Heat Shock Proteins

Volume 18

Series editors
Alexzander A. A. Asea
Professor, Department of Medicine and Director
Precision Therapeutics Proteogenomics Diagnostic Center
Eleanor N. Dana Cancer Center
University of Toledo College of Medicine and Life Sciences
Toledo, USA

Stuart K. Calderwood
Professor and Director, Division of Molecular and Cellular Radiation Oncology
Department of Radiation Oncology
Beth Israel Deaconess Medical Center and Harvard Medical School
Boston, USA

Heat Shock Proteins: key mediators of Health and Disease. Heat shock proteins (HSP) are essential molecules conserved through cellular evolution required for cells to survive the stresses encountered in the environment and in the tissues of the developing and aging organism. These proteins play the essential roles in stress of preventing the initiation of programmed cell death and repairing damage to the proteome permitting resumption of normal metabolism. Loss of the HSP is lethal either in the short-term in cases of acute stress or in the long-term when exposure to stress is chronic. Cells appear to walk a fine line in terms of HSP expression. If expression falls below a certain level, cells become sensitive to oxidative damage that influences aging and protein aggregation disease. If HSP levels rise above the normal range, inflammatory and oncogenic changes occur. It is becoming clear that HSP are emerging as remarkably versatile mediators of health and disease. The aim of this series of volumes is to examine how HSP regulation and expression become altered in pathological states and how this may be remedied by pharmacological and other interventions.

More information about this series at http://www.springer.com/series/7515

Alexzander A. A. Asea • Punit Kaur
Editors

Heat Shock Protein 60 in Human Diseases and Disorders

Springer

Editors
Alexzander A. A. Asea
Department of Medicine and Director,
Precision Therapeutics Proteogenomics
Diagnostic Center, Eleanor N. Dana
Cancer Center
University of Toledo, College of Medicine
and Life Sciences
Toledo, OH, USA

Punit Kaur
Department of Experimental
Radiation Oncology
MD Anderson Cancer Center
Houston, TX, USA

ISSN 1877-1246 ISSN 1877-1254 (electronic)
Heat Shock Proteins
ISBN 978-3-030-23153-8 ISBN 978-3-030-23154-5 (eBook)
https://doi.org/10.1007/978-3-030-23154-5

This Springer imprint is published by the registered company Springer Nature Switzerland AG.
The registered company address is: Gewerbestrasse 11, 6330 Cham, Switzerland

Preface

The sixty-kilo Dalton heat shock protein (HSP60) family is one of the oldest groups of proteins belonging to the molecular chaperones. HSP60 is present in all living organisms. HSP60 is a mitochondrial chaperonin that functions in the transportation and refolding of proteins from the cytoplasm into the mitochondrial matrix. In addition, HSP60 functions as a chaperonin to assist in folding linear amino acid chains into their respective three-dimensional structure. Through the extensive study of groEL, HSP60's bacterial homolog, HSP60 has been deemed essential in the synthesis and transportation of essential mitochondrial proteins from the cell's cytoplasm into the mitochondrial matrix. Further studies have linked HSP60 to diabetes, stress response, cancer and certain types of immunological disorders.

The book ***Heat Shock Protein 60 in Human Diseases and Disorders*** provides the most comprehensive review on contemporary knowledge on the role of HSP60 in human diseases and disorders. Using an integrative approach, the contributors provide a synopsis of novel mechanisms and signal transduction pathways. To enhance the ease of reading and comprehension the book has further been subdivided into various section including; Part I: Biomolecular Aspects of HSP60; Part II: HSP60 and Cancer; Part III: HSP60 and Inflammatory Diseases and Disorders; Part IV: HSP60 and Cardiovascular Diseases and Disorders; Part V: HSP60 and Neurological and Neurosciences; Part VI: HSP60 and Skeletal Muscle Diseases and Disorders; and Part VII: HSP60 in Human Health.

Key basic and clinical research laboratories from major universities, academic medical hospitals, biotechnology and pharmaceutical laboratories around the world have contributed chapters that review present research activity and importantly project the field into the future. The book is a must read for graduate students. medical students, basic science researchers and postdoctoral scholars in the fields of Translational Medicine, Clinical Research, Human Physiology, Biotechnology,

Neurology & Neuroscience, Oncology, Cardiovascular Disease, Skeletal Muscle Diseases and Disorders, Cell & Molecular Medicine, Pharmaceutical Scientists and Researchers involved in Drug Discovery.

Toledo, OH, USA Alexzander A. A. Asea
Houston, TX, USA Punit Kaur

Contents

About the Editors

Prof. Alexzander A. A. Asea is a highly innovative and accomplished world renowned clinical and basic research scientist and visionary executive leader who has exceptional experience spearheading clinical and basic science research, training, education, and commercialization initiatives within top ranked academic biomedical institutes. Prof. Dr. Asea's initial findings studying the effects of Hsp72 on human monocytes lead to the proposal of a novel paradigm that Hsp72, previously known to be an intracellular molecular chaperones, can be found in the extracellular milieu where it has regulatory effects on immuno-competent cells – a term now called chaperokine. Prof. Asea has authored over 255 scientific publications including peer-reviewed articles, reviews, books, book chapters, editorials, and news headliners in a wide range of biomedical-related disciplines. Prof. Asea is the series editor of the widely successful book series *Heat Shock Proteins* (Springer Nature Publishing) and is an editorial board member of numerous scientific peer-reviewed journals. Currently, Prof. Dr. Asea is at the University of Toledo College of Medicine and Life Sciences in Toledo, USA.

Dr. Punit Kaur is an expert in onco-proteogenomics, with extensive training and experience in quantitative mass spectrometry imaging, protein chemistry and biomarker discovery. Dr. Kaur's main research focus is on the use of heat-induced nanotechnology in combination with radiotherapy and chemotherapy in the cancer stem cell therapy. Dr. Kaur has published more than 40 scientific articles, book chapters, and reviews, and currently serves as editorial board member for the European *Journal of* Cancer Prevention and the *Journal of Proteomics and Bioinformatics*. Dr. Kaur is an editor of 9 books in the highly successful *Heat Shock Proteins* book series by Springer Nature Publishers. Currently, Dr. Kaur is a Visiting Scientist Professor at the University of Texas MD Anderson Cancer Center in Houston, USA.

Part I
Biomolecular Aspects of HSP60

Chapter 1
HSP60: A Story as Long as Life on the Earth

Francesco Cappello, Everly Conway de Macario, and Alberto J. L. Macario

Abstract In this Chapter, we briefly recount a few salient aspects of our personal experience and ideas about one of the most important anti-stress proteins in all cells, i.e., HSP60 (HSPD1 in humans). We outline the progression of HSP60 from gene to protein and its voyages inside and outside the cell, mentioning its various roles in health and disease. We correlate scientific data obtained by different experimental approaches to personal visions about the existence of life on Earth, its perpetuation, and the future of the human-species evolution. We believe that Science should not only be the application of technologies to find an answer to an unsolved question but also a source of philosophical considerations about ourselves and the future of our ecosystem, on Earth and elsewhere.

Keywords Chaperonins · Chaperonopathies · Chaperonotherapy · Endosymbiotic theory · Evolution · HSP60 · Molecular anatomy · Phylogenesis

F. Cappello (✉)
Department of Biomedicine, Neuroscience and Advanced Diagnostics, University of Palermo, Palermo, Italy

Euro-Mediterranean Institute of Science and Technology (IEMEST), Palermo, Italy

Department of Biology, College of Science and Technology, Temple University, Philadelphia, PA, USA
e-mail: francesco.cappello@unipa.it

E. Conway de Macario
Department of Microbiology and Immunology, University of Maryland at Baltimore-Institute of Marine and Environmental Technology (IMET), Baltimore, MD, USA

A. J. L. Macario
Euro-Mediterranean Institute of Science and Technology (IEMEST), Palermo, Italy

School of Medicine, Department of Microbiology and Immunology, University of Maryland at Baltimore-Institute of Marine and Environmental Technology (IMET), Columbus Center, Baltimore, MD, USA

A. A. A. Asea, P. Kaur (eds.), *Heat Shock Protein 60 in Human Diseases and Disorders*, Heat Shock Proteins 18,
https://doi.org/10.1007/978-3-030-23154-5_1

Abbreviations

CCT	Chaperonin containing TCP-1
Hsp	Heat shock protein
MIS	Mitochondrial import signal
TCP-1	T-complex protein 1
TRiC	TCP-1 ring complex

1.1 Introduction

When did the first cell ("protocell") appear on Earth? We may assume that when it happened, billions of years ago, it was immediately subjected to physical and chemical stressors, e.g. temperature, radiation, pH, etc. Therefore, it was necessary for it to develop anti-stress mechanisms. These mechanisms had to ensure not only that the cell was able to resist stress and therefore to survive, but also to reproduce itself. If today there is still life on Earth, in various forms and under a range of different conditions, including extreme ones (from very hot hydrothermal vents to icy poles), it is probably thanks to these anti-stress mechanisms that organisms, first unicellular and then multicellular, have been able to subsist and evolve. We can therefore say that it is a law of Nature that if an organism is able to fight stress it can survive, reproduce, and evolve. Vice versa, if it is not able to resist stress, it will succumb. In this short Chapter, we will attempt to summarize our own experience and views concerning the most studied cellular anti-stress mechanism, i.e., the chaperoning system, of which the molecular chaperones are the primary components, focusing on one of the most conserved of them throughout evolution, the mitochondrial chaperonin HSP60.

1.1.1 Molecular Chaperones

Molecular chaperones are among the phylogenetically older proteins. They are found in ancestral life forms (including Bacteria and Archaea) and are structurally much conserved (Macario and Conway de Macario 1997). The homologues present in the most complex and most evolved organisms, such as primates (including humans), are very similar to the most primitive forms (Gupta 1995; Macario and Conway de Macario 1997; Macario et al. 2004; Rowland and Robb 2017). This has allowed for decades studies on the structural (molecular anatomy) and functional (biochemistry) features of chaperones using very simple models (Conway de Macario et al. 2017). Many molecular chaperones but not all are heat shock proteins (Hsp), and vice versa, not all Hsp are chaperones, and they have been named and classified accordingly (Kampinga et al. 2009; Macario et al. 2013). We will focus on the HSP60 family, which is one of the oldest groups of proteins belonging to the molecular chaperones (Gupta 1995). These proteins present in all living species, including plants – are unique in their molecular characteristics, so that they have been called "chaperonins" in order to differentiate them from other chaperones (Hemmingsen et al. 1988).

1.1.1.1 Chaperonins

Classically, there are two groups of chaperonins. Group I are found in bacteria as well as in eukaryotic organelles of endosymbiotic origin (chloroplasts and mitochondria). In humans, it is known as HSP60 (or Cpn60, or HSPD1) and it works along with its co-chaperone HSP10 (or Cpn10, or HSPE1). Group II chaperonins are found in the eukaryotic cytosol and in archaea, and compared to Group I proteins they are less well characterized. In human cells, this group is represented by TRiC (TCP-1 Ring Complex), also called CCT (Chaperonin Containing TCP-1) (Macario et al. 2013). Recently a third group has been identified and it is under characterization (Rowland and Robb 2017). What both HSP60 and TRiC have in common is that they can form very large (~1 MDa) macromolecular complexes capable of accommodating other proteins inside them, the client proteins, i.e., proteins that must be folded to achieve the correct and functional final conformation (Koldewey et al. 2017; Skjærven et al. 2015). This unique morphological characteristic is the reason why they have been called "chaperonins" (Hemmingsen et al. 1988).

The fact that these macromolecular complexes form closed structures as if they were very small, miniature membrane-less cells, we can hypothesize that they could represent a first attempt to organize an isolated environment in which organic material (peptides) can flow and gather together to become functional. This would make them even more ancient and interesting, as membrane-less biological structures, than the protocells that first appeared on Earth. Most of the information we have about how human chaperonins are made and work derive from studies of their bacterial counterparts. Although phylogenetically they are much conserved proteins, there are billions of years of evolution in-between separating human proteins from their ancestors, which have led to modifications of some of their structural and functional characteristics.

For example, the bacterial homolog of the human protein HSP60 (Hsp60, Cpn60) is GroEL. In its quaternary structure, GroEL forms a heptameric ring, a toroidal shape, and two rings associate to form a sort of barrel with a cavity inside. The barrel can accommodate a protein to be folded, the substrate or client polypeptide. The barrel of GroEL works thanks to a "hood" formed by a heptamer of its co-chaperone, GroES (the homologous of eukaryotic HSP10 (Hsp10, Cpn10) and ATP. Two rings form a macromolecular complex in which each half works alternately to optimize the folding times: while one folds one protein, the other releases an already folded protein, and then vice versa, as in an assembly line (Nielsen and Cowan 1998;

Vilasi et al. 2018). Nevertheless, the heptameric ring of HSP60 can work even without forming a macro-complex with another ring and, surprisingly, even without ATP (Bhatt et al. 2018; Nielsen and Cowan 1998; Vilasi et al. 2018). HSP60 could have acquired these functions during phylogenesis to be able to work in emergency conditions, for example when there is not a sufficiently high concentration of HSP60 to form two rings that are near enough to each other to form a barrel, or when there is little or no ATP available. Even under these "extreme" conditions, HSP60 tries to perform its very important functions for human cells, often – but probably not always – succeeding.

1.1.2 HSP60 Impairments (Chaperonopathies)

HSP60 may not be able to function if it has undergone mutations, which may cause disease, i.e., chaperonopathies in homozygous and heterozygous conditions (Macario et al. 2013). Genetic chaperonopathies have been described in which HSP60 is mutated and this leads to premature death of the organism due to serious heart or nervous diseases (Bross et al. 2012; Macario et al. 2013). All the organs, and not only the heart and the neuraxis, suffer from the lack of functional HSP60 in their cells but the heart and the neuraxis are probably the first ones that are damaged as a result of HSP60 deficiency, which is probably the reason why the pathologies described as a result of these mutations affected predominantly these two anatomical districts.

It is likely that many fetuses or newborns affected by HSP60 mutations, not yet characterized, die prematurely and, consequently, it is also likely that a mutation affecting HSP60 would cause as yet unexplained early abortions. New studies are necessary to confirm or disprove this hypothesis. The confirmation that HSP60 is essential for life derives from experiments that tried to generate knock-out animal models for both *hsp60* alleles, e.g., in mice (Berger et al. 2016; Christensen et al. 2010), and zebrafish (personal data, unpublished); these experiments have been unsuccessful because the knock-out conditions were incompatible with life. Even experiments with *hsp60* knock-out cells often result in failure because the cells die (Tang et al. 2016). The explanation lies in the fact that HSP60, which in human cells is classically present in the mitochondria, serves to ensure the correct folding of virtually all other mitochondrial proteins, including those of the respiratory chain, which are essential for cell's life. Therefore, if HSP60 does not work well because it is defective due to mutation or other structural anomaly, for example aberrant post-translational modification, the cell, and in turn the whole organism, will die.

1.1.3 *HSP60 Odyssey*

The endosymbiotic theory postulates that mitochondria come from bacteria that have adapted to live within eukaryotic cells, supplying it with energy and receiving in return the molecular elements for their turn-over (Ku et al. 2015; Martin et al. 2015; Zimorski et al. 2014). Current versions of endosymbiotic theory have it that the host was an archaeon (an archaebacterium), not a eukaryote (Ku et al. 2015). It means that the Group I chaperonin HSP60 probably derives from bacteria, while the Group II chaperonin TRiC derives from archaea. The bacterial homolog of HSP60 is GroEL, which works along with its co-chaperone, GroES. However, the HSP60 and HSP10 genes in humans are nuclear and not mitochondrial. They are found on chromosome 2 and have a common promoter to which they are joined head-to-head (Hansen et al. 2003). The promoter's reading by the RNA polymerase in one direction or another determines the transcription of the HSP60 and HSP10 gene, respectively, and the mRNA formation that will go into the cytosol for translation into protein (Fig. 1.1A). The protein will then have to enter the mitochondrion. Differently from HSP10, the HSP60 transcribed into the cytosol possesses a mitochondrial import signal (MIS) that is lost when HSP60 finally enters the mitochondrion (Singh et al. 1990).

The fact that the HSP60 and HSP10 genes are located in the nucleus and not in the mitochondrion suggests that – according to the endosymbiotic theory – this fragment of DNA coding for both these proteins migrated to the nucleus, becoming integrated into the nuclear DNA. This forced the "Odyssey" that HSP60 and HSP10 must undertake to return to their "home of origin", the bacterium/mitochondrion (Fig. 1.1A). The mitochondrial DNA has some aspects that are still mysterious and not fully understood, but now we can ask a question: given the strategic importance these proteins have for the functioning of the mitochondrion, was it not simpler that the DNA for the HSP60 and HSP10 genes remained in this organelle? Why did it migrate to the nucleus? Geneticists argue that the nuclear environment is more stable than the mitochondrial environment for transcription. Even if this were true, the fact remains that other parts of the genetic code of these prokaryotic organisms stayed within them as mitochondrial DNA. However, one can argue that, moving to the nucleus and requiring a translation into the cytosol, not all copies of HSP60 and HSP10 produced should go into the mitochondria but some would remain in the cytosol to perform other functions. Is this possible? We are still interrogating ourselves on these issues. Nevertheless, we were able to show that human HSP60, containing its mitochondrial import signal, is able to form heptamers and tetradecamers remarkably stable over a wide range of concentrations (Vilasi et al. 2014). There are other "Odysseys" which HSP60 undertakes that will be discussed ahead.

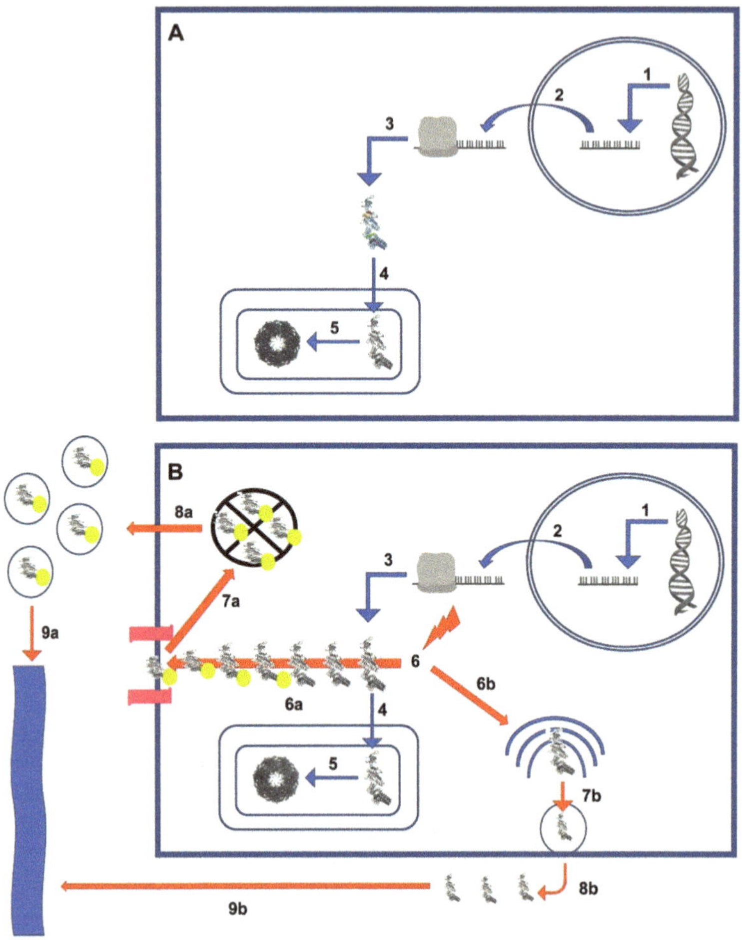

Fig. 1.1 HSP60 Odyssey. The endosymbiotic theory postulates that the DNA coding for HSP60 (and HSP10, not shown) in the endosymbiont moved to the nucleus of eukaryotic host-cells or their precursors. This resulted in a complex migration itinerary, a sort of Odyssey, for this molecule during both normal (**A**) and stressful (**B**) conditions. (**A**) After gene transcription (1), mRNA coding for HSP60 is transported to the cytosol (2) for translation (3) by ribosomes. The protein is then imported (4) into the mitochondrion where it works (5) as chaperonin. (**B**) Stress (6) causes HSP60 accumulation in the cytosol and the chaperonin undergoes post-translational modifications (yellow circle) that determine its migration (6a) to the plasma-cell membrane. Lipid rafts (horizontal purple bands) internalize HSP60 into multivesicular bodies (7a) from where it is released outside the cell via exosomes (8a). Another route consists of HSP60 reaching the Golgi's complex (6b), internalization in Golgi's vesicles (7b), and its secretion in its free (soluble) form (8b). In both cases, HSP60 bound to exosomes (9a) or free (9b) reaches the circulation (vertical undulating blue band) and, thereby, anatomical districts far from the site it was released. **Note:** it is possible that pathways shown in B occur not only in stressed cells but also under normal physiological conditions although at considerable lower levels than in stressed cells. See for further information (Campanella et al. 2012; Merendino et al. 2010)

1.1.4 *Extramitochondrial HSP60 (the HSP60 Odyssey, Part 2)*

Does the fact that HSP60, and its co-chaperonin HSP10, are translated into the cytosol represent an advantage or a disadvantage? Extracellular HSP10 has been characterized as EPF (Early Pregnancy Factor) since its concentrations increase after fertilization in mammals (Cavanagh 1996; Morton 1998). It is thought that this is due to the strong immunomodulatory properties of this protein: reducing the aggressiveness of immune cells, which would favor the implantation of the blastocyst and therefore the start of pregnancy. The immunomodulatory capacity of HSP10 is an additional function that obviously does not find any explanation in a single-cell organism that perpetuates simply by dividing. So, we can hypothesize that this function was acquired by HSP10 relatively late during phylogenesis and represents an advantage for the evolution of the species.

But can we say with certainty that the human being is the finished product of evolution on Earth? May be not, and may be that the many diseases – genetic and acquired – that occur in humans are unsuccessful attempts to produce a more evolved species. Even some studies about HSP60 give us some clues in this direction. For example, one thing we know is that HSP60 can accumulate in the cytosol (Cappello et al. 2013a, 2014, 2008; Pace et al. 2013; Vilasi et al. 2018). This occurs under particular stressful conditions. An example is carcinogenesis as originally shown by our group, both in vivo (Cabibi et al. 2016; Cappello et al. 2002, 2003a, b, 2005; Rappa et al. 2013, 2016) and in vitro (Campanella et al. 2008; Cappello et al. 2018; Caruso Bavisotto et al. 2017b; Gorska et al. 2013). The cancer cell must survive a series of stress conditions. This is the basis of the competition between the body and the tumor that develops in it. In tumor cells, there is over-transcription and -translation of HSP60, which accumulates in the cytosol (Fig. 1.1B). We do not know the cause, but we have proposed a possible explanation: following mitochondrial stress, HSP60 is released from the mitochondrion and, like cytochrome C, is able to activate apoptosis. In particular, HSP60 goes to activate pro-caspase 3, favoring its maturation into caspase 3, a pro-apoptotic molecule (Campanella et al. 2008; Chandra et al. 2007).

The HSP60 that is released from the mitochondrion, for the reasons we have explained before, is devoid of the MIS, which instead is present in that which has just been translated at the ribosomal level. Noteworthy, in the cytosol of the tumor cells both HSP60 with the MIS and without the MIS accumulate. The former has a function opposite to that of the latter with respect to pro-caspase 3, i.e., it prevents its activation. Thus, HSP60 without MIS is pro-apoptotic, while HSP60 with MIS is anti-apoptotic. To survive, the tumor cell needs to produce a lot of HSP60 so that the chaperonin – unable to enter the mitochondrion in its totality – accumulates in the cytosol, in which it has an anti-apoptotic effect. However, the cytosolic HSP60 is prone to undergo post-translational modifications, such as acetylation or nitration (Fig. 1.1B) (Campanella et al. 2016; Marino Gammazza et al. 2017). We hypothesize that these modifications increase the affinity of this protein with the plasma-cell membrane. In fact, in tumor cells, we find increased amounts of HSP60 at the plasma-cell membrane. But, when present at the plasma-cell membrane, HSP60 can turn into a self-antigen, as demonstrated with models of human atherosclerosis (Wick et al. 2014). Furthermore, bacterial infections can lead to the production of anti-GroEL antibodies capable of cross-reacting against human HSP60 (molecular mimicry) (Cappello et al. 2009). Therefore, for the tumor cell it is disadvantageous to have HSP60 at the plasma-cell membrane.

We were the first to show that when the HSP60 reaches the plasma-cell membrane of a tumor cell, the lipid rafts internalize it through endosomes in the multivesicular bodies from which exosomes loaded with HSP60 are released (Fig. 1.1B) (Campanella et al. 2012; Merendino et al. 2010). Further evidence lies in the fact that subjects affected by some forms of tumors that are known to overexpress HSP60 (such as colorectal carcinoma) have high circulating levels of exosomal HSP60, while these levels are drastically reduced when the tumor is removed (Campanella et al. 2015); these patterns suggest that HSP60 levels in circulation may have not only a role in tumorigenesis but also diagnostic and prognostic applications (Caruso Bavisotto et al. 2017a). The release of HSP60 through exosomes is the ending of the intracellular Odyssey of this protein but it is also the starting of a new Odyssey, the extracellular travel of this protein that – through exosomes and possibly also free (Fig. 1.1B) – may reach every district of the human body. However, we still do not know the effects of HSP60 on distant target organs. It is also unclear what are the actions and effects of HSP60 in the peri-tumoral microenvironment; for instance, we do not know yet if and how HSP60 can modify it, making it advantageous or disadvantageous for tumor progression. These all are open questions that deserve further investigation.

1.2 Conclusions

HSP60 has been dubbed *Proteus* and *Janus Bifrons* because of its multifaceted and changeable properties but it can also be said that it resembles *Odysseus* (or *Ulysses*). Its encoding gene and the gene product, the protein chaperone HSP60, are protagonists of complicated voyages inside and outside the cell. This Odyssey, from the times in which life originated to this day and today inside and outside cells is, like the Homeric epic poem, characterized by a variety of landing places and is pregnant with stressful situations and with the hero's encounters with local friends and foes. Maybe this is not unique to HSP60 but it is certainly one of its distinctive characteristics, perhaps indicating the chaperonin's extraordinary importance in evolution and, today, in health and disease. Indeed, HSP60 is one of the most important proteins for living cells and may have significantly affected the propagation and evolution of life on Earth. Without HSP60 there cannot be life and many types of mutations in HSP60 most likely cause fatal pathologies very early in embryo development. Other mutations cause more subtle deleterious effects and allow life to proceed to early adulthood: these are the "genetic chaperonopathies" that have been reported. Post-translational modifications can also affect HSP60 structural and functional properties and, most likely, are at the basis of cellular senescence as well as of degenerative events affecting many human tissues ("acquired chaperonopathies") (Cappello et al. 2013b; Macario et al. 2010, 2013). HSP60 seems to have contrasting effects (we discussed the pro- and anti-apoptotic effects, but elsewhere in this volume the pro- and anti-inflammatory effects will also be described) so much so that we have compared it to Proteus (Cappello and Zummo 2005), the Greek god of the sea whose "shape" could not be captured by anyone, while others compared it to the roman god Janus Bifrons (Henderson 2010). The truth is probably close to the fact that this molecule participates in the fine tuning of many homeostatic processes: very small deviations in the balance of the conditions under which it operates or is studied change its functions/effects and what we see. Also, assuming that the Human species is probably not the final product of the evolution of life on Earth, it is possible that HSP60 is still participating in this evolutionary process, also through its involvement in the pathogenesis of some human diseases (e.g., HSP60 chaperonopathies) that will be described in other chapters of this book. Given its plurality of roles in multiple places, and its voyages inside and outside cells, it can be surmised that HSP60 failure will cause morphological and functional lesions in virtually any tissue and organ of the human body. Consequently, HSP60 chaperonopathies are of interest to physicians and medical researchers across a wide range of specialties. As explained earlier, if the HSP60 damage is more than subtle the probability of survival is minimal, but minor abnormalities will produce phenotypes, namely patients with HSP60 chaperonopathies, as already shown by numerous investigators. Therefore, efforts should be made to develop means of diagnosis and treatment protocols targeting the chaperonin. HSP60 chaperonopathies can be by defect, excess or mistake (like other chaperonopathies pertaining to other chaperones) and, therefore, can in principle also be treated using the corresponding type of chaperonotherapy (Cappello et al. 2013a; Cappello et al. 2014). The future for research in developing chaperonotherapy strategies and tools to treat HSP60 chaperonopathies seems full of exciting possibilities.

Acknowledgements F.C. was partially supported by IEMEST and UniPA. A.J.L.M, and E.C. de M. were partially supported by IMET. This work was done under the agreement between IEMEST (Italy) and IMET (USA) (this is IMET contribution number IMET 19-004). The first draft of this work from FC was written during a few days FC spent in London at the Wellcome Foundation library, a place that relaxes and transmits serenity. For this reason, we are grateful to the Wellcome Foundation, which makes its Library available to visitors free of charge.

References

Agsteribbe E, Huckriede A, Veenhuis M, Ruiters MH, Niezen-Koning KE, Skjeldal OH, Skullerud K, Gupta RS, Hallberg R, van Diggelen OP, Scholte HR (1993) A fatal, systemic mitochondrial disease with decreased mitochondrial enzyme activities, abnormal ultrastructure of the mitochondria and deficiency of heat shock protein 60. Biochem Biophys Res Commun 193:146–154

Berger E, Rath E, Yuan D, Waldschmitt N, Khaloian S, Allgäuer M, Staszewski O, Lobner EM, Schöttl T, Giesbertz P, Coleman OI, Prinz M, Weber A, Gerhard M, Klingenspor M, Janssen KP, Heikenwalder M, Haller D (2016) Mitochondrial function controls intestinal epithelial stemness and proliferation. Nat Commun 7:13171

Bhatt JM, Enriquez AS, Wang J, Rojo HM, Molugu SK, Hildenbrand ZL, Bernal RA (2018) Single-ring intermediates are essential for some chaperonins. Front Mol Biosci 5:42

Bross P, Magnoni R, Bie AS (2012) Molecular chaperone disorders: defective Hsp60 in neurodegeneration. Curr Top Med Chem 12:2491–2503

Cabibi D, Conway de Macario E, Ingrao S, Porcasi R, Zucco F, Macario AJL, Cappello F, Rappa F (2016) CD1A-positive cells and HSP60 (HSPD1) levels in keratoacanthoma and squamous cell carcinoma. Cell Stress Chaperones 21:131–137

Campanella C, Bucchieri F, Ardizzone NM, Marino Gammazza A, Montalbano A, Ribbene A, Di Felice V, Bellafiore M, David S, Rappa F, Marasà M, Peri G, Farina F, Czarnecka AM, Conway de Macario E, Macario AJL, Zummo G, Cappello F (2008) Upon oxidative stress, the antiapoptotic Hsp60/procaspase-3 complex persists in mucoepidermoid carcinoma cells. Eur J Histochem 52:221–228

Campanella C, Bucchieri F, Merendino AM, Fucarino A, Burgio G, Corona DF, Barbieri G, David S, Farina F, Zummo G, Conway de Macario E, Macario AJL, Cappello F (2012) The odyssey of Hsp60 from tumor cells to other destinations includes plasma membrane-associated stages and Golgi and exosomal protein-trafficking modalities. PLoS One 7:e42008

Campanella C, Rappa F, Sciumè C, Marino Gammazza A, Barone R, Bucchieri F, David S, Curcurù G, Caruso Bavisotto C, Pitruzzella A, Geraci G, Modica G, Farina F, Zummo G, Fais S, Conway de Macario E, Macario AJL, Cappello F (2015) Heat shock protein 60 levels in tissue and circulating exosomes in human large bowel cancer before and after ablative surgery. Cancer 121:3230–3239

Campanella C, D'Anneo A, Marino Gammazza A, Caruso Bavisotto C, Barone R, Emanuele S, Lo Cascio F, Mocciaro E, Fais S, Conway de Macario E, Macario AJL, Cappello F, Lauricella M (2016) The histone deacetylase inhibitor SAHA induces HSP60 nitration and its extracellular release by exosomal vesicles in human lung-derived carcinoma cells. Oncotarget 7:28849–28867

Cappello F, Zummo G (2005) HSP60 expression during carcinogenesis: a molecular "proteus" of carcinogenesis? Cell Stress Chaperones 10:263–264

Cappello F, Bellafiore M, Palma A, Marcianò V, Martorana G, Belfiore P, Martorana A, Farina F, Zummo G, Bucchieri F (2002) Expression of 60-kD heat shock protein increases during carcinogenesis in the uterine exocervix. Pathobiology 70:83–88

Cappello F, Bellafiore M, Palma A, David S, Marcianò V, Bartolotta T, Sciumè C, Modica G, Farina F, Zummo G, Bucchieri F (2003a) 60KDa chaperonin (HSP60) is over-expressed during colorectal carcinogenesis. Eur J Histochem 47:105–110

Cappello F, Rappa F, David S, Anzalone R, Zummo G (2003b) Immunohistochemical evaluation of PCNA, p53, HSP60, HSP10 and MUC-2 presence and expression in prostate carcinogenesis. Anticancer Res 23:1325–1331

Cappello F, David S, Rappa F, Bucchieri F, Marasà L, Bartolotta TE, Farina F, Zummo G (2005) The expression of HSP60 and HSP10 in large bowel carcinomas with lymph node metastase. BMC Cancer 5:139

Cappello F, Conway de Macario E, Marasà L, Zummo G, Macario AJL (2008) Hsp60 expression, new locations, functions and perspectives for cancer diagnosis and therapy. Cancer Biol Ther 7:801–809

Cappello F, Conway de Macario E, Di Felice V, Zummo G, Macario AJL (2009) *Chlamydia trachomatis* infection and anti-Hsp60 immunity: the two sides of the coin. PLoS Pathog 5:e1000552

Cappello F, Angileri F, Conway de Macario E, Macario AJL (2013a) Chaperonopathies and chaperonotherapy. Hsp60 as therapeutic target in cancer: potential benefits and risks. Curr Pharm Des 19:452–457

Cappello F, Conway de Macario E, Marino Gammazza A, Bonaventura G, Carini F, Czarnecka AM, Farina F, Zummo G, Macario AJL (2013b) Hsp60 and human aging: Les liaisons dangereuses. Front Biosci (Landmark Ed) 18:626–637

Cappello F, Marino Gammazza A, Palumbo Piccionello A, Campanella C, Pace A, Conway de Macario E, Macario AJL (2014) Hsp60 chaperonopathies and chaperonotherapy: targets and agents. Expert Opin Ther Targets 18:185–208

Cappello F, Conway de Macario E, Rappa F, Zummo G, Macario AJL (2018) Immunohistochemistry of human Hsp60 in health and disease: from autoimmunity to cancer. Methods Mol Biol 1709:293–305

Caruso Bavisotto C, Cappello F, Macario AJL, Conway de Macario E, Logozzi M, Fais S, Campanella C (2017a) Exosomal HSP60: a potentially useful biomarker for diagnosis, assessing prognosis, and monitoring response to treatment. Expert Rev Mol Diagn 17:815–822

Caruso Bavisotto C, Nikolic D, Marino Gammazza A, Barone R, Lo Cascio F, Mocciaro E, Zummo G, Conway de Macario E, Macario AJL, Cappello F, Giacalone V, Pace A, Barone G, Palumbo Piccionello A, Campanella C (2017b) The dissociation of the Hsp60/pro-Caspase-3 complex by bis(pyridyl)oxadiazole copper complex (CubipyOXA) leads to cell death in NCI-H292 cancer cells. J Inorg Biochem 170:8–16

Cavanagh AC (1996) Identification of early pregnancy factor as chaperonin 10: implications for understanding its role. Rev Reprod 1:28–32

Chandra D, Choy G, Tang DG (2007) Cytosolic accumulation of HSP60 during apoptosis with or without apparent mitochondrial release: evidence that its pro-apoptotic or pro-survival functions involve differential interactions with caspase-3. J Biol Chem 282:31289–31301

Christensen JH, Nielsen MN, Hansen J, Fuchtbauer A, Fuchtbauer EM, West M, Corydon TJ, Gregersen N, Bross P (2010) Inactivation of the hereditary spastic paraplegia-associated Hspd1 gene encoding the Hsp60 chaperone results in early embryonic lethality in mice. Cell Stress Chaperones 15:851–863

Conway de Macario E, Robb FT, Macario AJL (2017) Prokaryotic chaperonins as experimental models for elucidating structure-function abnormalities of human pathogenic mutant counterparts. Front Mol Biosci 3:84

Gorska M, Marino Gammazza A, Zmijewski MA, Campanella C, Cappello F, Wasiewicz T, Kuban-Jankowska A, Daca A, Sielicka A, Popowska U, Knap N, Antoniewicz J, Wakabayashi T, Wozniak M (2013) Geldanamycin-induced osteosarcoma cell death is associated with hyperacetylation and loss of mitochondrial pool of heat shock protein 60 (hsp60). PLoS One 8:e71135

Gupta RS (1995) Evolution of the chaperonin families (Hsp60, Hsp10 and Tcp-1) of proteins and the origin of eukaryotic cells. Mol Microbiol 15:1–11

Hansen JJ, Bross P, Westergaard M, Nielsen MN, Eiberg H, Børglum AD, Mogensen J, Kristiansen K, Bolund L, Gregersen N (2003) Genomic structure of the human mitochondrial chaperonin genes: HSP60 and HSP10 are localised head to head on chromosome 2 separated by a bidirectional promoter. Hum Genet 112:71–77

Hemmingsen SM, Woolford C, Van der Vies SM, Tilly K, Dennis DT, Georgopoulos CP, Hendrix RW, Ellis RJ (1988) Homologous plant and bacterial proteins chaperone oligomeric protein assembly. Nature 333:330–334

Henderson B (2010) Integrating the cell stress response: a new view of molecular chaperones as immunological and physiological homeostatic regulators. Cell Biochem Funct 28:1–14

Kampinga HH, Hageman J, Vos MJ, Kubota H, Tanguay RM, Bruford EA, Cheetham ME, Chen B, Hightower LE (2009) Guidelines for the nomenclature of the human heat shock proteins. Cell Stress Chaperones 14:105–111

Koldewey P, Horowitz S, Bardwell JCA (2017) Chaperone-client interactions: non-specificity engenders multifunctionality. J Biol Chem 292:12010–12017

Ku C, Nelson-Sathi S, Roettger M, Garg S, Hazkani-Covo E, Martin WF (2015) Endosymbiotic gene transfer from prokaryotic pangenomes: inherited chimerism in eukaryotes. Proc Natl Acad Sci USA 112:10139–10146

Macario AJL, Conway de Macario E (1997) Mini-review; stress genes: an introductory overview. Stress 1:123–134

Macario AJL, Malz M, Conway de Macario E (2004) Evolution of assisted protein folding: the distribution of the main chaperoning systems within the phylogenetic domain archaea. Front Biosci 9:1318–1332

Macario AJL, Cappello F, Zummo G, Conway de Macario E (2010) Chaperonopathies of senescence and the scrambling of interactions between the chaperoning and the immune systems. Ann N Y Acad Sci 1197:85–93

Macario AJL, Conway de Macario E, Cappello F (2013) The Chaperonopathies. Diseases with defective molecular chaperones, Springer briefs in biochemistry and molecular biology. Springer, Dordrecht

Marino Gammazza A, Campanella C, Barone R, Caruso Bavisotto C, Gorska M, Wozniak M, Carini F, Cappello F, D'Anneo A, Lauricella M, Zummo G, Conway de Macario E, Macario AJL, Di Felice V (2017) Doxorubicin anti-tumor mechanisms include Hsp60 post-translational modifications leading to the Hsp60/p53 complex dissociation and instauration of replicative senescence. Cancer Lett 385:75–86

Martin WF, Garg S, Zimorski V (2015) Endosymbiotic theories for eukaryote origin. Philos Trans R Soc Lond Ser B Biol Sci 370:20140330

Merendino AM, Bucchieri F, Campanella C, Marcianò V, Ribbene A, David S, Zummo G, Burgio G, Corona DF, Conway de Macario E, Macario AJL, Cappello F (2010) Hsp60 is actively secreted by human tumor cells. PLoS One 5:e9247

Morton H (1998) Early pregnancy factor: an extracellular chaperonin 10 homologue. Immunol Cell Biol 76:483–496

Nielsen KL, Cowan NJ (1998) A single ring is sufficient for productive chaperonin-mediated folding in vivo. Mol Cell 2:93–99

Pace A, Barone G, Lauria A, Martorana A, Piccionello AP, Pierro P, Terenzi A, Almerico AM, Buscemi S, Campanella C, Angileri F, Carini F, Zummo G, Conway de Macario E, Cappello F, Macario AJL (2013) Hsp60, a novel target for antitumor therapy: structure-function features and prospective drugs design. Curr Pharm Des 19:2757–2764

Rappa F, Unti E, Baiamonte P, Cappello F, Scibetta N (2013) Different immunohistochemical levels of Hsp60 and Hsp70 in a subset of brain tumors and putative role of Hsp60 in neuroepithelial tumorigenesis. Eur J Histochem 57:e20

Rappa F, Pitruzzella A, Marino Gammazza A, Barone R, Mocciaro E, Tomasello G, Carini F, Farina F, Zummo G, Conway de Macario E, Macario AJL, Cappello F (2016) Quantitative patterns of Hsps in tubular adenoma compared with normal and tumor tissues reveal the value of Hsp10 and Hsp60 in early diagnosis of large bowel cancer. Cell Stress Chaperones 21:927–933

Rowland SE, Robb FT (2017) Structure, function and evolution of the Hsp60 chaperonins. In: Santosh Kumar CM and Mande SC (eds) Prokaryotic chaperonins. Multiple copies and multitude of functions, Springer Nature Singapore Pte Ltd. 129. Heat Shock Proteins 11, pp 3–20

Singh B, Patel HV, Ridley RG, Freeman KB, Gupta RS (1990) Mitochondrial import of the human chaperonin (HSP60) protein. Biochem Biophys Res Commun 169:391–396

Skjærven L, Cuellar J, Martinez A, Valpuesta JM (2015) Dynamics, flexibility, and allostery in molecular chaperonins. FEBS Lett 589:2522–2532

Tang H, Li J, Liu X, Wang G, Luo M, Deng H (2016) Down-regulation of HSP60 suppresses the proliferation of glioblastoma cells via the ROS/AMPK/mTOR pathway. Sci Rep 6:28388

Vilasi S, Carrotta R, Mangione MR, Campanella C, Librizzi F, Randazzo L, Martorana V, Marino Gammazza A, Ortore MG, Vilasi A, Pocsfalvi G, Burgio G, Corona D, Palumbo Piccionello A, Zummo G, Bulone D, Conway de Macario E, Macario AJL, San Biagio PL, Cappello F (2014) Human Hsp60 with its mitochondrial import signal occurs in solution as heptamers and tetradecamers remarkably stable over a wide range of concentrations. PLoS One 9:e97657

Vilasi S, Bulone D, Caruso Bavisotto C, Campanella C, Marino Gammazza A, San Biagio PL, Cappello F, Conway de Macario E, Macario AJL (2018) Chaperonin of group I: oligomeric spectrum and biochemical and biological implications. Front Mol Biosci 4:99

Wick G, Jakic B, Buszko M, Wick MC, Grundtman C (2014) The role of heat shock proteins in atherosclerosis. Nat Rev Cardiol 11:516–529

Zimorski V, Ku C, Martin WF, Gould SB (2014) Endosymbiotic theory for organelle origins. Curr Opin Microbiol 22:38–48

Chapter 2
Single-Ring Intermediates in the Catalytic Cycle of the Human Mitochondrial Hsp60

Jay M. Bhatt and Ricardo A. Bernal

Abstract The human mitochondrial heat shock protein 60 (hsp60) is composed to two heptameric rings stacked back-to-back. During the protein-folding cycle, these two rings dissociate to form single-ring intermediates that are catalytically active in-vitro and in-vivo. In this book chapter, we discuss the structural and catalytic features of hsp60 single-ring intermediates. We also provide a comparative analysis of the structural and mechanistic characteristics of single-ring intermediates observed in hsp60 versus ϕ-EL, OBP, and groEL chaperonin complexes that have also been reported to utilize single-rings within their catalytic cycle.

Keywords ϕ-EL · Catalytic cycle · GroEL · Hsp60 · Mitochondria · Single-ring

Abbreviations

hsp60	Heat shock protein 60
RUBISCO	Ribulose-1,5 bisphosphate carboxylase

2.1 Introduction

Proteostasis refers to a coordinated functioning of multiple biochemical pathways to maintain an active set of cellular proteins. These pathways regulate the synthesis, folding, transport, and degradation of nascent and unfolded proteins. Cellular proteostasis is achieved through the functional integration of several families of proteins that reside within and outside the cell. Chaperonins are one such family of

J. M. Bhatt (✉) · R. A. Bernal
Department of Chemistry and Biochemistry, The University of Texas at El Paso, El Paso, TX, USA
e-mail: jayuab@gmail.com

A. A. A. Asea, P. Kaur (eds.), *Heat Shock Protein 60 in Human Diseases and Disorders*, Heat Shock Proteins 18,
https://doi.org/10.1007/978-3-030-23154-5_2

proteins that modulate the correct folding of partially or completely unfolded proteins. Proteins that lose their 3-dimensional structure also lose their ability to function. These partially or completely unfolded proteins could form aggregates and lead to a wide myriad of diseases such as Parkinson's disease, Alzheimer's disease, Huntington's disease, cancer, cystic fibrosis, etc. (Chaudhuri and Paul 2006; Elborn 2016; Sweeney et al. 2017; Wang and Kaufman 2014). Therefore, it is not surprising that all cells have evolved intricate mechanisms to protect against protein aggregation that can lead to cell death.

Chaperonins are found in organisms ranging from bacteria to humans. In the human mitochondria, heat shock protein 60 (hsp60) is the predominant chaperonin that folds proteins in the mitochondrial matrix. Hsp60 along with its co-chaperonin, heat shock protein 10 (hsp10) have been known to be essential for the import and assembly of multimeric protein complexes in mitochondria (Cheng et al. 1989). Microarray analyses have shown that bacterial cells containing mutant hsp60/10 complexes display an up-regulation of genes that are essential for robust aerobic growth (Bross et al. 2008). This suggests that hsp60/10 plays an essential role in mitochondrial biogenesis and therefore, a functionally compromised hsp60/10 system leads to mitochondrial dysfunction causing early onset diabetes, juvenile rheumatoid arthritis, and neurodegenerative diseases (Hansen et al. 2002; Juwono and Martinus 2016; Magen et al. 2008; Vercoulen et al. 2009).

Hsp60 and hsp10 subunits arrange themselves into heptameric ring-like structures reminiscent of the bacterial homologues, chaperonin groEL and its co-chaperonins groES (together denotes as groEL/ES). Sequence alignment of hsp60 to groEL reveals that the hydrophilic residues that line the central cavity of groEL are conserved in hsp60 (Brocchieri and Karlin 2000). This sequence identity coupled with a similar mechanism for co-chaperonin binding indicates that hsp60 utilizes a hydrophilic central cavity to facilitate the burial of exposed hydrophobic patches in non-native substrate proteins. Despite similarities in the general ring structure, subunit architecture, and the nature of the central cavity, the hsp60/10 system significantly differs from groEL/ES in its mechanism of protein-folding. Hsp60/10 complexes have been shown to exist and function in single-ring and double-ring conformations. On the other hand, chimeras and mutations generated in groEL/ES have demonstrated that the bacterial system operates exclusively via a double-ring mechanism.

In this book chapter, we discuss the structural and catalytic features of single-ring intermediates observed in the protein-folding pathway of hsp60/10. Thus far, single-ring intermediates have also been identified in bacteriophage encoded chaperonins ϕ-EL and OBP (Molugu et al. 2016; Semenyuk et al. 2016). Recently, Yan et al. (2018) reported that groEL/ES might also be working using single-ring intermediates (Yan et al. 2018). We also provide a comparative analysis of the structural and mechanistic characteristics of single-ring intermediates observed across these various types of chaperonin complexes.

2.1.1 History of Hsp60/10 Single-Ring Intermediates

Early studies investigating the structural and biochemical nature of the hsp60/10 chaperonin suggested that homo-oligomeric single-rings composed of seven subunits could be expressed and purified from bacteria (Viitanen et al. 1992). These single-ring complexes were stable and could fold partially or completely unfolded substrate protein, RUBISCO (ribulose-1,5 bisphosphate carboxylase), in the presence of ATP. Similarly, chimeric hsp60/groEL chaperonin molecules that could not form double-ring complexes were demonstrated to have the ability to fold proteins in-vivo without the formation of double-ring intermediates (Nielsen and Cowan 1998). Hence, it was believed that hsp60/10 could maintain productive protein-folding without the use of double-ring complexes. In addition, expression of hsp60/10 in *E.coli* cells devoid of endogenous groEL/ES showed that the single-ring hsp60/10 complexes were able to compensate for the loss of groEL/ES and could carry out all the essential in-vivo functions that double-ring groEL/ES complexes would fulfill (Nielsen et al. 1999).

Most of what is currently known about the protein-folding mechanism of hsp60/10 is based on the knowledge derived from groEL/ES, its single-ring mutants, and the ϕ-EL chaperonin (Illingworth et al. 2011, 2015; Kovacs et al. 2010; Liu et al. 2009; Molugu et al. 2016; Sun et al. 2003). This is mainly attributed to the instability of the functional complexes in-vitro (Viitanen et al. 1992, 1998; Vilasi et al. 2014). Attempts to purify hsp60/10 complexes as tetradecamer were not successful since most groups could only isolate the monomers. The tetradecameric complexes had to be reconstituted in-vitro to study their architecture, however, such complexes would readily dissociate into monomers at low temperatures and in the presence of ATP (Viitanen et al. 1998). Therefore, unlike groEL/ES, the nucleotide induced conformational changes in the human mitochondrial hsp60 remained largely unknown. Negative stain electron microscopy studies indicated hsp60 favored a tetradecameric conformation in the presence of ATP. However, a high-resolution structure for the resulting ATP-hsp60/10 complex has not been determined (Levy-Rimler et al. 2001). Current knowledge allows us to postulate a catalytic cycle for hsp60 based on the conformational intermediates that have been observed under the electron microscope (see below). However, further studies are warranted to investigate the detailed mechanism by which hsp60/tenfold proteins along its catalytic pathway.

2.1.2 *Catalytic Cycle of GroEL/ES*

All chaperonins share a general architecture characterized by two rings stacked back-to-back. Each ring is composed of seven to nine subunits arranged in a homo- or hetero-oligomeric manner. The individual subunits are composed of equatorial, intermediate, and apical domains (Braig et al. 1994; Hildenbrand and Bernal 2012). Upon ATP binding and hydrolysis, these domains undergo conformational changes within each ring that allow for binding of substrate proteins and co-chaperonin to create an isolated environment to fold proteins, and finally release the newly folded protein back into the cellular milieu. However, the molecular details of these nucleotide-driven conformational changes are dramatically different across various chaperonin complexes.

The ATPase cycle of groEL/ES is divided into two phases. The first phase involves ATP binding to one of the heptameric groEL rings (Fig. 2.1). Seven molecules of ATP bind with positive cooperativity to the *cis*-ring followed by binding of the unfolded substrate and co-chaperonin groES. ATP hydrolysis drives the substrate folding within the internal cavity. ATP binding to the *trans*-ring stimulates the release of groES, ADP and folded substrate molecules in the *cis*-ring. This converts the *trans*-ring into the *cis*-ring conformation leading to the second phase of the cycle that involves sequential binding of ATP, unfolded substrate, and groES. The positive intra-ring cooperativity of groEL promotes the successive binding of ATP to all the subunits in the *cis*-ring. The negative inter-ring cooperativity in groEL

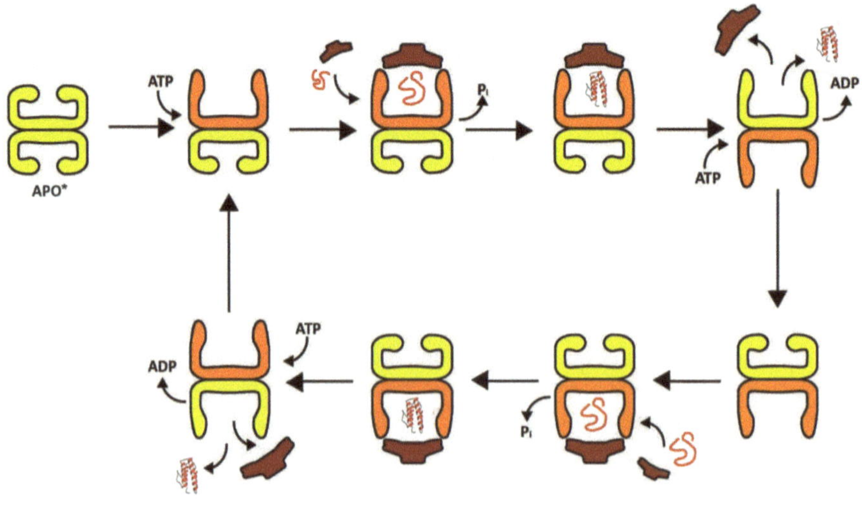

Fig. 2.1 GroEL/ES catalytic cycle: The APO conformation sequentially binds ATP molecules to the *cis*-ring, followed by substrate binding and capping by groES. ATP binding to the *trans*-ring leads to the release of folded substrate, ADP, and groES. The *trans*-ring transitions into the cis-ring conformation and the catalytic pathway continues in a cyclic manner such that the two rings alternate in their protein-folding responsibility

restricts the binding of ATP to both rings simultaneously. Therefore, the *cis*-ring is ATP-bound while the *trans*-ring is devoid of ATP. Hence, the groEL/ES mechanism resembles a "two-stroke engine" where protein-folding switches back and forth between the two heptameric rings. Yan et al. (2018) recently reported that groEL/ES might be forming single-ring intermediates (Yan et al. 2018). However, these observations need to be validated by further studies since they utilized groEL mutants and nucleotide analogs that in our hands lead to erroneous off-pathway conformations that do not match with natural (ATP and ADP) nucleotide-driven conformations.

2.1.3 Recent Advances in the Knowledge of ϕ-EL and Hsp60 Single-Ring Intermediates

We obtained cryo-EM reconstructions of the ϕ-EL chaperonin to show that nucleotide binding drives the conformational changes during its protein-folding catalytic pathway. The overall ϕ-EL subunit architecture and ring-arrangement was found to be like group I chaperonins. The ATP-bound conformation shows D7 symmetry and is 'open' at both ends such that the two chambers are fully accessible for substrate binding. ATP hydrolysis substantially diminishes the cavity opening since the apical domains rearrange themselves to form an in-built lid. The equatorial domains drop away from the cavity to allow the formation of an enlarged central cavity. ATP hydrolysis also results in the dissociation of the double-ring complexes into single-ring intermediates that can independently fold proteins in-vivo and in-vitro (Molugu et al. 2016). The ADP-bound single-ring 'closed' conformation of the ϕ-EL chaperonin has an expanded internal cavity, presumably to fold larger viral proteins that could not be accommodated within the host chaperonin. ϕ-EL has been shown to fold β-galactosidase, a 116-kDa substrate that would be too large to be folded by groEL/ES complexes (Molugu et al. 2016).

Upon release of ADP from the nucleotide binding pocket (APO conformation), the two heptameric rings come back together to form a tetradecameric double-ring conformation. The APO conformation is also a 'closed' conformation since the central cavity is not accessible from the cytosol. The ADP and APO conformations showed that the function of the co-chaperonin is circumvented by the conformational changes in the apical domain that effectively close the openings of the central cavity to provide an environment isolated from the cytosol to facilitate efficient protein-folding. This structural feature is similar to group II chaperonins that utilize apical domain rearrangements to facility ring closure and thereby preclude the requirement of a co-chaperonin (Hildenbrand and Bernal 2012). It is particularly interesting to note that ϕ-EL utilizes structural features of both group I and II chaperonins (reviewed in Bhatt et al. 2018). The catalytic cycle of ϕ-EL has been summarized in Fig. 2.2. In contrast with groEL's "two-stroke" mechanism, ϕ-EL utilizes a "one-stroke" mechanism where both the single-rings fold the proteins simultaneously.

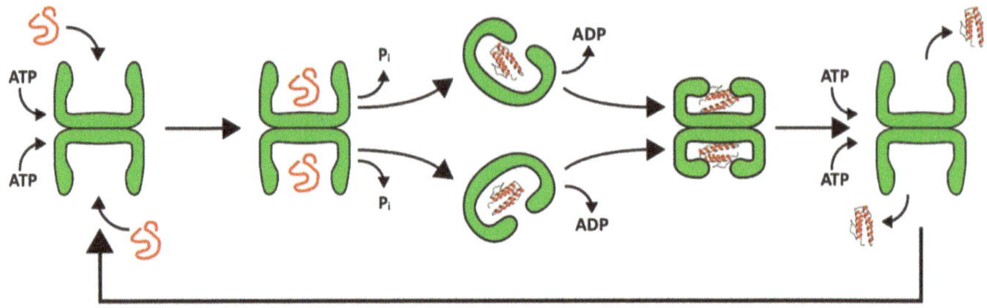

Fig. 2.2 Catalytic cycle of ϕ-EL: ATP binding followed by entry of the unfolded substrate yields the 'open' double-ring ATP-conformation. ATP hydrolysis results in a dramatic increase in the volume of the central cavity and separation of the two ADP-bound rings ('closed' single-ring conformation). These single-rings associate upon ADP release to form the 'closed' double-ring APO-conformation and release the folded substrate upon ATP binding

Like ϕ-EL, hsp60/10 has been known to utilize single-ring intermediates within its catalytic pathway to fold proteins. However, there are additional conformations that are observed within the hsp60/10 protein-folding pathway due to the requirement of co-chaperonin hsp10 for productive catalysis. ATP binding drives the association of hsp10 to the tetradecameric hsp60 complexes leading to the formation of unique "bullet" and "football" conformations (Levy-Rimler et al. 2001; Nisemblat et al. 2014, 2015). The asymmetric "bullet" conformation is composed of one heptameric hsp10 ring binding to only one heptameric hsp60 ring within the tetradecameric complex. The symmetric "American football" complex contains hsp10 heptamers bound to both rings of the hsp60 tetradecamer resulting in a fully enclosed chaperonin complex. The observation of "bullet" shaped hsp60/10 complexes in the presence of ATP indicate that the inter-ring negative cooperativity seen in groEL/ES is conserved in hsp60. However, the additional presence of "football" complexes suggests the inter-ring negative cooperativity is used to sequentially pack one ring with denatured substrate before the second ring can accommodate a substrate of its own. This is in stark contrast to groEL/ES that depends on the binding of ATP to the *trans*-ring to eject substrate and co-chaperonin form the *cis*-ring (Fig. 2.1). The presence of a "football" conformation and lack of a *trans*-ring indicates that hsp60/10 utilizes a different mechanism for ejecting substrate and co-chaperonin as compared to groEL/ES.

Upon ATP hydrolysis, the "football" complex dissociates into single-rings that can independently fold the substrate proteins. Recently, we reported that in the absence of nucleotide and substrate, hsp60 forms a stable tetradecameric conformation (Enriquez et al. 2017). Therefore, the two single-rings come back together to form a stable double-ring APO conformation after the release of ADP, hsp10 and folded substrate protein from both the rings. The protein-folding catalytic cycle of hsp60/10 is unique since it combines the mechanistic features of the "two-stroke" groEL/ES cycle and the "one-stroke" ϕ-EL cycle. However, the exact structural and

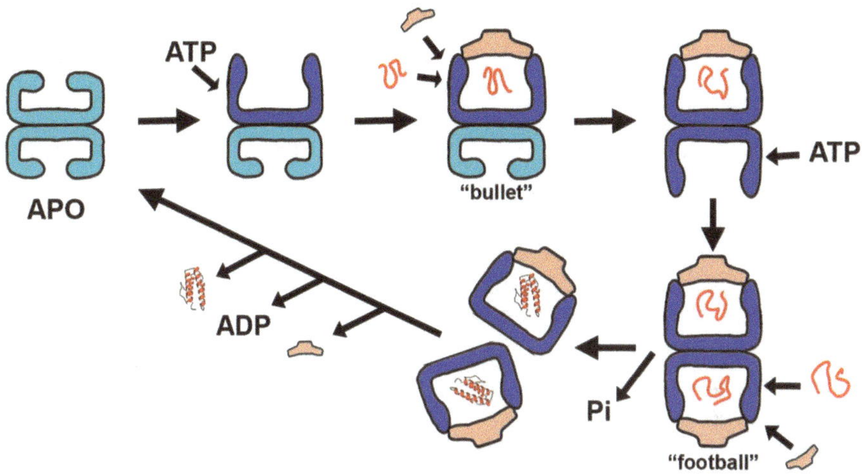

Fig. 2.3 Proposed hsp60/10 catalytic cycle: The APO conformation binds ATP, substrate, and hsp10 to form the "bullet" conformation. However, additional ATP binding leads to the formation of the "football" conformation with ATP, substrate, and hsp10 both to both the rings simultaneously. ATP hydrolysis leads to the separation of the double-ring complex. The single-ring conformation is probably where the substrate is folded as it has been noted that the single-ring is able to expand to a much larger internal chamber volume. The expulsion of ADP likely induces the removal of hsp10 and the folded substrate from both the rings and the return to the APO double-ring conformation

catalytic features of the hsp60/10 protein-folding cycle are yet to be elucidated since high-resolution structures of nucleotide-bound conformations have remain elusive. Based on the limited knowledge of the unique hsp60/10 conformations that have been observed under the electron microscope, we have proposed a catalytic cycle for hsp60/10 (Fig. 2.3).

Bioinformatics analyses have suggested that ϕ-EL and hsp60 form single-rings due to altered inter-ring salt bridge contacts as compared to groEL (Karlin and Brocchieri 2000; Molugu et al. 2016). These residues are critical for negative inter-ring allostery within groEL but are mutated in hsp60 and ϕ-EL. This observation has interesting implications for the overall architecture and subunit contacts across the two rings for ϕ-EL and hsp60. The inter-ring contacts between the groEL subunits are out-of-register (1:2); meaning one subunit contacts two subunits across the inter-ring interface. ϕ-EL demonstrates in-register (1:1) inter-ring contacts such that a subunit from one ring interacts with only one subunit from the other ring. One would expect hsp60/10 subunits to be arranged in an in-register (1:1) formation since it forms single-ring intermediates similar to ϕ-EL and have limited inter-ring negative cooperativity. However, our lab has shown that APO-hsp60 subunits display an out-of-register (1:2) organization (Enriquez et al. 2017). This reinforces the idea that hsp60/10 utilizes catalytic and architectural features from both the "two-stroke" groEL/ES cycle and the "one-stroke" ϕ-EL systems.

2.1.4 Unanswered Questions

The ADP-bound single-ring conformation of the ϕ-EL chaperonin has an expanded internal cavity, presumably to fold larger viral proteins that could not be accommodated within the host chaperonin, groEL/ES (Molugu et al. 2016). Likewise, an 80% increase in the volume of the internal cavity has also been reported for the single-ring mutant of groEL, SR398 (Chen et al. 2006). It has been postulated that single-ring hsp60/10 intermediate would also show an expanded central cavity to allow for larger substrate proteins to be sequestered by the chaperonin. This hypothesis is supported by the association between hsp60 and the 82 kDa mitochondrial Aconitase protein, a protein too large to be encapsulated by groEL/ES (Dubaquie et al. 1998; Wolf 2006). However, there is still no clear evidence to strongly support this hypothesis for hsp60/10 complexes.

Neurodegenerative disorders MitCHAP-60 and SPG13 have been associated with point mutations in hsp60. A D3G hsp60 mutation is a missense mutation that has been linked to MitCHAP-60 while a V72I mutation leads to SPG13 (Bross et al. 2008; Hansen et al. 2002; Magen et al. 2008; Parnas et al. 2009). MitCHAP-60 is inherited in an autosomal-recessive pattern and is characterized by neuronal hypomyelination and white matter degradation. Patients show psychomotor developmental delays, muscle weakness, limb spasticity, followed by death within the first two decades of life (Magen et al. 2008). SPG13 is a member of the hereditary spastic paraplegia group of neurodegenerative diseases. It is characterized by progressive muscle weakness, lower extremity spasticity, impaired vision, deafness, and cognitive deficits (Depienne et al. 2007; Lo Giudice et al. 2014; Morfini et al. 2009; Salinas et al. 2008). The exact molecular mechanism(s) by which the D3G and V72I mutations lead to these neurodegenerative disorders is unknown. It is possible that the single-ring intermediates might be involved in the pathogenesis of these diseases; however, more research needs to be done to uncover the molecular basis of these disorders.

Approximately 80% of the hsp60 complexes reside within the mitochondria while the remainder is scattered throughout the cell in the cytosol, nucleus, Golgi, plasma membrane, and peroxisomes (Campanella et al. 2012; Itoh et al. 1995, 2002; Khan et al. 1998; Kirchhoff et al. 2002; Meng et al. 2018; Piselli et al. 2000; Soltys and Gupta 1996). It is unclear if the extra-mitochondrial pool of hsp60 is capable of folding proteins and uses single-ring-intermediates. If yes, are single-ring intermediates relevant in pathological conditions such as infections, cancer, cellular stress, and apoptosis where a role of extra-mitochondrial hsp60 has been implicated? (Belles et al. 1999; Isidoro et al. 2005; Kirchhoff et al. 2002; Lin et al. 2007). The cellular conditions that regulate the formation of these single-ring intermediates, within and outside the mitochondria, are still unknown. Additional studies that elucidate the high-resolution structure of hsp60/10 conformations, along with biochemical, cell-based, and organismal experimental approaches are needed to understand the role of hsp60/10 single-ring intermediates in various biological processes.

2.2 Conclusions

Hsp60/10 has been known for a long time to operate via a protein-folding mechanism that utilizes both single-ring and double-ring intermediates. The hsp60/10 protein-folding cycle utilizes catalytic and architectural features from both the "two-stroke" groEL/ES cycle and the "one-stroke" ϕ-EL systems. However, there are still a number of unanswered questions revolving around the cellular, structural, and pathological details of the hsp60/10 single-ring intermediates. It is especially critical to resolve these questions since hsp60/10 is implicated in several biochemical pathways and disease conditions in humans.

Acknowledgements This work was made possible by the Welch Foundation award (AH-1649) and NIH-NIGMS (SC3GM113805) awarded to RAB. We would also like to acknowledge the UTEP BBRC and grant 5G12MD007592 from the National Institutes on Minority Health and Health Disparities (NIMHD), a component of the National Institutes of Health (NIH). JMB was supported by the NIH-NIGMS under linked Award Numbers RL5GM118969, TL4GM118971, and UL1GM118970.

References

Belles C, Kuhl A, Nosheny R, Carding SR (1999) Plasma membrane expression of heat shock protein 60 in vivo in response to infection. Infect Immun 67:4191–4200

Bhatt JM, Enriquez AS, Wang J et al (2018) Single-ring intermediates are essential for some chaperonins. Front Mol Biosci 5:42

Braig K, Otwinowski Z, Hegde R et al (1994) The crystal structure of the bacterial chaperonin GroEL at 2.8 A. Nature 371:578–586

Brocchieri L, Karlin S (2000) Conservation among HSP60 sequences in relation to structure, function, and evolution. Protein Sci 9:476–486

Bross P, Naundrup S, Hansen J et al (2008) The Hsp60-(p.V98I) mutation associated with hereditary spastic paraplegia SPG13 compromises chaperonin function both in vitro and in vivo. J Biol Chem 283:15694–15700

Campanella C, Bucchieri F, Merendino AM et al (2012) The odyssey of Hsp60 from tumor cells to other destinations includes plasma membrane-associated stages and Golgi and exosomal protein-trafficking modalities. PLoS One 7:e42008

Chaudhuri TK, Paul S (2006) Protein-misfolding diseases and chaperone-based therapeutic approaches. FEBS J 273:1331–1349

Chen DH, Song JL, Chuang DT, Chiu W, Ludtke SJ (2006) An expanded conformation of single-ring GroEL-GroES complex encapsulates an 86 kDa substrate. Structure 14:1711–1722

Cheng MY, Hartl FU, Martin J et al (1989) Mitochondrial heat-shock protein Hsp60 is essential for assembly of proteins imported into yeast mitochondria. Nature 337:620–625

Depienne C, Stevanin G, Brice A, Durr A (2007) Hereditary spastic paraplegias: an update. Curr Opin Neurol 20:674–680

Dubaquie Y, Looser R, Funfschilling U, Jeno P, Rospert S (1998) Identification of in vivo substrates of the yeast mitochondrial chaperonins reveals overlapping but non-identical requirement for hsp60 and hsp10. EMBO J 17:5868–5876

Elborn JS (2016) Cystic fibrosis. Lancet 388:2519–2531

Enriquez AS, Rojo HM, Bhatt JM, Molugu SK, Hildenbrand ZL, Bernal RA (2017) The human mitochondrial Hsp60 in the APO conformation forms a stable tetradecameric complex. Cell Cycle 16:1309–1319

Hansen JJ, Durr A, Cournu-Rebeix I et al (2002) Hereditary spastic paraplegia SPG13 is associated with a mutation in the gene encoding the mitochondrial chaperonin Hsp60. Am J Hum Genet 70:1328–1332

Hildenbrand ZL, Bernal RA (2012) Chaperonin-mediated folding of viral proteins. Adv Exp Med Biol 726:307–324

Illingworth M, Ramsey A, Zheng Z, Chen L (2011) Stimulating the substrate folding activity of a single ring GroEL variant by modulating the cochaperonin GroES. J Biol Chem 286:30401–30408

Illingworth M, Salisbury J, Li W, Lin D, Chen L (2015) Effective ATPase activity and moderate chaperonin-cochaperonin interaction are important for the functional single-ring chaperonin system. Biochem Biophys Res Commun 466:15–20

Isidoro A, Casado E, Redondo A et al (2005) Breast carcinomas fulfill the Warburg hypothesis and provide metabolic markers of cancer prognosis. Carcinogenesis 26:2095–2104

Itoh H, Kobayashi R, Wakui H et al (1995) Mammalian 60-kDa stress protein (chaperonin homolog). Identification, biochemical properties, and localization. J Biol Chem 270:13429–13435

Itoh H, Komatsuda A, Ohtani H et al (2002) Mammalian HSP60 is quickly sorted into the mitochondria under conditions of dehydration. Eur J Biochem 269:5931–5938

Juwono J, Martinus RD (2016) Does Hsp60 provide a link between mitochondrial stress and inflammation in diabetes mellitus? J Diabetes Res 2016:8017571

Karlin S, Brocchieri L (2000) Conservation among HSP60 sequences in relation to structure, function, and evolution. Protein Sci 9:476–486

Khan IU, Wallin R, Gupta RS, Kammer GM (1998) Protein kinase A-catalyzed phosphorylation of heat shock protein 60 chaperone regulates its attachment to histone 2B in the T lymphocyte plasma membrane. Proc Natl Acad Sci U S A 95:10425–10430

Kirchhoff SR, Gupta S, Knowlton AA (2002) Cytosolic heat shock protein 60, apoptosis, and myocardial injury. Circulation 105:2899–2904

Kovacs E, Sun Z, Liu H et al (2010) Characterisation of a GroEL single-ring mutant that supports growth of Escherichia coli and has GroES-dependent ATPase activity. J Mol Biol 396:1271–1283

Levy-Rimler G, Viitanen P, Weiss C et al (2001) The effect of nucleotides and mitochondrial chaperonin 10 on the structure and chaperone activity of mitochondrial chaperonin 60. Eur J Biochem 268:3465–3472

Lin L, Kim SC, Wang Y et al (2007) HSP60 in heart failure: abnormal distribution and role in cardiac myocyte apoptosis. Am J Phys Heart Circ Phys 293:H2238–H2247

Liu H, Kovacs E, Lund PA (2009) Characterisation of mutations in GroES that allow GroEL to function as a single ring. FEBS Lett 583:2365–2371

Lo Giudice T, Lombardi F, Santorelli FM, Kawarai T, Orlacchio A (2014) Hereditary spastic paraplegia: clinical-genetic characteristics and evolving molecular mechanisms. Exp Neurol 261:518–539

Magen D, Georgopoulos C, Bross P et al (2008) Mitochondrial hsp60 chaperonopathy causes an autosomal-recessive neurodegenerative disorder linked to brain hypomyelination and leukodystrophy. Am J Hum Genet 83:30–42

Meng Q, Li BX, Xiao X (2018) Toward developing chemical modulators of Hsp60 as potential therapeutics. Front Mol Biosci 5:35

Molugu SK, Hildenbrand ZL, Morgan DG et al (2016) Ring separation highlights the protein-folding mechanism used by the phage EL-encoded chaperonin. Structure 24:537–546

Morfini GA, Burns M, Binder LI et al (2009) Axonal transport defects in neurodegenerative diseases. J Neurosci 29:12776–12786

Nielsen KL, Cowan NJ (1998) A single ring is sufficient for productive chaperonin-mediated folding in vivo. Mol Cell 2:93–99

Nielsen KL, McLennan N, Masters M, Cowan NJ (1999) A single-ring mitochondrial chaperonin (Hsp60-Hsp10) can substitute for GroEL-GroES in vivo. J Bacteriol 181:5871–5875
Nisemblat S, Parnas A, Yaniv O, Azem A, Frolow F (2014) Crystallization and structure determination of a symmetrical 'football' complex of the mammalian mitochondrial Hsp60-Hsp10 chaperonins. Acta Crystallogr F Struct Biol Commun 70:116–119
Nisemblat S, Yaniv O, Parnas A, Frolow F, Azem A (2015) Crystal structure of the human mitochondrial chaperonin symmetrical football complex. Proc Natl Acad Sci U S A 112:6044–6049
Parnas A, Nadler M, Nisemblat S, Horovitz A, Mandel H, Azem A (2009) The MitCHAP-60 disease is due to entropic destabilization of the human mitochondrial Hsp60 oligomer. J Biol Chem 284:28198–28203
Piselli P, Vendetti S, Vismara D et al (2000) Different expression of CD44, ICAM-1, and HSP60 on primary tumor and metastases of a human pancreatic carcinoma growing in scid mice. Anticancer Res 20:825–831
Salinas S, Proukakis C, Crosby A, Warner TT (2008) Hereditary spastic paraplegia: clinical features and pathogenetic mechanisms. The Lancet Neurology 7:1127–1138
Semenyuk PI, Orlov VN, Sokolova OS, Kurochkina LP (2016) New GroEL-like chaperonin of bacteriophage OBP *Pseudomonas fluorescens* suppresses thermal protein aggregation in an ATP-dependent manner. Biochem J 473:2383–2393
Soltys BJ, Gupta RS (1996) Immunoelectron microscopic localization of the 60-kDa heat shock chaperonin protein (Hsp60) in mammalian cells. Exp Cell Res 222:16–27
Sun Z, Scott DJ, Lund PA (2003) Isolation and characterisation of mutants of GroEL that are fully functional as single rings. J Mol Biol 332:715–728
Sweeney P, Park H, Baumann M et al (2017) Protein misfolding in neurodegenerative diseases: implications and strategies. Transl Neurodegener 6:6
Vercoulen Y, van Teijlingen NH, de Kleer IM, Kamphuis S, Albani S, Prakken BJ (2009) Heat shock protein 60 reactive T cells in juvenile idiopathic arthritis: what is new? Arthritis Res Ther 11:231
Viitanen PV, Lorimer GH, Seetharam R et al (1992) Mammalian mitochondrial chaperonin 60 functions as a single toroidal ring. J Biol Chem 267:695–698
Viitanen PV, Lorimer G, Bergmeier W, Weiss C, Kessel M, Goloubinoff P (1998) Purification of mammalian mitochondrial chaperonin 60 through in vitro reconstitution of active oligomers. Methods Enzymol 290:203–217
Vilasi S, Carrotta R, Mangione MR et al (2014) Human Hsp60 with its mitochondrial import signal occurs in solution as heptamers and tetradecamers remarkably stable over a wide range of concentrations. PLoS One 9:e97657
Wang M, Kaufman RJ (2014) The impact of the endoplasmic reticulum protein-folding environment on cancer development. Nat Rev Cancer 14:581–597
Wolf SG (2006) Single-ring GroEL: an expanded view. Structure 14:1599–1600
Yan X, Shi Q, Bracher A et al (2018) GroEL ring separation and exchange in the chaperonin reaction. Cell 172:605–17 e11

Chapter 3
Hsp60 Inhibitors and Modulators

Antonio Palumbo Piccionello, Paola Marzullo, Silvestre Buscemi, and Andrea Pace

Abstract In this chapter, we focus on the 60 KDa Heat Shock Protein (Hsp60) and discuss some of its biological, molecular and pathological features. The structural and mechanistic aspect of the Hsp60 folding cycle will be also presented. We further illustrate how Hsp60 may be involved in many diseases and therefore considered as an effective therapeutic or theranostic target. Finally, the state-of-the-art on the development of Hsp60 and bacterial GroEL inhibitors and modulators of their expression will be illustrated. This is discussed in the light of a negative chaperonotherapy, and the consequent development of inhibitors, as well as positive chaperonotherapy, in the event its excessive activity is a disease-contrasting event.

Keywords Avrainvillamide · Carboranylphenoxyacetanilide · Cpn60 · Epolactaene · GroEL · Heat shock proteins · Hsp60 · Hsp60 inhibitors · HspD1 · Mizoribine · Pyrazolopyrimidine

Abbreviations

AD	alzheimer's disease
ADP	adenosine diphosphate
APP	amyloid precursor protein
ATP	adenosine triphosphate
AuTPP	Gold(III) *meso*-tetraphenylporphyrin chloride
Aβ	amyloid peptide
COPD	chronic obstructive pulmonary diseases
Cpn60	chaperonins
Cys	cysteine
DNA	deoxyribonucleic acid

A. Palumbo Piccionello (✉) · P. Marzullo · S. Buscemi · A. Pace
Department STEBICEF, University of Palermo, Palermo, Italy
e-mail: antonio.palumbopiccionello@unipa.it

A. A. A. Asea, P. Kaur (eds.), *Heat Shock Protein 60 in Human Diseases and Disorders*, Heat Shock Proteins 18,
https://doi.org/10.1007/978-3-030-23154-5_3

EM	electron microscopy
ESKAPE	*Enterococcus faecium, Staphylococcus aureus, Klebsiella pneumoniae, Acinetobacter baumannii, Pseudomonas aeruginosa* and *Enterobacter*
ETB	epolactaene tertiary butyl ester
GroEL	bacterial 60 KDa heat shock protein
GroES	bacterial 10 KDa heat shock protein
GTP	guanosine triphosphate
HNE	4-hydroxynonenal
HSP	heat shock proteins
Hsp10	10 KDa heat shock protein
Hsp40	40 KDa heat shock protein
Hsp60	(HspD1), 60 KDa heat shock protein.
Hsp70	70 KDa heat shock protein
Hsp90	90 KDa heat shock protein
Hsp110	110 KDa heat shock protein
HTS	high-throughput screening
ITC	isothermal titration calorimetry
MD	molecular dynamics
MIS	mitochondrial import sequence
NF-kB	nuclear factor kappa-light-chain-enhancer of activated B cells
MRSA	methicillin-resistant *Staphylococcus Aureus*
NMR	nuclear magnetic resonance
SAR	structure-activity relationship
SAXS	small-angle X-ray scattering
sHsp	small Hsp
SP	substrate protein
SPG13	hereditary spastic paraplegia
TEM	transmission electron microscopy

3.1 Introduction

The concept of the "chaperome" is related to a system that includes all molecular chaperones, co-chaperones and co-factors of an organism (Macario and Conway de Macario 2005). Chaperonology is the science that studies this system, which also includes the study of diseases due to chaperoning system malfunction (Macario and Conway de Macario 2007a). The chaperome is physiologically necessary for managing protein homeostasis, folding and trafficking in a highly regulated fashion, therefore, its failure results in a pathological condition (Macario and Conway de Macario 2005, 2007b). Heat Shock Proteins (HSP) are classical molecular chaperones and they play a crucial role in the biosynthesis, folding/unfolding, transport and assembly of other proteins (Finka and Goloubinoff 2013). Molecular chaperones have been traditionally classified on the basis of their molecular weight, into: heavy (Hsp110, Hsp90, Hsp70, Hsp60, Hsp40), and small Hsp (sHsp) (Macario et al.

2013). 60 KDa Heat Shock Proteins, usually known as chaperonins (Cpn60), mainly represented by the human Hsp60 and the bacterial homologue GroEL, are one of the main component of the chaperome and are classically described as mitochondrial proteins, constitutively expressed under normal conditions and induced by various types of stressors as heat shock, oxidative stress, and DNA damage (Czarnecka et al. 2006). Inside mitochondria, they act as a folding machine, together with co-chaperonin Hsp10 (GroES in bacteria), for the correct folding of several mitochondrial proteins (Bukau and Horwich 1998; Parnas et al. 2012). A new nomenclature has been proposed on the basis of the HUGO Gene Nomenclature Committee and used in the National Center of Biotechnology Information Entrez Gene database for the heat shock genes (Kampinga et al. 2009). Even if this terminology is not universally adopted in the scientific literature, Hsp60 is often referred as HspD1.

3.2 Hsp60 Structure and Functions

Hsp60/GroEL assists client protein's folding by forming a tetradecameric structure with a barrel shape obtained by two heptameric rings. (Horwich et al. 2006) The folding process is also assisted by the 10 kDa co-chaperonin Hsp10/GroES and by ATP to ADP hydrolysis. The monomeric chaperonin presents: (i) an equatorial domain responsible for ATP/ADP binding and inter-ring interaction; (ii) an intermediate domain and (iii) an apical domain, whose movement is responsible for substrate protein (SP) capture/release and interaction with co-chaperonin. Concerning the structure of the bacterial GroEL, for many years the active form of this supramolecular folding machine was assumed to be an asymmetric bullet-shaped $GroEL_{14}$:$GroES_7$ complex (Clare et al. 2012), while only recently a symmetric football-shaped $GroEL_{14}$:$(GroES_7)_2$ complex was also evidenced (Yang et al. 2013; Fei et al. 2014). In these models, the binding of the SP precedes the one of ATP. Therefore, the barrel-shaped tetradecamer in its T-state (ATP-unbound) is the basis for the recognition of SPs. Further studies revealed the role of the C-terminus of GroEL, at the bottom of the barrel chamber, useful to force SP entrance in, and final substrate release from, the folding chamber (Weaver and Rye 2014). On the other hand, the human homologue Hsp60 similarly works with its co-chaperone Hsp10 in a ATP-mediated process, even if Hsp60 shows less affinity for Hsp10 than the bacterial homologue, forming preferentially a single heptameric ring instead of the barrel-shaped double-ring (Parnas et al. 2009). The crystal structure of human Hsp60, in complex with Hsp10, was also obtained, showing a symmetric football-shaped assembly also for the mammalian form (Nisemblat et al. 2014, 2015). The human Hsp60 system behave differently from the bacterial GroEL; in fact, also single heptameric rings are described as stable oligomeric state together with tetradecameric double-rings (Enriquez et al. 2017), and the binding with ATP seems crucial for the Hsp10 recognition (Enriquez et al. 2017). A GTPase activity was also evidenced for Hsp60 (Okamoto et al. 2017). TEM experiments revealed the presence of a complex distribution of double-ring (football and bullet-type complexes) and single-ring complexes

(Ishida et al. 2018). In general, the folding cycle of Hsp60 seems more complex than that reported for GroEL. The growth of EM techniques boosted this research field and allowed to shed light into these complex mechanisms.

Furthermore, Naïve-Hsp60, which is characterized by the presence of a 26 peptides Mitochondrial Import Sequence (MIS) linked at the *N*-terminus, was also studied (Vilasi et al. 2014). The oligomeric states of naïve-Hsp60 was investigated, revealing the presence of an equilibrium between stable heptameric and tetradecameric forms (Spinello et al. 2015). Detailed quaternary structure in solution, under physiological conditions, was investigated by means of MD simulations and SAXS experimental data, analyzed by the QUAFIT method (Spinozzi and Beltramini 2012). The combination of MD simulations and the advanced SAXS data analysis provided the quaternary structure of both GroEL and Hsp60 in naïve-solution.

3.3 Hsp60 Involvement in Different Pathologies

The role of Hsp60 in many pathologies is evident and in particular in those cases related to the protein's malfunction due to mutations of the *Hsp60* (*HSPD1*) gene (Bross and Fernandez-Guerra 2016). Hereditary spastic paraplegia SPG13 is a rare and incurable neurodegenerative disorder related to the V98I mutation in Hsp60 (Bross et al. 2008), while MitCHAP-60 disease, which is associated with D27G or D3G mutations, is a neurodegenerative disorder that causes hypomyelination and leukodystrophy in the brain (Magen et al. 2008). Beside this direct involvement, Hsp60's role as "friend or foe" is currently under investigation for various conditions. In Cancer, the Hsp60's role has been widely assessed and its anti-apoptotic role in tumor-cell lifecycle unraveled (Wu et al. 2017; Cappello et al. 2014, Kondoh and Osada 2013; Pace et al. 2013). Tumor pro-survival effects are due, for example, to interactions with caspase-3 (Chandra et al. 2007) and nuclear factor-kB (NF-kB)-dependent pathway (Chun et al. 2010). In general, Hsp60 overexpression is a frequent pro-tumoral factor that could be used as diagnostic as well prognostic tool. In this field, also the interest toward the extracellular exosomial fraction is increasing (Campanella et al. 2015; Lv et al. 2012; Merendino et al. 2010).

Moreover, Hsp60 is a risk factor for atherosclerosis (Wick et al. 2014), diabete (Juwono and Martinus 2016) and cardiovascular diseases (Rizzo et al. 2011), such as myocardial injury (Kirchhoff et al. 2002).

Hsp60 is also involved in neurodegenerative diseases related to protein misfolding like Alzheimer's Disease (AD) (Campanella et al. 2018). Interactions of Hsp60 with both amyloid peptide (Aβ) (Veereshwarayya, et al. 2006; Mangione et al. 2016) and amyloid precursor protein (APP) (Walls et al. 2012) were reported but with controversial role regarding its protective role in neurons or as cause of mithocondrial stress in AD. Interestingly, an Aβ-Hsp60 conjugate vaccine was successfully tested in a mouse model of AD (Nemirovsky et al. 2011). Finally, due to strict homology of human Hsp60 with that of several bacteria, Hsp60 is also involved in different autoimmune diseases (van Eden et al. 2017) including inflammatory bowel diseases (Tomasello et al. 2011), chronic obstructive pulmonary diseases (COPD) (Cappello et al. 2011), Hashimoto's thyroiditis (Marino Gammazza et al. 2014) and myasthenia gravis (Marino Gammazza et al. 2012).

3.4 Hsp60 Inhibitors and Modulators

Despite various studies point toward targeting Hsp60 as a promising therapeutic strategy, only a few compounds have been characterized in some detail as Hsp60 inhibitors (Meng et al. 2018). Moreover, in many cases, the mechanism of action of such inhibitors needs to be clearly assessed. Therefore, both in the development of new Hsp60 inhibitors and in the study of their mechanism of action, a particular care should be paid to structural differences between eukaryotic Hsp60 and its more widely studied prokaryotic homolog, GroEL. For instance, GroEL lacks three cysteine residues that are instead present in the human Hsp60 (i.e. Cys237, Cys442 and Cys447). In fact, due to their nucleophilic character and easy oxidizability, such residues represent ideal sites for drug interaction (Cappello et al. 2014). Two approaches have been proposed so far to design Hsp60 inhibitors (Cappello et al. 2014). The first one targets the sites which are functional for hydrolysis and ATP binding, thus affecting those Hsp60's conformational changes which are ATP-dependent and crucial for inducing protein folding (Itoh et al. 1999; Chapman et al. 2008; Tanabe et al. 2012).

The second strategy targets Hsp60's cysteine residues as either oxidizable sites (Wang et al. 2013) or for covalent binding, likely by reaction with an electrophilic moiety on the drug candidate (Nagumo et al. 2004, 2005; Wulff et al. 2007). For instance, mizoribine (Fig. 3.1, left) is an immunosuppressant imidazole *N*-ribofuranoside which targets Hsp60's ATPase activity by forming a complex with Hsp60 and affecting its protein-folding activity (Itoh et al. 1999; Chapman et al. 2008). Mizoribine was shown to affect ATP hydrolysis thus slowing down the protein-folding cycle. Additionally, mizoribine's activity was hypothesized to inhibit the dissociation of the co-chaperonin Hsp10 from the Hsp60/Hsp10 complex. In this case, however, there was a relevant difference in the activity observed

Fig. 3.1 Modulators of Hsp60's ATP-dependent activity

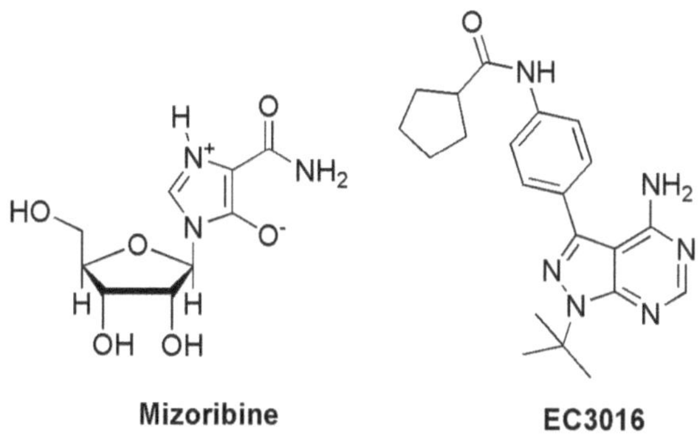

with prokaryotic and eukaryotic Hsp60, since the GroEL/GroES system was not significantly affected by mizoribine (Tanabe et al. 2012). Similarly, another aza-aromatic heterocycle, pyrazolopyrimidine EC3016 (Fig. 3.1, right), was able to affect the protein-folding function of Hsp60 by blocking ATP binding and hydrolysis (Chapman et al. 2008). Surprisingly, no further update on EC3016 use has appeared since the original report of its inhibitory activity against Hsp60's function. As for the second strategy, a different class of molecules has been reported to interact with of Hsp60's cysteine residues (Fig. 3.2) (Cappello et al. 2014).

One of these compounds, avrainvillamide, can alkylate the thiol function of cysteine residues through the 3-alkylidene-3H-indole 1-oxide electrophilic moiety (Wulff et al. 2007). However, its inhibitory effects on Hsp60 functions has not been unequivocally demonstrated so far. Other Hsp60-interacting molecules were identified found among natural compounds, such as epolactaene, which inhibits the chaperoning activity of human Hsp60 by specifically binding the Cys442 residue (Nagumo et al. 2005). Recently, other epolactaene derivatives, such as its tertiary butyl ester ETB, were proved to target mitochondrial transcription (Sun et al. 2012). A structure-activity relationship (SAR) analysis has been performed on a series of epolactaene derivatives demonstrating that both the α-β unsaturated ketone and the lactam are critical moieties to inhibit Hsp60's chaperone activity (Nagumo et al. 2004). However, by looking at the whole structure of ETB, one can recognize more than one electrophilic site able to covalently bind the thiol groups of cysteine residues. In particular, the epoxide moiety of epolactaene should be the most likely binding site for the nucleophilic cysteine residues. This was demonstrated in a study on the reactivity of epolactaene with a series of thiols, including cysteine, pointing out epolactaene's ability to favor oxidation to disulfide (Kuramochi et al. 2011). The mechanism of action of epolactaene was investigated by means of Molecular Dynamics (MD), revealing that epolactaene could bind Hsp60 at the ATP-binding site in the APO form and, more intriguing, a binding pocket in proximity of Cys442 after ATP binding (Spinello et al. 2016). Once in proximity of Cys 442, epolactaene covalently binds thiol moiety through attack at C14 and epoxide ring-opening. Interestingly, epolactaene binding

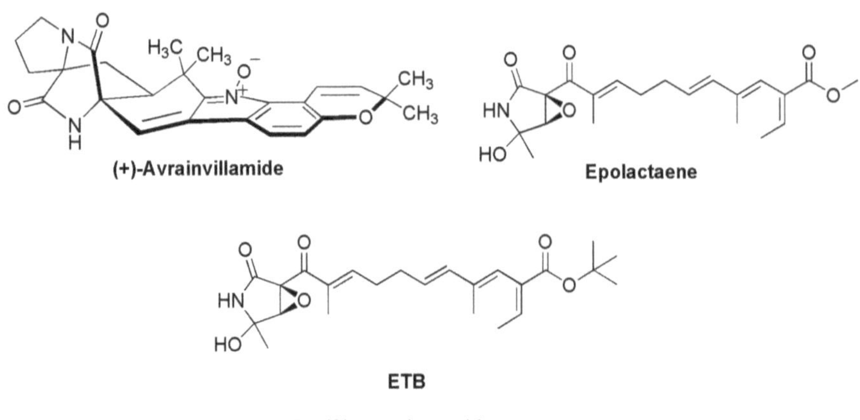

Fig. 3.2 Compounds targeting Hsp60's cysteine residues

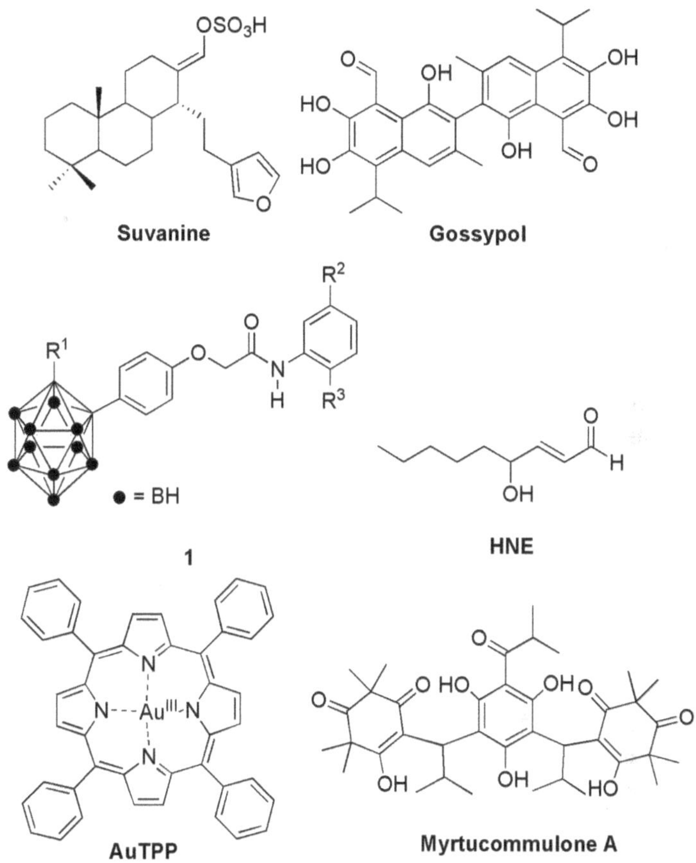

Fig. 3.3 Molecular structure of compounds interacting with Hsp60

hinders the dynamic conformational changes of the apical domain of the monomer, necessary for SP capture and release during folding process.

Another study, based on a proteomic screening of the interactions between several natural compounds and Hsp60, showed that suvanine (Fig. 3.3), a natural sesquiterpene of marine origin targets the chaperonin's cysteine residues for sulfation (Cassiano et al. 2012). On the other hand gossypol (Fig. 3.3), a polyphenolic drugs which induces apoptosis through oxidative stress, interacts with Hsp60 interfering with the typical thiol/disulfide redox reaction of its cysteine residues (Wang et al. 2013). Interestingly, a hypoxic-inducible factor 1 alpha inhibitor, containing an unusual (for pharmacological purposes) carboranyl moiety, was found to target Hsp60, although its actual binding site is still undefined (Ban et al. 2010). Recently, one of the carboranylphenoxyacetanilide derivatives (**1**) (Fig. 3.3; $R^1 = CH_2CH_3$, $R^2 = B(OH)_2$, $R^3 = OH$) showed a twofold chaperone inhibitory activity compared to ETB (Ban et al. 2012; Nakamura et al. 2013).

Another proteomic analysis performed on compounds targeting heat shock proteins involved in the stress response (Hsp60, Hsp70, Hsp90, and 78-kDa glucose regulated protein) revealed that 4-hydroxynonenal (HNE) (Fig. 3.3) targets Hsp60, with a dose-dependent increase in labeled proteins with increased sequence coverage at higher concentrations (Vila et al. 2008). Also in this case, the binding site was not discovered; however, a nucleophilic attack of the cysteine thiol group on the electrophilic α-β-unsaturated aldehyde moiety can be reasonably suggested as a mode of binding. Gold(III) *meso*-tetraphenylporphyrin chloride (AuTPP) (Fig. 3.3), is a well-known antitumoral that was revealed to inhibit Hsp60 through different experiments: chemoproteomic, NMR and folding assays (Hu et al. 2016). The presence of both Gold(III) and the porphyrin scaffold is necessary for activity as established through SAR studies; nevertheless, the mode of action is still unclear. Similarly, antitumoral drug myrtucommulone A (Fig. 3.3), was find to bind Hsp60 by means of a protein fishing approach (Wiechmann et al. 2017). The activity of myrtucommulone A was further confirmed by malate dehydrogenase protein refolding assay. Other representative molecules interacting with Hsp60 or affecting its expression with unknown mechanism of action ranges from copper complex (**2**), to matrine and curcumin (Fig. 3.4). For example, the exposure of tumor cells to some recently characterized copper complexes (Terenzi et al. 2010, 2011) showed antitumor activity that was correlated to decreased levels of Hsp60 and of the Hsp60/pro-Caspase-3-complex (Caruso Bavisotto et al. 2017). The neuroprotective effect of Matrine (Zhang et al. 2017) and Curcumin (Ding et al. 2016) was ascribed to reduced Hsp60 expression and therefore to lower microglia activation. Another interesting research field is the discovery of GroEL inhibitors as a new class of antibacterials. An HTS screening performed on 700,000 compounds allowed the individuation of compound (**3**) as hit inhibitor (Johnson et al. 2014), together with compounds (**4–5**) active toward ESKAPE pathogens (Abdeen et al. 2016). Further SAR optimization yields to compounds (**6–7**) as potent antibacterial versus resistant strains such as MRSA (Abdeen et al. 2018). The mode of action seem linked to the targeting of the ATP binding site, interestingly, from ITC experiments highlighted

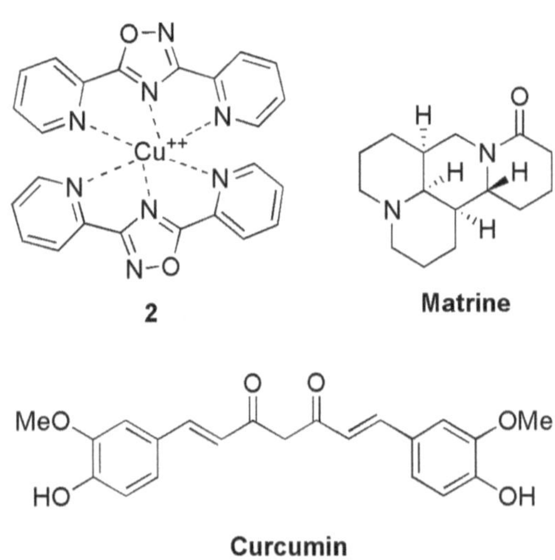

Fig. 3.4 Chemical structure of Hsp60 modulating molecules

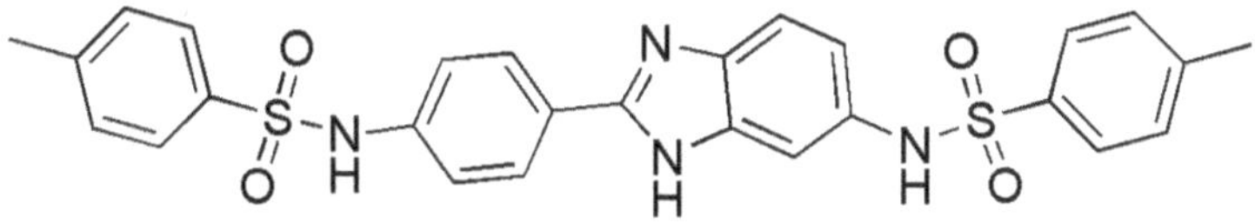

3

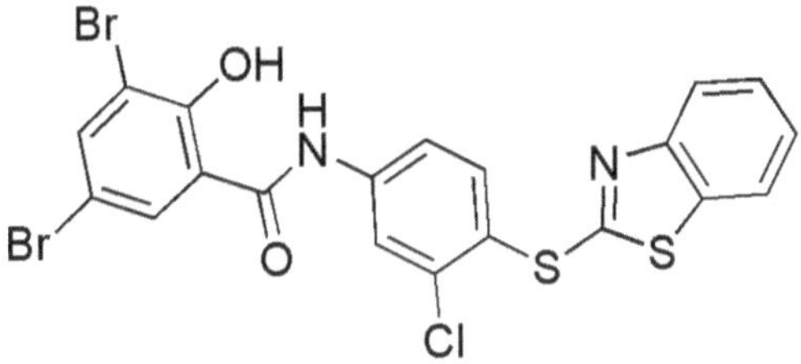

4

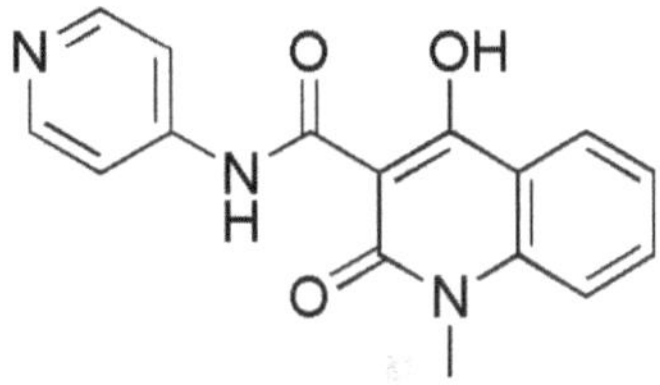

5

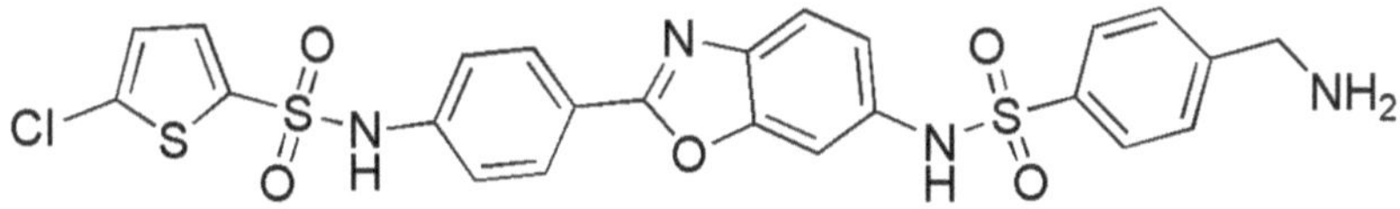

6

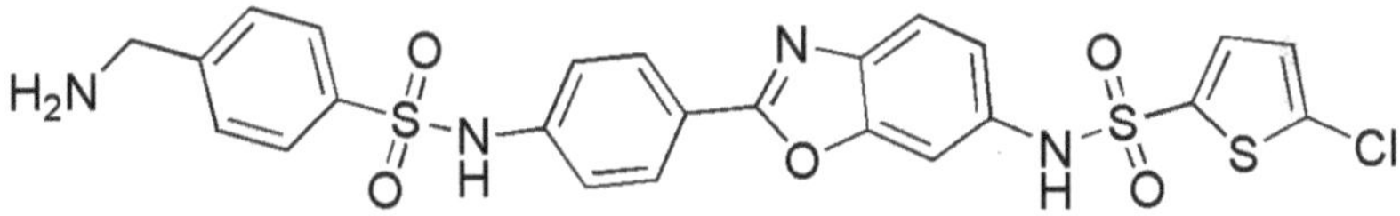

7

Fig. 3.5 Chemical structure of GroEL inhibitors

the binding of more than one molecule per monomer, opening the possibility to the presence of other bonding sites (Fig. 3.5).

3.5 Conclusions

In summary, current strategies to target chaperonin's activity focus on its ATP binding site or cysteine residues. Nevertheless, several mechanism of action need to be assessed and other regions of Hsp60 can be surveyed to develop novel inhibitors. For instance, crucial features for protein (re)folding activity, such as the site of interaction between the mitochondrial Hsp60 and Hsp10, could be suggested as target for new inhibitors/modulators. Ideally, a thorough study on the development of new Hsp60 inhibitors should test their efficacy in inhibiting ATP binding and hydrolysis and the chaperonin's protein folding activity, address the binding capability and identify the docking site (Cappello et al. 2014). Unfortunately, these aspects are rarely addressed in comprehensive studies (Nagumo et al. 2004; Tanabe et al. 2012). Therefore, there is still room for studies aiming at resolving the unanswered questions concerning current Hsp60 inhibitors. In this context, the lack of experimental information on the mechanism of action of several promising Hsp60-targeting drugs, remarks the importance and the supporting role of biomolecular computational studies. These can be used preliminarly to drug design, to explore other regions of Hsp60, including, for instance, the site of interaction between the mitochondrial Hsp60 and its co-chaperonin Hsp10. Additionally, the lack of consensus on the nature of the oligomers involved in the folding cycle and the lack of a co-crystallized structure with a known inhibitor, leave several unanswered questions concerning Hsp60 role in pathologies. Overall, the design of drugs targeting Hsp60 is a perspective growing field of research and its translation into potential therapies remains still unexploited.

Acknowledgements Financial support from the Italian MIUR within the "FIRB-Futuro in Ricerca 2012" Program - Project RBFR12SIPT is acknowledged.

References

Abdeen S, Salim N, Mammadova N et al (2016) GROEL/ES inhibitors as potential antibiotics. Bioorg Med Chem Lett 26:3127–3134

Abdeen S, Kunkle T, Salim N et al (2018) Sulfonamido-2-arylbenzoxazole GROEL/ES inhibitors as potent antibacterials against methicillin-resistant Staphylococcus aureus (MRSA). J Med Chem 61:7345–7357

Ban HS, Shimizu K, Minegishi H, Nakamura H (2010) Identification of HSP60 as a primary target of o-carboranylphenoxyacetanilide, an HIF-1α inhibitor. J Am Chem Soc 132:11870–11871

Ban HS, Shimizu K, Minegishi H, Nakamura H (2012) Identification of heat shock protein 60 as the regulator of the hypoxia-inducible factor subunit HIF-1α. Pure Appl Chem 84:2325–2337

Bross P, Naundrup S, Hansen J et al (2008) The Hsp60-(p.V98I) mutation associated with hereditary spastic paraplegia SPG13 compromises chaperonin function both in vitro and in vivo. J Biol Chem 283:15694–15700

Bross P, Fernandez-Guerra P (2016) Disease-associated mutations in the HSPD1 gene encoding the large subunit of the mitochondrial HSP60/HSP10 chaperonin complex. Front Mol Biosci 3:49

Bukau B, Horwich AL (1998) The Hsp70 and Hsp60 chaperone machines. Cell 92:351–366

Campanella C, Rappa F, Sciumé C et al (2015) Heat shock protein 60 levels in tissue and circulating exosomes in human large bowel cancer before and after ablative surgery. Cancer 121:3230–3209

Campanella C, Pace A, Caruso Bavisotto C, Marzullo P, Marino Gammazza A, Buscemi S, Palumbo Piccionello A (2018) Heat shock proteins in Alzheimer's disease: role and targeting. Int J Mol Sci 19:2603

Cappello F, Caramori G, Campanella C et al (2011) Convergent sets of data from in vivo and in vitro methods point to an active role of Hsp60 in chronic obstructive pulmonary disease pathogenesis. PLoS One 6:e28200

Cappello F, Marino Gammazza A, Palumbo Piccionello A et al (2014) Hsp60 chaperonopathies and chaperonotherapy: targets and agents. Expert Opin Ther Targets 18:185–208

Caruso Bavisotto C, Nikolic D, Marino Gammazza A et al (2017) The dissociation of the Hsp60/pro-caspase-3-complex by bis (pyrydil)oxadiazole complex (CubipyOXA) leads to cell death in NCl-H292 cancer cells. J Inorg Biochem 170:8–16

Cassiano C, Monti MC, Festa C, Zampella A, Riccio R, Casapullo A (2012) Chemical proteomics reveals heat shock protein 60 to be the main cellular target of the marine bioactive sesterterpene suvanine. Chembiochem 13:1953–1958

Chandra D, Choy G, Tang DG (2007) Cytosolic accumulation of HSP60 during apoptosis with or without apparent mitochondrial release: evidence that its pro-apoptotic or pro-survival functions involve differential interactions with caspase-3. J Biol Chem 282:31289–31301

Chapman E, Farr GW, Fenton WA, Johnson SM, Horwich AL (2008) Requirement for binding multiple ATPs to convert a GroEL ring to the folding-active state. Proc Natl Acad Sci U S A 105:19205–19210

Chun JN, Choi B, Lee KW et al (2010) Cytosolic Hsp60 is involved in the NF-kB dependent survival of cancer cells via IKK regulation. PLoS One 5:e9422

Clare DK, Vasishtan D, Stagg S et al (2012) ATP-triggered conformational changes delineate substrate-binding and -folding mechanics of the GroEL chaperonin. Cell 149:113

Czarnecka AM, Campanella C, Zummo G, Cappello F (2006) Mitochondrial chaperones in cancer: from molecular biology to clinical diagnostics. Cancer Biol Ther 5:714–720

Ding F, Li F, Li Y et al (2016) HSP60 mediates the neuroprotective effects of curcumin by suppressing microglial activation. Exp Ther Med 12:823–828

Enriquez AS, Rojo HM, Bhatta JM, Molugu SK, Hildenbrand ZL, Bernal RA (2017) The human mitochondrial Hsp60 in the APO conformation forms a stable tetradecameric complex. Cell Cycle 16:1309–1319

Fei X, Ye X, LaRonde NA, Lorimer GH (2014) Formation and structures of GroEL:GroES$_2$ chaperonin footballs, the protein-folding functional form. Proc Natl Acad Sci U S A 111:12775–12780

Finka A, Goloubinoff P (2013) Proteomic data from human cell cultures refine mechanisms of chaperone-mediated protein homeostasis. Cell Stress Chaperones 18:591–605

Horwich L, Farr GW, Fenton WA (2006) GroEL–GroES-mediated protein folding. Chem Rev 106:1917–1930

Hu D, Liu Y, Lai YT et al (2016) Anticancer gold(III) porphyrins target mitochondrial chaperone Hsp60. Angew Chem Int Ed Engl 55:1387–1391

Ishida R, Okamoto T, Motojima F et al (2018) Physicochemical properties of the mammalian molecular chaperone HSP60. Int J Mol Sci 19:489

Itoh H, Komatsuda A, Wakui H, Miura AB, Tashima Y (1999) Mammalian HSP60 is a major target for an immunosuppressant mizoribine. J Biol Chem 274:35147–35151

Johnson SM, Sharif O, Mak PA et al (2014) A biochemical screen for GROEL/GROES inhibitors. Bioorg Med Chem Lett 24:786–789

Juwono J, Martinus RD (2016) Does Hsp60 provide a link between mitochondrial stress and inflammation in diabetes mellitus? J Diabetes Res 2016:8017571

Kampinga HH, Hageman J, Vos MJ et al (2009) Guidelines for the nomenclature of the human heat shock proteins. Cell Stress Chaperones 14:105–111

Kirchhoff SR, Gupta S, Knowlton AA (2002) Cytosolic heat shock protein 60, apoptosis, and myocardial injury. Circulation 105:2899–28904

Kondoh Y, Osada H (2013) High-throughput screening identifies small molecule inhibitors of molecular chaperones. Curr Pharm Des 19:473–492

Kuramochi K, Sunoki T, Tsubaki K et al (2011) Transformation of thiols to disulfides by epolactaene and its derivatives. Bioorg Med Chem 19:4162–4172

Lv LH, Wan YL, Lin Y et al (2012) Anticancer drugs cause release of exosomes with heat shock proteins from human hepatocellular carcinoma cells that elicit effective natural killer cell antitumor responses in vitro. J Biol Chem 4:15874–15885

Macario AJL, Conway de Macario E (2005) Sick chaperones, cellular stress, and disease. N Engl J Med 353:1489–1501

Macario AJL, Conway de Macario E (2007a) Chaperonopathies and chaperonotherapy. FEBS Lett 581:3681–3688

Macario AJL, Conway de Macario E (2007b) Molecular chaperones: multiple functions, pathologies, and potential applications. Front Biosci 12:2588–2600
Macario AJL, Conway de Macario E, Cappello F (2013) The chaperonopathies diseases with defective molecular chaperones. Springer, Dordrecht/Heidelberg/New York. Available at: http://www.springer.com/biomed/book/978-94-007-4666-4)
Magen D, Georgopoulos C, Bross P et al (2008) Mitochondrial hsp60 chaperonopathy causes an autosomalrecessive neurodegenerative disorder linked to brain hypomyelination and leukodystrophy. Am J Hum Genet 83:30–42
Mangione MR, Vilasi S, Marino C et al (2016) Hsp60, amateur chaperone in amyloid-beta fibrillogenesis. Biochim Biophys Acta 1860:2474–2483
Marino Gammazza A, Bucchieri F, Grimaldi LM, Benigno A, Conway de Macario E, Macario AJL (2012) The molecular anatomy of human Hsp60 and its similarity with that of bacterial orthologs and acetylcholine receptor reveal a potential pathogenetic role of anti-chaperonin immunity in myasthenia gravis. Cell Mol Neurobiol 32:943–947
Marino Gammazza A, Rizzo M, Citarrella R et al (2014) Elevated blood Hsp60, its structural similarities and cross-reactivity with thyroid molecules, and its presence on the plasma membrane of oncocytes point to the chaperonin as an immunopathogenic factor in Hashimoto's thyroiditis. Cell Stress Chaperones 19:343–353
Meng Q, Li BX, Xiao X (2018) Toward developing chemical modulators of Hsp60 as potential therapeutics. Front Mol Biosci 5:35
Merendino AM, Bucchieri F, Campanella C et al (2010) Hsp60 is actively secreted by human tumor cells. PLoS One 5:e9247
Nagumo Y, Kakeya H, Yamaguchi J et al (2004) Structure–activity relationships of epolactaene derivatives: structural requirements for inhibition of Hsp60 chaperone activity. Bioorg Med Chem Lett 14:4425–4429
Nagumo Y, Kakeya H, Shoji M, Hayashi Y, Dohmae N, Osada H (2005) Epolactaene binds human Hsp60 Cys442 resulting in the inhibition of chaperone activity. Biochem J 387:835–840
Nakamura H, Yasui Y, Maruyama M, Minegishi H, Ban HS, Sato S (2013) Development of hypoxia-inducible factor (HIF)-1a inhibitors: effect of ortho-carborane substituents on HIF transcriptional activity under hypoxia. Bioorg Med Chem Lett 23:806–810
Nemirovsky A, Fisher Y, Baron R, Cohen IR, Monsonego A (2011) Amyloid beta-HSP60 peptide conjugate vaccine treats a mouse model of Alzheimer's disease. Vaccine 23:4043–4050
Nisemblat S, Parnas A, Yaniv O, Azem A, Frolow F (2014) Crystallization and structure determination of a symmetrical 'football' complex of the mammalian mitochondrial Hsp60-Hsp10 chaperonins. Acta Crystallogr F Struct Biol Cryst Commun 70:116–119
Nisemblat S, Yaniv O, Parnas A, Frolow F, Azem A (2015) Crystal structure of the human mitochondrial chaperonin symmetrical football complex. Proc Natl Acad Sci 112:6044–6049
Okamoto T, Yamamoto H, Kudo I et al (2017) HSP60 possesses a GTPase activity and mediates protein folding with HSP10. Sci Rep 7:16931
Pace A, Barone G, Lauria A et al (2013) Hsp60, a novel target for antitumor therapy: structure-function features and prospective drugs design. Curr Pharm Des 19:2757–2764
Parnas A, Nadler M, Nisemblat S, Horovitz A, Mandel H, Azem A (2009) The MitCHAP-60 disease is due to entropic destabilization of the human mitochondrial Hsp60 oligomer. J Biol Chem 284:28198–28203
Parnas A, Nisemblat S, Weiss C et al (2012) Identification of elements that dictate the specificity of mitochondrial Hsp60 for its co-chaperonin. PLoS One 7:e50318
Rizzo M, Macario AJL, Conway de Macario E et al (2011) Heat shock protein-60 and risk for cardiovascular disease. Curr Pharm Des 17:3662–3668
Spinello A, Ortore MG, Spinozzi F et al (2015) Quaternary structures of GROEL and naïve-HSP60 chaperonins in solution: a combined SAXS-MD study. RCS Adv 5:49871–49879
Spinello A, Barone G, Cappello F, Pace A, Buscemi S, Palumbo Piccionello A (2016) The binding mechanism of epolactaene to Hsp60 unveiled by in silico modelling. Chemistry Select 1:759–765

Spinozzi F, Beltramini M (2012) QUAFIT: a novel method for the quaternary structure determination from small-angle scattering data. Biophys J 103:511

Sun W, Wang L, Jiang H et al (2012) Targeting mitochondrial transcription in fission yeast with ETB, an inhibitor of HSP60, the chaperone that binds to the mitochondrial transcription factor Mtf1. Genes Cells 17:122–131

Tanabe M, Ishida R, Izuhara F et al (2012) The ATPase activity of molecular chaperone HSP60 is inhibited by immunosuppressant mizoribine. Am J Mol Biol 2:93–102

Terenzi A, Barone G, Palumbo Piccionello A et al (2010) Synthesis, characterization, cellular uptake and interaction with native DNA of a bis(pyridyl)-1,2,4-oxadiazole copper(II) complex. Dalton Trans 39:9140–9145

Terenzi A, Barone G, Palumbo Piccionello A, Giorgi G, Guarcello A, Pace A (2011) Synthesis and chemical characterization of CuII, NiII and ZnII complexes of 3,5-bis(2′-pyridyl)-1,2,4-oxadiazole and 3-(2′-pyridyl)5-(phenyl)-1,2,4-oxadiazole. Inorg Chim Acta 373:62–67

Tomasello G, Rodolico V, Zerilli M et al (2011) Changes in immunohistochemical levels and subcellular localization after therapy and correlation and colocalization with CD68 suggest a pathogenetic role of Hsp60 in ulcerative colitis. Appl Immunohistochem Mol Morphol 19:552–561

Van Eden W, Jansen MAA, Ludwig I, Van Kooten P, Van der Zee R, Broere F (2017) The enigma of heat shock proteins in immune tolerance. Front Immunol 8:1599

Veereshwarayya V, Kumar P, Rosen KM, Mestril R, Querfurth HW (2006) Differential effects of mitochondrial heat shock protein 60 and related molecular chaperones to prevent intracellular beta-amyloid-induced inhibition of complex IV and limit apoptosis. J Biol Chem 281:29468–29478

Vila A, Tallman KA, Jacobs AT, Liebler DC, Porter NA, Marnett LJ (2008) Identification of protein targets of 4-hydroxynonenal using click chemistry for ex vivo biotinylation of azido and alkynyl derivatives. Chem Res Toxicol 21:432–444

Vilasi S, Carrotta R, Mangione MR et al (2014) Human Hsp60 with its mitochondrial import signal occurs in solution as heptamers and tetradecamers remarkably stable over a wide range of concentrations. PLoS One 9:e97657

Walls KC, Coskun P, Gallegos-Perez JL et al (2012) Swedish Alzheimer mutation induces mitochondrial dysfunction mediated by HSP60 mislocalization of amyloid precursor protein (APP) and beta-amyloid. J Biol Chem 31:30317–33027

Wang J, Jin L, Li X et al (2013) Gossypol induces apoptosis in ovarian cancer cells through oxidative stress. Mol BioSyst 9:1489–1497

Weaver J, Rye HS (2014) The C-terminal tails of the bacterial chaperonin GroEL stimulate protein folding by directly altering the conformation of a substrate protein. J Biol Chem 289:23219–23232

Wick G, Jakic B, Buszko M, Wick MC, Grundtman C (2014) The role of heat shock proteins in atherosclerosis. Nat Rev Cardiol 11:516–529

Wiechmann K, Müller H, König S et al (2017) Mitochondrial chaperonin Hsp60 is the apoptosis related target for myrtucommulone. Cell Chem Biol 24:614–623

Wu J, Liu T, Rios Z, Mei Q, Lin X, Cao S (2017) Heat shock proteins and cancer. Trends Pharmacol Sci 38:226–256

Wulff JE, Herzon SB, Siegrist R, Myers AG (2007) Evidence for the rapid conversion of Stephacidin b into the electrophilic monomer avrainvillamide in cell culture. J Am Chem Soc 129:4898–4899

Yang D, Ye X, Lorimer GH (2013) Symmetric GroEL:GroES2 complexes are the protein-folding functional form of the chaperonin nanomachine. Proc Natl Acad Sci U S A 110:E4298

Zhang R, Li Y, Hou X, Miao Z, Wang Y et al (2017) Neuroprotective effect of heat shock protein 60 on matrine-suppressed microglial activation. Exp Ther Med 14:1832–1836

Chapter 4
HSP60 as Modulators of Apoptosis

Pratima Chaudhuri (Chattopadhyay) and Naira Rashid

Abstract Heat shock proteins are a significant class of proteins that play a pivotal role in cells undergoing stress (heat/chemical). They also perform certain important functions in cells during normal conditions. Majority of heat shock proteins execute the role of molecular chaperones by enabling the unfolded protein to fold properly. Heat shock protein 60 (HSP 60) plays a significant role in protein homeostasis inside the mitochondria. Aberrant expression of HSP 60 has been observed to result in various disorders in humans. It is also responsible for various cell survival and apoptotic pathways. A wide cell survival program is orchestrated by HSP 60 and this process can be particularly explored for carcinogenesis. Therefore it can act as a lucrative target for treatment of various types of cancers and other diseases.

Keywords Apoptosis · Cancer · Heat shock proteins · Lon protein · Molecular chaperones

Abbreviations

COX	cytochrome *c* oxidase
Cpn	chaperonin
Hip	HSP 70-interacting protein
Hop	HSP 70-HSP 90 organizing protein
HSP	heat shock proteins
MAP	mitogen activated protein
PARP	Poly (ADP ribose) polymerase

P. Chaudhuri (Chattopadhyay) (✉) · N. Rashid
Molecular Biophysics Lab, Amity Institute of Biotechnology, Amity University, Noida, Uttar Pradesh, India
e-mail: pchaudhuri@amity.edu

A. A. A. Asea, P. Kaur (eds.), *Heat Shock Protein 60 in Human Diseases and Disorders*, Heat Shock Proteins 18,
https://doi.org/10.1007/978-3-030-23154-5_4

SPG	spastic paraplegia
shRNA	short hairpin RNA
TUNEL	transferase-mediated dUTP nick-end labeling

4.1 Introduction

Molecular chaperones are ubiquitously present in all living organisms. They execute a significant role in protecting the cells from intrinsic and extrinsic stresses. In pathological conditions, these cyto-protective molecules assist in folding of proteins, repair, refolding of peptides that are misfolded, and also in degradation of proteins that cannot be repaired. They are constitutively expressed in ideal growth conditions; however, under heat shock or any other environmental insults, there is an up regulation of many of these chaperones. Thus such chaperones are also known as stress or heat shock proteins (Hsp). Their primary function is in establishing conformational quality control of the proteome. These molecules achieve the conformational stability of non native polypeptides by interacting with them. A number of other cellular functions have also been attributed to these molecular chaperones. Based on amino acid composition, designated function and molecular weight, these molecular chaperones have been classified into two groups; the high molecular weight and the small molecular weight heat shock proteins. The high molecular weight heat shock proteins have a molecular weight ranging between 60 and 110 kDa. They are dependent on ATP and primarily operate by binding and enabling the nascent polypeptides to fold through ATP-dependent allosteric organization. The small molecular weight heat shock proteins have a molecular weight ranging between 15 and 43 kDa (Creagh et al. 2000). They perform their function in an ATP independent manner. This type of molecular chaperones have been observed to act in various embryonic development pathways, formation of respiratory organs such as cardiac muscles, as biomarkers for formation of tumor, in exercise induced stress, as well as in protein folding .

The elemental capability of Heat shock proteins to perpetuate the longevity of cell is correlative to suppression of caspase activation and apoptosis that could usually be based on their chaperoning capabilities (Beere 2004). They are known to enhance the survival of cell by inhibition of mitochondrial cell death initiated by apoptosome, stabilizing the survival effectors, and inactivating p53. For example, the anti apoptotic activity of HSP 90 might become selectively exploited in case of carcinoma (Whitesell and Lindquist 2005) and may execute a pivotal part in maintenance of tumor cell (Ghosh et al. 2008; Isaacs et al. 2003).

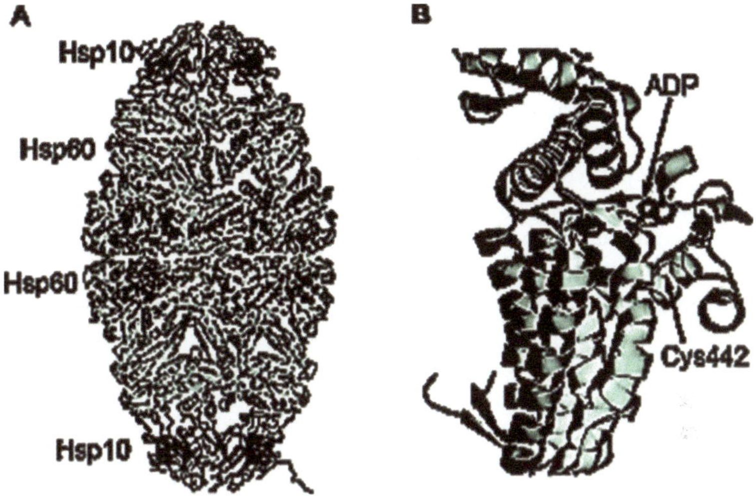

Fig. 4.1 HSP60. (**a**) Complex formed by HSP 60 and HSP 10. (**b**) ADP binding pocket and Cysteine residue 442 of HSP 60

4.2 Heat Shock Protein 60

Heat Shock Protein 60 (HSP 60), along with its co-chaperonin HSP10 (Zhao et al. 2002), is an evolutionarily conserved stress response chaperone system in *Eukaryotes*. Human HSP 60 is also known as the 60 kDa chaperonin (Cpn 60). It is a homolog of bacterial GroEL. It is usually localized within the mitochondria and has been referred to as the mitochondrial molecular chaperone. HSP 60 and HSP10 (Fig. 4.1) also are known to occupy extracellular space, cytosol, and nucleus. The mitochondrial compartmentalized HSP 60 and HSP 10 together form a heptameric ring assembly. A folding chamber is formed due to stacking of these heptameric rings (Nielsen et al. 1999; Nisemblat et al. 2014). HSP 60 executes cardinal functions such as organelle biogenesis and folding/refolding of imported pre-proteins. A number of studies carried out on HSP 60 have pointed out their role in a number of tumor series correlating with disease outcome. The role of heat shock proteins has been discerned in regulation of a significant cancer related gene called survivin. Survivin participates in defence against apoptosis. It is also responsible for controlling mitosis in transformed cells.

4.3 HSP60 and Human Diseases

HSP 60 plays a significant role in protein homeostasis inside the mitochondria (Meng et al. 2018). A number of diseases have been associated to mutations that occur in HSP 60 (also called HSPD1), e.g. a rare and non treatable hereditary neurodegenerative disorder called spastic paraplegia SPG13 has been discerned to occur due to a mutation V981 in HSP 60. This disorder exhibits its effect by causing spasticity and weakness of lower limbs (Meng et al. 2018). This mutation results in reduction of refolding ability of HSP 60 client proteins. MitCHAP-60 disease is another neurodegenerative disorder that is autosomal recessive in nature. It is designated by hypomyelination and leukodystrophy in the brain. This devitalizing disease occurs due to mutation in HSP 60 (Magen et al. 2012). The underlying mechanism contributing to this disease is the less stability of D3G mutant in formation of heptameric and tetradecameric oligomers in comparison to the wild type. The decreased stability is coupled with deterioration in the refolding ability and ATPase activity.

In addition to the mutations occurring in the HSP 60, aberrant expression of HSP 60 has also culminated in various diseases. The involvement of in various inflammatory responses and immune reactions HSP 60 has also been pointed out (Pockley 2003). Thus these physiopathological pathways can be modulated by the level of expression of HSP60. For instance, the expression of HSP 60 in skin allografts can mediate the rejection by host for these allografts. The increased expression in this case, results in augmented dismissal in non obese diabetic mice. HSP 60 is able to play the role as an auto-antigen and the HSP 60 autoimmunity could be mediated in non obese diagnostic mice by injecting mouse HSP 60 peptides subcutaneously. This vaccination protects against allograft rejection. Mechanistically, this HSP 60 vaccination pathway apparently is involved in shifting the phenotype of the T cell in response to self HSP 60 from a pro-inflammatory Th1 kind of reciprocation to a Th2 regulatory type of reciprocation. The idea of endogenously present HSP 60 being able to execute the role of auto-antigen which further produces anti-HSP 60 antibody has also been corroborated in humans. People suffering from spondyloarthritis or periodontitis possess higher concentration of HSP 60 antibody than normal healthy volunteers. Nevertheless, concentration of human serum anti-HSP 60 appears to be disconnected of predicting kidney allograft rejection (Meng et al. 2018). The autoimmunity opposing HSP 60 might perform as a preventive measure against the occurrence of atherosclerosis with ageing. The functioning of HSP 60 as an auto-antigen also executes a role in the progression of various other autoimmune diseases (e.g., Hashimoto's thyroiditis, myasthenia gravis, inflammatory bowel diseases, chronic obstructive pulmonary diseases (COPD)). The participation of HSP60 in various autoimmune disorders is extremely intriguing as one of the pharmacologically utilized immunosuppressant mizoribine targets HSP 60.

4.4 Heat Shock Proteins in Apoptosis

Apoptosis is a morphologically well defined, genetically programmed, cell death. It involves energy dependent biochemical mechanisms. This programmed cell death is a significant part of number of biochemical processes such as normal cell turnover, development and function of the immune system, hormone reliant atrophy, embryonic development and chemical induced cell death. Abnormal apoptosis i.e. too low or too high results in neurodegenerative diseases, ischemic damage, autoimmune disorders and numerous kinds of carcinomas in *Homo sapiens*. Numerous genes have been established that act as either positive or negative modulators of apoptosis. An increase in the production of inducible heat shock proteins results in hike in resistance to apoptosis that was induced by diversified cytotoxic agents and is engrossed in chemotherapeutic resistance of tumors and carcinogenesis (Creagh et al. 2000). Various studies in the past explored the underlying mechanism involved in apoptosis. The regulatory effects of heat shock proteins in apoptotic death are very well ascertained.

A strict modulation of proliferation, differentiation and death processes is essential for proper cell development and maintenance. Development of any abnormality in any of these processes culminates in serious complications including disorders like leukemia, autoimmunity, viral infections, allergic and neurodegenerative disorders. Aberrant expression of heat shock proteins has been implicated as one of the significant factors responsible for deregulation of development and maintenance of cell. The death of cells in eukaryotes can take place in two principally different ways; necrosis or apoptosis. Intense damage or harm to the cell leads to intense dysfunctioning of cell which subsequently leads to Necrosis. It is characterized by inert and disorderly processes that cause the cell to lose its control over the ionic transport. These disruptive processes results in cytolysis due to swelling of cell and the cellular organelles which takes place due to water uptake by the cell. A local inflammatory response is generated as the cell contents get released into the extracellular tissues present in the vicinity of the cell. On the other hand, apoptosis involves activation of intrinsic cascade leading to cell death. Apoptosis is genetically regulated and a number of external signals can modulate the cascade leading to cell death as shown in Fig. 4.2. This is a sequential step wise process that is initiated by shrinking of cells followed by blebbing of membranes, condensation of chromatin, fragmentation of inter-nucleosomal DNA and apoptotic body formation.

The occurrence of apoptosis or necrosis is dependent on the stress condition present in the cell. Necrosis takes place in harsh situations where the cell is unable to regulate the activation of stepwise programmed cell death i.e. the apoptosis. However, when the surrounding environment is not much harsh, the cells are able to initiate the cascade of apoptosis as shown in Fig. 4.3. Apoptosis is carried out by the aspartate specific cysteine proteases called caspases activity. These caspases split to make the target substrates either active or inactive. A cascade is initiated by the caspases during which preliminary caspases show an interaction with specific adaptor molecules which mediates their own autocatalytic processing. This further leads to cleavage and activation of the downstream caspases that execute the proteolytic disorganization of the cell (Fig. 4.2). Sometimes when the cell faces with a low enough level of stress, cells employ a distinct response pathway to sustain. This process is characterized by inhibition of synthesis of proteins inside the cells; however expression of heat shock proteins is simultaneously induced as a consequence of which the cell enters a thermo tolerant phase transiently. These proteins are clas-

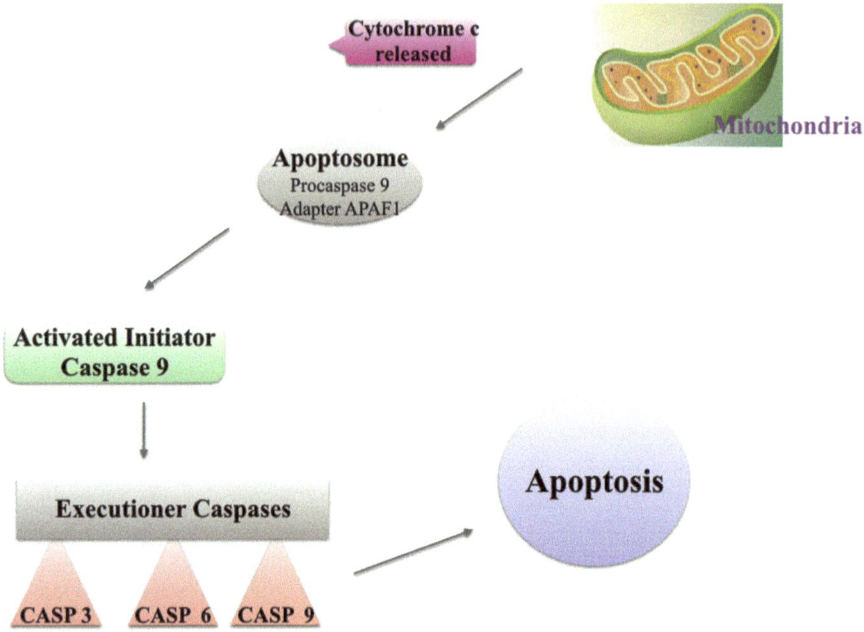

Fig. 4.2 Apoptotic pathway

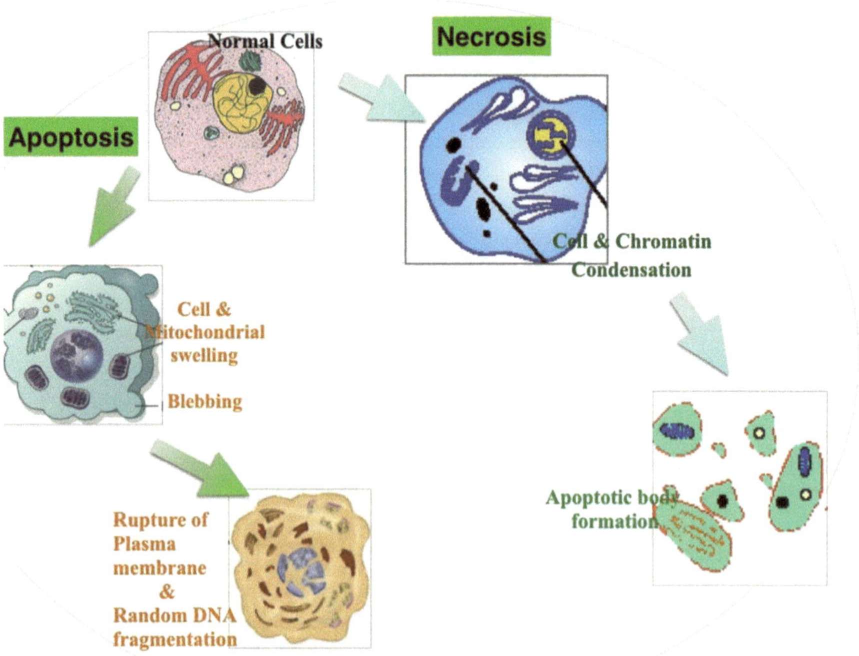

Fig. 4.3 Comparison between Necrosis and Apoptosis

sified on the basis of their size e.g. the HSP 27, HSP 60, HSP 70, HSP 90 and HSP 100 families (Beere 2004).

The role of heat shock protein in stress regulation is dependent on a single elemental feature which is their capacity of interaction with substrates that are proteins/polypeptides in nature. HSP 70 and HSP 90 proteins are composed of two parts: a highly conserved amino end that comprises the ATPase and a Carboxy end which constitutes the polypeptide-binding site. There are four amino acids, EEVD, at the carboxy end which is responsible for mediating the inter-domain interaction and binding ability of proteins, and are significant for modulating defence against heat induced stress. On the other hand, HSP27 does not possess an ATPase domain. Mitogen activated protein (MAP)-kinase dependent phosphorylation and self-oligomerization regulate the defense against stress in this case.

The chaperoning property of the heat shock proteins is regulated by a reaction cycle of binding of ATP, hydrolysis and nucleotide exchange to initiate a series of rapid association and dissociation cycles in-between the Heat shock proteins and their target polypeptides. The heat shock proteins bound to ATP, bind and release peptide quickly, which results in decreased overall affinity. On the other hand, the heat shock proteins bound to ADP, bind to the peptide at a slower pace but have more stable interactions. Presence of additional co factors/co chaperones responsible for catalyzing the conversion between ATP and ADP further regulates the capacity of these chaperones. HSP 40 (HDJ-1 and HDJ-2), HSP 70-interacting protein (Hip) and HSP 70-HSP 90 organizing protein (Hop) represent such classes of co-chaperones.

When the stress is eliminated from the cell surrounding, there is again a depreciation in the quantum of heat shock proteins and their number is maintained as present during the normal conditions of the cell. However, paradoxically the presence of high level of heat shock proteins during these stress conditions makes the cell inimical to the presence of various toxic agents. In case of tumor cells, the quantum of heat shock proteins is quite high as they are constitutively expressed at increased level. This increased level of heat shock proteins in tumor cells proves anomalous as these cells are protected from toxic agents which results in development of resistance in these tumor cells against chemotherapies and carcinogenesis. The elevated levels of heat shock proteins apparently lead to rapid growth of tumors that are resistant to therapies.

4.5 HSP60 in Apoptosis of Tumor Cells

HSP 60 coupled with its co chaperone HSP10, has been discerned to be evolutionarily conserved chaperone that is involved in defending cell against various stresses. It is mostly present inside mitochondria and executes a significant part in biogenesis of the cellular organelle and folding/refolding of proteins that are imported inside the mitochondria (Asea and Brown 2008). Thus it's able to prevent aggregating of proteins which is essential for the cells in stress environment. There is an upregulation of HSP 60 during different kind of human cancers. It displays anti-apoptotic properties and supports formation, progression, invasion and metastases of tumor, progression. It is also responsible for therapeutic resistance and decreased survival. HSP 60 is deposited on the exterior of mitochondria during carcinogenesis, in the cytosol, plasma membrane, and in secretory vesicles thereby imparting protection to tumor cells from external environmental stress, which further promotes proliferation of cells. HSP 60 also shows participation in permeabilization of mitochondrial membrane by interacting with the cyclophilin D which is responsible for regulation of mitochondrial permeability transition pore.

HSP 60 mediates a cyto-protective cascade based on imparting stability to the quantum of survivin and restraints the functioning of p53 as shown in Fig. 4.4. On the other hand, acute extraction of HSP 60 leads to depreciation in the survivin reserves present in the mitochondria. Survivin is particularly designated for inhibition of the programmed cell death i.e. apoptosis, simultaneous increase in p53 expression, and activating p53 dependent apoptosis in tumor cells. These cyto-protective properties of HSP 60 is carefully explored in tumors in vivo, wherein HSP 60 is subjected to selective upregulation, in comparison to normal cells, and deficit of HSP 60 in normal cells is not linked to dysfunctioning of mitochondria or death of cells.

Survivin is established as a cardinal carcinogenic gene with binary function in division of cell and reticence of apoptosis. A crucial prerequisite of this procedure

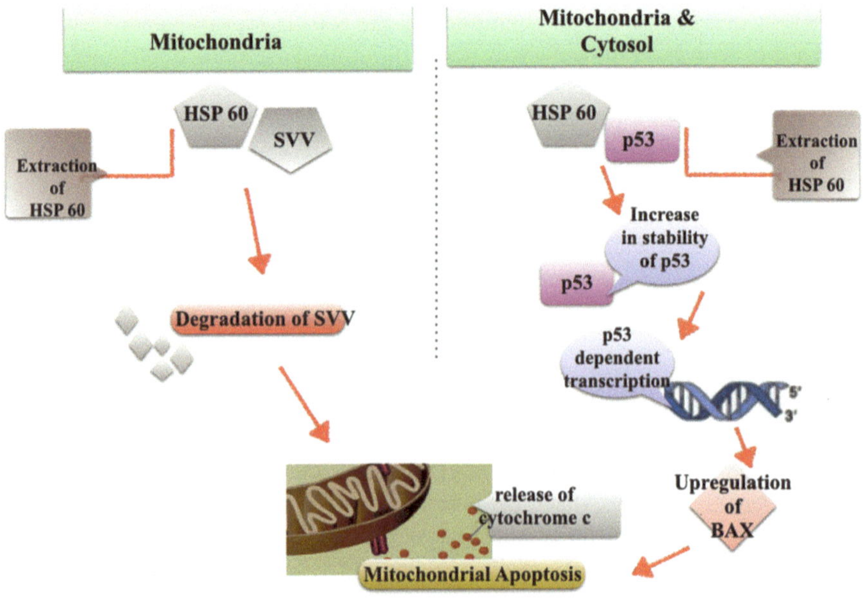

Fig. 4.4 Role of HSP60 in tumor cell Apoptosis

is the existence of a reserve of survivin occupying the mitochondria, generally in tumor cells, and excreted into the cytoplasm in return to stimulus of cell death. A number of evidences point out that a specific molecule designated for inhibition of apoptosis is provided by the survivin present in the mitochondria, thereby straightly exasperating the growth of tumor in vivo. This cascade is modulated by isolated phosphorylation taking place in mitochondria and associating differentially with the *X* chromosome-linked inhibitor of apoptosis which is an anti-apoptotic cofactor. Although there is an absence of a cleavable, amino end mitochondrial import sequence, survivin is vigorously transported to the mitochondrial compartment. A number of molecular chaperones associated to survivin occupying the cytoplasm might also be contributing to this cascade. These molecular chaperones include HSP 90 and/or AIP molecules that are involved in the import of pre-proteins into the mitochondria. Inside the mitochondria, survivin might need to associate with HSP 60 to form a complex in order to reinstate the optimum refolding after it's translocated across the membrane enveloping the mitochondria. This step of translocation includes unfolding of protein. Also, reports have pointed out that siRNA depletion of HSP 60 leads to destabilizing of quantum of survivin present and approximately total depletion of the deposits of survivin present in the mitochondria, thereby abolishing the comeback contrary to the apoptosis.

Besides, stabilizing the level of survivin present in the mitochondria, HSP 60 also imparts cyto-protection by another mechanism which includes the forming of a complex by association of HSP 60 with p53 that inhibits the functioning of p53 tumor cells. Molecular chaperones have precession in inhibiting functioning of p53. Mortalin, a mitochondrial HSP 70, has been reported to participate in binding to and sequestering p53 in the cytoplasm, thereby inhibiting its translocation into the nucleus and centrosomes, wherein such interplay invalidates a checkpoint for duplication of centrosomes that is dependent on p53. Modulation of p53 by HSP 60 is not implicated in bringing about modifications in the Mdm-2 which is a modulator of p53. Therefor the role of HSP 60 is different in comparison to role of inducible HSP 70 that is involved in counteracting the cell death brought about by p53. Based on their interaction with various pathways responsible for maintaining tumor (Whitesell and Lindquist 2005) and their frequent over-expression in carcinoma, molecular chaperones are robustly explored for new cancer diagnostics (Isaacs et al. 2003).

4.6 Regulation of Apoptosis by Mitochondrial Lon

Human Lon protease is a protein occupying the mitochondrial matrix. It performs a number of functions like degradation of proteins, binding of mitochondrial DNA, and chaperoning activity. Lon is a protease that has emerged as a significant modulator of tumorigenesis accorded by the mitochondria. There is increase in the expression of Lon in the cancer cells. It is an extremely conserved ATPase that is responsible for a number of cellular functions. It plays a cardinal role in ATP dependent proteolytic binding of DNA and chaperoning activity. Lon in eukaryotes participates in the protein quality control mechanism. It is also employed in significant cellular processes such as functioning of mitochondria, homeostasis and biogenesis. As already discussed, proteins are susceptible to getting inactivated by virtue of misfolding, unfolding, or aggregation when cells undergo stress conditions. In such circumstances, protein quality control mechanism, chaperones and proteases, protect the cell and enable its proper functioning. The two components coordinate to grant stability to the misfolded proteins and enable their refolding. They even expel these aggregated/misfolded proteins to evade the detrimental consequences of aggregation of proteins (Kao et al. 2015).

Mitochondria coordinates the life and death of cells. Thus, it employs a crucial regulation of signaling that enables the cell to survive, especially in the intrinsic pathway of senescence. The quantum of Lon present is responsible for regulating mitochondrial actions responsible for the fate of cell. Downregulation of Lon causes the mitochondria to lose its function. It also results in early embryonic mortality, decrease in proliferating cells, and programmed cell death. On the other hand, upregulation of Lon is significant for the cancer cells to survive. The upregulation of Lon also plays a role in tumorigenesis by modulating response to oxidation induced stress (Cheng et al. 2013; Quiros et al. 2014; Gibellini et al. 2014). The signature stress conditions of cancer cells including hypoxia, oxidative and mitochondrial unfolded protein induces the expression of Lon protein (Kao et al. 2015. Lon upregulation takes place in hypoxic conditions due to factor-1α (which is induced by hypoxia). It takes part in the pathway that responds to lower level of oxygen availability. The Lon protein is responsible for enabling cancer cells to get accustomed to hypoxia. Lon protein is also capable of showing chaperoning activity besides the proteolytic activity. Lon is also involved in promotion of assembling 4 to 1 subunits of cytochrome *c* oxidase (COX), (Hori et al. 2002; Fukuda et al. 2007; Ngo and Davies 2009) signifying the chaperoning properties of Lon protein in yeast and mammalian cells. Therefore, Lon protein occupying the mitochondria might be a protein chaperone to enable cells to sustain and accommodate to a number of stress conditions associated to oncogenesis.

4.7 Mitochondrial Lon in Regulation Apoptosis via the Interaction with HSP 60

Lon has been observed to increase the stability of proteins and levels of heat shock proteins HSP 60 and mitochondrial HSP 70 in response to cellular stress. Also, during increased stress the enhanced Lon expression was linked to lowering or enhancement in cleaved caspase 3 after hydrogen peroxide treatment, signifying that when Lon is upregulated it defends the cells from senescence/apoptosis and when Lon is downregulated it causes induction of programmed cell death while recovering from stress induced due to oxidation. It has been suggested when the Lon is upregulated, there is enhancement in the stability of proteins HSP

60–mitochondrial HSP 70 which subsequently provides protection to cell from apoptosis due to stresses, which is achieved when Lon binds to HSP 60 or mitochondrial HSP 70. The down regulated expression of HSP 60 by shRNA showed that the signals of pro apoptotic proteins such as Bax, cleaved caspase 3, cleaved PARP, p53, and phosphorylated p53Ser46, were subjected to activation after UV or hydrogen peroxide treatment in FADU cells, denoting that the treated cells underwent apoptosis. The quantum of pro apoptotic proteins decreases and apoptosis is suppressed in the cells that over-express Lon post stress that suggests the significance of Lon protein for regulating apoptosis under environmental stresses. On the other hand, knocking down the HSP 60 expression, increases the quantum of apoptotic proteins in the cells that over-expressing Lon protein in stress. These observations are indicative of the role played by HSP 60 during the stress conditions in the regulation of apoptotic processes which is mediated by Lon (Kao et al. 2015). The part played by HSP 60 in apoptosis modulated by the Lon protein has also been observed using terminal deoxynucleotidyl transferase-mediated dUTP nick-end labelling (TUNEL) assay. The cells that were TUNEL positive cells are lesser in number in cells where there is over-expression of Lon cells. Also, the TUNEL positive cells were greatly elevated in the cells that over-expressed Lon when HSP 60 was knocked down. These observations are indicative of the fact that increase in Lon protein defends cell from apoptotic pathway in stress conditions by virtue of interplay with and stabilizing HSP 60.

Mitochondrial Lon is responsible for modulating the organelle function, cell proliferation, and apoptosis. The increase in level of Lon is cardinal for survival of cells undergoing oxidative and hypoxic stress. Decrease in Lon results in decrease in number HSP 60 and mitochondrial HSP 70 in cell undergoing oxidative stress, depicting that the stability of proteins HSP 60–mitochondrial HSP 70 complex is reliant on Lon. Also, loss of HSP 60 results in the instability of Lon–mitochondrial HSP 70 interaction. Through binding with HSP 60–mitochondrial HSP 70 complex, increased mitochondrial Lon suppresses apoptosis and is conducive for survival of cell undergoing environment induced stresses, by which they are inclined to perform significant action in a cyto-protective chaperone network. On the basis of analysis of Lon related proteins purified from in solution digestion shotgun proteomics, these, significant mitochondrial proteins have been grouped into five functional classes: (1) mitochondrial chaperones, (2) cellular metabolism and energy, (3) Redox regulation, (4) cell death and survival, and (5) mitochondrial DNA stability. NDUFS8 (which is a NADH-ubiquinone iron-sulfur (Fe-S) 23 kDa subunit) HSP 60, and mitochondrial HSP70 among these are able to bind to Lon. Lon interplays with NDUFS8 of mitochondrial complex I that participates in the generation of ROS under induction by Lon.11 NDUFS8 is a subunit of mitochondrial NADH oxido-reductase (Complex I) that takes part in transferring of electrons from NADH to the respiratory chain in the mitochondrial inner membrane. Also, mitochondrial Fe-S proteins in complex I, II, and III, that is, SDHB (complex II) and Rieske (complex III), and COX 4–1 in complex IV are potential substrates of Lon protease. In case of yeast Lon modulates transport of electrons by disintegrating the subunits of complexes III–V in yeast.

4.8 Mechanism of Mitochondrial Lon Regulated Apoptosis

HSP60 coupled with mitochondrial HSP 70 is a binding associate. The quantum and stability of HSP 60 and mitochondrial HSP 70 is dependent on the quantum of Lon undergoing oxidative stress. Also, the capability of elevated Lon-deteriorated apoptosis significantly depends upon HSP 60. The mechanism involved in cell sustenance is modulated by Lon by the maintaining the stability of HSP 60 and mitochondrial HSP 70. HSP 60 displays anti-apoptotic as well as pro-apoptotic properties which is dependent on the background of cell type and situation. Mitochondrial HSP 60 is able to inhibit apoptotic pathway by enhancing the stability of survivin, refraining functioning of p53, opposing mitochondrial permeability transition which is cyclophilin D-dependent, and conserving generation of ATP for the complex IV. Also, mitochondrial HSP60 is an inducer of apoptotic pathway by increasing the maturity level of pro-caspase-3. Firstly, mitochondrial Lon stabilizes HSP60 and mitochondrial HSP 70 allowing them the execution of anti-apoptotic properties. Infact, over-expression of HSP 60 and mitochondrial HSP 70 in human tumor cells, has been depicted to oppose p53 and abrogate its apoptotic properties in cancer cells. Secondly, enhanced Lon stabilizes the complex made by association of HSP 60, mitochondrial HSP 70 and Lon to sequestrate HSP 60 in mitochondria, protecting it from activating pro-caspase-3 and cytoplasmic translocation that makes the cell sensitive to apoptotic pathway. Thirdly, in order to maintain the integrity of mitochondria, Lon protein depicts chaperoning properties in order to show cooperation with HSP 60 and mitochondrial HSP 70 complex and maintaining the protein homeostasis in mitochondria during stresses, which is backed by mitochondrial HSP 70 that might assist Lon chaperone misfolded proteins to conserved mitochondria functioning in yeast (Kao et al. 2015). Thus, the loss in the equilibrium between Lon and HSP 60 in the complex formed by Lon, HSP 60 and mitochondrial HSP 70 affects the apoptotic activation and the sustenance of cell. These quanta of Heat shock proteins are raised in different kinds of human tumors, which exhibits significant chaperone functioning to enhance the survival of stress in stress conditions.

4.9 Heat Shock Protein 60 Modulators and Inhibitors

As the complex formed by HSP 60 and HSP 10 chaperone is significant for maintenance of homeostasis in mitochondria. It executes a cardinal role in cardiovascular disorders including autoimmune diseases and carcinomas. Development of small molecules that can act as regulators of HSP 60 by targeting it can be used in various therapeutic treatment of diseases. Such small molecule regulators act as significant tools to further throw light on the biological functioning of HSP 60 in several contexts (Meng et al. 2018). A number of natural and synthetic compounds have flourished that can target HSP 60. Mechanistically, the HSP 60 inhibitors developed are grouped into two classes:

Type I: These inhibitors are responsible for blocking binding and hydrolysis of ATP. This affects the ATP-dépendent conformational changes as a result of which the refolding activity of complex formed by the HSP 60 and HSP 10 is inhibited.

Type II: These inhibitors incorporate compounds that show covalent reactions with particular cysteine residues in HSP 60.

Mizoribine is the first small organic molecule that has been known to inhibit HSP 60 inhibitor. It is an imidazole nucleoside antibiotic (Fig. 4.5) and has been isolated from *Eupenicillium brefeldianum*. Though it lacks anti-microbial activity, it does have potential of acting as immunosuppressor (Mizuno et al. 1974) as a result of which it is widely used in case of renal transplantation (Tajima et al. 1984). The direct binding of mizoribine inhibits the chaperoning properties of the binary complex formed by HSP 60 and HSP 10. Mizoribine also blocks the ATPase activity of HSP 60, which simultaneously results in more stable interaction of HSP 10 and HSP 60.

Epolactaene (Fig. 4.6) is another naturally found compound that inhibits the activity of HSP 60. Epolactaene (2, 2) has been isolated from the fungal strain Penicillium sp. BM 1689-P. It is capable of promoting outgrowth of neurites in SH-SY5Y cells (Meng et al. 2018). Another naturally found molecule capable of inhibiting the activity of HSP 60 is myrtucommulone (MC), (Fig. 4.7). It is a non-prenylated acylphloroglucinol possessing different bio-activities, such as anti-bacterial, anti-oxidant, anti-inflammatory, and anti-tumor properties. It has been shown to affect the mitochondria that have been isolated from human leukaemia

Fig. 4.5 Structure of Mizoribine

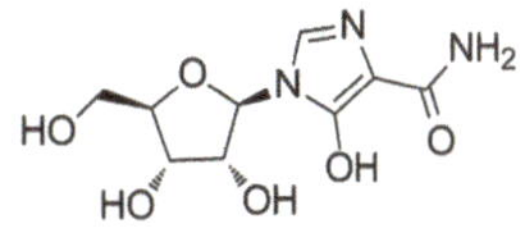

Fig. 4.6 Structure of Epolactene

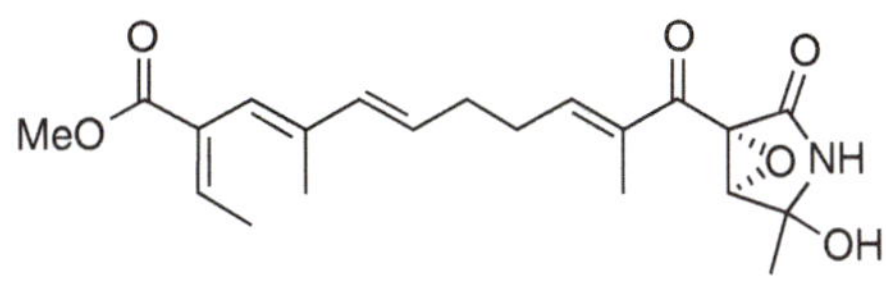

Fig. 4.7 Structure of Myrtucommulone

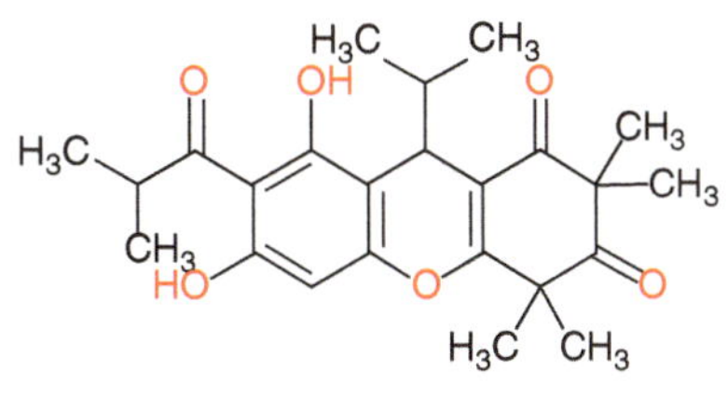

cells, and it also alters the mitochondrial functioning at sub-micromolar concentrations, which includes loss of mitochondrial membrane potential.

4.10 Conclusions

The cyto-protective property might be a general characteristic of various molecular chaperones, including HSP 60. This eventually results in enhancing an anti-apoptotic threshold in case of tumor cells in vivo. Interestingly, this property of molecular chaperones is particularly exploited in transformed cells and not in normal tissues. Additionally, a stark contrast in expression of the molecular chaperones, i.e., HSP 60, in case of oncogenesis in comparison to normal tissues in vivo, depicts that there are several other factors contributing to the selective use of this cascade in tumor cells. These might incorporate qualitative variations in chaperoning properties, as has been exhibited by HSP 90 ATPase function, or link with cancer genes distinctively expressed in cancer, e.g. for functional survivin-chaperone complexes . Despite the cyto-protection provided by these chaperones might enhance the sustenance of tumor cells and favor drug resistance, the distinctive expression or functional exploitation of this cascade in tumor cells might be suitaleble for wider, chaperone-administered anticancer approaches. This concept is further validated by molecular or pharmacologic targeting of complexes between survivin and HSP 90, mortalin and p53, and HSP 60 and survivin/p53 which is reconciled with selective induction of mitochondrial cell death in tumor cells sans any impact on normal cell types, including hematopoietic progenitor cells (Fortugno et al. 2003; Kang and Altieri 2006; Plescia et al. 2005). The concentration of HSP 60 and mitochondrial HSP 70 is reliant on Lon, and Lon-modulated programmed cell death is monitored by the HSP 60–mitochondria HSP 70 complex. Lon interactome, states that a Lon is a versatile protein involved in the regulation of mitochondrial chaperones, survival and death of cells, and mitochondrial DNA stability.

Acknowledgements The authors acknowledge the assistance and support provided by Amity Institute of Biotechnology, Amity University, Noida, Uttar Pradesh, India.

References

Asea AAA, Brown IR (2008) Heat shock proteins and the brain: Implications for neurodegenerative diseases and neuroprotection. Springer Nature

Beere HM (2004) The stress of dying: the role of heat shock proteins in the regulation of apoptosis. J Cell Sci 117:2641–2651

Cheng CW, Kuo CY, Fan CC, Fang WC, Jiang SS, Lo YK, Wang TY, Kao MC, Lee AYL (2013) Overexpression of Lon contributes to survival and aggressive phenotype of cancer cells through mitochondrial complex I-mediated generation of reactive oxygen species. Cell Death Dis 4:e681

Creagh EM, Sheehan D, Cotter TG (2000) Heat shock proteins – Modulators of apoptosis in tumor cells. Leukemia 14:1161–1173

Fortugno P, Beltrami E, Plescia J, Fontana J, Pradhan D, Marchisio PC, Sessa WC, Altieri DC (2003) Regulation of survivin function by Hsp90. Proc Natl Acad Sci U S A 100:13791–13796

Fukuda R, Zhang H, Kim JW, Shimoda L, Dang CV, Semenza GL (2007) HIF-1 regulates cytochrome oxidase subunits to optimize efficiency of respiration in hypoxic cells. Cell 129:111–122

Ghosh JC, Takehiko Dohi T, Kang BH, Altieri DC (2008) Hsp60 regulation of tumor cell apoptosis. J Biol Chem 283:5188–5194

Gibellini L, Pinti M, Boraldi F, Giorgio V, Bernardi P, Bartolomeo Nasi M, Biasi SD, Missiroli S, Carnevale G, Losi L, Tesei A, Pinton P, Quaglino D, Cossarizza A (2014) Silencing of mitochondrial Lon protease deeply impairs mitochondrial proteome and function in colon cancer cells. FASEB J 24:5122–5135

Hori O, Ichinoda F, Tamatani T, Yamaguchi A, Sato N, Ozawa K, Kitao Y, Miyazaki M, Harding HP, Ron D, Tohyama M, Stern DM, Ogawa S (2002) Transmission of cell stress from endoplasmic reticulum to mitochondria: enhanced expression of Lon protease. J Cell Biol 157:1151–1160

Isaacs JS, Xu W, Neckers L (2003) Heat shock protein 90 as a molecular target for cancer therapeutics. Cancer Cell 3:213–217

Kang BH, Altieri DC (2006) Regulation of survivin stability by the aryl hydrocarbon receptor-interacting protein. J Biol Chem 281:24721–24727

Kao T-Y, Chiu Y-C, Fang W-C, Cheng C-W, Kuo C-Y, Juan H-F, Wu S-H, Lee AY-L (2015) Mitochondrial Lon regulates apoptosis through the association with Hsp60–mtHsp70 complex. Cell Death Dis 6:e1642

Magen MJ, David JA, Michelhaugh SK, Schmidt CJ, Bannon MJ (2012) Increased heat shock protein 70 gene expression in the brains of cocaine-related fatalities may be reflective of post-drug survival and intervention rather than excited delirium. J Forensic Sci 57(6):1519–1523

Meng Q, Li BX, Xiangshu X (2018) Toward developing chemical modulators of Hsp60 as potential therapeutics. Front Mol Biosci 5:35

Mizuno K, Tsujino M, Takada M, Hayashi M, Atsumi K (1974) Studies on bredinin. I. Isolation, characterization and biological properties. J Antibiot 27:775–782

Ngo JK, Davies KJ (2009) Mitochondrial Lon protease is a human stress protein. Free Radic Biol Med 46:1042–1048

Nielsen KL, Lennan NM, Masters M, Cowan NJ (1999) A single-ring mitochondrial chaperonin (Hsp60-Hsp10) can substitute for GroEL-GroES in vivo. J Bacteriol 181:5871–5875

Nisemblat S, Parnas A, Yaniv O, Azem A, Frolow F (2014) Crystallization and structure determination of a symmetrical 'football' complex of the mammalian mitochondrial Hsp60–Hsp10 chaperonins. Acta Crystallogr F Struct Biol Commun 70:116–119

Plescia J, Salz W, Xia F, Pennati M, Zaffaroni N, Daidone MG, Meli M, Dohi T, Fortugno P, Nefedova Y, Gabrilovich DI, Colombo G, Altieri DC (2005) Rational design of shepherdin, a novel anticancer agent. Cancer Cell 7:457–468

Pockley AG (2003) Heat shock proteins as regulators of the immune response. Lancet 362:469–476

Quiros PM, Espanol Y, Acin-Perez R, Rodriguez F, Barcena C, Watanabe K, Calvo E, Loureiro M, FernAndez-Garcia MS, Fueyo A, VAzquez J, Enríquez JA, Lopez-Otin C (2014) ATP-dependent Lon protease controls tumor bioenergetics by reprogramming mitochondrial activity. Cell Rep 8:542–556

Tajima A, Hata M, Ohta N, Ohtawara Y, Suzuki K, Aso Y (1984) Bredinin treatment in clinical kidney allografting. Transplantation 38:116–118

Whitesell L, Lindquist SL (2005) HSP90 and the chaperoning of cancer. Nat Rev Cancer 5:761–772

Zhao Q, Wang J, Levichkin IV, Stasinopoulos S, Ryan MT, Hoogenraad NJ (2002) A mitochondrial cell response in mammalian cells. EMBO J 21:4411–4419

Chapter 5
Utilizing the Unique Architecture and Abilities of HSP60 in Drug Development

Tomohiro Mizobata and Yasushi Kawata

Abstract As a major active component in the maintenance of protein homeostasis, the HSP60 family of proteins have evolved numerous unique abilities that are useful in detecting, stabilizing, and facilitating the recovery of various proteins that have denatured under stress. From a technological viewpoint, many of the unique abilities of HSP60 may potentially be used to solve a variety of problems in drug development, as well as to serve as a scaffold for various applications that are relevant to the medical field. This section is an overview of recent efforts to harness the unique abilities of the HSP60 proteins in its role as a chaperonin, capable of preventing the aggregation of and stimulating the recovery of various proteins that have undergone stress-related denaturation.

Keywords Antibiotic target · Chaperonin · Molecular chaperone · Nanotechnological scaffold · Protein aggregation · Protein folding

Abbreviations

AD	Apical domain
CCT	Chaperonin containing T-complex polypeptide 1
GroEL-AD	The isolated apical domain of GroEL
HSP	Heat shock protein
PolyQ	Polyglutamine
TRiC	T-complex polypeptide 1 ring complex

T. Mizobata (✉) · Y. Kawata (✉)
Faculty of Engineering and Center for Research on Green Sustainable Chemistry, Tottori University, Tottori, Japan
e-mail: mizobata@tottori-u.ac.jp; kawata@tottori-u.ac.jp

A. A. A. Asea, P. Kaur (eds.), *Heat Shock Protein 60 in Human Diseases and Disorders*, Heat Shock Proteins 18,
https://doi.org/10.1007/978-3-030-23154-5_5

5.1 Introduction

Proteins in the cell are highly dynamic entities whose structures are responsive to various changes in the cellular environment and other stimuli when expressing their native activities. Our original understanding of the structural state of active proteins in the cell was that proteins were fairly rigid, with a clearly defined structure whose individual characteristics were indispensable to attain its biological activity. Recent new discoveries regarding the cellular environment have necessitated a reevaluation of this understanding, since research has found that many proteins in a typical eukaryotic cell retain large sequence regions that are in a highly dynamic, "intrinsically denatured" (Peng et al. 2015) state, and assume certain defined conformations only upon interaction with stimuli (Theillet et al. 2014), such as binding to other proteins, or posttranslational modifications such as phosphorylation (Collins et al. 2008; Iakoucheva et al. 2004). This present dynamic view of the cellular proteome suggests that almost the entire scope of the "folding landscape" of a given protein may be relevant to its actual function as a cellular component.

In this highly dynamic environment of the cellular proteome, the effects of the HSP60 family of molecular chaperones has proven to be highly complex and multifaceted. HSP60 proteins, together with the accessory molecular chaperones HSP10 in bacteria, form the family of chaperonin proteins, whose function is to detect and segregate proteins that are prone to irreversible aggregation, and also to promote the structural recovery of these proteins. HSP60 proteins are present in all three branches of the tree of life, and many species have more than one HSP60 gene; in an extreme example, *Bradyrhizobium japonicum* possesses seven homologous HSP60 genes in its genome (Kaneko et al. 2002).

Many HSP60 proteins are very similar in basic structure, consisting of a polypeptide chain folded into three clearly defined domains (Fig. 5.1), which then form a ring-like quaternary structure composed of 7–9 subunits (Yebenes et al. 2011). Two of these HSP60 rings are associated back-to-back to form a final "double-ring" quaternary structure. Each domain of the HSP60 subunit also possesses clearly defined functional roles. The apical domain is the binding site for unfolded proteins that are recognized through hydrophobic interactions. The equatorial domain contains a site for the binding and hydrolysis of ATP, an enzymatic reaction that acts as a timer for the overall functional mechanism of these proteins. The apical and equatorial domains are located at the top and bottom of the HSP60 subunit, respectively; binding these two domains together is the intermediate domain, which acts as both the functional and structural intermediary between the apical and intermediate domains.

A unique structural characteristic common to many members of the HSP60 family is a hole-like cavity that is located in the center of each ring of the HSP60 oligomer (Fig. 5.1). This cavity, termed the central cavity (Braig et al. 1994), is the location where HSP60 sequesters vulnerable protein molecules for a predetermined interval, after recognizing and binding to these unfolded proteins that are prone to aggregating irreversibly. This ability of sequester and subsequently re-release

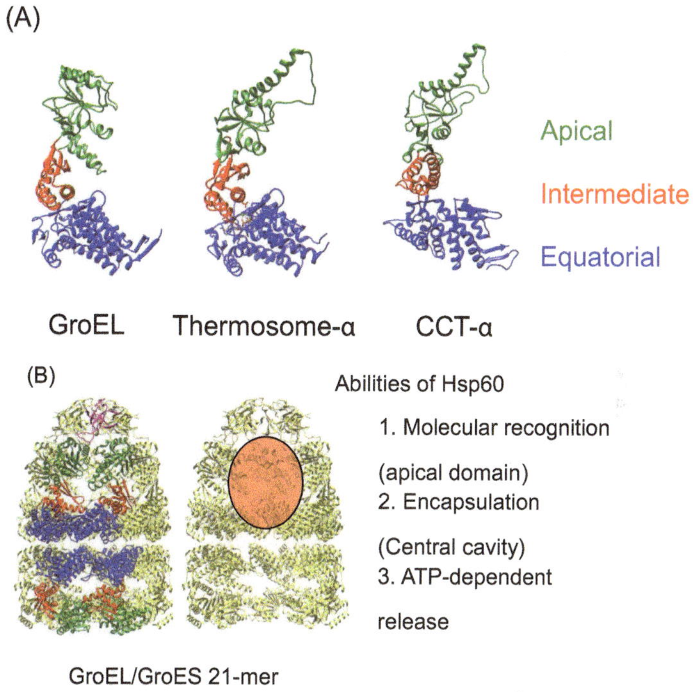

Fig. 5.1 Common structural characteristics of the HSP60 proteins. (**a**) Three representative examples of the HSP60 subunit. From left to right; *Escherichia coli* GroEL (PDB 1SVT Chaudhry et al. 2004), *Thermoplasma acidophilum* Thermosome alpha subunit (PDB 1A6E Ditzel et al. 1998), Yeast CCT alpha subunit (PDB 5GW5 Zang et al. 2016). Each subunit shows the nucleotide bound, or open form and the domains constituting each subunit are colored identically, with *green* indicating the apical domain, *red* the intermediate domain, and *blue* the equatorial domain. (**b**) Characteristics of the quaternary structure of HSP60, using the *E. coli* GroEL/ES system as example (PDB 1SVT Chaudhry et al. 2004). The overall architecture of HSP60 consists of a "double-ring" oligomeric structure with the apical domains of the HSP60 subunit situated at the two poles of the oligomer, and the equatorial domains at the equator (selected subunits are colored as in (**a**). The central cavity (*red shaded*) lies within each ring, with the size of the cavity changing depending on the state of bound nucleotide to the equatorial domains of each subunit that forms a given ring. Nucleotide binding and hydrolysis also govern the binding and release of the accessory chaperone HSP10, in bacterial HSP60 members (GroES in the figure, highlighted in *magenta*). The overall abilities of HSP60 stem from this unique quaternary structure and its dynamic behavior, and is summarized to the right of the main figure. Figures were generated using UCSF Chimera (Pettersen et al. 2004)

unstable protein molecules so that these proteins is a signature ability of the HSP60 proteins.

The functional mechanism of HSP60 generally consists of three phases: the initial phase involves the recognition and binding of denatured protein molecules by the apical domains through exposed hydrophobic regions on the target surface. The second phase is the "encapsulation" phase, where HSP60, sometimes in concert with the accessory molecular chaperone HSP10, moves the bound protein molecule to a position within the central cavity of the HSP60 ring oligomer. Once the molecule is placed into the central cavity, movements of the apical domains and/or the binding of HSP10 results in a completion of the encapsulated complex, where the protein molecule is placed in a state of infinite dilution and is prevented from aggregating. The third and final phase is the release phase, where the interactions that form the encapsulated complex are weakened, the chamber is opened, and the segregated protein molecule is free to diffuse into solution. The overall functional cycle of HSP60 is regulated by the binding and hydrolysis of the nucleotide ATP to the equatorial domains of the HSP60 subunit (Hayer-Hartl et al. 2016).

The HSP60 functional mechanism highlights three abilities of these proteins that may potentially be utilized in the development of reagents to treat and prevent various ailments (Fig. 5.1); hydrophobic recognition and binding to proteins, encapsulation within an enclosed chamber formed by its quaternary structure, and timed release of encapsulated proteins through ATP hydrolysis. In the following sections we summarize recent efforts to harness each of these unique abilities of HSP60 proteins in developing many medical applications.

5.1.1 HSP60 as Vehicle

The ability of HSP60 to encapsulate proteins of a certain size and segregate them from the bulk solution has drawn attention to the possibility of developing a nanometer-scale container that could house various bioactive molecules. The size of the central cavity of a representative member of the HSP60 proteins, GroEL from *E. coli*, is estimated from X-ray crystallographic analyses to be approx. 85,000 $Å^3$ in volume (Xu et al. 1997), which corresponds roughly to a protein with a relative molecular mass of 70 kDa (Xu et al. 1997). The central cavity expands roughly to twice its size (175,000 $Å^3$) when the accessory chaperone GroES is bound to the apical domains and forms a lid to close the cavity (Fig. 5.1). Additionally, opening and closing of this chamber is dictated by the binding and hydrolysis of the nucleotide ATP by the GroEL subunit, an attractive switchable mechanism that may be applied to developing signal responsive, controllable nanomedical applications.

Since the early 2000s, a concerted effort led by Professor Takuzo Aida of the University of Tokyo has attempted to modify the *E. coli* HSP60 GroEL for use as an environmentally sensitive supramolecular nanocarrier. The first positive results were reported in 2003 when the group succeed in sequestering CdS nanodots within the central cavity of *E. coli* GroEL and *Thermus thermophilus* cpn60, and also to release them in an ATP-dependent manner (Ishii et al. 2003). This result demonstrated that the encapsulating abilities of GroEL were not limited only to protein molecules, and that other compounds could be fitted into the central cavity, there to be released in a manner that was experimentally modulable and biologically relevant.

Subsequently, the group went on to modify the GroEL scaffold so that two cysteine residues were introduced into the apical domain, and then modified these cysteine residues with a photosensitive metal ion chelating group, merocyanine. The resultant modified GroEL formed long, unidirectional nanotubes in the presence of Mg^{2+} ion (which formed intermolecular chelates with merocyanine) that could be dissociated upon application of 400 nm light (which isomerized the merocyanyl group to spiropyran and abolished metal ion chelating abilities) (Biswas et al. 2009; Sendai et al. 2013). Further application of this concept was realized when the researchers loaded lactalbumin molecules that were esterified with a cyanine group into GroEL molecules (Biswas et al. 2013). The loaded GroEL molecules formed nanotubes in a manner similar to that of the Cys-modified versions, and subsequent surface modification allowed these nanotubes to be taken up into HeLa cells. Strikingly, once introduced into the cells, the concentration of ATP in the cells was sufficient to induce disruption of the nanotubes, and an intracellular esterase cleaved the cyanyl group from the construct. This represented the development of an ATP sensitive carrier of molecules that could conceivably be used as a potential drug application vector.

In a very recent study, a group led by Guangjun Nie of the University of Chinese Academy of Sciences was successful in applying GroEL for use as a carrier for doxorubicin, a hydrophobic antitumor molecule (Yuan et al. 2018). An interesting and fortunate development of this experimental system was that the construct carrying doxorubicin showed a strong affinity toward plectin, a cell surface protein that is strongly expressed in tumor cells. The inherent characteristics of GroEL served in this case to (1) Solubilize a hydrophobic antitumor drug so that the agent was protected during travel through the blood stream, (2) Enable the targeting of tumor cells through interaction with a specific cell surface protein, and (3) Allowed the release of the cargo molecule in an ATP-dependent manner.

As these studies demonstrate, the central cavity of GroEL and other HSP60 proteins holds great potential as a potential environmentally sensitive carrier of bioactive compounds.

5.1.2 HSP60 as Agent

The ability of HSP60 to bind to hydrophobic patches on a target protein's surface is an interesting example of a highly versatile yet specific molecular recognition mechanism. Estimates have been reported that about 250 proteins expressed in a typical *E. coli* cell are capable of stably binding to GroEL (Kerner et al. 2005). Additionally, GroEL is famous for its ability to assist the folding of proteins from other biological sources, indicating that the mechanism for molecular recognition in these molecular chaperones, especially those from bacterial sources, is versatile enough to be applied to many situations where molecular recognition through hydrophobic interactions is required.

As previously mentioned, the actual recognition mechanism for unfolded proteins is typically localized in the apical domain of the HSP60 subunit (Xu et al. 1997). In many cases, e. g. GroEL from *E. coli,* the apical domain may be taken out of the context of the subunit and expressed independently (Zahn et al. 1996). The resultant polypeptide is referred to in the literature as "minichaperones" or "apical domain (AD)". AD derived from HSP60 have been shown in many cases to assume a stable protein fold, retain its ability to bind to denatured polypeptides (Golbik et al. 1998; Zahn et al. 1996), and express a rudimentary "chaperoning" ability (Chatellier et al. 1998) to recognize and bind to denatured proteins, thereby preventing their aggregation.

A very intensely studied phenomenon in the protein structure field involves the irreversible aggregation of proteins to form regular structures termed protein fibrils, or amyloid deposits. These protein fibrils have been implicated in the pathology of many disorders, in particular, the neurological diseases that target the elderly and certain families. Cell biology studies have shown that for many protein fibril-forming reactions implicated in neuropathies such as Huntington's disease, Alzheimer's disease, and Parkinson's disease, various members of the heat shock protein family, for example HSP70 and HSP40 (Auluck et al. 2002; Muchowski et al. 2000; Warrick et al. 1999), are able to interact with aggregation-prone polypeptides, and suppress the irreversible formation of fibrils. The molecular recognition abilities of members of the HSP60 family have also been the focus of numerous studies regarding the suppression of amyloid fibrils, and numerous members of the HSP60 family are capable of suppressing fibrillogenesis of proteins implicated in a wide variety of debilitating syndromes.

The HSP60 protein in eukaryotic cells, T-complex polypeptide 1 ring complex (TRiC) (or chaperonin containing T-complex polypeptide 1, CCT) is a heterooligomeric ring complex of eight homologous subunits (8×2 = 16-mer) (Munoz et al. 2011). TRiC has been found to interact and suppress the fibril formation of huntingtin, the causative protein of Huntington's disease characterized by a long polyglutamine (PolyQ) repeat (Behrends et al. 2006; Kitamura et al. 2006; Tam et al. 2006). This suppressive effect is proposed to occur through binding of TRiC to soluble molecules of PolyQ-enriched huntingtin, and also by the capping of fibrils ends by the chaperone, preventing fibril extension (Shahmoradian et al. 2013). Effects are also seen in in vivo models of Huntington's disease (Carmichael et al. 2000, 2002). Further experiments have shown that an apical domain fragment from TRiC, ApiCCT1, is capable of suppressing the aggregation of huntingtin, both in vitro and in vivo (Sontag et al. 2013). This latter result is very promising to further development of an actual treating agent for Huntington's disease, as the ApiCCT1 polypeptide contains a sequence region that closely resembles the HIV Tat protein transduction domain, and is thereby capable of entering a eukaryotic cell on its own to localize at the cytosol.

TRiC has also been found to suppress the aggregation of α-synuclein, a 140-residue natively unfolded polypeptide of unclear cellular function whose aggregation and fibrillation is implicated in familial Parkinson's disease (Sot et al. 2017). In this case, the effects of TRiC were to bind to soluble α-synuclein molecules that have not associated, to lower the cellular concentration of these molecules and thereby inhibit aggregation.

In addition to the cytoplasmic HSP60 TRiC, eukaryotic cells also possess a mitochondrial version of HSP60. Mitochondrial HSP60 has also been demonstrated to bind to and suppress the fibril formation of Aβ 1–40 peptide, a small polypeptide whose fibrillation is correlated with the onset of Alzheimer's disease (Mangione et al. 2016). Mitochondrial HSP60 bound to this peptide and maintained it in a conformation resembling a random coil.

The bacterial HSP60, GroEL has also been shown in recent studies to possess fibril suppressing abilities. The isolated apical domain of GroEL (GroEL-AD) was successful in suppressing the fibrillation of multiple proteins that are implicated in amyloidosis, such as α-synuclein and Aβ 1–42 peptide (Ojha et al. 2016). This effect was not confined to pathological polypeptides, as the fibrillation of a non-pathogenic model protein (GroES) was also strongly suppressed by the presence of GroEL-AD. The recognition mechanism of GroEL-AD seemed to adapt versatilely to the affinities of each fibrillogenic polypeptide, as multiple modes of binding and fibril suppression were observed. The intact GroEL oligomer was also a potent suppressor of fibrillogenesis, as determined in a separate study (Fukui et al. 2016). Interestingly, in this study it was found that mutations to the GroEL subunit that changed the orientation of the important apical domain (Machida et al. 2009) resulted in a change in affinity toward various fibrillogenic polypeptides. It was possible to modulate, using point mutations localized on a single site in the GroEL subunit, the affinity between GroEL and target fibrillogenic protein to a surprising degree of fidelity, and these results offered the potential to lead to affinity-tailored HSP60 agents that may show differing specificities toward various pathogenic targets. The versatility of GroEL in recognizing and binding to proteins is however not without its caveats; a very recent study involving GroEL and the Het-s prion protein revealed that in this specific case, the actions of GroEL served to accelerate the formation of protofibrils and formed a unique construct that bound GroEL to the outer surface of the resultant fibrils (Walti et al. 2017). Also, it has been demonstrated that eukaryotic CCT promotes the fibrillar aggregation of α-tubulin in vivo and in vitro (Pouchucq et al. 2018). The effects of HSP60 proteins on fibrillogenic proteins most likely change according to the specific structural elements of each target and the conditions under which chaperone and target interact.

5.1.3 HSP60 as Target

In addition to the very prominent role as a molecular chaperone to unfolded polypeptides, members of the HSP60 family have been implicated in a large number of additional physiological processes, both in bacteria and in eukaryotes (Henderson et al. 2013). It has been hypothesized that the prevalence of HSP60 proteins in diverse cellular processes such as cell surface ligand binding, induction of apoptosis, stimulation of cytokine synthesis, and biofilm production is due partially to the evolution of multiple HSP60 genes, and further evolution of these HSP60 homologs into defined cellular roles. The HSP60 proteins have also been shown in recent studies to be transported out of the cell, and roles involving this extracellular transport mechanism have also been reported. In eukaryotes, HSP60 has been implicated in inflammation, immune response, and cancer, and as a prime example of a "moonlighting" protein with multiple and diverse set of functions in addition to its major role as a molecular chaperone, is a vast source of molecular targets that are currently being probed towards the development of numerous clinical applications (Henderson et al. 2013).

Apart from targeting the highly complex and numerous pathways that involve HSP60 as a moonlighting multifunctional protein, recently an effort to develop unique antibiotic compounds that target the original function of HSP60, that of molecular chaperone, have been reported. Since HSP60 is an essential molecular chaperone that is necessary for the survival of almost all bacterial species, and the sequence and function of HSP60 as molecular chaperone is highly conserved, low-molecular compounds targeting HSP60 function would be a highly potent way to prevent the spread of pathogenic bacteria.

Building on this concept, screenings of compounds that specifically bind to and suppress the activity of a reference HSP60 system, the *E. coli* GroEL/ES system, and a number of compounds that were effective in suppressing GroEL/ES activity were screened (Johnson et al. 2014). Interestingly, some of the screened compounds were effective as antibiotics to numerous pathogenic organisms, such as *Trypanosoma brucei* (Abdeen et al. 2016b), and multiple Gram-positive bacteria such as *Staphylococcus aureus* (Abdeen et al. 2016a, 2018). The potential to target HSP60 systems to develop broad-range antibiotics for battling infectious bacteria seems to be promising, and is an exciting area of research.

Mitochondrial HSP60 is the eukaryotic version of bacterial GroEL, and extensive efforts to discover low molecular weight compounds that inhibit its molecular chaperone activity are ongoing (Meng et al. 2018). In the case of mitochondrial HSP60, such compounds have long-reaching effects on a wide variety of cellular functions, owing to HSP60 proteins' characteristic as a moonlighting protein (Henderson et al. 2013) that is implicated in a large number of disorders ranging from neurodegenerative disease to autoimmune disorders to cancer. The reader is recommended to refer to the highly detailed review by Meng and colleagues regarding the search for various compounds that bind to and inhibit mitochondrial HSP60 (Meng et al. 2018); compounds highlighted in this review include mizoribine (Mizuno et al. 1974) (an immunosuppressant that binds to and suppresses the molecular chaperone (Itoh et al. 1995) and ATPase activity of mitochondrial HSP60 (Tanabe et al. 2012)), and epolactaene (Nagumo et al. 2005).

5.1.4 HSP60 as Facilitator

Recent advances in molecular biology and protein engineering have resulted in the development of numerous protein-based pharmaceuticals, and future medical breakthroughs will most likely be based upon the activity of a certain polypeptide.

The stability of this hypothetical bioactive polypeptide, therefore, will be an extremely important criteria in the large-scale development of numerous future pharmaceuticals. In this context, a number of studies have attempted to utilize the abilities of HSP60 proteins such as *E. coli* GroEL to identify ligands or experimental conditions that would improve the stability, and therefore the usefulness, of certain proteins with important biological activities. The role of HSP60 in these efforts would be to act as a "facilitator" in protein-based drug design, specifically, in the search for ligands and conditions that stabilize the final active agent.

The concept of using HSP60 as a facilitator in the search for protein stabilizing factors is based upon the principle that interactions with HSP60 proteins such as GroEL will increase when the target protein is destabilized (and prone to aggregation) and decrease when the target is stabilized. Generalization would then involve developing a sensitive method to detect these changes in the GroEL-target interaction. Recent efforts by the Fisher group have probed the feasibility of applying GroEL to such a generalized search platform (Naik et al. 2010), and their experiments have shown that utilizing the abilities of GroEL in such a manner is feasible, with additional benefits such as a highly promiscuous binding mechanism that recognizes many diverse target proteins of interest, and also a high rate of target recovery after the assay has been completed. Recently, the group has also successfully applied a sensitive detection method, bio-layer interferometry, to this system (Naik et al. 2014), which will allow the development of a highly sensitive and universally applicable method to detect the binding of protein targets to GroEL (which would subsequently be perturbed by the additional presence of promising stabilizing ligands). A robust technological platform, based on the detection and binding abilities of GroEL toward unstable protein molecules, may soon be realized and utilized in the development of protein-based pharmaceuticals with stable active protein reagents.

5.2 Conclusions

The HSP60 family of molecular chaperones have evolved to attain a unique set of functional characteristics. Its highly sensitive and yet versatile mechanism for recognizing protein molecules, its unique ability to encapsulate molecules within its structure, and the eventual timed release of these encapsulated molecules in response to changes in nucleotide concentration are all highly relevant to the development of numerous agents and scaffolds of medical applicability. Additionally, the versatility of HSP60 has presumably resulted in the great expansion of physiological roles of this protein in the cell, and has made this protein the foremost target in efforts to prevent and treat the many diseases that are harmful to the human body. Further efforts such as the examples listed in this chapter will certainly result in the development of potent and broadly applicable medical breakthroughs that will benefit mankind.

Acknowledgements The authors were funded by the Strategic Research Program for Brain Sciences from Japan Agency for Medical Research and development, AMED (JP18dm0107073).

References

Abdeen S, Salim N, Mammadova N, Summers CM, Frankson R, Ambrose AJ, Anderson GG, Schultz PG, Horwich AL, Chapman E, Johnson SM (2016a) GroEL/ES inhibitors as potential antibiotics. Bioorg Med Chem Lett 26:3127–3134

Abdeen S, Salim N, Mammadova N, Summers CM, Goldsmith-Pestana K, McMahon-Pratt D, Schultz PG, Horwich AL, Chapman E, Johnson SM (2016b) Targeting the HSP60/10 chaperonin systems of *Trypanosoma brucei* as a strategy for treating African sleeping sickness. Bioorg Med Chem Lett 26:5247–5253

Abdeen S, Kunkle T, Salim N, Ray AM, Mammadova N, Summers C, Stevens M, Ambrose AJ, Park Y, Schultz PG, Horwich AL, Hoang QQ, Chapman E, Johnson SM (2018) Sulfonamido-2-arylbenzoxazole GroEL/ES inhibitors as potent antibacterials against methicillin-resistant *Staphylococcus aureus* (MRSA). J Med Chem 61:7345–7357

Auluck PK, Chan HY, Trojanowski JQ, Lee VM, Bonini NM (2002) Chaperone suppression of alpha-synuclein toxicity in a Drosophila model for Parkinson's disease. Science 295:865–868

Behrends C, Langer CA, Boteva R, Bottcher UM, Stemp MJ, Schaffar G, Rao BV, Giese A, Kretzschmar H, Siegers K, Hartl FU (2006) Chaperonin TRiC promotes the assembly of polyQ expansion proteins into nontoxic oligomers. Mol Cell 23:887–897

Biswas S, Kinbara K, Oya N, Ishii N, Taguchi H, Aida T (2009) A tubular biocontainer: metal ion-induced 1D assembly of a molecularly engineered chaperonin. J Am Chem Soc 131:7556–7557

Biswas S, Kinbara K, Niwa T, Taguchi H, Ishii N, Watanabe S, Miyata K, Kataoka K, Aida T (2013) Biomolecular robotics for chemomechanically driven guest delivery fuelled by intracellular ATP. Nat Chem 5:613–620

Braig K, Otwinowski Z, Hegde R, Boisvert DC, Joachimiak A, Horwich AL, Sigler PB (1994) The crystal structure of the bacterial chaperonin GroEL at 2.8 Å. Nature 371:578–586

Carmichael J, Chatellier J, Woolfson A, Milstein C, Fersht AR, Rubinsztein DC (2000) Bacterial and yeast chaperones reduce both aggregate formation and cell death in mammalian cell models of Huntington's disease. Proc Natl Acad Sci U S A 97:9701–9705

Carmichael J, Vacher C, Rubinsztein DC (2002) The bacterial chaperonin GroEL requires GroES to reduce aggregation and cell death in a COS-7 cell model of Huntington's disease. Neurosci Lett 330:270–274

Chatellier J, Hill F, Lund PA, Fersht AR (1998) In vivo activities of GroEL minichaperones. Proc Natl Acad Sci U S A 95:9861–9866

Chaudhry C, Horwich AL, Brunger AT, Adams PD (2004) Exploring the structural dynamics of the E.coli chaperonin GroEL using translation-libration-screw crystallographic refinement of intermediate states. J Mol Biol 342:229–245

Collins MO, Yu L, Campuzano I, Grant SG, Choudhary JS (2008) Phosphoproteomic analysis of the mouse brain cytosol reveals a predominance of protein phosphorylation in regions of intrinsic sequence disorder. Mol Cell Proteomics 7:1331–1348

Ditzel L, Lowe J, Stock D, Stetter KO, Huber H, Huber R, Steinbacher S (1998) Crystal structure of the thermosome, the archaeal chaperonin and homolog of CCT. Cell 93:125–138

Fukui N, Araki K, Hongo K, Mizobata T, Kawata Y (2016) Modulating the effects of the bacterial chaperonin GroEL on fibrillogenic polypeptides through modification of domain hinge architecture. J Biol Chem 291:25217–25226

Golbik R, Zahn R, Harding SE, Fersht AR (1998) Thermodynamic stability and folding of GroEL minichaperones. J Mol Biol 276:505–515

Hayer-Hartl M, Bracher A, Hartl FU (2016) The GroEL-GroES chaperonin machine: a nano-cage for protein folding. Trends Biochem Sci 41:62–76

Henderson B, Fares MA, Lund PA (2013) Chaperonin 60: a paradoxical, evolutionarily conserved protein family with multiple moonlighting functions. Biol Rev Camb Philos Soc 88:955–987

Iakoucheva LM, Radivojac P, Brown CJ, O'Connor TR, Sikes JG, Obradovic Z, Dunker AK (2004) The importance of intrinsic disorder for protein phosphorylation. Nucleic Acids Res 32:1037–1049

Ishii D, Kinbara K, Ishida Y, Ishii N, Okochi M, Yohda M, Aida T (2003) Chaperonin-mediated stabilization and ATP-triggered release of semiconductor nanoparticles. Nature 423:628–632

Itoh H, Kobayashi R, Wakui H, Komatsuda A, Ohtani H, Miura AB, Otaka M, Masamune O, Andoh H, Koyama K, Sato Y, Tashima Y (1995) Mammalian 60-kDa stress protein (Chaperonin homolog).: identification, biochemical properties, and localization. J Biol Chem 270:13429–13435

Johnson SM, Sharif O, Mak PA, Wang HT, Engels IH, Brinker A, Schultz PG, Horwich AL, Chapman E (2014) A biochemical screen for GroEL/GroES inhibitors. Bioorg Med Chem Lett 24:786–789

Kaneko T, Nakamura Y, Sato S, Minamisawa K, Uchiumi T, Sasamoto S, Watanabe A, Idesawa K, Iriguchi M, Kawashima K, Kohara M, Matsumoto M, Shimpo S, Tsuruoka H, Wada T, Yamada M, Tabata S (2002) Complete genomic sequence of nitrogen-fixing symbiotic bacterium Bradyrhizobium japonicum USDA110. DNA Res 9:189–197

Kerner MJ, Naylor DJ, Ishihama Y, Maier T, Chang HC, Stines AP, Georgopoulos C, Frishman D, Hayer-Hartl M, Mann M, Hartl FU (2005) Proteome-wide analysis of chaperonin-dependent protein folding in *Escherichia coli*. Cell 122:209–220

Kitamura A, Kubota H, Pack CG, Matsumoto G, Hirayama S, Takahashi Y, Kimura H, Kinjo M, Morimoto RI, Nagata K (2006) Cytosolic chaperonin prevents polyglutamine toxicity with altering the aggregation state. Nat Cell Biol 8:1163–1170

Machida K, Fujiwara R, Tanaka T, Sakane I, Hongo K, Mizobata T, Kawata Y (2009) Gly192 at hinge 2 site in the chaperonin GroEL plays a pivotal role in the dynamic apical domain movement that leads to GroES binding and efficient encapsulation of substrate proteins. Biochim Biophys Acta 1794:1344–1354

Mangione MR, Vilasi S, Marino C, Librizzi F, Canale C, Spigolon D, Bucchieri F, Fucarino A, Passantino R, Cappello F, Bulone D, San Biagio PL (2016) Hsp60, amateur chaperone in amyloid-beta fibrillogenesis. Biochim Biophys Acta 1860:2474–2483

Meng Q, Li BX, Xiao X (2018) Toward developing chemical modulators of Hsp60 as potential therapeutics. Front Mol Biosci 5:35

Mizuno K, Tsujino M, Takada M, Hayashi M, Atsumi K (1974) Studies on bredinin. I. Isolation, characterization and biological properties. J Antibiot (Tokyo) 27:775–782

Muchowski PJ, Schaffar G, Sittler A, Wanker EE, Hayer-Hartl MK, Hartl FU (2000) Hsp70 and hsp40 chaperones can inhibit self-assembly of polyglutamine proteins into amyloid-like fibrils. Proc Natl Acad Sci U S A 97:7841–7846

Munoz IG, Yebenes H, Zhou M, Mesa P, Serna M, Park AY, Bragado-Nilsson E, Beloso A, de Carcer G, Malumbres M, Robinson CV, Valpuesta JM, Montoya G (2011) Crystal structure of the open conformation of the mammalian chaperonin CCT in complex with tubulin. Nat Struct Mol Biol 18:14–19

Nagumo Y, Kakeya H, Shoji M, Hayashi Y, Dohmae N, Osada H (2005) Epolactaene binds human Hsp60 Cys442 resulting in the inhibition of chaperone activity. Biochem J 387:835–840

Naik S, Haque I, Degner N, Kornilayev B, Bomhoff G, Hodges J, Khorassani AA, Katayama H, Morris J, Kelly J, Seed J, Fisher MT (2010) Identifying protein stabilizing ligands using GroEL. Biopolymers 93:237–251

Naik S, Kumru OS, Cullom M, Telikepalli SN, Lindboe E, Roop TL, Joshi SB, Amin D, Gao P, Middaugh CR, Volkin DB, Fisher MT (2014) Probing structurally altered and aggregated states of therapeutically relevant proteins using GroEL coupled to bio-layer interferometry. Protein Sci 23:1461–1478

Ojha B, Fukui N, Hongo K, Mizobata T, Kawata Y (2016) Suppression of amyloid fibrils using the GroEL apical domain. Sci Rep 6:31041
Peng Z, Yan J, Fan X, Mizianty MJ, Xue B, Wang K, Hu G, Uversky VN, Kurgan L (2015) Exceptionally abundant exceptions: comprehensive characterization of intrinsic disorder in all domains of life. Cell Mol Life Sci 72:137–151
Pettersen EF, Goddard TD, Huang CC, Couch GS, Greenblatt DM, Meng EC, Ferrin TE (2004) UCSF chimera – a visualization system for exploratory research and analysis. J Comput Chem 25:1605–1612
Pouchucq L, Lobos-Ruiz P, Araya G, Valpuesta JM, Monasterio O (2018) The chaperonin CCT promotes the formation of fibrillar aggregates of gamma-tubulin. Biochim Biophys Acta 1866:519–526
Sendai T, Biswas S, Aida T (2013) Photoreconfigurable supramolecular nanotube. J Am Chem Soc 135:11509–11512
Shahmoradian SH, Galaz-Montoya JG, Schmid MF, Cong Y, Ma B, Spiess C, Frydman J, Ludtke SJ, Chiu W (2013) TRiC's tricks inhibit huntingtin aggregation. elife 2:e00710
Sontag EM, Joachimiak LA, Tan Z, Tomlinson A, Housman DE, Glabe CG, Potkin SG, Frydman J, Thompson LM (2013) Exogenous delivery of chaperonin subunit fragment ApiCCT1 modulates mutant huntingtin cellular phenotypes. Proc Natl Acad Sci U S A 110:3077–3082
Sot B, Rubio-Munoz A, Leal-Quintero A, Martinez-Sabando J, Marcilla M, Roodveldt C, Valpuesta JM (2017) The chaperonin CCT inhibits assembly of alpha-synuclein amyloid fibrils by a specific, conformation-dependent interaction. Sci Rep 7:40859
Tam S, Geller R, Spiess C, Frydman J (2006) The chaperonin TRiC controls polyglutamine aggregation and toxicity through subunit-specific interactions. Nat Cell Biol 8:1155–1162
Tanabe M, Ishida R, Izuhara F, Komatsuda A, Wakui H, Sawada K, Otaka M, Nakamura N, Itoh H (2012) The ATPase activity of molecular chaperone HSP60 is inhibited by immunosuppressant mizoribine. Am J Mol Biol 02(02):4
Theillet FX, Binolfi A, Frembgen-Kesner T, Hingorani K, Sarkar M, Kyne C, Li C, Crowley PB, Gierasch L, Pielak GJ, Elcock AH, Gershenson A, Selenko P (2014) Physicochemical properties of cells and their effects on intrinsically disordered proteins (IDPs). Chem Rev 114:6661–6714
Walti MA, Schmidt T, Murray DT, Wang H, Hinshaw JE, Clore GM (2017) Chaperonin GroEL accelerates protofibril formation and decorates fibrils of the Het-s prion protein. Proc Natl Acad Sci U S A 114:9104–9109
Warrick JM, Chan HY, Gray-Board GL, Chai Y, Paulson HL, Bonini NM (1999) Suppression of polyglutamine-mediated neurodegeneration in Drosophila by the molecular chaperone HSP70. Nat Genet 23:425–428
Xu Z, Horwich AL, Sigler PB (1997) The crystal structure of the asymmetric GroEL–GroES–(ADP)7 chaperonin complex. Nature 388:741
Yebenes H, Mesa P, Munoz IG, Montoya G, Valpuesta JM (2011) Chaperonins: two rings for folding. Trends Biochem Sci 36:424–432
Yuan Y, Du C, Sun C, Zhu J, Wu S, Zhang Y, Ji T, Lei J, Yang Y, Gao N, Nie G (2018) Chaperonin-GroEL as a smart hydrophobic drug delivery and tumor targeting molecular machine for tumor therapy. Nano Lett 18:921–928
Zahn R, Buckle AM, Perrett S, Johnson CM, Corrales FJ, Golbik R, Fersht AR (1996) Chaperone activity and structure of monomeric polypeptide binding domains of GroEL. Proc Natl Acad Sci U S A 93:15024–15029
Zang Y, Jin M, Wang H, Cui Z, Kong L, Liu C, Cong Y (2016) Staggered ATP binding mechanism of eukaryotic chaperonin TRiC (CCT) revealed through high-resolution cryo-EM. Nat Struct Mol Biol 23:1083–1091

Chapter 6
Role of the Post-translational Modifications of HSP60 in Disease

Byron Baron

Abstract Heat Shock Protein 60 (HSP60) is primarily a chaperone protein responsible for refolding proteins in the mitochondria. Nevertheless it has numerous other pro- and anti-apoptotis and signalling functions which it achieves through a variety of post-translational modifications (PTMs). Increasing evidence indicates that in disease states from cancer to systemic inflammation, such modifications become dysregulated, and as a result HSP60 cannot perform its functions. Understanding the biological role of these PTMs in healthy and disease states, the context in which they are generated and removed, as well as the mechanisms by which they can be targetted and modulated extraneously will help to provide better therapeutic solutions to deal with a wide range of conditions driven by stress-related processes.

Keywords Acetylation · Glycosylation · HSP60 · Methylation · Nitration · Phosphorylation · Post-translational modifications · Sulfoxidation · Ubiquitination

Abbreviations

ADAM	A Disintegrin And Metalloproteinase
aDMA	asymmetric N,N-di-methylarginine
AdoMet	S-adenosyl-L-methioinine
Akt	protein kinase B
APCs	antigen-presenting cells
ATP	adenosine triphosphate
β-O-GlyNAcylation	O-linked N-acetylglucosamine
Bak	BCL2 Antagonist/Killer 1
Bax	BCL2-Associated X Protein

B. Baron (✉)
Centre for Molecular Medicine and Biobanking, Faculty of Medicine and Surgery, University of Malta, Msida, Malta
e-mail: byron.baron@um.edu.mt

A. A. A. Asea, P. Kaur (eds.), *Heat Shock Protein 60 in Human Diseases and Disorders*, Heat Shock Proteins 18,
https://doi.org/10.1007/978-3-030-23154-5_6

BCL2	B-Cell Chronic Lymphocytic Leukemia/Lymphoma 2
BCL2-L1	Bcl-2-Like Protein 1
CD	cluster of differentiation
CLEC	C-type lectin
DNA	deoxyribonucleic acid
Dox	Doxorubicin
E1s	ubiquitin-activating enzymes
E2s	ubiquitin-conjugating enzymes
E3s	ubiquitin ligases
ER	endoplasmic reticulum
ERB	eukaryotic ribosome biogenesis protein
ETF	electron transfer flavoprotein
ExAC	Exome Aggregation Consortium
GA	Geldanamycin
GPI anchor	Glycosylphosphatidylinositol anchor
H2B	histone 2B
HATs	histone acetyltransferases
HDACis	Histone deacetylase inhibitors
HDACs	histone deacetylases
HSE	heat shock response element
HSF1	heat shock factor 1
HSP	heat shock protein/s
IL-1b	interleukin-1b
iNOS	inducible nitric oxide synthase
K	lysine
KATs	lysine acetyltransferases
KDACs	lysine deacetylases
KO	knock-out
M	methionine
METTL	Methyltransferase-like
MHC	major histocompatibility complex
MIRKO	muscle insulin receptor knockout
MMPs	matrix metalloproteases
N	asparagine
NAC	N-acetyl-cysteine
NATs	N-terminal acetyltransferases
NF-kB	Nuclear Factor kappa-light-chain-enhancer of activated B cells
NKG2D	Natural killer group 2 member D receptor
NO	nitric oxide
NSP4	nonstructural protein 4
PKA-I	protein kinase A, type 1
PKMTs	protein lysine methyltransferases
PRL-3	phosphatase of Regenerating Liver 3
PRMTs	protein arginine methyltransferases
PTMs	post-translational modifications

RNS	reactive nitrogen species
ROS	reactive oxygen species
S	serine
SAHA	Suberoylanilide hydroxamic acid
SET	Su(var)3-9, Enhancer-of-zeste and Trithorax
SIRT	sirtuin
SNP	single nucleotide polymorphism
Src	sarcoma
T	threonine
TLR	toll-like-receptor
ULBP2	UL16 binding protein 2
Y	tyrosine

6.1 Introduction

Heat Shock Protein 60 (HSP60) is generally described as a mitochondrial chaperone, acting with its co-chaperone HSP10, to transport mitochondrial proteins from the cytoplasm into the mitochondrial matrix, facilitate the folding process of new proteins following import into the mitochondria or refold misfolded proteins within the mitochondria (Koll et al. 1992; Bukau and Horwich 1998; Ghosh et al. 2010). HSP60 also fulfils other functions within the mitochondria, including as a ROS sensor and scavenger, thus protecting cells against oxidative stress (Sarangi et al. 2013).

The HSP60 protein is composed of three structural domains (referred to as apical, intermediate, and equatorial) with distinct functional roles. The complex formed between HSP60 and HSP10 (HSP60-HSP10) is composed of two seven-meric rings of HSP60 stacked back-to-back, producing an inner cavity that is capped by a lid at each extremity composed of a seven-meric ring of HSP10 (Nisemblat et al. 2015). The HSP60-HSP10 complex undergoes functional folding cycles in which the unfolded or misfolded client protein binds to the HSP60 rings, the HSP10 rings enclose the protein by capping the complex, in an ATP-dependent process, promoting folding of proteins into their native shape, and then the HSP60 sub-units, HSP10 sub-units and the enclosed protein come apart (Nielsen and Cowan 1998; Levy-Rimler et al. 2001).

Despite the main localisation of this protein being within the mitochondria, HSP60 has also been isolated from other cellular compartments, such as the cytoplasm, endoplasmic reticulum, intracellular vesicles, peroxisomes, secretory granules and plasma membrane, as well as in the extracellular space of both normal and tumour cells (Soltys and Gupta 1996, 1997; Barazi et al. 2002; Chen et al. 2006; Chandra et al. 2007; Merendino et al. 2010). However, despite the various subcellular localisations of HSP60, this is not the result of splice variants of mRNA (Yogev and Pines 2011; Hayoun et al. 2012), directing the focus towards the application of PTMs to achieve this. An alternate mechanism has been proposed by which a single mRNA initiates translation from an additional non-AUG start codon, producing two variants of HSP60 (Touriol et al. 2003), where the non-AUG product lacks the mitochondrial targeting peptide, and can be directed to the ER membrane (Pyhtila et al. 2008), thus allowing HSP60 to undergo a variety of PTMs and relocate to the various sub-cellular compartments.

Apart from its main chaperoning role, HSP60 can fulfil other very diverse functions depending on the PTMs added and its sub-cellular localisation, as shall be described in the following sections. The best described of these functions is in the cytoplasm, where HSP60 has been reported to fulfil both anti- and pro-apoptotic functions depending on its tissue distribution, and the key factors it interacts with such as B-Cell Chronic Lymphocytic Leukemia/Lymphoma 2 (BCL2), BCL2-Associated X Protein (Bax), BCL2 Antagonist/Killer 1 (Bak), Bcl-2-Like Protein 1 (BCL2-L1) and caspase 3 (Chandra et al. 2007; Gupta and Knowlton 2002; Shan et al. 2003; Gupta and Knowlton 2005).

At the other end of the spectrum, the role of HSP60 on the plasma membrane is mainly as an immune regulator, being a ligand of Toll-like-receptor (TLR)2, TLR4, and major histocompatibility complex (MHC) proteins on immune cells such as neutrophils and macrophages and thus triggering both innate and adaptive immune responses (Quintana and Cohen 2011; Farrugia and Baron 2017).

Furthermore, despite not having an export sequence, HSP60 may be exported into the extracellular space or general circulation via microvesicles such as exosomes (Thery et al. 2009; Campanella et al. 2012, 2014; Greening et al. 2015). HSP60 first reaches the plasma membrane by a mechanism that involves lipid rafts and is then secreted outside the cell by an active mechanism in which HSP60 is carried inside of or on the membrane of exosomes (Macario and de Macario Conway 2007; Merendino et al. 2010; Campanella et al. 2012). In the extracellular space, it acts as a signalling molecule, mediating interactions between the secretory source (such as a tumour) and immune cells or other tissues, by interacting with cell-surface receptors, such as Cluster of Differentiation (CD)14, CD40, and TLRs, causing pro-inflammatory (Chen et al. 1999) or anti-inflammatory effects by promoting the secretion of suppressive cytokines from antigen-presenting cells (APCs), with subsequent activation of T-cells (Osterloh et al. 2004; Zanin-Zhorov et al. 2005; Van Eden et al. 2007).

6.2 Role in Disease

The mutation of the HSP60 gene is known to be linked only to two serious human genetic disorders, a dominant form of hereditary spastic paraplegia and a recessively inherited white matter disorder called MitCHAP60 disease (Magen et al. 2008; Hansen et al. 2002). Despite this, in most cases it is the dysregulation of HSP60 at the protein level which is related to a harmful condition or disease state. While most studies have traditionally looked at the total HSP60 expression and its localisation, advances in protein separation and peptide enrichment techniques, together with the ever-increasing sensitivity of mass spectrometric technology, are now giving a more in-depth view of the changes to HSP60 at a post-translational level and how this translates to alterations in biochemistry, which are associated with disease (Baron 2015, 2018). The function of HSP60 related to cytoprotection against oxidative stress underlies its role in stress response and survival under such stressful conditions as well as some of the most common diseases such as cancer, diabetes, chronic inflammatory conditions, and neurological disorders such as Alzheimer's disease (Koeck et al. 2009).

6.2.1 Role in Cancer

Through the stages of tumour progression, in various types of cancers, intracellular, pericellular or circulating HSP60 levels, have been found to increase, as a consequence of some dysregulated process, accompanied by uncontrolled proliferation and neoplastic transformation and metastasis (Barazi et al. 2002; Cappello et al. 2005; Czarnecka et al. 2006; Macario and de Macario Conway 2007; Ghosh et al. 2008; Tsai et al. 2009; Campanella et al. 2012; Rappa et al. 2012; Li et al. 2014a; Lianos et al. 2015). The intracellular role of HSP60 in the process of carcinogenesis spans from the promotion of cell transformation via regulation of proto-oncogenes such as c-MYC (Tsai et al. 2008), to evasion of apoptosis and cell survival by interacting with proteins such as p53 and survivin (Gupta and Knowlton 2002; Ghosh et al. 2008, 2010). Once released in exosomes, HSP60 can also plays an extracellular role in cancer progression through its modulatory effect on the immune system, acting as a mediator of tumour immunity or conversely blocking or misdirecting the immune system (Macario and de Macario Conway 2007; Merendino et al. 2010; Campanella et al. 2012).

The presence of total HSP60 in the blood, urine, and other biological fluids has been proposed as a diagnostic and prognostic marker of cancer. For example, both HSP10 and HSP60 are over-expressed during colorectal carcinogenesis (Cappello et al. 2003, 2009) and the levels of HSP60-containing exosomes in the blood of colon cancer patients decreased from significantly high levels before operation to normal levels within 1 week from surgical resection of the tumour mass (Campanella et al. 2015).

6.2.2 Role in Chronic Inflammation, Type 2 Diabetes Mellitus and Autoimmune Diseases

The role of HSP60 in chronic inflammation, Type 2 Diabetes Mellitus and autoimmune diseases stems from its ability to interact with both the innate and the adaptive immune system, acting as a foreign antigen (resulting in the induction of the immune system by infectious agents), a TLR ligand and a carrier of other functional molecules (providing a means of communication between immune cells and other cells in the body) (Quintana and Cohen 2011) as well as a self-antigen. The latter is due to the high degree of structural similarity between eukaryotic and prokaryotic HSP60, and this similarity is the basis of the failure of the mechanism discriminating self from non-self, which leads to induction of autoimmunity (Campanella et al. 2009) and inflammation (Tomasello et al. 2011) and, thereby, to chronic inflammatory disorders including Type 2 Diabetes Mellitus.

6.3 Post-translational Modifications of HSP60

6.3.1 Acetylation

Acetylation occurs either at the N-terminus of a protein or on the ε-amino group of a lysine (K) residue and this reaction uses acetyl-coenzyme A as the acetyl group donor. N-terminal acetylation is considered irreversible and the enzymes catalysing such reactions are called N-terminal acetyltransferases (NATs), while lysine acetylation is reversible, with acetyl groups being added by lysine acetyltransferases (KATs; a.k.a. histone acetyltransferases - HATs) and removed by lysine deacetylases (KDACs; a.k.a. histone deacetylases - HDACs). There are 22 KATs and 18 KDACs in humans, which can act on both histone and non-histone proteins (Drazic et al. 2016). Based on their catalytic mechanisms, KDACs are divided into two groups, namely Zn^{2+}-dependent HDACs (HDAC1–11) and NAD+ -dependent sirtuins (SIRT1–7) (Glozak et al. 2005).

The activity of these KDACs can be inhibited by a group of compounds called Histone deacetylase inhibitors (HDACis), that cause accumulation of acetylated proteins (Marks and Xu 2009). One such HDACi, Suberoylanilide hydroxamic acid (SAHA; Vorinostat), presenting anti-cancer effects (Butler et al. 2000; Lauricella et al. 2012; Chen et al. 2013), has been administered alone or in combination with other anti-cancer drugs (etoposide, gemcitabine, cisplatin, paclitaxel, doxorubicin, cyclophosphamide, gefitinib) to cancer patients (Kim and Bae 2011; Kirschbaum et al. 2014; Gueugnon et al. 2014; Tu et al. 2014; Han et al. 2015; Straus et al. 2015). HDACis can hyperacetylate HSP40, HSP70 and HSP90, altering functions regulated by acetylation (Fiskus et al. 2007; Hageman et al. 2010; Rao et al. 2012; Yang et al. 2013; Ha et al. 2014; Haaland et al. 2014; Huang et al. 2015). In the case of HSP90 hyperacetylation, HDACis inhibit HDAC6, which brings about the dissociation of client proteins ERB1, ERB2, and Akt (Fiskus et al. 2007). HDACi drugs are also known to produce reactive oxygen species (ROS), inducing oxidative stress and resulting in cell death, in a variety of cancer cell types (Yang et al. 2013; Ha et al. 2014; De Bellis et al. 2014). For HSP60 however, hyperacetylation was not observed after SAHA treatment (Campanella et al. 2016)

HSP60 has been shown to be a substrate of Sirtuin 3 (Sirt3), with increased acetylation contributing to increased ROS and an altered response to oxidative stress (Lombard et al. 2007; Sol et al. 2012). Sirt3 is a member of the sirtuin family, classified as Class III HDACs, located in the mitochondria (Schwer et al. 2002; Lombard et al. 2007), and its over-expression affects the expression of genes involved in mitochondrial function, leading to the mitochondrial stress response and ROS generation (Shi et al. 2005; Jing et al. 2011). Sirt3 has been found to play an important role in diabetes, being down-regulated in response to a high-fat diet, in conjugation with an increased rate of fatty acid oxidation and mitochondrial protein acetylation in Sirt3 knock-out mice (Alrob et al. 2014). It was also detected at significantly decreased expression in the muscles of muscle insulin receptor knockout (MIRKO) mice, an insulin-deficient diabetic mouse model (Yechoor et al. 2004). Moreover, Sirt3 expression was found to be altered in the skeletal muscles of both type 1 and type 2 diabetic mouse models. The alterated Sirt3 expression regulated mitochondrial oxidation and ROS production, which in turn altered insulin signaling, resulting in insulin resistance in skeletal muscle (Jing et al. 2011). On the other hand, Sirt3 is upregulated in white and brown adipose tissue in response to caloric restriction (Shi et al. 2005).

Cellular senescence (cell cycle arrest in the G2/M phase) can also be triggered by oxidative stress, such as following chemotherapy, making its induction an attractive aim in anticancer treatment to limit cell proliferation (Braig et al. 2005; Guo et al. 2009). Senescence induced by DNA-damaging anticancer drugs is one of the major factors of successful chemotherapy (Kikuchi et al. 2010). Doxorubicin (Dox), an anthracycline drug widely used as a chemotherapeutic agent, induces senescence at low doses and apoptosis at high doses (Song et al. 2005; Spallarossa et al. 2009; Altieri et al. 2012). Dox intercalates with DNA inhibiting topoisomerase II progression and generates ROS. Dox induces the expression of the tumour suppressor protein p53, which mediates the DNA damage response (Kurz et al. 2004; Spallarossa et al. 2009), as well as up-regulation of p21, a transcriptional target of p53 involved in the induction of senescence (Roninson 2002). After human-lung mucoepidermoid carcinoma cells underwent Dox-treatment, HSP60 gene expression levels significantly increased, while total protein levels decreased and acetylation levels increased. Hyperacetylation of Hsp60 was possibly due to the inhibition of sirtuins since Dox is known to induce senescence via down-regulation of Sirt1 and Sirt2 (Eren et al. 2015). This was linked to a significant increase in HSP60 ubiquitination, which is thought to have tagged the hypermethylated HSP60 for proteosomal degradation. The ubiquitination and degradation of acetylated HSP60, impaired the anti-apoptotic effect of HSP60, leading to tumour cell death. On the other hand the extracellular release of HSP60 was not consistent with increasing doses (Gammazza et al. 2017). Furthermore, the amount of p53 in the HSP60:p53 complex significantly decreased after Dox-treatment in a concentration-dependent manner, probably favoured by HSP60 acetylation, suggesting that HSP60 acetylation may interfere with the formation of the HSP60:p53 complex and/or promote its dissociation, both causing an increase in the levels of free p53, which can then activate cell senescence (Gammazza et al. 2017). The disruption of the HSP60:p53 complex and the activation of senescence possibly occurs via the p53-p21 pathway (Mirzayans et al. 2012).

A similar outcome was observed following treatment of osteosarcoma cells with Geldanamycin (GA), which induced 2.4-fold upregulation of HSP60 gene expression, but decreased both total and hyperacetylated HSP60 protein levels at 6 h post-treatment, compared to control (Gorska et al. 2013). Geldanamycin (GA) is a benzoquinone ansamycin antibiotic, known to be a potent inhibitor of HSP90 via its N-terminal ATP binding site, destabilising the complex between HSP90 and its client proteins, and leading to proteasomal degradation (Powers and Workman 2007). GA also appears to influence the expression of HSP by stabilising the trimeric complex between Heat Shock Factor 1 (HSF1; the HSP transcription factor) and the heat shock response element (HSE), stimulating HSP gene expression (Kim et al. 1999; Lai et al. 2003; Chao et al. 2008). Apart from changes in HSP60 levels, GA treatment of osteosarcoma cells brought about an overall decrease in cell viability and increased cancer cell death, suggesting that hyperacetylation of HSP60 is associated with the anti-cancer activity of GA (Gorska et al. 2013). HSP60 knockdown is known to contribute to osteosarcoma cell arrest (Kaul et al. 2006; Cappello and Zummo 2005), while HSP60 was shown to be acetylated in osteosarcoma cells with Sirt3 knockout (Sol et al. 2012). Thus it has been proposed that since hyperacetylation affects HSP function (Nahleh et al. 2012), GA-induced HSP60 hyperacetylation could act to deregulate HSP60 activity, disturbing protein folding (Cappello and Zummo 2005), and causing proteotoxic stress, leading to a switch from pro-survival signaling to cell death within cancer cells (Karbowski et al. 2001).

Moreover, the acetylation status of HSP10, the co-chaperone of HSP60, may play a role in the stabilisation and function of the HSP60-HSP10 complex. It has been shown that HSP10 is also a functional substrate of Sirt3 deacetylation (Fritz et al. 2012; Rardin et al. 2013; Lu et al. 2015). Proteomic analysis of Sirt3 KO mice identified four differentially acetylated sites on HSP10, namely K28, K40, K54 and K56 and further mutational analysis confirmed that K56 alters HSP60-HSP10 binding, conformation, and protein folding. Maintained acetylation of HSP10 K56 thus reduced HSP60-HSP10 binding affinity, prevented the accumulation of misfolded proteins, while enhancing the efficiency of protein folding by the HSP60-HSP10 complex (Lu et al. 2015).

6.3.2 Nitration

The nitration of tyrosine (Y) residues to 3-nitrotyrosine is one of a variety of oxidative PTMs related to the generation of ROS (Radi 2004, 2012; Monzani et al. 2004; Pacher et al. 2007). Reactive nitrogen species (RNS) such as peroxynitrite are produced from nitric oxide (NO) or NO-derived metabolites in the presence of elevated levels of ROS (Pacher et al. 2007), which when produced in the mitochondria, can lead to increased tyrosine nitration of mitochondrial proteins. Following nitration by RNS, key properties such as the structure and function of target proteins can be modified (Monzani et al. 2004; Radi 2012; Abdelmegeed and Song 2014), which in turn may alter cell homeostasis (Abello et al. 2009). Such nitration can be delaterious to mitochondrial functions, impeding cell maintenance and leading to cell death (Davis et al. 2010). The pathological conditions in which tyrosine nitration has been detected include cardiovascular disorders, diabetes, and hepatic and neurodegenerative diseases (Thomson 2015; Chavarría and Souza 2013).

Nitration of HSP60 most likely occurs on one or both of the highly conserved tyrosines Y223 and Y227, found in the apical domain, which is crucial for the binding of the co-chaperonin HSP10 and substrates (Campanella et al. 2016). HSP60 nitration has also been shown to bring about a significant decrease in the ATP-hydrolysis activity of the chaperonin as well as disturbing the interaction of HSP60 with its co-chaperonin HSP10 and to a lesser extent, substrate proteins (Koeck et al. 2009). In the case of HSP60 nitration in lung cancer cells, after treatment with SAHA, the cell cycle was interrupted at the G2/M phase, ROS over-production caused oxidative stress together with increased mitochondrial damage, and HSP60 levels were significantly reduced, leading to cell death. Once nitrated, HSP60 was released into the extracellular space via exosomes. The increase in nitrated HSP60 brought about by SAHA treatment could be off-set by the antioxidant N-acetylcysteine (NAC), indicating that nitration was directly linked to the oxidative stress induced by SAHA (Campanella et al. 2016).

In the context of hyperglycaemia, when rat insulinoma cells and rat islets of Langerhans were exposed to elevated glucose levels, a significant glucose concentration-dependent increase in 3-nitrotyrosine immunoreactivity was observed specifically in the mitochondria, indicating that HSP60 nitration was organelle-based (Koeck et al. 2009). ROS are known to activate NF-kB and the expression of pro-inflammatory cytokines such as interleukin-1b (IL-1b), which is known to be a major cytokine in the pathogenesis of Type 2 Diabetes Mellitus (Pacher et al. 2007; Alexandraki et al. 2006). In turn, IL-1b is a primary mediator of inducible nitric oxide synthase (iNOS) expression and nitric NO generation by B-cells (Corbett and McDaniel 1995). Hyperglycemia is known to induce IL-1b production, iNOS expression, and activation of neuronal NOS in the pancreatic islets (Pacher et al. 2007; Maedler et al. 2002; Jimenez-Feltstrom et al. 2005), and an NO-dependent occurrence of tyrosine nitration has been reported in the islets of Langerhans (Suarez-Pinzon et al. 1997; Lakey et al. 2001; Szabó et al. 2002). The processing of proinsulin to produce active insulin is thought to include a step in which it binds to HSP60 (Brudzynski and Martinez 1993; Arias et al. 2000). However it has been shown that the ability of ATP hydrolysis and binding efficiency of HSP60 to proinsulin are significantly decreased following nitration, advocating the possibility that HSP60 nitration could bring about a decrease in insulin secretion (Koeck et al. 2009), with increased levels of nitro-tyrosine having been correlated with reduced glucose-stimulated insulin secretion in human type 2 diabetic islets (Lupi et al. 2007). This further indicates that even small changes in glucose levels can bring about physiological changes through a variety of PTMs, that if not corrected early on, can lead to chronic disease such as Type 2 Diabetes Mellitus.

With respect to alcohol consumption, HSP60 nitration is part of the outcome of mitochondrial dysfunction and protein nitration in the liver (Barone et al. 2016). In both the liver and the pancreas, HSP60 is considered to be a major part of the defence mechanism against cellular damage resulting from ethanol consumption (Rakonczay et al. 2003). Apart from this context, HSP60 in the liver is known to be involved in hepatocellular carcinoma, chronic active hepatitis, and primary biliary cirrhosis (Cappello et al. 2014).

6.3.3 Sulfoxidation

The sulfoxidation of methionine (M) residues to methionine sulfoxide (or methionine sulfone) is another of the oxidative PTMs related to the generation of ROS (Stadtman 2006). The oxidation of methionine to methionine sulfoxide and eventual reduction back by methionine sulfoxide reductase (Msr) is a critical pathway for repairing oxidatively damaged proteins and eliminating ROS (Lee et al. 2009; Moskovitz and Oien 2010). HSP60 has a methionine-rich C-terminal motif, which is highly conserved in different species and has been proposed as a ROS-acceptor, although the exact function is as yet unknown (Brocchieri and Karlin 2000; Li et al. 2014b). Using mass spectrometry, it was shown that between one and five of the methionines in the C-terminus of HSP60 were oxidised and these co-localised with biotinylated lysines (Li et al. 2014b). Lysine biotinylation is catalysed by holocarboxylase synthetase (Chapman-Smith and Cronan 1999), which is localised in the cytoplasm, nucleus and mitochondria (Narang et al. 2004; Chew et al. 2006; Bailey et al. 2010; Bao et al. 2011). Furthermore, fibroblasts stressed by biotin depletion, produced significantly higher levels of ROS (>9-fold compared to controls using biotin-supplemented medium), resulting in higher levels of methionine sulfoxidation, which in turn brought about cell cycle arrest in the G0/G1 and S phases (Li et al. 2014b). This led to the postulation of a mechanism in which lysine biotinylation increased the chance that an adjacent methionine undergoes sulfoxidation, facilitating the scavenging and elimination of ROS through the methionine reduction pathway, which is involved in the function of HSP60 as a means of defense against ROS (Li et al. 2014b).

6.3.4 Phosphorylation

Phosphorylation is a PTM which can occur on tyrosine (Y), serine (S) and threonine (T). Phosphorylation is performed by cellular kinases, while dephosphorylation is performed by phosphatases. HSP60 can be tyrosine phosphorylated at Y227 and Y243 (Rikova et al. 2007; Gu et al. 2011a) and HSP60 tyrosine phosphorylation is required for its surface activation (Asquith et al. 2004). HSP60 present in the plasma membrane of a human leukaemic T-cell line was phosphorylated by protein kinase A, type 1 (PKA-I). PKA-I is known to phosphorylate serine or threonine residues in the general consensus sequence R-R-X-S/T or K-R/X-S/T (Kennelly and Krebs 1991). HSP60 has six potential PKA-I phosphorylation sites: K72-V73-T74,

K130-I131-S132, K157-Q158-S159, K250-I251-S252, K396-L397-S398, and K469-R470-T471 (Jindal et al. 1989). It was shown that HSP60 binds to histone 2B (H2B) in the cytosol, when both molecules were unphosphorylated. HSP60 then chaperones H2B into the plasma membrane, where PKA-I specifically catalyses the phosphorylation of both HSP60 and H2B (Khan et al. 1998). One of the functions of HSP60 phosphorylation thus appears to be in regulating the binding affinity to its targets such as H2B, since it brings about the dissociation of H2B from HSP60 and release of H2B from the plasma membrane of the T-cells, into the extracellular space, while HSP60 is retained (Khan et al. 1998).

The tyrosine phosphorylation of HSP60, regulated by Phosphatase of Regenerating Liver 3 (PRL-3; a.k.a. Protein tyrosine phosphatase 4A3), was found to be essential for the protein maturation and surface expression of UL16 binding protein 2 (ULBP2). Newly synthesised ULBP2 (27 kDa) is glycosylated to become a mature 37-kDa glycoprotein (Cosman et al. 2001) and complete maturation of ULBP2 requires the activity of PRL-3 (Leung et al. 2015). Treating cancer cells with PRL-3 inhibitor I (PRL3-I) reduced the amount of phosphorylated HSP60 bound to ULBP2, which suppressed ULBP2 post-translational maturation, involving the formation of a GPI anchor and glycosylation, resulting in retention of the immature protein in the ER and reducing ULBP2 expression on the surface of cancer cells (Leung et al. 2015). ULBP2 is a ligand of the Natural killer group 2 member D (NKG2D) receptor, a C-type lectin (CLEC) expressed on the surface of human NK cells and T-cells. The interaction between this receptor and its ligands in situations of stress or infection triggers NK cells to secrete cytokines, which induce apoptosis of target cells (Raulet et al. 2013). However, the NKG2D ligands present on the cell surface of many cancer cells are proteolytically shed by the single-pass transmembrane and secreted A Disintegrin And Metalloproteinase (ADAM) matrix metalloproteases (MMPs) (Salih et al. 2002; Waldhauer and Steinle 2006, 2008), in order to evade immunosurveillance by NK cells and CD8 T-cells and serum levels of ULBP2 have been correlated with cancer prognosis (Hilpert et al. 2012; Yamaguchi et al. 2012). PRL-3 has also been associated with cancer progression and metastasis (Bessette et al. 2008; Al-Aidaroos and Zeng 2010; Semba et al. 2010), with inhibition of PRL-3 being able to block migration and invasion of metastatic cancer cells both *in vitro* and *in vivo* (Guo et al. 2008; Wang et al. 2009).

HSP60 phosphorylation was also found to play a role in rotavirus infection. Rotaviruses are double stranded RNA viruses, which cause gastroenteritis in children. During the early stage of rotavirus infection, HSP60 tyrosine-phosphorylation increased. This led to increased ubiquitinated HSP60 and proteasomal degradation, with an overall decreased expression level of HSP60. HSP60 was found to be phosphorylated at Y227 by an activated form of Src kinase, and this was critical for degradation. HSP60 phosphorylation and subsequent degradation could be inhibited by silencing the Src gene or using a Src kinase inhibitor. HSP60 assists the translocation of the rotaviral nonstructural protein 4 (NSP4) into the mitochondria and since NSP4 has a pro-apoptotic function, transient degradation of HSP60 during early hours of rotavirus infection, inhibits premature mitochondrial import of NSP4, preventing premature apoptosis, acting as one of several strategies to increase viral replication and propagation via a lower yield of viral progeny (Chattopadhyay et al. 2017). NSP4 is know to act by dissipating the mitochondrial potential, cytosolic release of cytochrome c and caspase activation (Martin-Latil et al. 2007; Bhowmick et al. 2012), together with mobilisation of Ca^{2+} ions from the ER, triggering calcium-mediated apoptosis (Tian et al. 1995; Zambrano et al. 2008). One important aspect of phosphorylation is its crosstalk with ubiquitination, with phosphorylation positively or negatively regulating ubiquitination, in a wide variety of cellular contexts. Another important occurrence of phosphorylation crosstalk is its competition with O-linked N-acetylglucosamine (O-GlcNacylation or O-GlcNac) on serine and threonine, where O-GlcNac prevents phosphorylation at a specific residue or adjacent sites (Hunter 2007).

6.3.5 Ubiquitination

Ubiquitination is the process by which one (mono-ubiquitination) or multiple (poly-ubiquitination) ubiquitin monomers are added to the lysine, cysteine, serine, threonine or N-terminus of the target protein, marking it for degradation via the proteasome, altering the cellular location, affecting activity, and modulating protein-protein interactions (Schnell and Hicke 2003; Komander and Rape 2012; McDowell and Philpott 2013). Ubiquitination involves three main steps: activation, conjugation, and ligation, performed by ubiquitin-activating enzymes (E1s), ubiquitin-conjugating enzymes (E2s), and ubiquitin ligases (E3s), respectively (Glickman and Ciechanover 2002; Pickart and Eddins 2004).

Proteomic analysis showed that in monocytes undergoing necrosis due to treatment with 5-azacytidine (as a ROS-generator), cytosolic and nuclear chaperones are down-regulated, while ER and mitochondrial chaperones are up-regulated. HSP60 was found to be the most abundant up-regulated mitochondrial chaperone, with expression after 5-azacytidine treatment increasing sevenfold. Moreover HSP60 in necrotic cells was found to be ubiquitinated at K396, whereas ubiquitinated HSP60 was not found in the untreated sample. While the role of ubiquitination at K396 in stress-induced monocyte necrosis was not determined, it was speculated that upregulation of HSP60 may serve as the first line of defense against cytotoxicity elicited by 5-azacytidine-induced cell stress (Tang et al. 2013).

Contrarily, following SAHA treatment (which was also used as a ROS generator), the level of ubiquitinated HSP60 in lung cancer cells was lower compared to untreated cells. The addition of MG132, a proteasome inhibitor, did not prevent the reduction in HSP60 levels induced by SAHA, indicating that the ubiquitin-proteasome system is probably not involved in this reduction (Campanella et al. 2016). However in this case, the position of the ubiquitinated residue/s involved was not determined.

6.3.6 Glycosylation

Glycosylation of proteins in the majority of cases involves the addition of glycans to the nitrogen on asparagine (N) within an N-X-S/T motif, where X can be any amino acid except proline (N-linked glycosylation) or the hydroxyl oxygen of serine, threonine, tyrosine, hydroxylysine, or hydroxyproline side-chains (O-linked glycosylation) (Spiro 2002). In tumours, as well as in normal cells under stress, N-glycosylated HSP60 is expressed on the cell surface or secreted extracellularly (Barazi et al. 2002), with about 10% of total HSP60 undergoing glycosylation (Avidan et al. 2009; Hayoun et al. 2012). The first part of the N-linked glycans (blocks of 14 glycan monomers) is added to the asparagine in the N-X-S/T motif in the ER (Helenius and Aebi 2001). HSP60 has three potential N-linked glycosylation sites, N103, N230 and N426. Subsequently, the glycan side chains are extensively modified as the glycoproteins matures through the ER and the Golgi complex (Helenius and Aebi 2001) into bi- and tri-antennary structures, including a fraction of high mannose glycans (with 5–9 mannosyl residues) as well as the addition of phosphorylation to the high mannose. This phosphorylation possibly occurs during the synthesis and trafficking of the glycosylated protein to the cell surface (Hayes et al. 1995; Hayoun et al. 2012). Once the modified HSP60 reaches the surface of the tumour, it can affect the immunological properties of the proteins in the tumour microenvironment, having many implications due to its ability to modulate the immune response within the tumour microenvironment.

HSP60 can also undergo addition of O-linked N-acetylglucosamine (β-O-GlyNAcylation), a cytosolic glycosylation (unlike N- and O-glycosylation), at serine or threonine residues, making it more akin to phosphorylation (Comer and Hart 2001; Copeland et al. 2008; Whelan et al. 2008). O-GlcNAcylation is commonly observed to be altered under high glucose conditions (hyperglycaemia), as found in various cell types in diabetic patients (Liu et al. 2000; Akimoto et al. 2001; Walgren et al. 2003). Hyperglycaemia is known to affect the phosphorylation status of O-GlcNAcylated proteins, where O-GlcNAcylation and serine or threonine phosphorylation have been shown to inversely affect each other while tyrosine phosphorylation has been found to facilitate O-GlcNAcylation (Cheng et al. 2000; Kamemura et al. 2002; Dentin et al. 2008; Ande et al. 2009; Whelan et al. 2008; Gu et al. 2011b). Moreover, increased O-GlcNAcylation has been strongly correlated with the development of insulin resistance (Buse 2006; Copeland et al. 2008).

Under hyperglycaemic conditions, O-GlcNAcylation and serine phosphorylation of HSP60 were found to be up-regulated whereas threonine phosphorylation was down-regulated (Gu et al. 2011b). Levels of cytosolic O-GlcNAcylated HSP60 were found to increase two-fold under hyperglycemic compared to normoglycemic conditions, with total cytosolic HSP60 levels remaining unchanged. This lead to the release of Bax from O-GlcNAcylated HSP60, which translocated to the mitochondria, triggering cytochrome c release and caspase-3 activation, leading to pancreatic β-cell death. Thus it appears that under normoglycemic conditions HSP60 prevents apoptosis by sequestering Bax in the cytoplasm. (Kim et al. 2006).

6.3.7 Arginine Methylation

Arginine methylation involves the addition of one or two methyl groups on one (asymmetric) or both (symmetrical) of the terminal nitrogen atoms of the arginine side chain. This reaction is catalysed by a family of enzymes called protein arginine methyltransferases (PRMTs), which use S-adenosyl-L-methioinine (AdoMet) as the methyl donor. So far, nine mammalian PRMTs (named PRMT1 to PRMT9) have been isolated (Wolf 2009), and are classified into two sub-families based on the end-product of the reaction they catalyse. Following the production of N-mono-methylarginine as an intermediate, type I enzymes (PRMT1, PRMT3, PRMT4/CARM1, PRMT6, and PRMT8) produce asymmetric N,N-di-methylarginine (aDMA), while type II enzymes (PRMT5, PRMT7, and PRMT9) produce symmetric N,N-di-methylarginine (Pahlich et al. 2006). It has been shown that there is an inversely proportional relationship between the level of arginine methylation and the cellular age or proliferation for proteins with molecular masses ranging between 50 and 100 kDa, with asymmetric arginine di-methylation being lower in senescent cells compared to low-passage cells (Lim et al. 2008). In agreement with this, asymmetric arginine di-methylation of HSP60 was found to be lower in senescent fibroblasts compared to low-passage fibroblasts, with no significant difference in asymmetric dimethyl-arginine expression levels between immortalised and low-passage fibroblasts, which was consistent with the reduction of total HSP60 observed in senescent cells (Lim et al. 2010). This implies that arginine asymmetric di-methylation of HSP60 is correlated with the proliferation potential of cells and might be useful as a marker of cellular senescence at least in some cell types.

6.3.8 Lysine Methylation

Lysine methylation involves the addition of one, two or three methyl groups on the terminal nitrogen atom of the lysine side chain. This reaction is catalysed by a family of enzymes called protein lysine methyltransferases (PKMTs), which use AdoMet as the methyl donor. The number of isolated mammalian PKMTs is over 100 (Letunic et al. 2004; Bateman et al. 2004), and are classified into nine subclasses, based on the presence of the Su(var)3–9, Enhancer-of-zeste and Trithorax (SET) and some other defined protein domain or homologous sequence. Most, but not all PKMTs, have a SET domain. The two families lacking the SET-domain are the KMT4/DOT1L and the Methyltransferase-like (METTL) family proteins, which are structurally closer to PRMTs (Cloutier et al. 2013). A small number of METTL family enzymes preferentially methylate chaperone proteins, regulating their activity (Cloutier et al. 2013). The first evidence of HSP60 lysine methylation pointed at an interaction with METTL20 (Cloutier et al. 2013). Data from various cell lines indicated that one functional methylation present on HSP60 is mono-methylated lysine 490 (K490me1) (Cao et al. 2013). Using quantitative mass spectrometry with Stable Isotope Labeling by Amino acids in Cell culture (SILAC)-labelling it was found that there is a possibility that METTL20 interacts with HSP60, while electron transfer flavoprotein (ETF)α and ETFβ were found to be strong binding partners of METTL20. Further experiments however did not support the finding that METTL20 methylates HSP60 and the interaction was deemed as being unlikely (Rhein et al. 2014). On the other hand, data generated from in vitro assays using GroEL, the bacterial homologue to mammalian HSP60, suggests tight binding of METTL23 to the chaperone protein (Bernkopf et al. 2014). Yeast-Two Hybrid and mass spectrometric data generated in our lab supports the findings that METTL20 presents no interaction with HSP60 but rather our data suggest an interaction with HSP27. Moreover our data does not support the interaction of METTL23 with HSP60.

6.4 Inferences from SNPs

The HSP60 and HSP10 proteins are encoded by the HSPD1 and HSPE1 genes respectively. The two genes are situated in a head-to-head configuration on chromosome 2, at chromosome locus 2q33.1, with a common bi-directional promoter (Bross and Fernandez-Guerra 2016). Complete deletion of the HSPD1 gene is embryonic lethal and mutations at most positions have deleterious or strong dominant negative effects, also resulting in embryonic lethality (Christensen et al. 2010; Bross and Fernandez-Guerra 2016). As mentioned previously, two very low frequency missense mutations in HSPD1, V98I and D29G, have been found to be associated with disease (Hansen et al. 2002; Magen et al. 2008). On the other hand, deletion of the HSPE1 gene is not embryonic lethal but a potentially disease-associated mutation, L73F, has been identified in a patient with a neurological and developmental disorder (Suarez et al. 2014; Bie et al. 2016). Nevertheless, within the human HSPD1 gene, a number of non-synonymous SNPs (altering the amino acid sequence) are known which may produce a less functionally efficient or viable HSP60 protein product (Bross and Fernandez-Guerra 2016). These changes affect function either because the variant amino acid is chemically different and therefore affects the overall chemistry of the protein or else, possibly, because a functionally critical PTM can no longer be added at that position. For example, when K156, found to be acetylated in the human acute myeloid leukemia cell line MV4–11 (Choudhary et al. 2009) and in mouse liver (Rardin et al. 2013) is replaced by the arginine variant (K156R), despite being classified as a benign variation due to the conservation of the positive charge, the acetylation site is lost and this may in turn affect the function or regulation of HSP60.

Other than K156R, numerous benign non-synonymous mutations can be found on HSPD1 in the Exome Aggregation Consortium (ExAC) database (Lek et al. 2016), namely K157R and K205R, which are chemically similar and K192E, K369N, K418T and K473Q which are chemically different. New lysines are also introduced via R16K, E129K, R142K, Q461K. Considering the number of PTMs which lysine can undergo and their importance in regulation, these mutations open a myriad of dysregulatory options of the HSP60 protein.

6.5 Conclusions

Very little is as yet known about the position and function of numerous HSP60 PTMs, but the emerging picture is that this protein covers a diversity of roles both inside the mitochondria and in other sub-cellular compartments, with all these functions being controlled by the small chemical additions to the protein, which are dependent on the cellular context. Besides the ability of HSP60 to produce functional variants on its own, PTMs on its co-chaperone HSP10 would also play a role in the HSP60-HSP10 complex function. Understanding the function and context of these PTMs will improve the prospect of identifying therapies for treating conditions involving HSP60 dysregulation.

References

Abdelmegeed MA, Song BJ (2014) Functional roles of protein nitration in acute and chronic liver diseases. Oxidative Med Cell Longev 2014:149627

Abello N, Kerstjens HA, Postma DS, Bischoff R (2009) Protein tyrosine nitration: selectivity, physicochemical and biological consequences, denitration, and proteomics methods for the identification of tyrosine-nitrated proteins. J Proteome Res 8(7):3222–3238

Akimoto Y, Kreppel LK, Hirano H, Hart GW (2001) Hyperglycemia and the O-GlcNAc transferase in rat aortic smooth muscle cells: elevated expression and altered patterns of O-GlcNAcylation. Arch Biochem Biophys 389(2):166–175

Al-Aidaroos AQO, Zeng Q (2010) PRL-3 phosphatase and cancer metastasis. J Cell Biochem 111(5):1087–1098

Alexandraki K, Piperi C, Kalofoutis C, Singh J, Alaveras A, Kalofoutis A (2006) Inflammatory process in type 2 diabetes. Ann N Y Acad Sci 1084(1):89–117

Alrob OA, Sankaralingam S, Ma C, Wagg CS, Fillmore N, Jaswal JS, Sack MN, Lehner R, Gupta MP, Michelakis ED, Padwal RS (2014) Obesity-induced lysine acetylation increases cardiac fatty acid oxidation and impairs insulin signalling. Cardiovasc Res 103(4):485–497

Altieri P, Spallarossa P, Barisione C, Garibaldi S, Garuti A, Fabbi P, Ghigliotti G, Brunelli C (2012) Inhibition of doxorubicin-induced senescence by PPARδ activation agonists in cardiac muscle cells: cooperation between PPARδ and Bcl6. PLoS One 7(9):e46126

Ande SR, Moulik S, Mishra S (2009) Interaction between O-GlcNAc modification and tyrosine phosphorylation of prohibitin: implication for a novel binary switch. PLoS One 4(2):e4586

Arias AE, Vélez-Granell CS, Mayer G, Bendayan M (2000) Colocalization of chaperone Cpn60, proinsulin and convertase PC1 within immature secretory granules of insulin-secreting cells suggests a role for Cpn60 in insulin processing. J Cell Sci 113(11):2075–2083

Asquith KL, Baleato RM, McLaughlin EA, Nixon B, Aitken RJ (2004) Tyrosine phosphorylation activates surface chaperones facilitating sperm-zona recognition. J Cell Sci 117(16):3645–3657

Avidan A, Perlmutter M, Tal S, Oraki O, Kapp T, Krelin Y, Elkabets M, Dotan S, Apte RN, Lichtenstein RG (2009) Differences in the sialylation patterns of membrane stress proteins in chemical carcinogen-induced tumors developed in BALB/c and IL-1α deficient mice. Glycoconj J 26(9):1181

Bailey LM, Wallace JC, Polyak SW (2010) Holocarboxylase synthetase: correlation of protein localisation with biological function. Arch Biochem Biophys 496(1):45–52

Bao B, Wijeratne SS, Rodriguez-Melendez R, Zempleni J (2011) Human holocarboxylase synthetase with a start site at methionine-58 is the predominant nuclear variant of this protein and has catalytic activity. Biochem Biophys Res Commun 412(1):115–120

Barazi HO, Zhou L, Templeton NS, Krutzsch HC, Roberts DD (2002) Identification of heat shock protein 60 as a molecular mediator of α3β1 integrin activation. Cancer Res 62(5):1541–1548
Baron B (2015) Lysine methylation of non-histone proteins. Biochem Mod Appl 1:1–2
Baron B (2018) The diagnostic and prognostic application of heat shock proteins and their post-translational modifications from liquid biopsies. IntechOpen [Online First]
Barone R, Rappa F, Macaluso F, Bavisotto CC, Sangiorgi C, Di Paola G, Tomasello G, Di Felice V, Marcianò V, Farina F, Zummo G (2016) Alcoholic liver disease: a mouse model reveals protection by *Lactobacillus fermentum*. Clin Transl Gastroenterol 7(1):e138
Bateman A, Coin L, Durbin R, Finn RD, Hollich V, Griffiths-Jones S, Khanna A, Marshall M, Moxon S, Sonnhammer EL, Studholme DJ (2004) The Pfam protein families database. Nucleic Acids Res 32(Suppl 1):D138–D141
Bernkopf M, Webersinke G, Tongsook C, Koyani CN, Rafiq MA, Ayaz M, Müller D, Enzinger C, Aslam M, Naeem F, Schmidt K (2014) Disruption of the methyltransferase-like 23 gene METTL23 causes mild autosomal recessive intellectual disability. Hum Mol Genet 23(15):4015–4023
Bessette DC, Qiu D, Pallen CJ (2008) PRL PTPs: mediators and markers of cancer progression. Cancer Metastasis Rev 27(2):231–252
Bhowmick R, Halder UC, Chattopadhyay S, Chanda S, Nandi S, Bagchi P, Nayak MK, Chakrabarti O, Kobayashi N, Chawla-Sarkar M (2012) Rotaviral enterotoxin nonstructural protein 4 targets mitochondria for activation of apoptosis during infection. J Biol Chem 287(42):35004–35020
Bie AS, Fernandez-Guerra P, Birkler RI, Nisemblat S, Pelnena D, Lu X, Deignan JL, Lee H, Dorrani N, Corydon TJ, Palmfeldt J (2016) Effects of a mutation in the HSPE1 gene encoding the mitochondrial co-chaperonin HSP10 and its potential association with a neurological and developmental disorder. Front Mol Biosci 3:65
Braig M, Lee S, Loddenkemper C, Rudolph C, Peters AH, Schlegelberger B, Stein H, Dörken B, Jenuwein T, Schmitt CA (2005) Oncogene-induced senescence as an initial barrier in lymphoma development. Nature 436(7051):660
Brocchieri L, Karlin S (2000) Conservation among HSP60 sequences in relation to structure, function, and evolution. Protein Sci 9(3):476–486
Bross P, Fernandez-Guerra P (2016) Disease-associated mutations in the HSPD1 gene encoding the large subunit of the mitochondrial HSP60/HSP10 chaperonin complex. Front Mol Biosci 3:49
Brudzynski K, Martinez V (1993) Synaptophysin-containing microvesicles transport heat-shock protein hsp60 in insulin-secreting beta cells. Cytotechnology 11(1):23–33
Bukau B, Horwich AL (1998) The Hsp70 and Hsp60 chaperone machines. Cell 92(3):351–366
Buse MG (2006) Hexosamines, insulin resistance, and the complications of diabetes: current status. Am J Physiol-Endocrinol Metab 290(1):E1–E8
Butler LM, Agus DB, Scher HI, Higgins B, Rose A, Cordon-Cardo C, Thaler HT, Rifkind RA, Marks PA, Richon VM (2000) Suberoylanilide hydroxamic acid, an inhibitor of histone deacetylase, suppresses the growth of prostate cancer cells in vitro and in vivo. Cancer Res 60(18):5165–5170
Campanella C, Marino Gammazza A, Mularoni L, Cappello F, Zummo G, Di Felice V (2009) A comparative analysis of the products of GROEL-1 gene from chlamydia trachomatis serovar D and the HSP60 var1 transcript from *Homo sapiens* suggests a possible autoimmune response. Int J Immunogenet 36(1):73–78
Campanella C, Bucchieri F, Merendino AM, Fucarino A, Burgio G, Corona DF, Barbieri G, David S, Farina F, Zummo G, de Macario EC (2012) The odyssey of Hsp60 from tumor cells to other destinations includes plasma membrane-associated stages and Golgi and exosomal protein-trafficking modalities. PLoS One 7(7):e42008
Campanella C, Bavisotto CC, Gammazza AM, Nikolic D, Rappa F, David S, Cappello F, Bucchieri F, Fais S (2014) Exosomal heat shock proteins as new players in tumour cell-to-cell communication. J Circulating Biomark 3:4

Campanella C, Rappa F, Sciumè C, Marino Gammazza A, Barone R, Bucchieri F, David S, Curcurù G, Caruso Bavisotto C, Pitruzzella A, Geraci G (2015) Heat shock protein 60 levels in tissue and circulating exosomes in human large bowel cancer before and after ablative surgery. Cancer 121(18):3230–3239

Campanella C, D'Anneo A, Gammazza AM, Bavisotto CC, Barone R, Emanuele S, Cascio FL, Mocciaro E, Fais S, De Macario EC, Macario AJ (2016) The histone deacetylase inhibitor SAHA induces HSP60 nitration and its extracellular release by exosomal vesicles in human lung-derived carcinoma cells. Oncotarget 7(20):28849

Cao XJ, Arnaudo AM, Garcia BA (2013) Large-scale global identification of protein lysine methylation in vivo. Epigenetics 8(5):477–485

Cappello F, Zummo G (2005) HSP60 expression during carcinogenesis: a molecular "Proteus" of carcinogenesis? Cell Stress Chaperones 10(4):263

Cappello F, Bellafiore M, David S, Anzalone R, Zummo G (2003) Ten kilodalton heat shock protein (HSP10) is overexpressed during carcinogenesis of large bowel and uterine exocervix. Cancer Lett 196(1):35–41

Cappello F, David S, Rappa F, Bucchieri F, Marasà L, Bartolotta TE, Farina F, Zummo G (2005) The expression of HSP60 and HSP10 in large bowel carcinomas with lymph node metastase. BMC Cancer 5(1):139

Cappello F, Bellafiore M, Palma A, David S, MarcianÃ V, Bartolotta T, SciumÃ C, Modica G, Farina F, Zummo G, Bucchieri F (2009) 60KDa chaperonin (HSP60) is over-expressed during colorectal carcinogenesis. Eur J Histochem 47(2):105–110

Cappello F, Marino Gammazza A, Palumbo Piccionello A, Campanella C, Pace A, Conway de Macario E, Macario AJ (2014) Hsp60 chaperonopathies and chaperonotherapy: targets and agents. Expert Opin Ther Targets 18(2):185–208

Chandra D, Choy G, Tang DG (2007) Cytosolic accumulation of Hsp60 during apoptosis with or without apparent mitochondrial release evidence that its pro-apoptotic or pro-survival functions involve differential interactions with caspase-3. J Biol Chem 282(43):31289–31301

Chao CC, Sun FC, Wang CH, Lo CW, Chang YS, Chang KC, Chang MDT, Lai YK (2008) Concerted actions of multiple transcription elements confer differential transactivation of HSP90 isoforms in geldanamycin-treated 9L rat gliosarcoma cells. J Cell Biochem 104(4):1286–1296

Chapman-Smith A, Cronan JE Jr (1999) Molecular biology of biotin attachment to proteins. J Nutr 129(2):477S–484S

Chattopadhyay S, Mukherjee A, Patra U, Bhowmick R, Basak T, Sengupta S, Chawla-Sarkar M (2017) Tyrosine phosphorylation modulates mitochondrial chaperonin Hsp60 and delays rotavirus NSP4-mediated apoptotic signaling in host cells. Cell Microbiol 19(3):e12670

Chavarría C, Souza JM (2013) Oxidation and nitration of α-synuclein and their implications in neurodegenerative diseases. Arch Biochem Biophys 533(1–2):25–32

Chen W, Syldath U, Bellmann K, Burkart V, Kolb H (1999) Human 60-kDa heat-shock protein: a danger signal to the innate immune system. J Immunol 162(6):3212–3219

Chen W, Wang J, Shao C, Liu S, Yu Y, Wang Q, Cao X (2006) Efficient induction of antitumor T cell immunity by exosomes derived from heat-shocked lymphoma cells. Eur J Immunol 36(6):1598–1607

Chen S, Zhao Y, Gou WF, Zhao S, Takano Y, Zheng HC (2013) The anti-tumor effects and molecular mechanisms of suberoylanilide hydroxamic acid (SAHA) on the aggressive phenotypes of ovarian carcinoma cells. PLoS One 8(11):e79781

Cheng X, Cole RN, Zaia J, Hart GW (2000) Alternative O-glycosylation/O-phosphorylation of the murine estrogen receptor β. Biochemistry 39(38):11609–11620

Chew YC, Camporeale G, Kothapalli N, Sarath G, Zempleni J (2006) Lysine residues in N-terminal and C-terminal regions of human histone H2A are targets for biotinylation by biotinidase. J Nutr Biochem 17(4):225–233

Choudhary C, Kumar C, Gnad F, Nielsen ML, Rehman M, Walther TC, Olsen JV, Mann M (2009) Lysine acetylation targets protein complexes and co-regulates major cellular functions. Science 325(5942):834–840

Christensen JH, Nielsen MN, Hansen J, Füchtbauer A, Füchtbauer EM, West M, Corydon TJ, Gregersen N, Bross P (2010) Inactivation of the hereditary spastic paraplegia-associated Hspd1 gene encoding the Hsp60 chaperone results in early embryonic lethality in mice. Cell Stress Chaperones 15(6):851–863

Cloutier P, Lavallée-Adam M, Faubert D, Blanchette M, Coulombe B (2013) A newly uncovered group of distantly related lysine methyltransferases preferentially interact with molecular chaperones to regulate their activity. PLoS Genet 9(1):e1003210

Comer FI, Hart GW (2001) Reciprocity between O-GlcNAc and O-phosphate on the carboxyl terminal domain of RNA polymerase II. Biochemistry 40(26):7845–7852

Copeland RJ, Bullen JW Jr, Hart GW (2008) Crosstalk between GlcNAcylation and phosphorylation: roles in insulin resistance and glucose toxicity. Am J Physiol-Endocrinol Metab 295:E17–E28

Corbett JA, McDaniel ML (1995) Intraislet release of interleukin 1 inhibits beta cell function by inducing beta cell expression of inducible nitric oxide synthase. J Exp Med 181(2):559–568

Cosman D, Müllberg J, Sutherland CL, Chin W, Armitage R, Fanslow W, Kubin M, Chalupny NJ (2001) ULBPs, novel MHC class I–related molecules, bind to CMV glycoprotein UL16 and stimulate NK cytotoxicity through the NKG2D receptor. Immunity 14(2):123–133

Czarnecka AM, Campanella C, Zummo G, Cappello F (2006) Mitochondrial chaperones in cancer: from molecular biology to clinical diagnostics. Cancer Biol Ther 5(7):714–720

Davis CW, Hawkins BJ, Ramasamy S, Irrinki KM, Cameron BA, Islam K, Daswani VP, Doonan PJ, Manevich Y, Madesh M (2010) Nitration of the mitochondrial complex I subunit NDUFB8 elicits RIP1-and RIP3-mediated necrosis. Free Radic Biol Med 48(2):306–317

De Bellis F, Carafa V, Conte M, Rotili D, Petraglia F, Matarese F, Françoijs KJ, Ablain J, Valente S, Castellano R, Goubard A (2014) Context-selective death of acute myeloid leukemia cells triggered by the novel hybrid retinoid-HDAC inhibitor MC2392. Cancer Res 74(8):2328–2339

Dentin R, Hedrick S, Xie J, Yates J, Montminy M (2008) Hepatic glucose sensing via the CREB coactivator CRTC2. Science 319(5868):1402–1405

Drazic A, Myklebust LM, Ree R, Arnesen T (2016) The world of protein acetylation. Biochim Biophys Acta (BBA)-Proteins Proteomics 1864(10):1372–1401

Eren MK, Kilincli A, Eren Ö (2015) Resveratrol induced premature senescence is associated with DNA damage mediated SIRT1 and SIRT2 down-regulation. PLoS One 10(4):e0124837

Farrugia M, Baron B (2017) The role of Toll-like receptors in autoimmune diseases through failure of the self-recognition mechanism. Int J Inflamm 2017:8391230

Fiskus W, Ren Y, Mohapatra A, Bali P, Mandawat A, Rao R, Herger B, Yang Y, Atadja P, Wu J, Bhalla K (2007) Hydroxamic acid analogue histone deacetylase inhibitors attenuate estrogen receptor-α levels and transcriptional activity: a result of hyperacetylation and inhibition of chaperone function of heat shock protein 90. Clin Cancer Res 13(16):4882–4890

Fritz KS, Galligan JJ, Hirschey MD, Verdin E, Petersen DR (2012) Mitochondrial acetylome analysis in a mouse model of alcohol-induced liver injury utilizing SIRT3 knockout mice. J Proteome Res 11(3):1633–1643

Gammazza AM, Campanella C, Barone R, Bavisotto CC, Gorska M, Wozniak M, Carini F, Cappello F, D'Anneo A, Lauricella M, Zummo G (2017) Doxorubicin anti-tumor mechanisms include Hsp60 post-translational modifications leading to the Hsp60/p53 complex dissociation and instauration of replicative senescence. Cancer Lett 385:75–86

Ghosh JC, Dohi T, Kang BH, Altieri DC (2008) Hsp60 regulation of tumor cell apoptosis. J Biol Chem 283(8):5188–5194

Ghosh JC, Siegelin MD, Dohi T, Altieri DC (2010) Heat shock protein 60 regulation of the mitochondrial permeability transition pore in tumor cells. Cancer Res 70(22):8988–8993

Glickman MH, Ciechanover A (2002) The ubiquitin-proteasome proteolytic pathway: destruction for the sake of construction. Physiol Rev 82(2):373–428

Glozak MA, Sengupta N, Zhang X, Seto E (2005) Acetylation and deacetylation of non-histone proteins. Gene 363:15–23

Gorska M, Gammazza AM, Zmijewski MA, Campanella C, Cappello F, Wasiewicz T, Kuban-Jankowska A, Daca A, Sielicka A, Popowska U, Knap N (2013) Geldanamycin-induced osteosarcoma cell death is associated with hyperacetylation and loss of mitochondrial pool of heat shock protein 60 (hsp60). PLoS One 8(8):e71135

Greening DW, Gopal SK, Xu R, Simpson RJ, Chen W (2015) Exosomes and their roles in immune regulation and cancer. Semin Cell Dev Biol 40:72–81. Academic Press

Gu TL, Deng X, Huang F, Tucker M, Crosby K, Rimkunas V, Wang Y, Deng G, Zhu L, Tan Z, Hu Y (2011a) Survey of tyrosine kinase signaling reveals ROS kinase fusions in human cholangiocarcinoma. PLoS One 6(1):e15640

Gu Y, Ande SR, Mishra S (2011b) Altered O-GlcNAc modification and phosphorylation of mitochondrial proteins in myoblast cells exposed to high glucose. Arch Biochem Biophys 505(1):98–104

Gueugnon F, Cartron PF, Charrier C, Bertrand P, Fonteneau JF, Gregoire M, Blanquart C (2014) New histone deacetylase inhibitors improve cisplatin antitumor properties against thoracic cancer cells. Oncotarget 5(12):4504

Guo K, Li J, Ping Tang J, Bobby Tan CP, Wang H, Wang H, Zeng Q (2008) Monoclonal antibodies target intracellular PRL phosphatases to inhibit cancer metastases in mice. Cancer Biol Ther 7(5):750–757

Guo X, Keyes WM, Papazoglu C, Zuber J, Li W, Lowe SW, Vogel H, Mills AA (2009) TAp63 induces senescence and suppresses tumorigenesis in vivo. Nat Cell Biol 11(12):1451

Gupta S, Knowlton AA (2002) Cytosolic heat shock protein 60, hypoxia, and apoptosis. Circulation 106(21):2727–2733

Gupta S, Knowlton AA (2005) HSP60, Bax, apoptosis and the heart. J Cell Mol Med 9(1):51–58

Ha K, Fiskus W, Choi DS, Bhaskara S, Cerchietti L, Devaraj SG, Shah B, Sharma S, Chang JC, Melnick AM, Hiebert S (2014) Histone deacetylase inhibitor treatment induces 'BRCAness' and synergistic lethality with PARP inhibitor and cisplatin against human triple negative breast cancer cells. Oncotarget 5(14):5637

Haaland I, Opsahl JA, Berven FS, Reikvam H, Fredly HK, Haugse R, Thiede B, McCormack E, Lain S, Bruserud Ø, Gjertsen BT (2014) Molecular mechanisms of nutlin-3 involve acetylation of p53, histones and heat shock proteins in acute myeloid leukemia. Mol Cancer 13(1):116

Hageman J, Rujano MA, Van Waarde MA, Kakkar V, Dirks RP, Govorukhina N, Oosterveld-Hut HM, Lubsen NH, Kampinga HH (2010) A DNAJB chaperone subfamily with HDAC-dependent activities suppresses toxic protein aggregation. Mol Cell 37(3):355–369

Han JY, Lee SH, Lee GK, Yun T, Lee YJ, Hwang KH, Kim JY, Kim HT (2015) Phase I/II study of gefitinib (Iressa®) and vorinostat (IVORI) in previously treated patients with advanced non-small cell lung cancer. Cancer Chemother Pharmacol 75(3):475–483

Hansen JJ, Dürr A, Cournu-Rebeix I, Georgopoulos C, Ang D, Nielsen MN, Davoine CS, Brice A, Fontaine B, Gregersen N, Bross P (2002) Hereditary spastic paraplegia SPG13 is associated with a mutation in the gene encoding the mitochondrial chaperonin Hsp60. Am J Hum Genet 70(5):1328–1332

Hayes GR, Williams A, Costello CE, Enns CA, Lucas JJ (1995) The critical glycosylation site of human transferrin receptor contains a high-mannose oligosaccharide. Glycobiology 5(2):227–232

Hayoun D, Kapp T, Edri-Brami M, Ventura T, Cohen M, Avidan A, Lichtenstein RG (2012) HSP60 is transported through the secretory pathway of 3-MCA-induced fibrosarcoma tumour cells and undergoes N-glycosylation. FEBS J 279(12):2083–2095

Helenius A, Aebi M (2001) Intracellular functions of N-linked glycans. Science 291(5512):2364–2369

Hilpert J, Grosse-Hovest L, Grünebach F, Buechele C, Nuebling T, Raum T, Steinle A, Salih HR (2012) Comprehensive analysis of NKG2D ligand expression and release in leukemia: implications for NKG2D-mediated NK cell responses. J Immunol 189(3):1360–1371

Huang YC, Huang FI, Mehndiratta S, Lai SC, Liou JP, Yang CR (2015) Anticancer activity of MPT0G157, a derivative of indolylbenzenesulfonamide, inhibits tumor growth and angiogenesis. Oncotarget 6(21):18590

Hunter T (2007) The age of crosstalk: phosphorylation, ubiquitination, and beyond. Mol Cell 28(5):730–738

Jimenez-Feltstrom J, Lundquist I, Salehi A (2005) Glucose stimulates the expression and activities of nitric oxide synthases in incubated rat islets: an effect counteracted by GLP-1 through the cyclic AMP/PKA pathway. Cell Tissue Res 319(2):221–230

Jindal S, Dudani AK, Singh BHAG, Harley CB, Gupta RS (1989) Primary structure of a human mitochondrial protein homologous to the bacterial and plant chaperonins and to the 65-kilodalton mycobacterial antigen. Mol Cell Biol 9(5):2279–2283

Jing E, Emanuelli B, Hirschey MD, Boucher J, Lee KY, Lombard D, Verdin EM, Kahn CR (2011) Sirtuin-3 (Sirt3) regulates skeletal muscle metabolism and insulin signaling via altered mitochondrial oxidation and reactive oxygen species production. Proc Natl Acad Sci 108(35):14608–14613

Kamemura K, Hayes BK, Comer FI, Hart GW (2002) Dynamic interplay between O-glycosylation and O-phosphorylation of nucleocytoplasmic proteins alternative glycosylation/phosphorylation of THR-58, a known mutational hot spot of c-Myc in lymphomas, is regulated by mitogens. J Biol Chem 277(21):19229–19235

Karbowski M, Spodnik JH, Teranishi MA, Wozniak M, Nishizawa Y, Usukura J, Wakabayashi T (2001) Opposite effects of microtubule-stabilizing and microtubule-destabilizing drugs on biogenesis of mitochondria in mammalian cells. J Cell Sci 114(2):281–291

Kaul Z, Yaguchi T, Kaul SC, Wadhwa R (2006) Quantum dot-based protein imaging and functional significance of two mitochondrial chaperones in cellular senescence and carcinogenesis. Ann N Y Acad Sci 1067(1):469–473

Kennelly PJ, Krebs EG (1991) Consensus sequences as substrate specificity determinants for protein kinases and protein phosphatases. J Biol Chem 266(24):15555–15558

Khan IU, Wallin R, Gupta RS, Kammer GM (1998) Protein kinase A-catalyzed phosphorylation of heat shock protein 60 chaperone regulates its attachment to histone 2B in the T lymphocyte plasma membrane. Proc Natl Acad Sci 95(18):10425–10430

Kikuchi I, Nakayama Y, Morinaga T, Fukumoto Y, Yamaguchi N (2010) A decrease in cyclin B1 levels leads to polyploidization in DNA damage-induced senescence. Cell Biol Int 34(6):645–653

Kim HJ, Bae SC (2011) Histone deacetylase inhibitors: molecular mechanisms of action and clinical trials as anti-cancer drugs. Am J Transl Res 3(2):166

Kim HR, Kang HS, Do Kim H (1999) Geldanamycin induces heat shock protein expression through activation of HSF1 in K562 erythroleukemic cells. IUBMB Life 48(4):429–433

Kim HS, Kim EM, Lee J, Yang WH, Park TY, Kim YM, Cho JW (2006) Heat shock protein 60 modified with O-linked N-acetylglucosamine is involved in pancreatic β-cell death under hyperglycemic conditions. FEBS Lett 580(9):2311–2316

Kirschbaum M, Gojo I, Goldberg SL, Bredeson C, Kujawski LA, Yang A, Marks P, Frankel P, Sun X, Tosolini A, Eid JE (2014) A phase 1 clinical trial of vorinostat in combination with decitabine in patients with acute myeloid leukaemia or myelodysplastic syndrome. Br J Haematol 167(2):185–193

Koeck T, Corbett JA, Crabb JW, Stuehr DJ, Aulak KS (2009) Glucose-modulated tyrosine nitration in beta cells: targets and consequences. Arch Biochem Biophys 484(2):221–231

Koll H, Guiard B, Rassow J, Ostermann J, Horwich AL, Neupert W, Hartl FU (1992) Antifolding activity of hsp60 couples protein import into the mitochondrial matrix with export to the intermembrane space. Cell 68(6):1163–1175

Komander D, Rape M (2012) The ubiquitin code. Annu Rev Biochem 81:203–229

Kurz EU, Douglas P, Lees-Miller SP (2004) Doxorubicin activates ATM-dependent phosphorylation of multiple downstream targets in part through the generation of reactive oxygen species. J Biol Chem 279(51):53272–53281

Lai MT, Huang KL, Chang WM, Lai YK (2003) Geldanamycin induction of grp78 requires activation of reactive oxygen species via ER stress responsive elements in 9L rat brain tumour cells. Cell Signal 15(6):585–595

Lakey JR, Suarez-Pinzon WL, Strynadka K, Korbutt GS, Rajotte RV, Mabley JG, Szabó C, Rabinovitch A (2001) Peroxynitrite is a mediator of cytokine-induced destruction of human pancreatic islet β cells. Lab Investig 81(12):1683

Lauricella M, Ciraolo A, Carlisi D, Vento R, Tesoriere G (2012) SAHA/TRAIL combination induces detachment and anoikis of MDA-MB231 and MCF-7 breast cancer cells. Biochimie 94(2):287–299

Lee BC, Dikiy A, Kim HY, Gladyshev VN (2009) Functions and evolution of selenoprotein methionine sulfoxide reductases. Biochim Biophys Acta (BBA)-Gen Subj 1790(11):1471–1477

Lek M, Karczewski KJ, Minikel EV, Samocha KE, Banks E, Fennell T, O'Donnell-Luria AH, Ware JS, Hill AJ, Cummings BB, Tukiainen T (2016) Analysis of protein-coding genetic variation in 60,706 humans. Nature 536(7616):285

Letunic I, Copley RR, Schmidt S, Ciccarelli FD, Doerks T, Schultz J, Ponting CP, Bork P (2004) SMART 4.0: towards genomic data integration. Nucleic Acids Res 32(suppl_1):D142–D144

Leung WH, Vong QP, Lin W, Bouck D, Wendt S, Sullivan E, Li Y, Bari R, Chen T, Leung W (2015) PRL-3 mediates the protein maturation of ULBP2 by regulating the tyrosine phosphorylation of HSP60. J Immunol 194(6):2930–2941

Levy-Rimler G, Viitanen P, Weiss C, Sharkia R, Greenberg A, Niv A, Lustig A, Delarea Y, Azem A (2001) The effect of nucleotides and mitochondrial chaperonin 10 on the structure and chaperone activity of mitochondrial chaperonin 60. Eur J Biochem 268(12):3465–3472

Li XS, Xu Q, Fu XY, Luo WS (2014a) Heat shock protein 60 overexpression is associated with the progression and prognosis in gastric cancer. PLoS One 9(9):e107507

Li Y, Malkaram SA, Zhou J, Zempleni J (2014b) Lysine biotinylation and methionine oxidation in the heat shock protein HSP60 synergize in the elimination of reactive oxygen species in human cell cultures. J Nutr Biochem 25(4):475–482

Lianos GD, Alexiou GA, Mangano A, Mangano A, Rausei S, Boni L, Dionigi G, Roukos DH (2015) The role of heat shock proteins in cancer. Cancer Lett 360(2):114–118

Lim Y, Lee E, Lee J, Oh S, Kim S (2008) Down-regulation of asymmetric arginine methylation during replicative and H2O2-induced premature senescence in WI-38 human diploid fibroblasts. J Biochem 144(4):523–529

Lim Y, Hong E, Kwon D, Lee E (2010) Proteomic identification and comparative analysis of asymmetrically arginine-methylated proteins in immortalized, young and senescent cells. Electrophoresis 31(23–24):3823–3833

Liu K, Paterson AJ, Chin E, Kudlow JE (2000) Glucose stimulates protein modification by O-linked GlcNAc in pancreatic β cells: linkage of O-linked GlcNAc to β cell death. Proc Natl Acad Sci 97(6):2820–2825

Lombard DB, Alt FW, Cheng HL, Bunkenborg J, Streeper RS, Mostoslavsky R, Kim J, Yancopoulos G, Valenzuela D, Murphy A, Yang Y (2007) Mammalian Sir2 homolog SIRT3 regulates global mitochondrial lysine acetylation. Mol Cell Biol 27(24):8807–8814

Lu Z, Chen Y, Aponte AM, Battaglia V, Gucek M, Sack MN (2015) Prolonged fasting identifies heat shock protein 10 as a Sirtuin 3 substrate elucidating a new mechanism linking mitochondrial protein acetylation to fatty acid oxidation enzyme folding and function. J Biol Chem 290(4):2466–2476

Lupi R, Del Guerra S, Mancarella R, Novelli M, Valgimigli L, Pedulli GF, Paolini M, Soleti A, Filipponi F, Mosca F, Boggi U (2007) Insulin secretion defects of human type 2 diabetic islets are corrected in vitro by a new reactive oxygen species scavenger. Diabetes Metab 33(5):340–345

Macario AJ, de Macario Conway E (2007) Molecular chaperones: multiple functions, pathologies, and potential applications. Front Biosci 12:2588–2600

Maedler K, Sergeev P, Ris F, Oberholzer J, Joller-Jemelka HI, Spinas GA, Kaiser N, Halban PA, Donath MY (2002) Glucose-induced β cell production of IL-1β contributes to glucotoxicity in human pancreatic islets. J Clin Invest 110(6):851–860

Magen D, Georgopoulos C, Bross P, Ang D, Segev Y, Goldsher D, Nemirovski A, Shahar E, Ravid S, Luder A, Heno B (2008) Mitochondrial hsp60 chaperonopathy causes an autosomal-recessive neurodegenerative disorder linked to brain hypomyelination and leukodystrophy. Am J Hum Genet 83(1):30–42

Marks PA, Xu WS (2009) Histone deacetylase inhibitors: potential in cancer therapy. J Cell Biochem 107(4):600–608

Martin-Latil S, Mousson L, Autret A, Colbère-Garapin F, Blondel B (2007) Bax is activated during rotavirus-induced apoptosis through the mitochondrial pathway. J Virol 81(9):4457–4464

McDowell GS, Philpott A (2013) Non-canonical ubiquitylation: mechanisms and consequences. Int J Biochem Cell Biol 45(8):1833–1842

Merendino AM, Bucchieri F, Campanella C, Marcianò V, Ribbene A, David S, Zummo G, Burgio G, Corona DF, de Macario EC, Macario AJ (2010) Hsp60 is actively secreted by human tumor cells. PLoS One 5(2):e9247

Mirzayans R, Andrais B, Hansen G, Murray D (2012) Role of in replicative senescence and DNA damage-induced premature senescence in p53-deficient human cells. Biochem Res Int 2012:951574

Monzani E, Roncone R, Galliano M, Koppenol WH, Casella L (2004) Mechanistic insight into the peroxidase catalyzed nitration of tyrosine derivatives by nitrite and hydrogen peroxide. Eur J Biochem 271(5):895–906

Moskovitz J, Oien DB (2010) Protein carbonyl and the methionine sulfoxide reductase system. Antioxid Redox Signal 12(3):405–415

Nahleh Z, Tfayli A, Najm A, El Sayed A, Nahle Z (2012) Heat shock proteins in cancer: targeting the 'chaperones'. Future Med Chem 4(7):927–935

Narang MA, Dumas R, Ayer LM, Gravel RA (2004) Reduced histone biotinylation in multiple carboxylase deficiency patients: a nuclear rôle for holocarboxylase synthetase. Hum Mol Genet 13(1):15–23

Nielsen KL, Cowan NJ (1998) A single ring is sufficient for productive chaperonin-mediated folding in vivo. Mol Cell 2(1):93–99

Nisemblat S, Yaniv O, Parnas A, Frolow F, Azem A (2015) Crystal structure of the human mitochondrial chaperonin symmetrical football complex. Proc Natl Acad Sci 112(19):6044–6049

Osterloh A, Meier-Stiegen F, Veit A, Fleischer B, von Bonin A, Breloer M (2004) Lipopolysaccharide-free heat shock protein 60 activates T cells. J Biol Chem 279(46):47906–47911

Pacher P, Beckman JS, Liaudet L (2007) Nitric oxide and peroxynitrite in health and disease. Physiol Rev 87(1):315–424

Pahlich S, Zakaryan RP, Gehring H (2006) Protein arginine methylation: cellular functions and methods of analysis. Biochim Biophys Acta (BBA)-Proteins Proteomics 1764(12):1890–1903

Pickart CM, Eddins MJ (2004) Ubiquitin: structures, functions, mechanisms. Biochim Biophys Acta (BBA)-Mol Cell Res 1695(1–3):55–72

Powers MV, Workman P (2007) Inhibitors of the heat shock response: biology and pharmacology. FEBS Lett 581(19):3758–3769

Pyhtila B, Zheng T, Lager PJ, Keene JD, Reedy MC, Nicchitta CV (2008) Signal sequence-and translation-independent mRNA localization to the endoplasmic reticulum. RNA 14(3):445–453

Quintana FJ, Cohen IR (2011) The HSP60 immune system network. Trends Immunol 32(2):89–95

Radi R (2004) Nitric oxide, oxidants, and protein tyrosine nitration. Proc Natl Acad Sci 101(12):4003–4008

Radi R (2012) Protein tyrosine nitration: biochemical mechanisms and structural basis of functional effects. Acc Chem Res 46(2):550–559

Rakonczay Z Jr, Boros I, Jarmay K, Hegyi P, Lonovics J, Takacs T (2003) Ethanol administration generates oxidative stress in the pancreas and liver, but fails to induce heat-shock proteins in rats. J Gastroenterol Hepatol 18(7):858–867

Rao R, Fiskus W, Ganguly S, Kambhampati S, Bhalla KN (2012) HDAC inhibitors and chaperone function. Adv Cancer Res 116:239–262. Academic Press

Rappa F, Farina F, Zummo G, David S, Campanella C, Carini F, Tomasello G, Damiani P, Cappello F, De Macario EC, Macario AJ (2012) HSP-molecular chaperones in cancer biogenesis and tumor therapy: an overview. Anticancer Res 32(12):5139–5150

Rardin MJ, Newman JC, Held JM, Cusack MP, Sorensen DJ, Li B, Schilling B, Mooney SD, Kahn CR, Verdin E, Gibson BW (2013) Label-free quantitative proteomics of the lysine acetylome in mitochondria identifies substrates of SIRT3 in metabolic pathways. Proc Natl Acad Sci 110(16):6601–6606

Raulet DH, Gasser S, Gowen BG, Deng W, Jung H (2013) Regulation of ligands for the NKG2D activating receptor. Annu Rev Immunol 31:413–441

Rhein VF, Carroll J, He J, Ding S, Fearnley IM, Walker JE (2014) Human METTL20 methylates lysine residues adjacent to the recognition loop of the electron transfer flavoprotein in mitochondria. J Biol Chem 289(35):24640–24651

Rikova K, Guo A, Zeng Q, Possemato A, Yu J, Haack H, Nardone J, Lee K, Reeves C, Li Y, Hu Y (2007) Global survey of phosphotyrosine signaling identifies oncogenic kinases in lung cancer. Cell 131(6):1190–1203

Roninson IB (2002) Oncogenic functions of tumour suppressor p21Waf1/Cip1/Sdi1: association with cell senescence and tumour-promoting activities of stromal fibroblasts. Cancer Lett 179(1):1–14

Salih HR, Rammensee HG, Steinle A (2002) Cutting edge: down-regulation of MICA on human tumors by proteolytic shedding. J Immunol 169(8):4098–4102

Sarangi U, Singh MK, Abhijnya KVV, Reddy LPA, Prasad BS, Pitke VV, Paithankar K, Sreedhar AS (2013) Hsp60 chaperonin acts as barrier to pharmacologically induced oxidative stress mediated apoptosis in tumor cells with differential stress response. Drug Target Insights 7:DTI-S12513

Schnell JD, Hicke L (2003) Non-traditional functions of ubiquitin and ubiquitin-binding proteins. J Biol Chem 278(38):35857–35860

Schwer B, North BJ, Frye RA, Ott M, Verdin E (2002) The human silent information regulator (Sir) 2 homologue hSIRT3 is a mitochondrial nicotinamide adenine dinucleotide–dependent deacetylase. J Cell Biol 158(4):647–657

Semba S, Mizuuchi E, Yokozaki H (2010) Requirement of phosphatase of regenerating liver-3 for the nucleolar localization of nucleolin during the progression of colorectal carcinoma. Cancer Sci 101(10):2254–2261

Shan YX, Liu TJ, Su HF, Samsamshariat A, Mestril R, Wang PH (2003) Hsp10 and Hsp60 modulate Bcl-2 family and mitochondria apoptosis signaling induced by doxorubicin in cardiac muscle cells. J Mol Cell Cardiol 35(9):1135–1143

Shi T, Wang F, Stieren E, Tong Q (2005) SIRT3, a mitochondrial sirtuin deacetylase, regulates mitochondrial function and thermogenesis in brown adipocytes. J Biol Chem 280(14):13560–13567

Sol EM, Wagner SA, Weinert BT, Kumar A, Kim HS, Deng CX, Choudhary C (2012) Proteomic investigations of lysine acetylation identify diverse substrates of mitochondrial deacetylase sirt3. PLoS One 7(12):e50545

Soltys BJ, Gupta RS (1996) Immunoelectron microscopic localization of the 60-kDa heat shock chaperonin protein (Hsp60) in mammalian cells. Exp Cell Res 222(1):16–27

Soltys BJ, Gupta RS (1997) Cell surface localization of the 60 kDa heat shock chaperonin protein (hsp60) in mammalian cells. Cell Biol Int 21(5):315–320

Song YS, Lee BY, Hwang ES (2005) Dinstinct ROS and biochemical profiles in cells undergoing DNA damage-induced senescence and apoptosis. Mech Ageing Dev 126(5):580–590

Spallarossa P, Altieri P, Aloi C, Garibaldi S, Barisione C, Ghigliotti G, Fugazza G, Barsotti A, Brunelli C (2009) Doxorubicin induces senescence or apoptosis in rat neonatal cardiomyocytes by regulating the expression levels of the telomere binding factors 1 and 2. Am J Phys Heart Circ Phys 297:H2169–H2181

Spiro RG (2002) Protein glycosylation: nature, distribution, enzymatic formation, and disease implications of glycopeptide bonds. Glycobiology 12(4):43R–56R

Stadtman ER (2006) Protein oxidation and aging. Free Radic Res 40(12):1250–1258

Straus DJ, Hamlin PA, Matasar MJ, Palomba ML, Drullinsky PR, Zelenetz AD, Gerecitano JF et al (2015) Phase I/II trial of vorinostat with rituximab, cyclophosphamide, etoposide and prednisone as palliative treatment for elderly patients with relapsed or refractory diffuse large B-cell lymphoma not eligible for autologous stem cell transplantation. Br J Haematol 168(5):663–670

Suarez J, Wang H, Ouyang K, Scott B (2014) Cardiac myocyte-specific deletion of heat shock protein 10 results in altered cytosolic and mitochondrial protein expression (699.6). FASEB J 28(1_supplement):699–696

Suarez-Pinzon WL, Szabó C, Rabinovitch A (1997) Development of autoimmune diabetes in NOD mice is associated with the formation of peroxynitrite in pancreatic islet β-cells. Diabetes 46(5):907–911

Szabó C, Mabley JG, Moeller SM, Shimanovich R, Pacher P, Virág L, Soriano FG, Van Duzer JH, Williams W, Salzman AL, Groves JT (2002) Part I: pathogenetic role of peroxynitrite in the development of diabetes and diabetic vascular complications: studies with FP15, a novel potent peroxynitrite decomposition catalyst. Mol Med 8(10):571–580

Tang H, Tian E, Liu C, Wang Q, Deng H (2013) Oxidative stress induces monocyte necrosis with enrichment of cell-bound albumin and overexpression of endoplasmic reticulum and mitochondrial chaperones. PLoS One 8(3):e59610

Théry C, Ostrowski M, Segura E (2009) Membrane vesicles as conveyors of immune responses. Nat Rev Immunol 9(8):581

Thomson L (2015) 3-nitrotyrosine modified proteins in atherosclerosis. Dis Markers 2015:708282

Tian P, Estes MK, Hu Y, Ball JM, Zeng CQ, Schilling WP (1995) The rotavirus nonstructural glycoprotein NSP4 mobilizes Ca2+ from the endoplasmic reticulum. J Virol 69(9):5763–5772

Tomasello G, Rodolico V, Zerilli M, Martorana A, Bucchieri F, Pitruzzella A, Gammazza AM, David S, Rappa F, Zummo G, Damiani P (2011) Changes in immunohistochemical levels and subcellular localization after therapy and correlation and colocalization with CD68 suggest a pathogenetic role of Hsp60 in ulcerative colitis. Appl Immunohistochem Mol Morphol 19(6):552–561

Touriol C, Bornes S, Bonnal S, Audigier S, Prats H, Prats AC, Vagner S (2003) Generation of protein isoform diversity by alternative initiation of translation at non-AUG codons. Biol Cell 95(3–4):169–178

Tsai YP, Teng SC, Wu KJ (2008) Direct regulation of HSP60 expression by c-MYC induces transformation. FEBS Lett 582(29):4083–4088

Tsai YP, Yang MH, Huang CH, Chang SY, Chen PM, Liu CJ, Teng SC, Wu KJ (2009) Interaction between HSP60 and β-catenin promotes metastasis. Carcinogenesis 30(6):1049–1057

Tu Y, Hershman DL, Bhalla K, Fiskus W, Pellegrino CM, Andreopoulou E, Makower D, Kalinsky K, Fehn K, Fineberg S, Negassa A (2014) A phase I-II study of the histone deacetylase inhibitor vorinostat plus sequential weekly paclitaxel and doxorubicin-cyclophosphamide in locally advanced breast cancer. Breast Cancer Res Treat 146(1):145–152

Van Eden W, Wick G, Albani S, Cohen I (2007) Stress, heat shock proteins, and autoimmunity. Ann N Y Acad Sci 1113(1):217–237

Waldhauer I, Steinle A (2006) Proteolytic release of soluble UL16-binding protein 2 from tumor cells. Cancer Res 66(5):2520–2526

Waldhauer I, Goehlsdorf D, Gieseke F, Weinschenk T, Wittenbrink M, Ludwig A, Stevanovic S, Rammensee HG, Steinle A (2008) Tumor-associated MICA is shed by ADAM proteases. Cancer Res 68(15):6368–6376

Walgren JL, Vincent TS, Schey KL, Buse MG (2003) High glucose/insulin promote O-GlcNAc modification of proteins including α-tubulin. Am J Physiol-Endocrinol Metab 284:E424–E434

Wang L, Shen Y, Song R, Sun Y, Xu J, Xu Q (2009) An anticancer effect of curcumin mediated by down-regulating phosphatase of regenerating liver-3 expression on highly metastatic melanoma cells. Mol Pharmacol 76(6):1238–1245

Whelan SA, Lane MD, Hart GW (2008) Regulation of the O-linked β-N-acetylglucosamine transferase by insulin signaling. J Biol Chem 283(31):21411–21417

Wolf SS (2009) The protein arginine methyltransferase family: an update about function, new perspectives and the physiological role in humans. Cell Mol Life Sci 66(13):2109–2121

Yamaguchi K, Chikumi H, Shimizu A, Takata M, Kinoshita N, Hashimoto K, Nakamoto M, Matsunaga S, Kurai J, Miyake N, Matsumoto S (2012) Diagnostic and prognostic impact of serum-soluble UL 16-binding protein 2 in lung cancer patients. Cancer Sci 103(8):1405–1413

Yang Y, Fiskus W, Yong B, Atadja P, Takahashi Y, Pandita TK, Wang HG, Bhalla KN (2013) Acetylated hsp70 and KAP1-mediated Vps34 SUMOylation is required for autophagosome creation in autophagy. Proc Natl Acad Sci 110(17):6841–6846

Yechoor VK, Patti ME, Ueki K, Laustsen PG, Saccone R, Rauniyar R, Kahn CR (2004) Distinct pathways of insulin-regulated versus diabetes-regulated gene expression: an in vivo analysis in MIRKO mice. Proc Natl Acad Sci 101(47):16525–16530

Yogev O, Pines O (2011) Dual targeting of mitochondrial proteins: mechanism, regulation and function. Biochim Biophys Acta (BBA)-Biomembr 1808(3):1012–1020

Zambrano JL, Díaz Y, Peña F, Vizzi E, Ruiz MC, Michelangeli F, Liprandi F, Ludert JE (2008) Silencing of rotavirus NSP4 or VP7 expression reduces alterations in Ca2+ homeostasis induced by infection of cultured cells. J Virol 82(12):5815–5824

Zanin-Zhorov A, Bruck R, Tal G, Oren S, Aeed H, Hershkoviz R, Cohen IR, Lider O (2005) Heat shock protein 60 inhibits Th1-mediated hepatitis model via innate regulation of Th1/Th2 transcription factors and cytokines. J Immunol 174(6):3227–3236

Part II
HSP60 and Cancer

Chapter 7
Evaluation of Heat Shock Protein 60 (HSP60) Chaperonin in Oncology

İsmail Ağababaoğlu

Abstract Heat shock proteins (Hsp) are a group of chaperonin that are increased production at cellular level in cellular stress situations. These stress include such as heat, infection, inflammation, many toxins such as ethanol, arsenic, some metals, ultraviolet light, some oncogenes. Under physiological conditions, Hsp helps the newly produced proteins in the cell in folding correctly. They provide stabilization of the mitochondria as a mechanism that prevents the cell from apoptosis. They function by interacting with proteins called chaperonin in the cell cycle and stabilization of mitochondria. And it protects the cell from apoptosis and directs it to the process of carcinogenesis. Therefore, it has great potential in cancer studies in many stages. Thus, our chapter aims to briefly evaluate up to date knowledge for Hsp in oncological field.

Keywords Biomarker · Cancer vaccines · Chaperonin · Carcinogenesis · Heat shock protein 60 · Target therapy

Abbreviations

2D gel	Two-dimensional gel
5-FU	5- fluorouracil
17AAG	17-Allylamino-Demethoxygeldanamycin
ATPase	adenosine triphosphate
CA-19-9	cancer Antigen 19-9
CD8	cytotoxic T cells surface antigen
CEA	carcinoembryonic antigen

İ. Ağababaoğlu (✉)
Department of Thoracic Surgery, Yıldırım Beyazıt University Yenimahalle Research Hospital, Ankara, Turkey

A. A. A. Asea, P. Kaur (eds.), *Heat Shock Protein 60 in Human Diseases and Disorders*, Heat Shock Proteins 18,
https://doi.org/10.1007/978-3-030-23154-5_7

c-MYC	myelocytomatosis oncogene
Cyp-D	peptidylprolyl isomerase D (*cyclophilin D)*
E-Box	enhancer box
ELISA	Enzyme-linked immunosorbent assay
HCV	hepatitis C virus
HSP	heat shock protein
kDa	kilodaltons
MALDI-TOF	matrix assisted laser desorption ionization time of flight mass spectrometry
Mrna	messenger ribonucleic acid
PSA	prostate specific antigen
PCR	polymerase chain reaction
SW80	human colorectal cancer cell line
TNF-Alpha	tumor necrosis factor-alpha

7.1 Introduction

Hsp are a large family of proteins that perform many functions within the cell. They function in the most important response of living beings in the cellular dimension to stress, and the mechanism of evolutionary adaptation. They help to preserve the structure of cellular proteins that have lost their stability or are about to be denatured. They are particularly sensitive to heat and are named after these sources of stress. In this respect they work as intracellular chaperons (Liu et al. 2012). The molecular weights of this particular protein group range from 7 to 110 kDa and are classified according to this weight. It was demonstrated by the researchers that Hsp have important functions in cell differentiation, proliferation, intracellular transport, secretion, folding of protein structures and regulation of cell cycle with the functions they perform at cellular level (Ciocca and Calderwood 2005). They provide stabilization of the mitochondria as a mechanism that prevents the cell from apoptosis. They function by interacting with proteins called chaperonin in the cell cycle and stabilization of mitochondria. Molecular chaperonins allow the proteins to remain in the correct configuration (Faried et al. 2004; Ritossa 1996). Denaturation of protein structures in the cell is one of the factors that induce apoptosis.

Hsp show antiapoptotic effect with their functions protecting protein structures. This plays a role in protection and immortality of damaged cells during malignant transformation (Ghosh et al. 2008).As an important member of the HSP, HSP60 can be found in the large oligomer structure, in the mitochondria, in the cytoplasm, in the intercellular matrix or even in peripheral blood. In particular, it interacts with mitochondria proteins to ensure that they are in the correct spatial configuration. It functions in the combination of larger proteins and in cellular respiration. Even if it functions weakly in cell signaling pathways, it shows ATPase activity. HSP60 interacts considerably with HSP 10 and HSP70 during its cellular level function and provides its functions in this way (Cappello et al. 2008). It is located on HSP60 protein and HSP10 chromosome 2 with which it works together (Cappello et al. 2013; Chaiwatanasirkul and Sala 2011; Chang et al. 2012; Ciocca and Calderwood 2005; Desmetz et al. 2008; Faried et al. 2004; Ghosh et al. 2008, 2010; Gorska et al. 2013; Hamelin et al. 2011; Hamrita et al. 2008; Hjerpe et al. 2013; Hoang et al. 2000; Hwang et al. 2009; Kang et al. 2009; Liffers et al. 2011). It has a triple protein structure and is bound to Y-box binding protein and creatine 23 during its physiological activities (Sigler et al. 1998). They are antigenically identified by B and T lymphocytes and form ligand structures for toll-like receptors in antigen-presenting cells (Lv et al. 2012).

Among the Hsp, HSP 27, HSP70 and HSP90 are the most studied in the oncological field. There is much more limited literature on HSP 60. HSP60 protein is equally important due to its fundamentally similar function and its role in the same processes, and it deserves to be investigated. It was shown to function in all phases of the carcinogenesis stage. Therefore, it has the potential to be used in every stage of the clinical approach in oncology. It has the potential to be used in all stages of diagnosis, treatment, follow-up and prognosis.

7.2 Carcinogenesis

HSP60 is the first evaluated heat shock protein in studies on cancer. The synthesis of this protein increases in cellular stress conditions as in other heat shock proteins. One of the most important cases of cellular stress is carcinogenesis. Therefore, in many studies, HSP60 protein was shown to increase in cancer tissue and is associated with many types of cancer (Ağababaoğlu et al. 2017; Arya et al. 2007). In the study performed with MALDI-TOF mass spectrometry in subgroup patients with lung adenocarcinoma, HSP 60 protein was highly expressed and its exact mechanism could not be evaluated but it was found to increase in brain tumors especially in glioblastoma and neuroblastoma patients (Chaiwatanasirkul and Sala 2011).

HSP60 protein is effective in tumorigenesis, i.e. tumor growth and transformation. It was shown to be effective in the metastasis process as it supports angiogenesis in tumor tissue (Tsai et al. 2009). Increased metastasis rates in certain types of cancers with over-expressed HSP60 protein have been demonstrated in both in vitro and in vivo studies (Hoang et al. 2000).But the main mechanism of action of HSP60 in carcinogenesis is anti-apoptotic effect on mitochondria and inhibition of cell death. It accomplishes this by reducing the mitochondrial permeability in apoptosis. HSP60 protein is directly associated with cyclophilin D (CypD), a component of the mitochondrial permeability transition pore. Inhibition of HSP60 triggered cyclophilin D-dependent mitochondrial permeability transition. Therefore, HSP60 is a novel regulator of mitochondrial permeability transition and antagonizes Cyp D-dependent cell death in tumors (Ghosh et al. 2010). In this process, it is critical for the cancer cell to gain immortality in carcinogenesis (Nakamura and

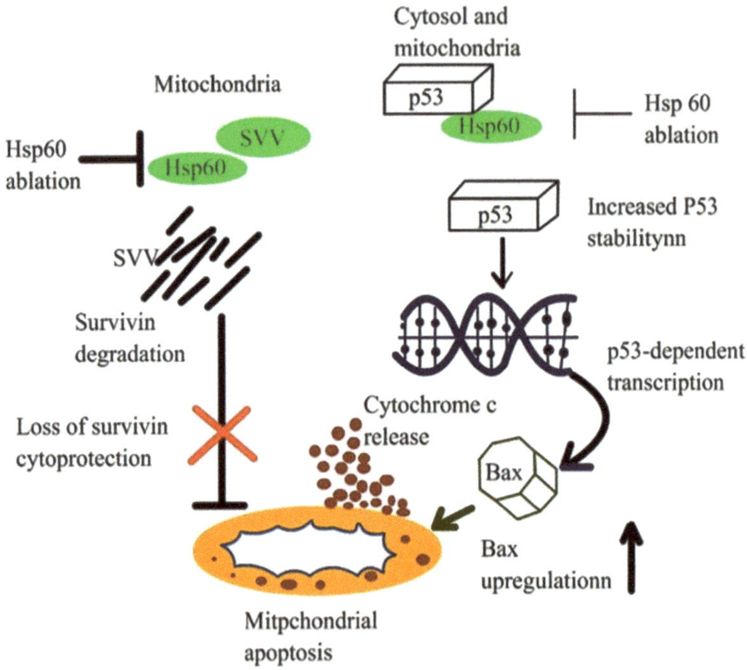

Fig. 7.1 Hsp60-regulated cytoprotection in cancer (Hwang et al. 2009)

Minegishi 2013) (Fig. 7.1) (Ghosh et al. 2008). Similar to this mechanism, it plays a role in the development of resistance in chemotherapeutic agents. In particular, it was shown to play a role in the development of resistance to cisplatin and oxaliplatin and 5-FU chemotherapies (Abu-hadid et al. 1997; Ağababaoğlu et al. 2017; Arya et al. 2007; Cappello et al. 2008, 2013; Chaiwatanasirkul and Sala 2011; Chang et al. 2012; Ciocca and Calderwood 2005; Desmetz et al. 2008; Faried et al. 2004; Ghosh et al. 2008, 2010; Gorska et al. 2013; Hamelin et al. 2011; Hamrita et al. 2008; Hjerpe et al. 2013; Hoang et al. 2000; Hwang et al. 2009; Kang et al. 2009; Liffers et al. 2011; Liu et al. 2012; Lv et al. 2012; Nakamura and Minegishi 2013; Peng et al. 2005; Ritossa 1996; Sigler et al. 1998; Skvortsov et al. 2011; Srivastava 2002; Su et al. 2012; Tsai et al. 2008; Tsai et al. 2009; Wong et al. 2008). Kang et al. demonstrated the effect of Hsp60 overexpression on apoptosis in the liver disease with chronic HCV infection by reducing the production of reactive oxygen radicals. As a result of their studies, they found a decrease in the production of reactive oxygen radicals as a result of an increase in HSP 60, and they found a decrease in the direction of damaged hepatocytes towards apoptosis through TNF-alpha. In light of these results, they on apoptosis the progress towards the oncogenic process in hepatocyte cells and the possibility of hepatoma development may increase (Kang et al. 2009).

7.3 Role of HSP60 in Diagnosis

Hsp60 has the potential to be used in every stage of the diagnosis process of cancer cases due to its role in the cell. It can be used as an early diagnosis and screening biomarker in almost every type of cancer. It has the potential to be used in diagnostic tests or combined diagnostic tests. It has the potential to be used in the investigation of recurrence in treatment follow-up. HSP 60 protein was shown to increase in many types of cancers as previously mentioned. In lung adenocarcinoma, tumor tissues were observed to be significantly increased compared to the conjugated tissues of the patients (Ağababaoğlu et al. 2017). In addition, it was observed that tumor tissue increases in colorectal, gastric, pancreatic, head and neck cancers and many other cancers compared to normal healthy tissue samples (Desmetz et al. 2008).

In the literature, it was shown that the increase in HSP60 protein expression was an independent prognostic factor for disease-free survival by immunohistochemical staining for HSP60 protein on 103 specimens in the study of Xu X. et al. on HSA60 protein and lung adenocarcinoma. It was reported that it may be an important biomarker in determining the prognosis in patients with pulmonary adenocarcinoma in clinical practice (Xu et al. 2011). In addition to lung cancer, HSP60 protein was found to be an important biomarker for the prognosis and follow-up of the response to treatment in serous ovarian carcinomas. In this study of Hjerpe et al. on patients with ovarian carcinoma, average survival was 31 months in patients with HSP60 protein expression and 60 months in patients without expression. In addition, one of the important results of this study was that chemotherapy treatments targeting the HSP60 protein could increase survival (Hjerpe et al. 2013).

The studies to be carried out with Elisa or serum samples have the potential to provide clinicians with much information in early diagnosis, diagnosis and treatment follow-up and detection of recurrences with HSP60. In addition, immunohistochemistry studies for HSP 60 protein may be very useful in the diagnosis stage considering the newly developed immunological diagnostic methods. HSP60 protein was found to be associated with cancer progression in the literature. This is related to the grade and stage of the tumor (Skvortsov et al. 2011). Lung cancer can be used to predict the clinical prognosis of many cancers such as head and neck tumors, gastric cancers, colorectal, and prostate cancer. Of course, into a large series of studies are required for its use in both the diagnosis and the prognosis of the patient and its entry to the clinical routine.

7.4 HSP60 in Treatment and Prognosis

Cancer vaccines are important for targeted immunotherapies due to their antigenic structure and effects in oncogenic processes. It was also shown that it can be used in terms of recurrence and prognosis, which is an important part of the treatment in oncology (Chaiwatanasirkul and Sala 2011). In some studies, it was found that it is associated with a resistance to chemotherapy agents. Thus, it is valuable to show the increased rate in the patients in the decision-making stage of the treatment agent in clinical use.

Among the HSP, HSP90 and HSP70 are the most studied structures for targeted therapies and immunotherapy in oncology. HSP90 protein was shown to be a suitable target for HSP 90 inhibitors of the ATPase domain. In this respect, Geldanamycin derivative is the first drug to progress to 17 AAG Phase II activity trials. HSP90 protein inhibitor geldanamycin was shown to be highly selective in tumor cells and not effective in the intact cells while showing effect on the cancer tissue (Peng et al. 2005). Even its protective effect on the normal cells was defined.It was shown in the studies on HSP 70 that it can be used for tumor immunotherapy and HSP antigen structures associated with carcinogenesis were started to be used in immunotherapy of cancers (Srivastava 2002).

HSP60 contains similar potentials in this respect, as in all HSP members. It was shown in the literature that HSP60 inhibitor agents can be used in some types of cancers in studies on HSP60 protein. In the osteosarcoma cell culture study conducted by Magdalena et al. it was demonstrated that HSP60 may be a potential target for developing new anti-cancer strategies. Geldanamycin induced upregulation in HSP60 gene exspresion, and loss of hyrepacetylated mitochondrial HSP60 pool resulting in decreased cancer cell viability and increased cancer cell death. Hyperacetylation of HSP60 is associated with chemotherapeutic activity of geldanamycin (Gorska et al. 2013).

Bortezomib, a proteosome inhibitor, was researched in a study of ovarian cancer animal model and cell culture. Bortezomib chemotherapy, a CD8 + T cell-mediated cell growth inhibitor, enhances cell surface expression of HSP60 and increases the immunogenicity of tumor tissue. Thus, it was shown to increase dentritic cell-mediated phagocytosis and to produce an immunity mediated anti-tumor effect. Accordingly, it was shown that bortezomib chemotherapy will benefit significantly in ovarian cancer patients with ovarian expression of HSP 60 protein (Chang et al. 2012).In another study on melanoma cell culture, it was shown that in melanoma cells, sinularin inhibits cell proliferation, stop cell migration and induce apoptosis. In addition, proteomic analysis of the cells that could be and could not be treated with sinularin was performed in this study. HSP60 was shown to be significantly lower in cell cultures treated with sinularin and it was found that the effect of this chemotherapeutic agent on melanoma takes place through HSP60 (Su et al. 2012).

In addition to targeted agents, Hsp60 was shown to be associated with resistance to chemotherapeutic agents in cancer types with overexpression. There are cell culture studies on the patients with bladder carcinoma and ovarian carcinoma. A strong association was found between resistance to platinum-based chemotherapies and increased mRNA expression encoding HSP60 in vitro trials. In particular, it was found to be associated with resistance to cisplatin and oxaliplatin chemotherapies in ovarian carcinoma and bladder carcinoma (Ghosh et al. 2010). Again, in a study on colorectal cancer cell cultures, HSP60 was associated with resistance to 5-FU chemotherapy. In the study conducted on SW480 cell passage, 2000 30 proteins were evaluated and it was observed that 8 proteins were down-regulated. One of the down-regulated proteins, HSP60, was thought to play a role in the resistance to 5-FU. Considering the fact that 5-FU is the first choice agent in colorectal cancers and 80% resistance to this agent, immunohistochemical evaluation of HSP60 expression level in the determination of chemotherapy agent will be guiding in the selection of treatment (Nakamura and Minegishi 2013).

Another important stage in the treatment of oncologic routine is to determine prognosis and recurrence. In the study on cervical cancer patients, the relationship between HSP 60 and carcinogenesis and the fact that it is an independent prognostic factor were determined. In the study using 2d gel electrophoresis and western blot technique, the expression of Hsp60 was shown to be significantly increased in cancerous cervical tissue compared to normal cervical tissue. In the same study, Hsp60 was shown to increase at mRNA level by real-time PCR. Both the increase in protein level and increased mRNA correlate with the clinical stages of the cases and can be used to determine prognosis in this type of cancer (Hwang et al. 2009).Similarly, in a study on colorectal cancers, similar results were found in cervical cancers. In the study using 2d electrophoresis in the tissues of the cases and western blot techniques in the serum samples, HSP60 levels were increased, and they were shown to be increased more in advanced stage cases. In addition, the results of immunohistochemistry in a wide series of tissues were similar. When combined with CEA and CA-19-9, its sensitivity as a biomarker for early-stage diagnosis or for the follow-up of treatment was shown to increase up to 0.77 (Hamelin et al. 2011).

In a study including 107 patients, Castilla et al. evaluated HSP60 protein in prostate cancer patients. The patients were divided into two groups as early and late stage. Immunohistochemistry and tissue samples of the patients were also studied with western blotting. There were high levels of HSP 60 in patients with high Gleason score and high serum PSA levels and in patients with lymph node metastases and they demonstrated that the elevation of HSP60 after ablation treatment in 50 local advanced patients was associated with early recurrence. They found that protein markers such as HSP60 would be useful in clinical use related to prostate cancer (Ghosh et al. 2010). Hamrita et al. studied HSP60 protein in breast cancer patients with the same purpose. They organized an antibody study in serum samples of 42 patients diagnosed with breast cancer, at the time of diagnosis and 42 patients clinically proven to have no breast cancer, and they found significantly increased antibody levels for HSP60 antigen. These results show that serum antibody levels for HSP60 can be used for diagnostic purposes (Hamrita et al. 2008).

In another study evaluating the relationship of HSP60 with invasion and metastasis, the regulation of the HSP60 level with c-MYC mechanism was investigated. It was shown that c-MYC directly activates HSP60 expression through an E-box region in the proximal promoter of HSP60 gene, and HSP60 also increases the concentration of beta-catenin protein. Beta-catenin was shown to be associated with invasion and metastasis, particularly associated with different types of metastatic cancers, such as colorectal, lung, prostate, and ovarian cancer. In addition, it was determined in this study that this increase in HSP60 at the cellular level can be used to predict metastasis and bad prognosis in head and neck cancer patients when it is accompanied by beta-catenin (Tsai et al. 2009).

In light of this information, HSP60 has a great potential in clinical use in oncology. As with all heat shock proteins, as HS60 is better understood, the pathophysiology of oncological diseases with unknown facts will be understood and it will present very valuable information to clinicians in all processes of screening, diagnosis, treatment and all subsequent processes.

7.5 Conclusions

Cancer is as old as human history. The term cancer was first used by Hippocrates in the third century BC. Cancer has been a major health problem since then. The process of transforming a normal cell into a cancerous cell is called carcinogenesis and various studies done to elucidate this process. Studies on heat shock proteins provide important information for carcinogenesis. Hsp help to preserve the structure of cellular proteins that have lost their stability or are about to be denatured. HSP60 provide this stabilization of the mitochondria as a mechanism that prevents the cell from apoptosis. Thus Hsp show antiapoptotic effect with their functions protecting protein structures in mitochondria. This plays a role in protection and immortality of damaged cells during malignant transformation. So, HSP60 has the potential to be used in every stages of cancer studies. They have potential for early diagnosis and screening biomarker in almost every type of cancer and can be used in areas for cancer diagnosis as well. Hsp60 proteins have been shown to be associated with angiogenesis and metastasis and this is used in the staging treatment follow-up and prognosis. Furthermore, Bortezomib chemotherapy, a CD8 + T cell-mediated cell growth inhibitor, enhances cell surface expression of HSP60 and increases the immunogenicity of tumor tissue, increase dentritic cell-mediated phagocytosis and to produce an immunity mediated anti-tumor effect. Accordingly, expression of HSP 60 protein can have potential target for developing new anti-cancer strategies.

Acknowledgments I wish to thank Safiye Aktaş, Cahit Koçak and Efe Özgür Serinan for contributing many ideas that improved the chapter.

References

Abu-hadid M, Wilkes JD, Elakawi Z, Pendyala L, Perez RP (1997) Relationship between heat shock protein 60 (HSP60) mRNA expression and resistance to platinum analogues in human ovarian and bladder carcinoma cell lines. Cancer Lett 119:63–70

Ağababaoğlu İ, Demir AB, Erçetin AP, Altun ZS, Ayça Pamukoğlu, Aktaş S, Ersöz H, Akkoçlu A, Önen A (2017) Chaperonin (HSP60) and annexin 2 are candidate biomarkers for early diagnosis of non small cell lung carcinoma. Med J 96(6)

Arya R, Mallik M, Lakhotia SC (2007) Heat shock genes – integrating cell survival and death. J Biosci 32:595–610

Cappello F, Conway de Macario E, Marasà L, Zummo G, Macario AJ (2008) Hsp60 expression, new locations, functions and perspectives for cancer diagnosis and therapy. Cancer Biol Ther 7:801–809

Cappello F, Conway de Macario E, Marino Gammazza A, Bonaventura G, Carini F, Czarnecka AM, Farina F, Zummo G, Macario AJ (2013) Hsp60 and human aging: les liaisons dangereuses. Front Biosci 18:626–637
Chaiwatanasirkul KA, Sala A (2011) The tumour-suppressive function of CLU is explained by its localisation and interaction with HSP60. Cell Death Dis 2:e219
Chang CL, Hsu YT, Wu CC, Yang YC, Wang C, Wu TC, Hung CF (2012) Immune mechanism of the antitumor effects generated by bortezomib. J Immunol 189:3209–3220
Ciocca DR, Calderwood SK (2005) Heat shock proteins in cancer: diagnostic, prognostic, predictive, and treatment implications. Cell Stress Chaperones 10:86–103
Desmetz C, Bibeau F, Boissière F, Bellet V, Rouanet P, Maudelonde T, Mangé A, Solassol J (2008) Proteomics-based identification of HSP60 as a tumor-associated antigen in early stage breast cancer and ductal carcinoma in situ. J Proteome Res 7:2830–2837
Faried A, Sohda M, Nakajima M, Miyazaki T, Kato H, Kuwano H (2004) Expression of heat-shock protein Hsp60 correlated with the apoptotic index and patient prognosis in human esophageal squamous cell carcinoma. Eur J Cancer 40:2804–2811
Ghosh JC, Dohi T, Kang BH, Altieri DC (2008) Hsp60 regulation of tumor cell apoptosis. J Biol Chem 283:5188–5194
Ghosh JC, Siegelin MD, Dohi T, Altieri DC (2010) Heatshock protein 60 regulation of the mitochondrial permeability transition pore in tumor cells. Cancer Res 15:8988–8993
Gorska M, Gammazza AM, Zmijewski MA, Campanella C, Cappello F, Wasiewicz T, Kuban-Jankowska A, Daca A, Sielicka A, Popowska U, Knap N, Antoniewicz J, Wakabayashi T, Wozniak M (2013) Geldanamycin-induced osteosarcoma cell death is associated with hyperacetylation and loss of mitochondrial pool of heat shock protein 60 (hsp60). PLoS One 8:e71135
Hamelin C, Cornut E, Poirier F, Pons S, Beaulieu C, Charrier J-P, Haïdous H, Cotte E, Lambert C, Piard F, Ataman-Önal Y, Choquet-Kastylevsky G (2011) Identification and verification of heat shock protein 60 as a potential serum marker for colorectal cancer. FEBS J 278:4845–4859
Hamrita B, Chahed K, Kabbage M, Guillier CL, Trimeche M, Chaïeb A, Chouchane L (2008) Identification of tumor antigens that elicit a humoral immune response in breast cancer patients' sera bysero- logical proteome analysis (SERPA). Clin Chim Acta 393:95–102
Hjerpe E, Egyhazi S, Carlson J, Stolt MF, Schedvins K, Johansson H, Shoshan M, Avall-Lundqvist E (2013) Int J Gynecol Cancer 23(3):448–455. https://doi.org/10.1097/IGC.0b013e318284308b
Hoang AT, Huang J, Rudra-Ganguly N et al (2000) A novel association between the human heat shock transcription factor 1 (HSF1) and prostate adenocarcinoma. Am J Pathol 156(3):857–864
Hwang YJ, Soon PL, Suk YK, Young HC, Min JK, Choong HL, Joo YL, Dae YK (2009) Expression of heats hock protein 60 kDa is upregulated in cervical cancer. Yonsei Med J 50:399–406
Kang SM, Kim SJ, Kim JH, Lee W, Kim GW, Lee KH, Choi KY, Oh JW (2009) Interaction of hepatitis C virus core protein with Hsp60 triggers the production of reactive oxygen species and enhances TNF-alpha-mediated apoptosis. Cancer Lett 279:230–237
Liffers ST, Maghnouj A, Munding JB, Jackstadt R, Herbrand U, Schulenborg T, Marcus K, Klein-Scory S, Wolff Schmiegel W, Irmgard Schwarte-Waldhoff I, Meyer HE, Stühler K, Hahn SA (2011) Keratin 23, a novel DPC4/Smad4 target gene which binds 14-3-3e. BMC Cancer 11:137
Liu T, Daniels CK, Cao S (2012) Comprehensive review on the HSC70 functions, interactions with related molecules and involvement in clinical diseases and therapeutic potential. Pharmacol Ther 136:354–374
Lv LH, Wan YL, Lin Y, Zhang W, Yang M, Li GL, Lin HM, Shang CZ, Chen YJ, Min J (2012) Anticancer drugs cause release of exosomes with heat shock proteins from human hepatocellular carcinoma cells that elicit effective natural killer cell antitumor responses in vitro. J Biol Chem 287:15874–15885
Nakamura H, Minegishi H (2013) Curr Pharm Des 19(3):441–451. Review
Peng X, Guo X, Borkan SC, Bharti A, Kuramochi Y, Calderwood S, Sawyer DB (2005) Heat shock protein 90 stabilization of ErbB2 expression is disrupted by ATP depletion in myocytes. J Biol Cem 280:13148–13152

Ritossa F (1996) Discovery of the heat shock response. Cell Stress Chaperones 1(2):97–98
Sigler PB, Xu Z, Rye HS, Burston SG, Fenton WA, Horwich AL (1998) Structure and function in GroEL-mediated protein folding. Annu Rev Biochem 67:581–608
Skvortsov S, Schafer G, Stasyk T, Fuchsberger C, Bonn GK, Bartsch G, Klocker H, Huber LA (2011) Proteomics profiling of micro dissected low- and high-grade prostate tumors identifies Lamin A as a discriminatory biomarker. J Proteome Res 10:259–268
Srivastava P (2002) Interaction of heat shock proteins with peptides and antigen presenting cells: chaperoning of the innate and adaptive immune responses. Annu Rev Immunol 20:395–425
Su TR, Lin JJ, Chiu CC, Chen JY, Su JH, Cheng ZJ, Hwang WI, Huang HH, Wu YJ (2012) Proteomic investigation of anti-tumor activities exerted by sinularin against A2058 melanoma cells. Electrophoresis 33:1139–1152
Tsai YP, Teng SC, Wu KJ (2008) Direct regulation of HSP60 expression by c-MYC induces. Transformation 582(29):4083–4088. https://doi.org/10.1016/j.febslet.2008.11.004. Epub 2008 Nov 20
Tsai YP, Yang MH, Huang CH, Chang SY, Chen PM, Liu CJ, Teng SC, Wu KJ (2009) Interaction between HSP60 and beta-catenin promotes metastasis. Carcinogenesis 30:1049–1057
Wong CS, Wong VW, Chan CM, Ma BB, Hui EP, Wong MC, Lam MY, Au TC, Chan WH, Cheuk W, Chan AT (2008) Identification of 5-fluorouracil response proteins in colorectal carcinoma cell line SW480 by two-dimensional electrophoresis and MALDI-TOF mass spectrometry. Oncol Rep 20:89–98
Xu X, Wang W, Shao W, Yin W, Chen H, Qiu Y, Mo M, Zhao J, Deng Q, He J (2011) J Surg Oncol 104(6):598–603. https://doi.org/10.1002/jso.21992. Epub 2011Jun13

Chapter 8
Exosomal Hsp60: A Tumor Biomarker?

Stefano Fais, Mariantonia Logozzi, Giusi Alberti, and Claudia Campanella

Abstract Exosomes (EXs) are extracellular vesicles containing proteins, DNA, mRNA, non-coding RNAs, such as miRNAs, and lipid. The EXs can be easily isolated from different biological fluids and their content is considered a potential biomarker in various diseases, such as cancer. EXs play an important role in intercellular communication, permitting cells to exchange proteins, lipids, and genetic material in normal and pathological conditions. New data have shown that tumor cells-derived EXs contribute to cancer progression through the modulation of tumor microenvironment. Heat shock proteins 60 kDa (Hsp60) is classically considered mitochondrial proteins with different biological roles. In recent years, many studies have focused on the extracellular roles played by Hsp60 that appear to be involved in cancer development and immune system stimulation. Hsp60 is localized on the surface of EXs, secreted by cells and could be a key player in intercellular cross-talk during the course of different diseases. Therefore, exosomal Hsp60 has a great potential for clinical applications, including its use as biomarker for diagnostics, assessing prognosis, and monitoring disease progression and response to treatment, particularly in cancer.

S. Fais
Department of Therapeutic Research and Medicines Evaluation, National Institute of Health, Rome, Italy

Euro-Mediterranean Institute of Science and Technology (IEMEST), Palermo, Italy

M. Logozzi
Department of Therapeutic Research and Medicines Evaluation, National Institute of Health, Rome, Italy

G. Alberti
Department of Biomedicine, Neuroscience and Advanced Diagnostics (BIND), Section of Human Anatomy, University of Palermo, Palermo, Italy

C. Campanella (✉)
Euro-Mediterranean Institute of Science and Technology (IEMEST), Palermo, Italy

Department of Biomedicine, Neuroscience and Advanced Diagnostics (BIND), Section of Human Anatomy, University of Palermo, Palermo, Italy
e-mail: claudia.campanella@unipa.it

A. A. A. Asea, P. Kaur (eds.), *Heat Shock Protein 60 in Human Diseases and Disorders*, Heat Shock Proteins 18,
https://doi.org/10.1007/978-3-030-23154-5_8

Keywords Exosomes · Heat shock proteins · Hsp · Hsp60 · Tumor biomarker

Abbreviations

ESCRT	endosomal sorting complex
EVs	extracellular vesicles
EXs	exosomes
Hsp	heat shock proteins
Hsp60	heat shock protein 60 kDa
ILVs	intraluminal vesicles
MVBs	multivesicular bodies

8.1 Introduction

EXs are a subtype of extracellular vesicles (EVs) with an average diameter of 30–100 nm that derive from intraluminal endosomal vesicles (Campanella et al. 2014). They are released by all type of cells and they are present in all body fluid such as blood urine, cerebral liquid etc. (Cappello et al. 2017). EXs mediate intercellular communication under physiological and pathophysiological conditions, including different types of cancer (Zhao et al. 2017). EXs contain proteins, nucleic acids (DNA, mRNA, miRNA) and lipids (Zhao et al. 2017). According to their molecular signature, they can reach selected target cells at local or distance sites (Campanella et al. 2014). EXs contain cell-specific profile, and they have been proposed as biomarkers in a variety of diseases (Caruso Bavisotto et al. 2017a, b). Recent studies have shown that EXs, derived from different types of tumor, such as pancreatic (Samandari et al. 2018), colorectal (Campanella et al. 2015), play a pivotal role in tumor progression (Cappello et al. 2017). They are able to determine changes in tumor microenvironment favoring their growth and dissemination (Li et al. 2018). Recent studies have validated the hypothesis that EXs may provide a secretory pathway, allowing cells to actively release specific Heat Shock Proteins (Hsp) (Campanella et al. 2014). Hsp are evolutionary conserved proteins, with diverse biological roles (Campanella et al. 2018; Vilasi et al. 2018). They play an important role in cellular homeostasis and cell viability (Czarnecka et al. 2006a). Moreover, many Hsp are defined molecular chaperones by their capacity to recognize and bind substrate proteins that are in an unstable or inactive state (Rappa et al. 2012). In particular, the Hsp act as molecular chaperones assisting in protein transport, oligomeric proteins and protein complexes assembly, refolding of misfolded proteins and triggers of degradation by proteasome (Czarnecka et al. 2006b). Furthermore, the Hsp are involved in many physiological processes in normal cells, such as DNA replication and regulation of gene expression (Pockley and Multhoff 2008). Hsp60 is a mitochondrial chaperone that collaborates with its co-chaperonine Hsp10 for the correct folding of mitochondrial proteins (Czarnecka et al. 2006b). In recent years, many studies have focused on the extracellular roles played by Hsp60 that appear to be involved in cancer development and immune system stimulation (Caruso Bavisotto et al. 2017a). Hsp60 has been found in extra-mitochondrial sites such as cytosol, interstitial space, on the cellular membrane and also in biological fluids in stable, cell free form or associated with EXs (Cappello et al. 2008). Many data demonstrated that its expression increased in different types of tumors and that its presence was often associated with a poor prognosis (Rappa et al. 2012). Campanella and coll. (2012, 2015) have found that Hsp60 is released by tumor cells, in vitro and in vivo, but much less or not at all by normal cells, and a possible secretion pathway, involving lipid rafts and EXs, has been proposed. In this chapter, we describe exosomal characteristic in normal and tumor cells and the role of exosomal Hsp60 in cancer.

8.2 Exosomes Biogenesis and Characteristic

The mechanisms involved in the EXs biogenesis have not yet been fully identified, even if some authors have proposed the possible EXs formation and release as intraluminal vesicles (ILVs) during multivesicular bodies (MVBs) maturation (Campanella et al. 2014; Gould and Raposo 2013; Colombo et al. 2014). ILVs formation involves the endocytosis of transmembrane proteins directed to early endosomes, whereas a second membrane invagination lead to ILVs formation in the late endosomes, generating MVBs (Cocucci and Meldolesi 2015). The fate of ILVs inside MVBs to the plasma membrane or to the lysosomal lumen is related to the presence of specific surface proteins. MVBs, expressing proteins, such as HD-PTP, the HOP complex, and the GTPase Rab-7, are destined to lysosomes compartments, with a subsequent degradation of ILVs. On the contrary, the absence of these proteins leads MVBs to fuse with plasma membrane, allowing ILVs release as EXs (Cocucci and Meldolesi 2015). In the biogenesis of the ILV and MVB are involved the components of the endosomal sorting complex required for transport (ESCRT). In particular, ESCRTs is composed of approx 20 proteins that assemble into four complexes (ESCRT-0, -I, -II and -III) and specific proteins associated (VPS4, VTA1, ALIX), which are conserved from yeast to mammals (Henne et al. 2011). ESCRT-0 consists of hepatocyte growth factor–regulated tyrosine kinase substrate (HRS) that recognizes the mono-ubiquitylated cargo proteins (STAM, Eps15 and clathrin). After that, HRS recruits TSG101 of the ESCRT-I complex. The latter is involved in the recruitment of ESCRT-III, through ESCRT-II or ALIX, an ESCRT-accessory protein. To date, it is unclear whether ESCRT-II has a direct role in ILVs biogenesis or whether its function is limited to a particular cargo (Bowers et al. 2006; Malerød et al. 2007; Colombo et al. 2014). Lack subunits belonging to the four ESCRT complexes does not totally impair the formation of MVBs, indicating that a possibly additional mechanisms may operate in the formation of ILVs and thereby of EXs (Babst 2011). In the literature, it is reported a second mechanism of EXs secretion independent of ESCRT based on the involvement of the sphingolipid ceramide. In particular it was demonstrated that ceramide plays a role in membrane microdomain coalescence, receptor clustering, vesicle formation, membrane fusion/fission and vesicular trafficking (Trajkovic et al. 2008). Another ESCRT-independent mechanism may involve CD63 trafficking that belong to tetraspanins (TSPANs) family. However, other TSPANs, such as CD9, CD81, CD151, and intercellular adhesion molecule-1, CD51 and CD61, integrin, Alix, externalized phosphatidylserine, milk fat globule-E8/lactoferrin, CD80, CD86, CD96, Rab-5b, and major histocompatibility complex class I and class II complexes, are commonly used as molecular marker for EXs (Fais et al. 2016). They are transmembrane proteins that may also be involved in endosomal sorting pathways (van Niel et al. 2011). These data have highlighted that there are different subpopulations of MVBs using different mechanisms for their biogenesis, leads to the generation of different types of EXs. It is important to underline that the EXs are defined by specific molecular features, including the phenotypic expression of specific surface molecules and furthermore their composition can be distinct from the originated cells due to the selective sorting of the cargo into EXs (Mathivanan et al. 2010; Li et al. 2017). In addition to proteins and lipids, EXs contain large amounts of nucleic acids, such as mRNA, microRNA, circular RNA, long non-coding RNA and DNA, which are protected from degradation due to the double lipid membrane (Li et al. 2015). Besides, the EXs have a different molecular characteristic in physiological and pathological conditions (Li et al. 2017). In fact, the EXs are involved in initiation, growth, progression and drug-resistance of cancers through interactions with the microenvironment by transferring oncogenic proteins and nucleic acids (Zhang et al. 2015; Li et al. 2017).

8.3 Exosomes in Tumor

EXs are emerging as major players able to transfer information locally within the tumor microenviroment as well as systemically to distant tissue sites (Li et al. 2018). They are distributed throughout all body fluids and they readily access all sites creating a communication network (Fais et al. 2016). Tumor-derived EXs carry information from the tumor to distant tissues and participate in the activation of the immune system and in pathological events, including malignant transformation (Zhao et al. 2017). Many data demonstrated that tumor-derived EXs can facilitate tumor progression and development malignant growth and metastatization (Cappello et al. 2017). Li et al. (2017) have been demonstrated that oncogenic proteins enclosed in EXs can lead to spread of tumor to the adjacent tissues through the transfer of oncogenic signals from one tumor cell to another, or from tumor cells to normal cells. For example, colon carcinoma cells show a mutanted K-RAS, which is transported by the EXs together a growth-promoting proteins as EGFR, SRC family kinases and integrins (Demory Beckler et al. 2013). EXs derived from glioblastoma expressing activating EGFRvIII mutation, transfer this receptor to glioblastoma cells lacking this mutation and convert the recipient cells to a more malignant phenotype (Choi et al. 2018). Intercellular transfer of molecules containing oncogenic mutations to normal or malignant recipient cells includes activated oncoproteins, their transcripts, oncogenic DNA sequences and oncogenic micro-RNAs, and leads to reprogramming of cellular pathways, especially those responsible for growth factor production (Skog et al. 2008; Caruso Bavisotto et al. 2018). Autocrine signals mediated by EXs, between different cancer cell lines, increased cellular proliferation by Akt phosphorylation and extracellular signal regulated kinase, which, with other downstream molecules, are associated with cellular proliferation (Khalyfa et al. 2016). In recent years, it has been highlighted that EXs could promote anti-tumor functions of immune cells, especially of dendritic cells and lymphocytes. On the other hand, in more advanced cancers, which have escaped immune surveillance, EXs carry an immunosuppressive cargo and become active participants in the tumor escape from the host immune system. Tumor derived EXs cells ready interact with blood vessels, stromal elements and immune cells in order to establish a pre-metastatic niche (Zhang and Grizzle 2014). Melanoma-derived EXs were shown to accumulate in sentinel lymph nodes, stimulate angiogenesis, re-model extracellular matrix and induce melanoma cells recruitment to the lymph nodes (Hood et al. 2011). Tumor cells secrete millions of EXs in order to re-program their surroundings to tumor-promoting microenvironment (Cappello et al. 2017). Experimental data demonstrated that the lipid, nucleic acid, and protein content of EXs are tumor-specific (Fais et al. 2016). A recent report provided evidence that prostate cancer patients show significantly higher levels of plasmatic exosomes as compared to both patients with benign prostate hypertrophy and healthy individuals (Logozzi et al. 2017). Tumor-derived EXs are abundant in the body fluid, are tumor-specific and their content correlates with tumor staging and treatment outcome, for this reason they can be considered potential biomarkers for a cancer (Cappello et al. 2017). EXs are readily accessible in nearly all body fluids including blood, urine, saliva, and ascites and contain bioactive molecules that reflect the pathological state of the originated cells, thus providing an enriched source of biomarkers (Fais et al. 2016; Caruso Bavisotto et al. 2013; Li et al. 2018). Indeed, they are very stable under various conditions such as freezing, thawing, and cold-storage (Zhao et al. 2017; Cappello et al. 2017).

8.4 Exosomal Hsp60

Several years ago, some researchers hypothesized that EXs may provide a secretory pathway, permitting cancer cells to actively release specific Hsp (Campanella et al. 2014). Recent studies have confirmed this hypothesis by demonstrating that many members of the HSP family can be secreted by cancerous cells via the exocytotic pathway (Table 8.1). These proteins may inhibit apoptosis and increase cellular proliferation so they provide a strong stimulus to the microenvironment that can facilitate the growth of cancers (Rappa et al. 2012; Caruso Bavisotto et al. 2017b).

Table 8.1 HSPs identified in the tumor exosomes

Exosomal Hsps	Tumoural cells/cancer origin	Function	References
Hsp70	Myeloid-derived suppressor cells	Induction of proinflammation cytokines, tumor growth factors of and tumor progression	Diao et al. (2015) and Kumar et al. (2016)
	Breast, lung, and ovarian cancer	Anti-tumor immune response	Gobbo et al. (2016)
	Human pancreas, colon and hepatocellular carcinoma cells	Stimulate migratory and cytolytic activity of natural killer cells	Gastpar et al. (2005)
Hsp90α	Brain glioblastoma cells, fibrosarcoma cells, human breast adenocarcinoma cells	Activation of plasminogen cell adhesion	McCready et al. (2010)
	Human hepatocellular carcinoma cells	Stimulate migratory and cytolytic activity of natural killer cells	Lv et al. (2012)
Hsp27	Ovarian Cancer cells	Controls the tumor microenvironment	Stope et al. (2017)
Gpr78	Colon cancer cells	Tumour cell proliferation and mesenchymal stem cell differentiation	Li et al. (2016)
Hsp60	Mucoepidermoid cancer cells	Imunnomodulatory function	Campanella et al. (2012, 2014)
	Colon cancer		
	Human hepatocellular carcinoma cells	Stimulate migratory and cytolytic activity of natural killer cells	

Various data support the hypothesis that Hsp60 favors carcinogenesis and show a correlation between high levels of Hsp60 and different types of cancer, thus the immunopositivity for Hsp60 can be considered as biomarker useful for the diagnosis and monitoring these types of malignancies (Rappa et al. 2012; Campanella et al. 2015). Moreover, some researchers are evaluating the use of potential Hsp60 inhibitors in the treatment of certain diseases, such as neurodegenerative diseases or cancer (Cappello et al. 2014; Campanella et al. 2018). In recent years, increasing evidence supporting the idea that Hsp60 can be localize in extracellular site and, when it is outside cells, may be involved in intercellular cross talk in normal and pathological conditions (Pockley and Henderson 2018; Campanella et al. 2016). Circulating Hsp60 may have a stimulator or inhibitor role on the immune system depending on the type of interaction between the chaperonin and immune system components (Huber et al. 2005). It can bind a variety of receptors present on the surface of plasma cells, such as TLR, CD14, CD40, and CD91, and can activate macrophages and neutrophils (Huber et al. 2005; Cappello et al. 2011). We have

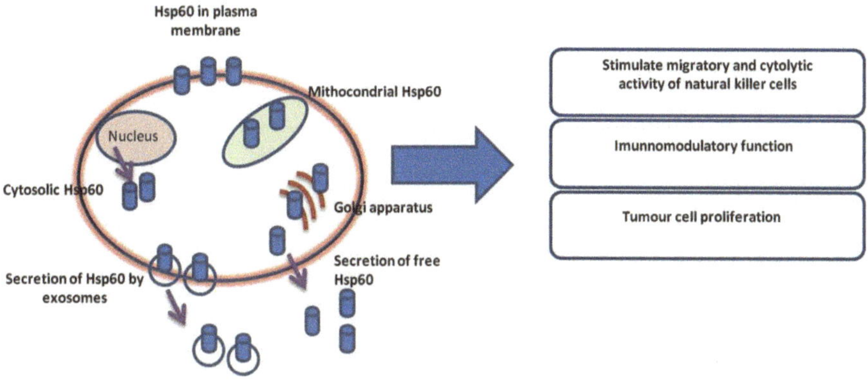

Fig. 8.1 Proposed pathway of HSP60 secretion via exosomes. HSP60 is overexpressed by cancer cells and it is classically considered an intracellular chaperone. It may have non-canonical localizations such as Golgi apparatus, cytoplasm and in plasma membrane. HSP60 is released by tumour cells by an unconventional secretion mechanism, i.e., the lipid-raft exosome pathway. The exosomal HSP60 would reach other cells near and far through the circulation, contributing to carcinogenesis

shown that in the blood of patients with Hashimoto's thyroiditis there were high levels of Hsp60, demonstrating that extracellular Hsp60 may be involved in HT pathogenesis via an antibody-mediated immune mechanism (Marino Gammazza et al. 2014).Recently, we reported, in vitro, that Hsp60 is released by tumor cells by an unconventional secretion mechanism, namely the lipid raft-exosome pathway (Merendino et al. 2010). We shown that Hsp60 is present also in the membrane of tumor cells and it is associated with lipid rafts, and from there ends up in the exosomal membrane, "Fig. (8.1)" (Campanella et al. 2012). Successively, we demonstrated in vivo that Hsp60 is increased in tumor-tissue samples from colorectal adenocarcinoma compared with controls and it is localized not only in the mitochondria but also in extramitochondrial sites, such as cytosol and on membrane (Campanella et al. 2015). After, we focused on the levels of Hsp60 in EXs obtained from the blood of colorectal adenocarcinoma patients before and after surgery. Hsp60 was present in EXs membranes from both time points, but at different levels. Particularly, the Hsp60 levels in the EXs from patients before surgery were significantly higher than in the EXs from the same patients after tumor ablation (Campanella et al. 2015). Moreover, we shown that the histone deacetylase inhibitor, suberoylanilide hydroxamic acid (SAHA) determine, in tumor cells, a decrease of intracellular levels of Hsp60 and also undergoes a Hsp60 post-translational modification. In particular, SAHA induced the Hsp60 nitration, which his exported towards intercellular space via EXs. We hypothesize that exosomal nitrate Hsp60 may have a role in activating the immune system. Many data indicates that protein nitration is linked to tumor cell evasion from T lymphocyte mediated immune response (Campanella et al. 2016).

8.5 Conclusions

All of these data point towards the potential of exosomal Hsp60, secreted by cancer cells, as a novel biomarker for diagnostic and therapeutic applications. As diagnostic and disease monitoring device, exosomal Hsp60 has the advantage that Hsp60 can be obtained from blood samples with minimal distress for the patient. All the results from clinical studies evaluating levels of exosomal Hsp60 in plasma or other body fluids, in tumor patients and controls may will provide a definitive answer to the question: Is exosomal Hsp60 a tumor biomarker?

Acknowledgements C.C. and G.A. C.C.B. were partially supported by UniPA.

References

Babst M (2011) MVB vesicle formation: ESCRT-dependent, ESCRT-independent and everything in between. Curr Opin Cell Biol 23(4):452–457

Bowers K, Piper SC, Edeling MA et al (2006) Degradation of endocytosed epidermal growth factor and virally ubiquitinated major histocompatibility complex class I is independent of mammalian ESCRTII. J Biol Chem 281:5094–5105

Campanella C, Bucchieri F, Merendino AM et al (2012) The odyssey of Hsp60 from tumor cells to other destinations includes plasma membrane-associated stages and Golgi and exosomal protein-trafficking modalities. PLoS One 7:e42008

Campanella C, Caruso Bavisotto C, Marino Gammazza A et al (2014) Exosomal heat shock proteins as new players in TumourCell-to-cell communication. J Circ Biomark 3:5

Campanella C, Rappa F, Sciumè C et al (2015) Heat shock protein 60 levels in tissue and circulating exosome s in human large bowel cancer before and after ablative surgery. Cancer 15:3230–3239

Campanella C, D'Anneo A, Gammazza M et al (2016) The histone deacetylase inhibitor SAHA induces HSP60 nitration and its extracellular release by exosomal vesicles in human lung-derived carcinoma cells. Oncotarget 17:28849–28867

Campanella C, Pace A, Caruso Bavisotto C et al (2018) Heat shock proteins in Alzheimer's disease: role and targeting. Int J Mol Sci 19:2603

Cappello F, Conway de Macario E, Marasà L, Zummo G, Macario AJ (2008) Hsp60 expression, new locations, functions and perspectives for cancer diagnosis and therapy. Cancer Biol Ther 7(6):801–809

Cappello F, Caramori G, Campanella C et al (2011) Convergent sets of data from in vivo and in vitro methods point to an active role of Hsp60 in chronic obstructive pulmonary disease pathogenesis. PLoS One 6(11):e28200

Cappello F, Marino Gammazza A, Palumbo Piccionello A et al (2014) Hsp60 chaperonopathies and chaperonotherapy: targets and agents. Expert Opin Ther Targets 18(2):185–208

Cappello F, Logozzi M, Campanella C et al (2017) Reprint of "exosomes levels in human body fluids: a tumor marker by themselves?". Eur J Pharm Sci 15:64–69

Caruso Bavisotto C, Marino Gammazza A et al (2013) Exosomes: can doctors still ignore their existence? Euromediterranean Biomed J 8:137–139

Caruso Bavisotto C, Cappello F, Macario AJL et al (2017a) Exosomal HSP60: a potentially useful biomarker for diagnosis, assessing prognosis, and monitoring response to treatment. Expert Rev Mol Diagn 17:815–822

Caruso Bavisotto C, Nikolic D, Marino Gammazza A et al (2017b) The dissociation of the Hsp60/pro-Caspase-3 complex by bis(pyridyl)oxadiazole copper complex (CubipyOXA) leads to cell death in NCI-H292 cancer cells. J Inorg Biochem 170:8–16

Caruso Bavisotto C, Graziano F, Rappa F et al (2018) Exosomal chaperones and miRNAs in gliomagenesis: state-of-art and theranostics perspectives. Int J Mol Sci 5:pii:E2626

Choi D, Montermini L, Kim DK, Meehan B, Roth FP, Rak J (2018) The impact of oncogenic EGFRvIII on the proteome of extracellular vesicles released from glioblastoma cells. Mol Cell Proteomics 13. pii: mcp.RA118.000644

Cocucci E, Meldolesi J (2015) Ectosomes and exosomes: shedding the confusion between extracellular vesicles. Trends Cell Biol 25:364–372

Colombo M, Raposo G, Théry C (2014) Biogenesis, secretion, and intercellular interactions of exosomes and other extracellular vesicles. Annu Rev Cell Dev Biol 30:255–289

Czarnecka AM, Campanella C, Zummo G, Cappello F (2006a) Heat shock protein 10 and signal transduction: a “capsula eburnea” of carcinogenesis? Cell Stress Chaperones 11:287–294

Czarnecka AM, Campanella C, Zummo G, Cappello F (2006b) Mitochondrial chaperones in cancer: from molecular biology to clinical diagnostics. Cancer Biol Ther 5:714–720

Demory Beckler M, Higginbotham JN, Franklin JL et al (2013) Proteomic analysis of exosomes from mutant KRAS colon cancer cells identifies intercellular transfer of mutant KRAS. Mol Cell Proteomics 12:343–355

Diao J, Yang X, Song X et al (2015) Exosomal Hsp70 mediates immunosuppressive activity of the myeloid-derived suppressor cells via phosphorylation of Stat3. Med Oncol 32(2):453

Fais S, O’Driscoll L, Borras FE et al (2016) Evidence-based clinical use of nanoscale extracellular vesicles in nanomedicine. ACS Nano 26:3886–3899

Gastpar R, Gehrmann M, Bausero MA et al (2005) Heat shock protein 70 surface-positive tumor exosomes stimulate migratory and cytolytic activity of natural killer cells. Cancer Res. **15** 65(12):5238–5247

Gobbo J, Marcion G, Cordonnier M et al (2016) Restoring anticancer immune response by targeting tumor-derived exosomes with a HSP70 peptide aptamer. J Natl Cancer Inst 108:3

Gould SJ, Raposo G (2013) As we wait: coping with an imperfect nomenclature for extracellular vesicles. J Extracell Vesic 2:20389

Henne WM, Buchkovich NJ, Emr SD (2011) The ESCRT pathway. Dev Cell 21:77–91

Hood JL, San RS, Wickline SA (2011) Exosomes released by melanoma cells prepare sentinel lymph nodes for tumor metastasis. Cancer Res 71(11):3792–3801

Huber V, Fais S, Iero M et al (2005) Human colorectal cancer cells induce T-cell death through release of proapoptotic microvesicles: role in immune escape. Gastroenterology 128:1796–1804

Khalyfa A, Almendros I, Gileles-Hillel A et al (2016) Circulating exosome potentiate tumor malignant properties in a mouse model of chronic sleep fragmentation. Oncotarget 7(34):54676–54690

Kumar S, Stokes J III, Singh UP et al (2016) Targeting Hsp70: a possible therapy for cancer. Cancer Lett 374(1):156–166

Li Y, Zheng Q, Bao C et al (2015) Circular RNA is enriched and stable in exosomes: a promising biomarker for cancer diagnosis. Cell Res 25:981–984

Li Z, Zhuang M, Zhang L, Zheng X, Yang P, Li Z (2016) Acetylation modification regulates GRP78 secretion in colon cancer cells. Sci Rep 27:30406

Li W, Li C, Zhou T, Liu X et al (2017) Role of exosomal proteins in cancer diagnosis. Mol Cancer 29:145

Li K, Chen Y, Li A, Tan C, Liu X (2018) Exosomes play roles in sequential processes of tumor metastasis. Int J Cancer 144:1486–1495

Logozzi M, Angelini DF, Iessi E et al (2017) Increased PSA expression on prostate cancer exosomes in in vitro condition and in cancer patients. Cancer Lett 10:318–329

Lv LH, Wan YL, Lin Y et al (2012) Anticancer drugs cause release of exosomes with heat shock proteins from human hepatocellular carcinoma cells that elicit effective natural killer cell antitumor responses in vitro. J Biol Chem 4:5874–5885

Malerød L, Stuffers S, Brech A, Stenmark H (2007) Vps22/EAP30 in ESCRT-II mediates endosomal sorting of growth factor and chemokine receptors destined for lysosomal degradation. Traffic 8:1617–1629
Marino Gammazza A, Rizzo M, Citarrella R et al (2014) Elevated blood Hsp60, its structural similarities and cross-reactivity with thyroid molecules, and its presence on the plasma membrane of oncocytes point to the chaperonin as an immunopathogenic factor in Hashimoto's thyroiditis. Cell Stress Chaperones 19:343–353
Mathivanan S, Ji H, Simpson RJ (2010) Exosomes: extracellular organelles important in intercellular communication. J Proteome 10:1907–1920
McCready J, Sims JD, Chan D, Jay DG (2010) Secretion of extracellular hsp90alpha via exosomes increases cancer cell motility: a role for plasminogen activation. BMC Cancer 16:294
Merendino AM, Bucchieri F, Campanella C et al (2010) Hsp60 is actively secreted by human tumor cells. PLoS One 16:e9247
Pockley AG, Henderson B (2018) Extracellular cell stress (heat shock) proteins-immune responses and disease: an overview. Philos Trans R Soc Lond B Biol Sci 373(1738)
Pockley AG, Multhoff G (2008) Cell stress proteins in extracellular fluids: friend or foe? Novartis Found Symp 291:86–95
Rappa F, Farina F, Zummo G et al (2012) HSP-molecular chaperones in cancer biogenesis and tumor therapy: an overview. Anticancer Res 32:5139–5150
Rodriguez GM, Galpin KJC, McCloskey CW, Vanderhyden BC (2018) The tumor microenvironment of epithelial ovarian Cancer and its influence on response to immunotherapy. Cancers (Basel) 10:pii:E242
Samandari M, Juliá MG, Rice A, Chronopoulos A, Del Río Hernández AE (2018) Liquid biopsies for management of pancreatic cancer. Transl Res pii: S1931-5244(18)30109-9
Skog J, Wurdinger T, van Rijn S et al (2008) Glioblastoma microvesicles transport RNA and proteins that promote tumour growth and provide diagnostic biomarkers. Nat Cell Biol 10:1470–1476
Stope MB, Klinkmann G, Diesing K, Koensgen D, Burchardt M, Mustea A (2017) Heat shock protein HSP27 secretion by ovarian Cancer cells is linked to intracellular expression levels, occurs independently of the endoplasmic reticulum pathway and HSP27's phosphorylation status, and is mediated by exosome liberation. Dis Markers 2017:1575374
Trajkovic K, Hsu C, Chiantia S, Rajendran L, Wenzel D, Wieland F et al (2008) Ceramide triggers budding of exosome vesicles into multivesicular endosomes. Science 29:1244–1247
van Niel G, Charrin S, Simoes S et al (2011) The tetraspanin CD63 regulates ESCRT-independent and -dependent endosomal sorting during melanogenesis. Dev Cell 21:708–721
Vilasi S, Bulone D, Caruso Bavisotto C et al (2018) Chaperonin of group I: oligomeric Spectrum and biochemical and biological implications. Front Mol Biosci 25:99
Zhang HG, Grizzle WE (2014) Exosomes: a novel pathway of local and distant intercellular communication that facilitates the growth and metastasis of neoplastic lesions. Am J Pathol 184:28–41
Zhang X, Yuan X, Shi H, Wu L, Qian H, Xu W (2015) Exosomes in cancer: small particle, big player. J Hematol Oncol 10(8):83
Zhao H, Achreja A, Iessi E et al (2017) The key role of extracellular vesicles in the metastatic process. Biochim Biophys Acta 1869:64–77

Chapter 9
Hsp60 in Cancer Immunity: Biological Basis, Diagnostic Potential and Therapeutic Opportunities

Christian R. Gomez

Abstract Hsp60 is involved in tumor immune mechanisms leading to recognition of transformed cells and inhibition of growth of neoplastic tissue. As tumors grow, Hsp60 participates in disease progression through its involvement in immunoescape. In this chapter, the interactions between Hsp60 and the tumor microenvironment are discussed as they offer key elements to illustrate the context-dependent functions of Hsp60 in tumor immunomodulation. Next, the applicability of Hsp60 as a diagnostic, prognostic, and marker for therapy response is discussed. Finally, the prospect of targeting Hsp60 and its immunomodulatory effects in tumors is explored. A critical component of self- and nonself-recognition, Hsp60 has value as a tumor marker and therapeutic agent. Study of its context-dependent immune functions offer the prospect of delivering better diagnostic and personalized therapeutic approaches.

Keywords Biomarker · Cancer · Hsp60 · Immune response · Therapy · Tumor microenvironment

Abbreviations

CTL	Cytotoxic T lymphocytes
HSP	Heat shock proteins
IHC	Immunohistochemistry

C. R. Gomez (✉)
Department of Pathology, University of Mississippi Center for Clinical and Translational Science, University of Mississippi Medical Center, Jackson, MS, USA

Department of Radiation Oncology, University of Mississippi Center for Clinical and Translational Science, University of Mississippi Medical Center, Jackson, MS, USA

Translational and Clinical Research Unit, University of Mississippi Center for Clinical and Translational Science, University of Mississippi Medical Center, Jackson, MS, USA
e-mail: crgomez@umc.edu

A. A. A. Asea, P. Kaur (eds.), *Heat Shock Protein 60 in Human Diseases and Disorders*, Heat Shock Proteins 18,
https://doi.org/10.1007/978-3-030-23154-5_9

MC	Myrtucommulone A
MDSC	Myeloid-derived suppressor cells
NK	Natural killer
TAA	Tumor-associated antigens
TAM	Tumor-associated macrophages
TME	Tumor microenvironment
Treg	Regulatory T cells

9.1 Introduction

Heat shock proteins (HSP) are induced not only in response to thermal stress, but in response to a variety of stressors such as infectious agents, diverse forms of radiation, intracellular stress, and stressing metabolic conditions. As such, HSP functions are dependent of the situation in triggering their expression as well as the compartment in which they are present (Cappello et al. 2018). Hsp60, originally described as a mitochondrial protein (Soltys and Gupta 1996), has been localized in the cytosol (Chandra et al. 2007), mitochondrial membrane (Meinhardt et al. 1995; Soltys and Gupta 1996), cell surface (Soltys and Gupta 1997), intracellular vesicles (Soltys and Gupta 1996), extracellular space (Campanella et al. 2008; Campanella et al. 2012; Merendino et al. 2010), and even in the circulatory compartment (Campanella et al. 2012; Desmetz et al. 2008; Hamelin et al. 2011).

In the extracellular milieu, presence of Hsp60 is considered a danger signal and can lead to the activation of immune responses (Pockley and Henderson 2018; Quintana and Cohen 2011). Under non-pathological conditions cell surface Hsp60, as well as its extracellular and bloodstream forms trigger innate and adaptive immunity responses. Rather than attributing multiple functions for Hsp60, it has been proposed that the immune system has evolved to recognize it (Pockley and Henderson 2018). Therefore, actions of the immune system on Hsp60 seem to operate over the basis of their localization rather than the context defined by a given physiological function.

Hsp60 has been involved in the pathogenesis of numerous diseases (Cappello et al. 2014; Kakkar et al. 2014). Since the role of this cell, stress protein in physiopathology relates to its aberrant expression and/or distribution, discussion on this matter must include consideration of its context-specific functions. Expression of Hsp60 is increased in various cancers (Chatterjee and Burns 2017; Wu et al. 2017b) and correlates with disease progression and poor prognosis (Chatterjee and Burns 2017). However, in some types of cancers, decreased levels have been associated with aggressive tumors and poor outcomes (Chatterjee and Burns 2017). Just like under normal conditions and in different compartments, Hsp60 mediates context-specific tumor immunity. How beneficial or detrimental are these immune responses to tumor Hsp60 is a subject of controversy.

Context-associated functions of Hsp60 contribute to define antitumor immune responses. Understanding these interactions is key due to the relevance of this chaperonin as a biomarker and as a therapeutic target. In this chapter, the involvement of Hsp60 in tumor immunity is reviewed. A special focus will be on the role of Hsp60, as a component of the tumor microenvironment (TME), on immunoevassion. The value of Hsp60 as a biomarker of antitumor immunity is also discussed, as it pertains to disease prognosis and therapy responses. Finally, targeting Hsp60 in context of TME is explored. We propose this approximation as a strategy to overcome tumor immunoescape and promote development of innovative antitumor therapeutic approaches.

9.2 Immunoregulatory Properties of the Tumor Microenvironment

The TME includes the complex and dynamic interactions between cancer cells, and other local components of the tumor stroma such as cancer associated fibroblasts, cancer stem cells, and other local or bone-marrow derived stem and progenitor cells (Batlle and Clevers 2017; Buck et al. 2017; Elkhattouti et al. 2015; Valkenburg et al. 2018). In addition to resident constituents, the TME includes immunocytes, T and B cells (Fridman et al. 2012), natural killer (NK) and NKT cells (Sznurkowski et al. 2014), dendritic cells (DC) (Gabrilovich et al. 2012), tumor-associated macrophages (TAM) (Galdiero et al. 2013), and myeloid-derived suppressor cells (MDSC) (Gabrilovich and Nagaraj 2009) among others. These diverse cell types deposit and remodel extracellular matrix, release cytokines, chemokines, adhesion molecules, growth factors, cellular matrix proteins, small regulatory RNAs, DNA, and metabolites. The interactions between cellular and non-celullar components of the stroma determine tumor growth, invasion, and metastasis (Arendt et al. 2010; Patel et al. 2018; Schiavoni et al. 2013), contribute to the natural selection of distinct clones, and to the establishment of aggressive tumors.

Emergence of neoantigens encoded by tumor-specific mutated genes is a relevant mechanism leading to cell transformation within the TME. The immune system can recognize tumor antigens resulting in elimination of emergent tumor cells (Burnet 1970). However, some cancer cells escape immunosurveillance by limiting their expression. Known as immunoediting, this mechanism of active immune tolerance (immunoescape) (Schreiber et al. 2011) conduces to selection of tumor cells and leads to malignant progression. Immune cells, which represent a substantial percentage of the TME (Fridman et al. 2012), are among the factors tumors exploit to avoid immune responses (Vinay et al. 2015). For example, tumor-derived $CD4^{+}CD25^{+}$ $FoxP3^{+}$ Regulatory T cells (Treg) have comparatively higher suppressive activity than naturally occurring Treg (Onizuka et al. 1999; Shimizu et al. 1999). The TME produces immune suppressive mediators. For example MDSC, DC, and TAM secrete cytokines, growth factors, and other products which mediate tumor initiation, angiogenesis, and metastasis, suppress apoptosis on immunosupressive MDSC, or delete tumor-specific CTL through apoptosis (Albini et al. 2018; Quail and Joyce 2013). In addition to effects on immune cells, tumors cause anergy or tolerance in effector cells by not expressing costimulatory molecules (Driessens et al. 2009), by expressing inhibitory molecules [e.g., programmed cell death ligand-1 (PD-L1)/B7-H1] (Flies and Chen 2007), or by down regulation of death receptors, which prevents the killing actions of tumor cells by cytotoxic T lymphocytes (CTL), and NK (French and Tschopp 2002). Thus, by multiple mechanisms, some of them directly operating within the TME, cancer cells act on immunocytes and lead to tumor immunoescape.

9.3 Role of Hsp60 in Tumorigenesis

Hsp60 is aberrantly expressed in tumors (Castle et al. 2005; Cornford et al. 2000; Ghosh et al. 2008; Thomas et al. 2005; Zhou et al. 2018) and has contributing roles in tumor initiation, progression, and metastasis (Chatterjee and Burns 2017; Ciocca and Calderwood 2005; Wu et al. 2017b). The pro-tumorigenic functions of Hsp60 include direct binding and stabilization of the antiapoptotic protein survivin (Ghosh et al. 2008), formation of a complex with Tumor Protein P53, which overrides the checkpoint of centrosomal duplication (Ghosh et al. 2008), interaction and stabilization of the anti-apoptotic mediator clusterin, (Chaiwatanasirikul and Sala 2011), and interaction with cyclophilin D, a component of the mitochondrial permeability transition pore (Ghosh et al. 2010). Independent of its anti-apoptotic role, Hsp60 has been recently reported to promote pancreatic ductal adenocarcinoma by regulating the generation of mitochondrial ATP, critically relevant for activation of extracellular signal-regulated kinase (Erk) 1/2 (Zhou et al. 2018). Similarly, in cervical adenocarcinoma HeLa S3 cells, cytosolic Hsp60 mediated enhanced nuclear factor-κB (NF-κB) activation by direct interaction and modulation of serine-dependent phosphorylation of IκB kinase (IKK) (Chun et al. 2010). The survival program activated by this mechanism, also demonstrated in vivo, involves control of mitochondrial-derived reactive oxygen species (ROS), critically relevant to tumor progression through accumulation of cellular damage (Chun et al. 2010). Hsp60 also has been identified as a downstream regulator on signaling pathways triggered by tumor suppressor insulin-like growth factor binding protein 7 (IGFBP7) in colon cancer cells (Ruan et al. 2010), and as a promoter of pro-metastatic effects via interaction with β-catenin (Tsai et al. 2009). Thus, Hsp60 through signaling pathways responsive to stress, contributes to the survival of cancer cells and to tumor progression.

9.4 Immunomodulatory Effects of Hsp60

Hsp60 is a critical participant in immune responses. Its immune functions include its role as a link between immunocytes and other cell types (Gammazza et al. 2017; Quintana and Cohen 2011), as an antigen (self or foreign) (Pockley and Henderson 2018), and as a ligand for immune-related signaling (Gammazza et al. 2017; Quintana and Cohen 2011). The effects of Hsp60 on immune function depend on its concentration, its antigenic nature (self vs. non-self Hsp60), and the particular constitution of the milieu where Hsp60 acts (Quintana and Cohen 2011). Physiological levels, antigenicity, and a normal milieu of Hsp60 will contribute to the steady state of immune function. The aberrant level of Hsp60, its antigenic nature, or Hsp60's interactions with components of the TME will promote immune responses consistent with disease progression.

Hsp60 has a role in the mechanism of immunosurveillance. In patients with colon cancer, Hsp60 present in the pericellular interstitium, in conjunction with TAM, and NK contributed to tumor removal, through a mechanism involving secretion of cytokines by the immunocytes in the TME (Campanella et al. 2015). As the tumor progresses, Hsp60 contributes to immunoescape by promoting an immunosuppressive environment. The different mechanisms involved in this process seem to depend on location of Hsp60. In Jurkat cells, a human T lymphocytic cell line utilized to study acute T-cell leukemia, mitochondrial pro-caspase-3 was found as part of a complex with Hsp60 and Hsp10 (Samali et al. 1999). Induction of apoptosis was associated with dissociation of active caspase-3 from the complex with the Hsp60 and Hsp10 (Samali et al. 1999). This evidence indicates that intracellular mechanisms of cell survival mediated by Hsp60 are involved in immunoscape.

When present on the cell surface of stressed apoptotic tumor cells, Hsp60 provided a danger signal that stimulated DC and induced a potent cytotoxic response in T cells (Feng et al. 2002). These findings suggest that Hsp60 present on the tumor cell surface may promote effective antitumor responses mediated by antigen presenting cells. In the extracellular space, Hsp60 released by the B16 murine melanoma cell line, increased expression of cytokines IL-6, IL-10, IL-13, TGF-β1, chemokine CCL-2, and chemokine receptor CCR8 through signaling via the Toll-like receptor (TLR2) and activation of the transcription factor signal transducer and activator of transcription 3 (STAT3) (Yang et al. 2009). Consistent with the involvement of Hsp60 as a factor of tumor aggressiveness, B16 cells, which are highly metastatic, released higher levels of Hsp60, had higher expression of TLR2 and had elevated baseline activation of STAT3, relative to low metastatic B16-F1 cells (Yang et al. 2009). These findings provide mechanistic evidences to the role of extracellular Hsp60 in promoting a TME suitable for the development of tumor immune tolerance and tumor progression.

9.5 Hsp60 as a Marker of Tumor Immune Function

Hsp60, as a key constituent and mediator of immune networks (Quintana and Cohen 2011), provides a unique perspective as a biomarker of immune function. Tumor-associated expression of Hsp60 has been found correlated in most reports with cancer progression and poor prognosis (Ge et al. 2018). From these studies, it also has been revealed that the heterogeneity on Hsp60's localization, is important for its use as a marker (Cappello et al. 2018). In many cases, using immunohistochemistry (IHC) the authors have explored the value of cytosolic Hsp60 as a prognostic marker. For example, Hsp60 predicted for poor survival in tumors of the prostate (Beyene et al. 2018; Castilla et al. 2010; Cornford et al. 2000), breast (Desmetz et al. 2008), lung (Ghosh et al. 2008), stomach (Giaginis et al. 2009; Li et al. 2014), bowel (Campanella et al. 2015; Hamelin et al. 2011; Li et al. 2017; Rappa et al. 2016), and ovary (Hjerpe et al. 2013). In other studies, high Hsp60 has been associated with good prognosis. In esophageal squamous cell carcinoma, survival was higher in patients with high Hsp60 expression, relative to those with low Hsp60 (Faried et al. 2004). Similarly, a significantly better prognosis was observed in ovarian cancer patients with high Hsp60-expression (Schneider et al. 1999). Lower cytoplasmic Hsp60 has been found in some tumors. Examples include tumors such as those of the tongue (Ito et al. 1998), bladder (Lebret et al. 2003), bronchia (Cappello et al. 2006), and liver (Zhang et al. 2016). In this last tumor site, cytoplasmic Hsp60, reduced in the tumor correlated with poorer overall survival (Zhang et al. 2016).

Location-dependent intracellular expression of Hsp60 provides differential diagnostic and prognostic value. To explain these discrepancies, it has been proposed that aberrant expression of proteins involved in control of Hsp60 expression can lead to its tumor-specific expression. For example, pro-survival or pro-death functions of cytosolic Hsp60 were found to be dependent on differential interactions with caspase-3, triggered by different apoptotic inducers (Chandra et al. 2007). More research on this will expand the potential of intracellular Hsp60 as a marker. Hsp60 carried in exosomes, mediates interactions between tumor cells and immunocytes in the TME and in other tissues (Campanella et al. 2012; Gammazza et al. 2017). Since Hsp60 contained in exosomes can reach out to other tissues and the circulatory compartment, its presence in biological fluids may have promising clinical applications in diagnosis, prognosis, and as a mean to monitor therapy response (Caruso Bavisotto et al. 2017).

Serum Hsp60 has proven valuable clinical utility for cancer management (Hamelin et al. 2011). Detected by ELISA assay, in a cohort of colorectal cancer patients, Hsp60 was elevated when compared to healthy volunteers (Hamelin et al. 2011). The increase in serum Hsp60 was observed across different stages, but was more pronounced in patients with stage IV cancer (Hamelin et al. 2011). This evidence suggests that serum Hsp60, rather than suited for early detection, could assess the prognosis in patients with advanced disease stage. Due to the facts indicating that Hsp60 expression is initiated early during the carcinogenesis process (Cappello et al. 2003), these findings are rather surprising. As noted by the authors of this report (Cappello et al. 2003), this limitation most likely is due to the sensitivity of their ELISA assay. Certainly, the utility of serum Hsp60 for early detection may benefit, in part, by development of assay methods with increased analytical sensitivity. Consistent with this notion, Hsp60 was not detected in the sera of breast cancer patients, despite the gradual increase in cytoplasmic Hsp60 from normal breast ductal epithelium through in situ ductal carcinoma to invasive ductal carcinoma (Desmetz et al. 2008). Noteworthy, as a highly-sensitive ELISA assay used by the authors of this publication, failed to detect serum Hsp60, independent of quality, other factors such as protein stability may be relevant to the limitations of this detection in early state tumors.

Critically relevant for the immunosurveillance mechanism, is the balance between tumor-derived antibodies and their corresponding antigens (Wu et al. 2017a). As tumors progress and the natural immune repertoire of autoantibodies to antigens is disrupted, Tumor-Associated Antigens (TAA), presented in the surface of tumor cells lead to synthesis of immunoglobulins. In many cases, TAA may represent proteins expressed in the surface of stressed tumor cells (Multhoff et al. 1995). In other cases, TAA are neoantigens, such as tumor-specific mutated proteins (Braunlein and Krackhardt 2017). Since their discovery tumor-associated autoantibodies have been explored as cancer markers. Identified in different kinds of tumors [reviewed in (Wu et al. 2017a)], circulating autoantibodies have attractive characteristics for their use as tumor makers (Anderson and LaBaer 2005). In the early stages of carcinogenesis, tumors express antigens that are recognized as non-self by the immune system. Systemic release of the autoantibodies to TAA represents the end-stage of the humoral mechanism of destruction for transformed cells containing neoantigens (Zaenker et al. 2016). Presence of cancer-specific autoantibodies provides biological amplification and allows for early detection prior to the presence of clinical signs (Casiano et al. 2006). Since access to premalignant lesions is limited, assessment of humoral response provides an alternative approach for identification of early-stage tumor markers. Autoantibodies are produced in higher qualities than TAAs. Additionally, antibodies are very stable proteins and have a long-lasting half-life. Finally, due to our vast knowledge of the biochemical properties of antibodies, assay development typically attempted with approaches such as serological analysis of recombinant cDNA expression libraries, multiplex autoantibody tests, enzyme linked immunosorbent assays, and antigen microarrays is achievable (Macdonald et al. 2017).

Autoantibodies have been utilized as markers for early-stage cancer in tumors of diverse origin such as breast (Anderson et al. 2011; Desmetz et al. 2009), prostate (Ma et al. 2015; Wang et al. 2005), lung (Chapman et al. 2011), liver (Hong et al. 2015), ovary (Anderson et al. 2015; Bodzek et al. 2014), and oral cavity (Wu et al. 2014). Relevant to our discussion, reactivity against Hsp60 has been of interest for early cancer detection (Anderson et al. 2011; Bodzek et al. 2014; Desmetz et al. 2009; Hong et al. 2015; Ma et al. 2015; Wu et al. 2014). Autoantibodies against Hsp60 were identified in early stage breast cancer sera (Anderson et al. 2011; Desmetz et al. 2008, 2009). Serum levels of Hsp60 correlated with increase in disease severity, suggesting that Hsp60 expression may have anticipated the development of the immune response. In another study, a panel of antibodies including five TAAs, peptidylprolyl isomerase A (PPIA), peroxiredoxin 2 (PRDX2), HSP-binding immunophilin (FKBP52), mucin 1 (MUC1), and Hsp60 was utilized as a method to distinguish primary breast cancer and carcinoma *in situ* from healthy controls in women under the age of 50 years (Desmetz et al. 2009). This ELISA-serum based multi analyte screening test demonstrated utility for detection of early-stage cancer in young women with potential to develop aggressive disease (Desmetz et al. 2009). This approach, not only in this study, but also in other reports (Hong et al. 2015; Lacombe et al. 2014) has demonstrated the improved sensitivity or specificity of autoantibody panels relative to assays based on a single autoantibody.

Cancer progression benefits from the selective conditions present in the TME. Among them, tumor hypoxia is a main effector. Since oxygen levels can directly or indirectly influence the function of almost all immune cell types, the role of tumor hypoxia on immune responses is determinant for tumor development (Petrova et al. 2018). We explored microenvironmental hypoxia as a variable to reveal an antigenic landscape potentially relevant to tumor aggressiveness (Ma et al. 2015). Low oxygen in the tumor contributes to tumor initiation (Taiakina et al. 2014; Vaupel et al. 2001) and therapy resistance (Tredan et al. 2007; Westhoff et al. 2009; Wouters et al. 2007), in part due to its effects on innate and adaptive immunity (Bosco et al. 2006; Imtiyaz and Simon 2010). Following culture of prostate cancer cells LNCaP and VCaP in low (2%) or normal (20%) oxygen, circulating prostate cancer-specific autoantibodies were identified by two-dimensional gel electrophoresis (2DGE) and immunoblotting followed by MALDI-TOF mass spectrometry (Ma et al. 2015). Among selected TAA, anti- Hsp60 and anti- Hsp70 antibodies, detected in plasma obtained from prostate cancer patients, were further elevated when lysates from cells cultured under hypoxia were used as source of TAA (Ma et al. 2015). Consistent with 2DGE and Western blot analysis, when assessed at the individual patient level, ELISA analysis showed higher frequency of Hsp60 autoantibodies in the circulation of prostate cancer patients relative to healthy controls (Ma et al. 2015). The data showing elevated Hsp60 autoantibody level in the plasma from cancer patients, highlight the immunogenic nature of tumors, and support a role of hypoxia as an immunomodulatory factor in the TME. Under the assumption that tumor cells cultured under hypoxia are akin to those present in tumors in situ, our (Ma et al. 2015) and other studies (Olin et al. 2010, 2011) may help to design a protocol to develop TME-relevant tumor markers.

9.6 Opportunities for Therapeutic Applicability

A challenge of current cancer screening tests refers to their applicability in early diagnosis and prediction of response to therapy. As we have described HSP, in particular Hsp60, are valuable when screening patients at risk to develop aggressive tumors. HSP may have a utility as markers with ability to anticipate treatment response (Chatterjee and Burns 2017). For example, exosomal Hsp60 obtained from the blood of colorectal cancer patients was higher before surgery compared to post surgery levels (Campanella et al. 2015). These findings suggest measurements of exosomal Hsp60 may provide a non-invasive tool to identify patients prone to develop aggressive disease. Likewise, the measurements would have value to anticipate treatment response.

In a retrospective study, IHC was used to evaluate cytoplasmic expression of Hsp27, Hsp60, Hsp70, Hsp90, and p53 in biopsies obtained from patients with aggressive bladder cancer (Urushibara et al. 2007). Pathological response to treatment was ascertained prior to low-dose neoadjuvant chemoradiotherapy, followed by radical or partial cystectomy (Urushibara et al. 2007). For all HSP, cytoplasmic expression was detected prior to chemoradiotherapy, however using a multivariate analysis, only positive expression of Hsp60 marginally predicted for a good response to chemoradiotherapy (Urushibara et al. 2007). Similarly, and also in a marginally significant manner, Hsp60 predicted for better overall survival (Urushibara et al. 2007). Despite the limitation of this study, based on a low number of cases (N = 54), this work provides additional support to the use of Hsp60 to predict for therapy responses. Use of Hsp60 as a predictive marker may help to personalize cancer treatment.

Due to the role of Hsp60 on mitochondrial homeostasis and its involvement in apoptosis and cell survival signaling pathways, this chaperonin has been proposed as a good drug target for cancer treatment (Chatterjee and Burns 2017; Meng et al. 2018). This notion has been further supported by experimental evidence showing Hsp60's involvement in the resistance of cancer cells to chemotherapeutic drugs (Gorska et al. 2013; Wong et al. 2008). In relation to the development of Hsp60 inhibitors with potential as therapeutic targets in cancer, most of the research has been focused on testing the applicability of known natural or synthetic compounds with described bioactive properties (Meng et al. 2018). In leukemia HL-60 cells, Hsp60 was recently reported as a direct mitochondrial target of myrtucommulone A (MC), a non-prenylated acylphloroglucinol naturally present in myrtle (Wiechmann et al. 2017). Myrtucommulone A is a promising therapeutic agent since it selectively induces apoptosis in cancer cells over normal cells (Tretiakova et al. 2008). By direct binding to Hsp60, MC inhibited its chaperone activity, and induced aggregation of mitochondrial stress proteins Lon protease-like protein (LONP) and leucine-rich PPR motif-containing protein (LRP130) (Wiechmann et al. 2017). The interference of MC on Hsp60's role on protein transport and stability under stress conditions can be proposed as an underlining mechanism for induction of the intrinsic mitochondrial pathway of cancer cell apoptosis.

Applicability of MC and other natural of currently available synthetic compounds based on targeting Hsp60, will need to overcome challenges. Some include off target effects, and other challenges related to the intrinsic structural properties which limit the development of more potent derivatives (Meng et al. 2018). While recognizing these limitations, arguably we point to the need of further investigations focused on unraveling the biological roles of Hsp60, particularly those relevant to conditions of the TME. Progress into that direction will enable a better understanding of the potential for Hsp60 as a drug target for anti-cancer therapy. For example, Hsp60 is a primary molecular target of the synthetic compound *o*-carboranylphenoxyacetanilide 8 (Ban et al. 2010). Mechanisms of action for this compound include inhibition of Hsp60-Hsp10's refolding activity and Hsp60's ATPase activity (Ban et al. 2010). As a critical downstream effect, Ban and collaborators found that *o*-carboranylphenoxyacetanilide 8 suppressed hypoxia-induced HIF-1α activation (Ban et al. 2010).

Providing further support to the relevance of targeting Hsp60 in the context of the TME, our group reported on tumor expression of HIF-1α, Hsp60, and the angiogenic factor vascular endothelial growth factor (VEGF), a HIF-1α target associated with tumor progression (Espinoza et al. 2016). The relationship between expression of Hsp60 and HIF-1α may be further studied in tumors as a histological subrogate of tumor hypoxia and as a marker of aggressive disease. From the therapeutic point of view, because of the role of HIF-1α on the cellular response of tumor cells to hypoxia (Petrova et al. 2018), the functional association reported by Ban (Ban et al. 2010) may have applicability for treatment of hypoxic areas, prevalent in the microenvironment of aggressive tumors.

As we have illustrated during our discussion, the expression of Hsp60 in the cell surface and in the extracellular environment defines the role of this protein as a connection between immune cells, as a coordinator of immune function, and as a target for immune-based therapy. Full length Hsp60 and derived peptides have immunoregulatory roles on the inflammatory response. This property has been utilized at the preclinical and clinical levels for therapeutic use to treat diseases such as type 1 diabetes (Fischer et al. 2010; Quintana and Cohen 2011; Raz et al. 2014), rheumatoid arthritis (Quintana et al. 2008), and for the development of anti-microbial vaccines (Bajzert et al. 2018; Kaur et al. 2015).

At a preclinical level, chemotherapeutic drugs have shown immune mechanisms involving Hsp60 in their antitumor effects. Bortezomib is the first FDA-approved proteasome inhibitor used in multiple myeloma (Mateos et al. 2006), and mantle-cell lymphoma (Goy et al. 2009). In tumor cells, this drug exerts its actions via the induction of apoptosis and autophagy (Hu et al. 2014; Selimovic et al. 2013; Zheng et al. 2015). Despite the advanced knowledge on the therapeutic effects of bortezomib, its effects on tumor cell immunogenicity have yet to be fully explored. In a murine ovarian ([C57BL/6×C3/He] F1) OV-HM cancer model with intact host immunity, antitumor effects mediated by bortezomib were observed (Chang et al. 2012). However, these effects were abolished in nude mice, suggesting immune involvement. As a strategy to develop a tumor cell-based immunotherapy, the authors injected mice with bortezomib-treated tumor cells. Animals receiving this vaccine had high numbers of $CD8^+$ tumor-infiltrating lymphocytes, demonstrating the immunological relevance of this approach. Tumor cells treated with bortezomib had upregulation of Hsp60 and Hsp90 on their cell surface, and were more susceptible to phagocytosis by DC (Chang et al. 2012).

Hsp60-dependent protective immunity involved in bortezomib antitumor effect represents groundwork not only for improving the therapeutic efficacy of this chemotherapeutic drug, but also for revealing its potential to improve the design of syngeneic therapeutic cancer vaccines. The composition of current cancer vaccines ranges from purified single antigens to complex antigen mixtures, such as whole cancer cells. Whole cells are believed to raise protective immunity against tumors much more effectively than tumor cell lysates (Strome et al. 2002).The relative success of whole cell cancer vaccines (Michael et al. 2005) could originate in the presentation of more relevant antigenic determinants in comparison to individual antigens or their combinations (Small et al. 2007). Nonetheless, the success of whole-cell cancer vaccines is still limited (Michael et al. 2005). One plausible reason for the incomplete response is that vaccine cells, prepared under standard tissue culture conditions, can drastically differ in expression of macromolecules from cognate cells growing in the lab, and thus may immunize against less pertinent antigen spectrum. Many differences between conditions in cell culture and in situ have been recognized as important modifiers of cell biology and phenotype. Among these are the ability of cells to form three–dimensional cell–cell contacts and maintain in situ–like shape, flow of nutrients and metabolites, hydrodynamic pressure, access to macromolecules, etc. Protein expression and antigenic signature could be more akin to that of cells in tumors when components of the TME, such as hypoxia are included in the preparation of whole cell cancer vaccines (Bardos and Ashcroft 2005; Bosco et al. 2006). Relevant to this discussion, Hsp60 was identified by us as a hypoxia-reactive TAA (Ma et al. 2015). These findings underscore the contribution of Hsp60 as a relevant antigenic determinant with potential for therapeutic development, additionally they suggest that hypoxia profoundly affects the antigenic signature of cancer cells. Cancer cells deemed largely independent of oxygen (the Warburg effect), are nonetheless sensitive to changes in oxygen levels. Tumor hypoxia, as a major modifier of tumor cell properties (Vaupel 2008) provides a variable of the TME with effects on immunity functions mediated by Hsp60 (Gabrilovich et al. 1996) and potential for therapeutic development.

9.7 Conclusions

We have summarized the involvement of Hsp60 in tumor immunity and described how context-dependent functions of this chaperonin protein define its involvement in immunosurveillance and immunoescape. Research on the tumor-specific intracellular signaling involved in the immunomodulatory roles of Hsp60, immune interactions of extracellular Hsp60, cell populations, and variables of the TME, and involvement of Hsp60 in tumor progression is needed. Derived knowledge will broaden the scope of Hsp60 as a component of tumor immune response and will expand its therapeutic potential. Hsp60 and in particular cancer-specific autoantibodies to this protein have utility to anticipate early disease. In many cases, circulating levels of autoantibodies to Hsp60 precede those of other tumor markers by several months to years. The increased sensitivity and specificity of this detection, added to the applicability of autoantibody panels represents a refinement with diagnostic, prognostic, and theragnostic applications. When applied to biomarker driven clinical trial design, Hsp60 will help to improve personalized clinical decisions and disease management. Not only related to stress responses, Hsp60 has been associated to the aggressiveness of tumors. A critical component of self-versus non-self-recognition, this protein drives tumor progression and has value as a marker and therapeutic agent. Study of context-dependent functions of Hsp60 in tumor immunomodulation offers the prospect of delivering better diagnostic and therapeutic approaches.

Acknowledgements The author thanks Drs. Venkat Mannam, Ingrid Espinoza, and Amit Reddy for valuable discussions and critical reading of the manuscript. This work was supported in part by Grant DOD PC094680 and PCF Creativity Award.

References

Albini A, Bruno A, Noonan DM, Mortara L (2018) Contribution to tumor angiogenesis from innate immune cells within the tumor microenvironment: implications for immunotherapy. Front Immunol 9:527

Anderson KS, LaBaer J (2005) The sentinel within: exploiting the immune system for cancer biomarkers. J Proteome Res 4:1123–1133

Anderson KS, Sibani S, Wallstrom G, Qiu J, Mendoza EA, Raphael J, Hainsworth E, Montor WR, Wong J, Park JG, Lokko N, Logvinenko T, Ramachandran N, Godwin AK, Marks J, Engstrom P, Labaer J (2011) Protein microarray signature of autoantibody biomarkers for the early detection of breast cancer. J Proteome Res 10:85–96

Anderson KS, Cramer DW, Sibani S, Wallstrom G, Wong J, Park J, Qiu J, Vitonis A, LaBaer J (2015) Autoantibody signature for the serologic detection of ovarian cancer. J Proteome Res 14:578–586

Arendt LM, Rudnick JA, Keller PJ, Kuperwasser C (2010) Stroma in breast development and disease. Semin Cell Dev Biol 21:11–18

Bajzert J, Gorczykowski M, Galli J, Stefaniak T (2018) The evaluation of immunogenic impact of selected bacterial, recombinant Hsp60 antigens in DBA/2J mice. Microb Pathog 115:100–111

Ban HS, Shimizu K, Minegishi H, Nakamura H (2010) Identification of HSP60 as a primary target of o-carboranylphenoxyacetanilide, an HIF-1alpha inhibitor. J Am Chem Soc 132:11870–11871

Bardos JI, Ashcroft M (2005) Negative and positive regulation of HIF-1: a complex network. Biochim Biophys Acta 1755:107–120

Batlle E, Clevers H (2017) Cancer stem cells revisited. Nat Med 23:1124–1134

Beyene DA, Naab TJ, Kanarek NF, Apprey V, Esnakula A, Khan FA, Blackman MR, Brown CA, Hudson TS (2018) Differential expression of Annexin 2, SPINK1, and Hsp60 predict progression of prostate cancer through bifurcated WHO Gleason score categories in African American men. Prostate 78:801–811

Bodzek P, Partyka R, Damasiewicz-Bodzek A (2014) Antibodies against Hsp60 and Hsp65 in the sera of women with ovarian cancer. J Ovarian Res 7:30

Bosco MC, Puppo M, Santangelo C, Anfosso L, Pfeffer U, Fardin P, Battaglia F, Varesio L (2006) Hypoxia modifies the transcriptome of primary human monocytes: modulation of novel immune-related genes and identification of CC-chemokine ligand 20 as a new hypoxia-inducible gene. J Immunol 177:1941–1955

Braunlein E, Krackhardt AM (2017) Identification and characterization of neoantigens as well as respective immune responses in cancer patients. Front Immunol 8:1702

Buck MD, Sowell RT, Kaech SM, Pearce EL (2017) Metabolic instruction of immunity. Cell 169:570–586

Burnet FM (1970) The concept of immunological surveillance. Prog Exp Tumor Res 13:1–27
Campanella C, Bucchieri F, Ardizzone NM, Marino Gammazza A, Montalbano A, Ribbene A, Di Felice V, Bellafiore M, David S, Rappa F, Marasa M, Peri G, Farina F, Czarnecka AM, Conway de Macario E, Macario AJ, Zummo G, Cappello F (2008) Upon oxidative stress, the antiapoptotic Hsp60/procaspase-3 complex persists in mucoepidermoid carcinoma cells. Eur J Histochem 52:221–228
Campanella C, Bucchieri F, Merendino AM, Fucarino A, Burgio G, Corona DF, Barbieri G, David S, Farina F, Zummo G, de Macario EC, Macario AJ, Cappello F (2012) The odyssey of Hsp60 from tumor cells to other destinations includes plasma membrane-associated stages and Golgi and exosomal protein-trafficking modalities. PLoS One 7:e42008
Campanella C, Rappa F, Sciume C, Marino Gammazza A, Barone R, Bucchieri F, David S, Curcuru G, Caruso Bavisotto C, Pitruzzella A, Geraci G, Modica G, Farina F, Zummo G, Fais S, Conway de Macario E, Macario AJ, Cappello F (2015) Heat shock protein 60 levels in tissue and circulating exosomes in human large bowel cancer before and after ablative surgery. Cancer 121:3230–3239
Cappello F, Bellafiore M, Palma A, David S, Marciano V, Bartolotta T, Sciume C, Modica G, Farina F, Zummo G, Bucchieri F (2003) 60KDa chaperonin (HSP60) is over-expressed during colorectal carcinogenesis. Eur J Histochem 47:105–110
Cappello, F. Di Stefano, A. David, S. Rappa, F. Anzalone, R. La Rocca, G. D'Anna, S.E. Magno, F. Donner, C.F. Balbi, B. and Zummo, G. (2006) Hsp60 and Hsp10 down-regulation predicts bronchial epithelial carcinogenesis in smokers with chronic obstructive pulmonary disease. Cancer 107, 2417–2424
Cappello F, Marino Gammazza A, Palumbo Piccionello A, Campanella C, Pace A, Conway de Macario E, Macario AJ (2014) Hsp60 chaperonopathies and chaperonotherapy: targets and agents. Expert Opin Ther Targets 18:185–208
Cappello F, Conway de Macario E, Rappa F, Zummo G, Macario AJL (2018) Immunohistochemistry of human Hsp60 in health and disease: from autoimmunity to Cancer. Methods Mol Biol 1709:293–305
Caruso Bavisotto C, Cappello F, Macario AJL, Conway de Macario E, Logozzi M, Fais S, Campanella C (2017) Exosomal HSP60: a potentially useful biomarker for diagnosis, assessing prognosis, and monitoring response to treatment. Expert Rev Mol Diagn 17:815–822
Casiano CA, Mediavilla-Varela M, Tan EM (2006) Tumor-associated antigen arrays for the serological diagnosis of cancer. Mol Cell Proteomics 5:1745–1759
Castilla C, Congregado B, Conde JM, Medina R, Torrubia FJ, Japon MA, Saez C (2010) Immunohistochemical expression of Hsp60 correlates with tumor progression and hormone resistance in prostate cancer. Urology 76:1017 e1–1017 e6
Castle PE, Ashfaq R, Ansari F, Muller CY (2005) Immunohistochemical evaluation of heat shock proteins in normal and preinvasive lesions of the cervix. Cancer Lett 229:245–252
Chaiwatanasirikul KA, Sala A (2011) The tumour-suppressive function of CLU is explained by its localisation and interaction with HSP60. Cell Death Dis 2:e219
Chandra D, Choy G, Tang DG (2007) Cytosolic accumulation of HSP60 during apoptosis with or without apparent mitochondrial release: evidence that its pro-apoptotic or pro-survival functions involve differential interactions with caspase-3. J Biol Chem 282:31289–31301
Chang CL, Hsu YT, Wu CC, Yang YC, Wang C, Wu TC, Hung CF (2012) Immune mechanism of the antitumor effects generated by bortezomib. J Immunol 189:3209–3220
Chapman CJ, Thorpe AJ, Murray A, Parsy-Kowalska CB, Allen J, Stafford KM, Chauhan AS, Kite TA, Maddison P, Robertson JF (2011) Immunobiomarkers in small cell lung cancer: potential early cancer signals. Clin Cancer Res 17:1474–1480
Chatterjee S, Burns TF (2017) Targeting heat shock proteins in cancer: a promising therapeutic approach. Int J Mol Sci 18
Chun JN, Choi B, Lee KW, Lee DJ, Kang DH, Lee JY, Song IS, Kim HI, Lee SH, Kim HS, Lee NK, Lee SY, Lee KJ, Kim J, Kang SW (2010) Cytosolic Hsp60 is involved in the NF-kappaB-dependent survival of cancer cells via IKK regulation. PLoS One 5:e9422

Ciocca DR, Calderwood SK (2005) Heat shock proteins in cancer: diagnostic, prognostic, predictive, and treatment implications. Cell Stress Chaperones 10:86–103
Cornford PA, Dodson AR, Parsons KF, Desmond AD, Woolfenden A, Fordham M, Neoptolemos JP, Ke Y, Foster CS (2000) Heat shock protein expression independently predicts clinical outcome in prostate cancer. Cancer Res 60:7099–7105
Desmetz C, Bibeau F, Boissiere F, Bellet V, Rouanet P, Maudelonde T, Mange A, Solassol J (2008) Proteomics-based identification of HSP60 as a tumor-associated antigen in early stage breast cancer and ductal carcinoma in situ. J Proteome Res 7:3830–3837
Desmetz C, Bascoul-Mollevi C, Rochaix P, Lamy PJ, Kramar A, Rouanet P, Maudelonde T, Mange A, Solassol J (2009) Identification of a new panel of serum autoantibodies associated with the presence of in situ carcinoma of the breast in younger women. Clin Cancer Res 15:4733–4741
Driessens G, Kline J, Gajewski TF (2009) Costimulatory and coinhibitory receptors in anti-tumor immunity. Immunol Rev 229:126–144
Elkhattouti A, Hassan M, Gomez CR (2015) Stromal fibroblast in age-related cancer: role in tumorigenesis and potential as novel therapeutic target. Front Oncol 5:158
Espinoza I, Sakiyama MJ, Ma T, Fair L, Zhou X, Hassan M, Zabaleta J, Gomez CR (2016) Hypoxia on the expression of hepatoma upregulated protein in prostate cancer cells. Front Oncol 6:144
Faried A, Sohda M, Nakajima M, Miyazaki T, Kato H, Kuwano H (2004) Expression of heat-shock protein Hsp60 correlated with the apoptotic index and patient prognosis in human oesophageal squamous cell carcinoma. Eur J Cancer 40:2804–2811
Feng H, Zeng Y, Graner MW, Katsanis E (2002) Stressed apoptotic tumor cells stimulate dendritic cells and induce specific cytotoxic T cells. Blood 100:4108–4115
Fischer B, Elias D, Bretzel RG, Linn T (2010) Immunomodulation with heat shock protein DiaPep277 to preserve beta cell function in type 1 diabetes – an update. Expert Opin Biol Ther 10:265–272
Flies DB, Chen L (2007) The new B7s: playing a pivotal role in tumor immunity. J Immunother 30:251–260
French LE, Tschopp J (2002) Defective death receptor signaling as a cause of tumor immune escape. Semin Cancer Biol 12:51–55
Fridman WH, Pages F, Sautes-Fridman C, Galon J (2012) The immune contexture in human tumours: impact on clinical outcome. Nat Rev Cancer 12:298–306
Gabrilovich DI, Nagaraj S (2009) Myeloid-derived suppressor cells as regulators of the immune system. Nat Rev 9:162–174
Gabrilovich DI, Chen HL, Girgis KR, Cunningham HT, Meny GM, Nadaf S, Kavanaugh D, Carbone DP (1996) Production of vascular endothelial growth factor by human tumors inhibits the functional maturation of dendritic cells. Nat Med 2:1096–1103
Gabrilovich DI, Ostrand-Rosenberg S, Bronte V (2012) Coordinated regulation of myeloid cells by tumours. Nat Rev 12:253–268
Galdiero MR, Bonavita E, Barajon I, Garlanda C, Mantovani A, Jaillon S (2013) Tumor associated macrophages and neutrophils in cancer. Immunobiology 218:1402–1410
Gammazza AM, Caruso C, David S, Barone R, Rappa F, Campanella C, Conway de Macario E, Cappello F, Macario AJL (2017) HSP60 is a ubiquitous player in the physiological and pathogenic interactions between the chaperoning and the immune systems. Curr Immunol Rev 13:44–55
Ge H, Yan Y, Guo L, Tian F, Wu D (2018) Prognostic role of HSPs in human gastrointestinal cancer: a systematic review and meta-analysis. Oncol Targets Ther 11:351–359
Ghosh JC, Dohi T, Kang BH, Altieri DC (2008) Hsp60 regulation of tumor cell apoptosis. J Biol Chem 283:5188–5194
Ghosh JC, Siegelin MD, Dohi T, Altieri DC (2010) Heat shock protein 60 regulation of the mitochondrial permeability transition pore in tumor cells. Cancer Res 70:8988–8993
Giaginis C, Daskalopoulou SS, Vgenopoulou S, Sfiniadakis I, Kouraklis G, Theocharis SE (2009) Heat shock protein-27, -60 and -90 expression in gastric cancer: association with clinicopathological variables and patient survival. BMC Gastroenterol 9:14

Gorska M, Marino Gammazza A, Zmijewski MA, Campanella C, Cappello F, Wasiewicz T, Kuban-Jankowska A, Daca A, Sielicka A, Popowska U, Knap N, Antoniewicz J, Wakabayashi T, Wozniak M (2013) Geldanamycin-induced osteosarcoma cell death is associated with hyperacetylation and loss of mitochondrial pool of heat shock protein 60 (hsp60). PLoS One 8:e71135

Goy A, Bernstein SH, Kahl BS, Djulbegovic B, Robertson MJ, de Vos S, Epner E, Krishnan A, Leonard JP, Lonial S, Nasta S, O'Connor OA, Shi H, Boral AL, Fisher RI (2009) Bortezomib in patients with relapsed or refractory mantle cell lymphoma: updated time-to-event analyses of the multicenter phase 2 PINNACLE study. Ann Oncol 20:520–525

Hamelin C, Cornut E, Poirier F, Pons S, Beaulieu C, Charrier JP, Haidous H, Cotte E, Lambert C, Piard F, Ataman-Onal Y, Choquet-Kastylevsky G (2011) Identification and verification of heat shock protein 60 as a potential serum marker for colorectal cancer. FEBS J 278:4845–4859

Hjerpe E, Egyhazi S, Carlson J, Stolt MF, Schedvins K, Johansson H, Shoshan M, Avall-Lundqvist E (2013) HSP60 predicts survival in advanced serous ovarian cancer. Int J Gynecol Cancer 23:448–455

Hong Y, Long J, Li H, Chen S, Liu Q, Zhang B, He X, Wang Y, Li H, Li Y, Zhang T, Lu C, Yan H, Zhang M, Li Q, Cao B, Bai Z, Wang J, Zhang Z, Zhu S, Zheng J, Ou X, Ma H, Jia J, You H, Wang S, Huang J (2015) An analysis of immunoreactive signatures in early stage hepatocellular carcinoma. EBioMedicine 2:438–446

Hu X, Xuan H, Du H, Jiang H, Huang J (2014) Down-regulation of CD9 by methylation decreased bortezomib sensitivity in multiple myeloma. PLoS One 9:e95765

Imtiyaz HZ, Simon MC (2010) Hypoxia-inducible factors as essential regulators of inflammation. Curr Top Microbiol Immunol 345:105–120

Ito T, Kawabe R, Kurasono Y, Hara M, Kitamura H, Fujita K, Kanisawa M (1998) Expression of heat shock proteins in squamous cell carcinoma of the tongue: an immunohistochemical study. J Oral Pathol Med 27:18–22

Kakkar V, Meister-Broekema M, Minoia M, Carra S, Kampinga HH (2014) Barcoding heat shock proteins to human diseases: looking beyond the heat shock response. Dis Model Mech 7:421–434

Kaur G, Sts C, Nimker C, Bansal A (2015) rIL-22 as an adjuvant enhances the immunogenicity of rGroEL in mice and its protective efficacy against S. Typhi and S. Typhimurium. Cell Mol Immunol 12:96–106

Lacombe J, Mange A, Bougnoux AC, Prassas I, Solassol J (2014) A multiparametric serum marker panel as a complementary test to mammography for the diagnosis of node-negative early-stage breast cancer and DCIS in young women. Cancer Epidemiol Biomarkers Prev 23:1834–1842

Lebret T, Watson RW, Molinie V, O'Neill A, Gabriel C, Fitzpatrick JM, Botto H (2003) Heat shock proteins HSP27, HSP60, HSP70, and HSP90: expression in bladder carcinoma. Cancer 98:970–977

Li XS, Xu Q, Fu XY, Luo WS (2014) Heat shock protein 60 overexpression is associated with the progression and prognosis in gastric cancer. PLoS One 9:e107507

Li G, Li M, Liang X, Xiao Z, Zhang P, Shao M, Peng F, Chen Y, Li Y, Chen Z (2017) Identifying DCN and HSPD1 as potential biomarkers in colon cancer using 2D-LC-MS/MS combined with iTRAQ technology. J Cancer 8:479–489

Ma T, Schreiber CA, Knutson GJ, Khattouti AE, Sakiyama MJ, Hassan M, Charlesworth MC, Madden BJ, Zhou X, Vuk-Pavlovic S, Gomez CR (2015) Effects of oxygen on the antigenic landscape of prostate cancer cells. BMC Res Notes 8:687

Macdonald IK, Parsy-Kowalska CB, Chapman CJ (2017) Autoantibodies: opportunities for early cancer detection. Trends Cancer 3:198–213

Mateos MV, Hernandez JM, Hernandez MT, Gutierrez NC, Palomera L, Fuertes M, Diaz-Mediavilla J, Lahuerta JJ, de la Rubia J, Terol MJ, Sureda A, Bargay J, Ribas P, de Arriba F, Alegre A, Oriol A, Carrera D, Garcia-Larana J, Garcia-Sanz R, Blade J, Prosper F, Mateo G, Esseltine DL, van de Velde H, San Miguel JF (2006) Bortezomib plus melphalan and prednisone in elderly untreated patients with multiple myeloma: results of a multicenter phase 1/2 study. Blood 108:2165–2172

Meinhardt A, Parvinen M, Bacher M, Aumuller G, Hakovirta H, Yagi A, Seitz J (1995) Expression of mitochondrial heat shock protein 60 in distinct cell types and defined stages of rat seminiferous epithelium. Biol Reprod 52:798–807

Meng Q, Li BX, Xiao X (2018) Toward developing chemical modulators of Hsp60 as potential therapeutics. Front Mol Biosci 5:35

Merendino AM, Bucchieri F, Campanella C, Marciano V, Ribbene A, David S, Zummo G, Burgio G, Corona DF, Conway de Macario E, Macario AJ, Cappello F (2010) Hsp60 is actively secreted by human tumor cells. PLoS One 5:e9247

Michael A, Ball G, Quatan N, Wushishi F, Russell N, Whelan J, Chakraborty P, Leader D, Whelan M, Pandha H (2005) Delayed disease progression after allogeneic cell vaccination in hormone-resistant prostate cancer and correlation with immunologic variables. Clin Cancer Res 11:4469–4478

Multhoff G, Botzler C, Wiesnet M, Muller E, Meier T, Wilmanns W, Issels RD (1995) A stress-inducible 72-kDa heat-shock protein (HSP72) is expressed on the surface of human tumor cells, but not on normal cells. Int J Cancer 61:272–279

Olin MR, Andersen BM, Zellmer DM, Grogan PT, Popescu FE, Xiong Z, Forster CL, Seiler C, SantaCruz KS, Chen W, Blazar BR, Ohlfest JR (2010) Superior efficacy of tumor cell vaccines grown in physiologic oxygen. Clin Cancer Res 16:4800–4808

Olin MR, Andersen BM, Litterman AJ, Grogan PT, Sarver AL, Robertson PT, Liang X, Chen W, Parney IF, Hunt MA, Blazar BR, Ohlfest JR (2011) Oxygen is a master regulator of the immunogenicity of primary human glioma cells. Cancer Res 71:6583–6589

Onizuka S, Tawara I, Shimizu J, Sakaguchi S, Fujita T, Nakayama E (1999) Tumor rejection by in vivo administration of anti-CD25 (interleukin-2 receptor alpha) monoclonal antibody. Cancer Res 59:3128–3133

Patel H, Nilendu P, Jahagirdar D, Pal JK, Sharma NK (2018) Modulating secreted components of tumor microenvironment: a masterstroke in tumor therapeutics. Cancer Biol Ther 19:3–12

Petrova V, Annicchiarico-Petruzzelli M, Melino G, Amelio I (2018) The hypoxic tumour microenvironment. Oncogenesis 7:10

Pockley AG, Henderson B (2018) Extracellular cell stress (heat shock) proteins-immune responses and disease: an overview. Philos Trans R Soc Lond Ser B Biol Sci 373

Quail DF, Joyce JA (2013) Microenvironmental regulation of tumor progression and metastasis. Nat Med 19:1423–1437

Quintana FJ, Cohen IR (2011) The HSP60 immune system network. Trends Immunol 32:89–95

Quintana FJ, Mimran A, Carmi P, Mor F, Cohen IR (2008) HSP60 as a target of anti-ergotypic regulatory T cells. PLoS One 3:e4026

Rappa F, Pitruzzella A, Marino Gammazza A, Barone R, Mocciaro E, Tomasello G, Carini F, Farina F, Zummo G, Conway de Macario E, Macario AJ, Cappello F (2016) Quantitative patterns of Hsps in tubular adenoma compared with normal and tumor tissues reveal the value of Hsp10 and Hsp60 in early diagnosis of large bowel cancer. Cell Stress Chaperones 21:927–933

Raz I, Ziegler AG, Linn T, Schernthaner G, Bonnici F, Distiller LA, Giordano C, Giorgino F, de Vries L, Mauricio D, Prochazka V, Wainstein J, Elias D, Avron A, Tamir M, Eren R, Peled D, Dagan S, Cohen IR, Pozzilli P, Group, D.-A.W (2014) Treatment of recent-onset type 1 diabetic patients with DiaPep277: results of a double-blind, placebo-controlled, randomized phase 3 trial. Diabetes Care 37:1392–1400

Ruan W, Wang Y, Ma Y, Xing X, Lin J, Cui J, Lai M (2010) HSP60, a protein downregulated by IGFBP7 in colorectal carcinoma. J Exp Clin Cancer Res 29:41

Samali A, Cai J, Zhivotovsky B, Jones DP, Orrenius S (1999) Presence of a pre-apoptotic complex of pro-caspase-3, Hsp60 and Hsp10 in the mitochondrial fraction of jurkat cells. EMBO J 18:2040–2048

Schiavoni G, Gabriele L, Mattei F (2013) The tumor microenvironment: a pitch for multiple players. Front Oncol 3:90

Schneider J, Jimenez E, Marenbach K, Romero H, Marx D, Meden H (1999) Immunohistochemical detection of HSP60-expression in human ovarian cancer. Correlation with survival in a series of 247 patients. Anticancer Res 19:2141–2146

Schreiber RD, Old LJ, Smyth MJ (2011) Cancer immunoediting: integrating immunity's roles in cancer suppression and promotion. Science (New York, NY) 331:1565–1570

Selimovic D, Porzig BB, El-Khattouti A, Badura HE, Ahmad M, Ghanjati F, Santourlidis S, Haikel Y, Hassan M (2013) Bortezomib/proteasome inhibitor triggers both apoptosis and autophagy-dependent pathways in melanoma cells. Cell Signal 25:308–318

Shimizu J, Yamazaki S, Sakaguchi S (1999) Induction of tumor immunity by removing CD25+CD4+ T cells: a common basis between tumor immunity and autoimmunity. J Immunol 163:5211–5218

Small EJ, Sacks N, Nemunaitis J, Urba WJ, Dula E, Centeno AS, Nelson WG, Ando D, Howard C, Borellini F, Nguyen M, Hege K, Simons JW (2007) Granulocyte macrophage colony-stimulating factor--secreting allogeneic cellular immunotherapy for hormone-refractory prostate cancer. Clin Cancer Res 13:3883–3891

Soltys BJ, Gupta RS (1996) Immunoelectron microscopic localization of the 60-kDa heat shock chaperonin protein (Hsp60) in mammalian cells. Exp Cell Res 222:16–27

Soltys BJ, Gupta RS (1997) Cell surface localization of the 60 kDa heat shock chaperonin protein (hsp60) in mammalian cells. Cell Biol Int 21:315–320

Strome SE, Voss S, Wilcox R, Wakefield TL, Tamada K, Flies D, Chapoval A, Lu J, Kasperbauer JL, Padley D, Vile R, Gastineau D, Wettstein P, Chen L (2002) Strategies for antigen loading of dendritic cells to enhance the antitumor immune response. Cancer Res 62:1884–1889

Sznurkowski JJ, Zawrocki A, Biernat W (2014) Subtypes of cytotoxic lymphocytes and natural killer cells infiltrating cancer nests correlate with prognosis in patients with vulvar squamous cell carcinoma. Cancer Immunol Immunother 63(3):297–303

Taiakina D, Pra AD, Bristow RG (2014) Intratumoral hypoxia as the genesis of genetic instability and clinical prognosis in prostate cancer. Adv Exp Med Biol 772:189–204

Thomas X, Campos L, Mounier C, Cornillon J, Flandrin P, Le QH, Piselli S, Guyotat D (2005) Expression of heat-shock proteins is associated with major adverse prognostic factors in acute myeloid leukemia. Leuk Res 29:1049–1058

Tredan O, Galmarini CM, Patel K, Tannock IF (2007) Drug resistance and the solid tumor microenvironment. J Natl Cancer Inst 99:1441–1454

Tretiakova I, Blaesius D, Maxia L, Wesselborg S, Schulze-Osthoff K, Cinatl J Jr, Michaelis M, Werz O (2008) Myrtucommulone from Myrtus communis induces apoptosis in cancer cells via the mitochondrial pathway involving caspase-9. Apoptosis : Int J Program Cell Death 13:119–131

Tsai YP, Yang MH, Huang CH, Chang SY, Chen PM, Liu CJ, Teng SC, Wu KJ (2009) Interaction between HSP60 and beta-catenin promotes metastasis. Carcinogenesis 30:1049–1057

Urushibara M, Kageyama Y, Akashi T, Otsuka Y, Takizawa T, Koike M, Kihara K (2007) HSP60 may predict good pathological response to neoadjuvant chemoradiotherapy in bladder cancer. Jpn J Clin Oncol 37:56–61

Valkenburg KC, de Groot AE, Pienta KJ (2018) Targeting the tumour stroma to improve cancer therapy. Nat Rev Clin Oncol 15:366–381

Vaupel P (2008) Hypoxia and aggressive tumor phenotype: implications for therapy and prognosis. Oncologist 13(Suppl 3):21–26

Vaupel P, Kelleher DK, Hockel M (2001) Oxygen status of malignant tumors: pathogenesis of hypoxia and significance for tumor therapy. Semin Oncol 28:29–35

Vinay DS, Ryan EP, Pawelec G, Talib WH, Stagg J, Elkord E, Lichtor T, Decker WK, Whelan RL, Kumara H, Signori E, Honoki K, Georgakilas AG, Amin A, Helferich WG, Boosani CS, Guha G, Ciriolo MR, Chen S, Mohammed SI, Azmi AS, Keith WN, Bilsland A, Bhakta D, Halicka D, Fujii H, Aquilano K, Ashraf SS, Nowsheen S, Yang X, Choi BK, Kwon BS (2015) Immune evasion in cancer: mechanistic basis and therapeutic strategies. Semin Cancer Biol 35(Suppl):S185–SS98

Wang X, Yu J, Sreekumar A, Varambally S, Shen R, Giacherio D, Mehra R, Montie JE, Pienta KJ, Sanda MG, Kantoff PW, Rubin MA, Wei JT, Ghosh D, Chinnaiyan AM (2005) Autoantibody signatures in prostate cancer. N Engl J Med 353:1224–1235

Westhoff MA, Kandenwein JA, Karl S, Vellanki SH, Braun V, Eramo A, Antoniadis G, Debatin KM, Fulda S (2009) The pyridinylfuranopyrimidine inhibitor, PI-103, chemosensitizes glioblastoma cells for apoptosis by inhibiting DNA repair. Oncogene 28:3586–3596

Wiechmann K, Muller H, Konig S, Wielsch N, Svatos A, Jauch J, Werz O (2017) Mitochondrial chaperonin HSP60 is the apoptosis-related target for myrtucommulone. Cell Chem Biol 24:614–23 e6

Wong CS, Wong VW, Chan CM, Ma BB, Hui EP, Wong MC, Lam MY, Au TC, Chan WH, Cheuk W, Chan AT (2008) Identification of 5-fluorouracil response proteins in colorectal carcinoma cell line SW480 by two-dimensional electrophoresis and MALDI-TOF mass spectrometry. Oncol Rep 20:89–98

Wouters A, Pauwels B, Lardon F, Vermorken JB (2007) Review: implications of in vitro research on the effect of radiotherapy and chemotherapy under hypoxic conditions. Oncologist 12:690–712

Wu CC, Chang YT, Chang KP, Liu YL, Liu HP, Lee IL, Yu JS, Chiang WF (2014) Salivary autoantibodies as noninvasive diagnostic markers of oral cavity squamous cell carcinoma. Cancer Epidemiol, Biomarkers Prev 23:1569–1578

Wu J, Li X, Song W, Fang Y, Yu L, Liu S, Churilov LP, Zhang F (2017a) The roles and applications of autoantibodies in progression, diagnosis, treatment and prognosis of human malignant tumours. Autoimmun Rev 16:1270–1281

Wu J, Liu T, Rios Z, Mei Q, Lin X, Cao S (2017b) Heat shock proteins and cancer. Trends Pharmacol Sci 38:226–256

Yang HZ, Cui B, Liu HZ, Mi S, Yan J, Yan HM, Hua F, Lin H, Cai WF, Xie WJ, Lv XX, Wang XX, Xin BM, Zhan QM, Hu ZW (2009) Blocking TLR2 activity attenuates pulmonary metastases of tumor. PLoS One 4:e6520

Zaenker P, Gray ES, Ziman MR (2016) Autoantibody production in Cancer – the humoral immune response toward autologous antigens in cancer patients. Autoimmun Rev 15:477–483

Zhang J, Zhou X, Chang H, Huang X, Guo X, Du X, Tian S, Wang L, Lyv Y, Yuan P, Xing J (2016) Hsp60 exerts a tumor suppressor function by inducing cell differentiation and inhibiting invasion in hepatocellular carcinoma. Oncotarget 7:68976–68989

Zheng RP, Wang W, Wei CD (2015) Bortezomib inhibits cell proliferation in prostate cancer. Exp Ther Med 10:1219–1223

Zhou C, Sun H, Zheng C, Gao J, Fu Q, Hu N, Shao X, Zhou Y, Xiong J, Nie K, Zhou H, Shen L, Fang H, Lyu J (2018) Oncogenic HSP60 regulates mitochondrial oxidative phosphorylation to support Erk1/2 activation during pancreatic cancer cell growth. Cell Death Dis 9:161

Chapter 10
Hsp60 Involvement During Carcinogenesis

Francesca Rappa, Francesco Carini, and Pierre Schembri Wismayer

Abstract The implication of Hsp60 in cancer development is due to its participation in many metabolic and biomolecular mechanisms in cancer cells. Hsp60 interacts with various molecules that are responsible of apoptosis, cell proliferation and other mechanisms involved when a normal cell becomes malignant. Hsp60 expression was found to be increased in many types of cancer but in same tumors of different anatomical district was found decreased. The mechanism of action of Hsp60 is different depending on the type of tumor. Its involvement in the carcinogenetic process of some tumors, such as large bowel carcinoma or cervical carcinoma, seems to occur in the very early stages of disease. Hsp60 participates in the mechanism of modulation of the immune response the cancer cells use to invade surrounding tissues, and expand the tumor mass.

Keywords Apoptosis · Biomolecular · Cancer cell · Carcinogenesis · Hsp60 · Tumor progression

Abbreviations

APCs	Antigen-presenting cells
Bax	Bcl-2-associated X protein
CD	Cluster of differentiation
c-myc	Cancer myelocytomatosis

F. Rappa (✉) · F. Carini
Department of Biomedicine, Neuroscience and Advanced Diagnostics (BIND), Section of Human Anatomy, University of Palermo, Palermo, Italy

P. Schembri Wismayer
Department of Anatomy, Centre for Molecular Medicine and Biobanking, Faculty of Medicine and Surgery, University of Malta, Msida, Malta

A. A. A. Asea, P. Kaur (eds.), *Heat Shock Protein 60 in Human Diseases and Disorders*, Heat Shock Proteins 18,
https://doi.org/10.1007/978-3-030-23154-5_10

COPD	Chronic obstructive pulmonary disease
DCs	Dendritic cells
G1	Grading 1
G2	Grading 2
G3	Grading 3
Hsp	Heat shock protein
IAP	Inhibitors of apoptosis protein
IFNγ	Interferon gamma
IKK	IκB kinase
IL	Interleukin
KA	Keratoacantomas
MMP9	Matrix metallopeptidase 9
mtHsp	Mithocondrial Heat shock protein
MyD88	Myeloid differentiation primary response 88
NCI-H292	Human lung mucoepidermoid cell
NF-kB	Nuclear factor kappa-light-chain-enhancer of activated B cells
NK	Natural killer
pC3	Pro-caspase-3
PIN	Prostatic intraepithelial lesions
ROS	Reactive oxygen species
SCC	Squamous cell carcinomas
SIL	Squamous intraepithelial lesion
Th1	Type 1 T helper
TLR	Toll-like receptor
VEGF	Vascular endothelial growth factor

10.1 Introduction

Hsp60 is involved in the carcinogenic process and in the progression of different types of human cancer. Its implication during carcinogenesis is due to its interactions with many metabolic and biomolecular mechanisms of cells that concluded with neoplastic transformation. The carcinogenic role of Hsp60 has been evaluated in many experiments that also examined the cellular and extracellular localization of this protein. Typically, in normal cells, Hsp60 is localized in and around the mitochondria, but, in cancer cells, it is localized in the cytosol, very close to, and in, the cell membrane (Campanella et al. 2008, 2012). When it accumulates inside the cell, Hsp60 can be actively released into the extracellular space and the general circulation and, thus elicit an immune response. Hsp60 also occurs in exosomes released by cancer cells through an active secretion mechanism not related to apoptosis or necrosis (Merendino et al. 2010). The identification of intracellular and extracellular Hsp60 localization is therefore important because its intracellular accumulation is related to many cancerous changes and its extracellular secretion is associated with the cell to cell cross- talk.

10.1.1 Hsp60 Involvement in Biomolecular Mechanism of Cancer Cells

Among the many functions of Hsp60, it is important to remember its potential involvement in mechanisms of programmed cell death, cellular proliferation, invasiveness, regulation of angiogenesis and immune system regulation. In some such mechanisms cited, Hsp60 is beneficial for the cancerous cell and, as such, pathogenic for the organism (Macario and Conway de Macario 2007). HSP60 favors the survival of certain types of tumors and in some cases, it may even be essential for tumor cell growth.

10.1.1.1 Hsp60 and Programmed Cell Death and Cell Proliferation

A number of studies showed that Hsp60 has dual roles in cell apoptosis, pro- and anti-apoptotic, depending on the cellular situation (Chandra et al. 2007). The opinions about the role of this chaperone in apoptosis differ and it is not yet clear when and how Hsp60 is pro-apoptotic (i.e., it acts as an anti-cancer factor) or the contrary, it works in favor of the tumor by interfering with apoptosis (Ghosh et al. 2008; Campanella et al. 2008; Caruso Bavisotto et al. 2017). Thus, Hsp60 can have opposing effects and may be defined as "a molecular Proteus" of tumor cell survival (Cappello and Zummo 2005). Some studies, using *in vitro* models, demonstrated that Hsp60 has a pro-apoptotic role. Hsp60 in combination with the co-chaperone Hsp10 may regulate caspase-3 activation by facilitating the maturation of cleaved caspase-3 (Samali et al. 1999; Xanthoudakis et al. 1999; Rappa et al. 2012). On the other hand several reports in literature, support Hsp60's survival-promoting role. In many experiments, the Hsp60 plays, in tumor cells, a cytoprotective role centered on stabilization of levels of Survivin, a member of Inhibitors of Apoptosis Protein family (IAP), that inhibits caspase and blocks cell death (Ghosh et al. 2008). Hsp60 also interact with p53 protein function through the formation of Hsp60-p53 complexes, which inhibited p53 pro-apoptotic function in tumor cells (Ghosh et al. 2008). Moreover, Hsp60 may form a complex with the pro-apoptotic protein Bax, that may prevent apoptosis (Lanneau et al. 2008). Recently, it has been shown that Hsp60 interacts with human Lon protease, a mitochondrial matrix protein which has emerged lately as a regulator of mitochondrial-contributed tumorigenesis. Lon has an anti-apoptotic function and contributes to cell survival by association with the Hsp60-mtHsp70 complex. Consequently, Hsp60 is essential to maintain apoptosis inhibition preserved by Lon overexpression (Kao et al. 2015). Other authors have shown that Hsp60 interacts with Pro-Caspase-3 (pC3) in the mucoepidermoid carcinoma cell line NCI-H292 with this association persisting after the induction of oxidative stress and maintaining an anti-apoptotic function (Campanella et al. 2008). Stressing cancer cells affects the Hsp60 levels (Gorska et al. 2013; Campanella et al. 2015) and it is known that this chaperone plays a crucial role in apoptosis.

10.1.1.2 Hsp60 and Invasiveness and Regulation of Angiogenesis

Cancer cells that overexpress Hsp60 may show an increased tendency to invade the peri-tumoral environment and metastasize. The interaction of Hsp60 and β-catenin increases β-catenin protein levels and predicts, in this manner, the metastasis occurring in head and neck squamous cell carcinoma (Tsai et al. 2009). B-catenin, in fact, promotes tumorigenesis and metastasis in turn. It also enhances the expression of molecules implicated in the cell cycle (c-myc, Cyclin D1), in matrix degradation (MMP9) and in vascular regeneration (VEGF) (Zhang et al. 2017). Hsp60 has been considered a molecular mediator of activation of α3β1 integrin (Barazi et al. 2002), which is associated with metastasis in breast carcinoma and with cell proliferation and angiogenesis by stimulating cell proliferation (Chandrasekaran et al. 2000).

10.1.1.3 Hsp60 and Immune System Regulation

The presence of Hsp60 in the plasma membrane and its transit to the extracellular space and into the peritumoral environment, allow it to elicit an immune response. Hsp60 may bind to the receptors present on the surface of inflammatory cells, such as macrophages and NK cells. It has been suggested that Hsp60 can activate cells of the innate immune system, determining the release of Th1 cytokines, as well as acting as a danger signal and promoting maturation of dendritic cells (DCs) (Flohé et al. 2003), important for the tissue immunity. Hsp60 can also stimulate B cells via TLR4-MyD88 signaling and thereby, promote their proliferation, the expression of costimulatory molecules, and the secretion of Th2 cytokines (Cohen-sfady et al. 2005; Pasare and Medzhitov 2005). Moreover, Hsp60 is also able to interact with TLR2 on T cells, in turn inhibiting the cytoskeletal rearrangement and the chemotaxis induced by the Stromal cell-derived factor-1α chemokine (Zanin-Zhorov et al. 2003; Zanin-Zhorov et al. 2005). Hsp60 is also a ligand of the CD14 receptor lipopolysaccharide-binding sites, stimulating an immune response. When Hsp60 binds the CD14 co-receptor and the TLR4 signaling receptor, it stimulates at the same time, the production of IL-12 by antigen-presenting cells (APCs) and of IFNγ by T cells, whose activation depends on the presence of professional APCs, such as DCs (Osterloh et al. 2007). Cytosolic Hsp60 interacts with kinases (such as IKK) that activate the NF-kB survival pathway, thus inducing the expression of two genes (Bfl1/A1 and MnSOD) implicated in the control of mitochondrial-derived ROS. This molecular and genetic mechanism, mediated by Hsp60, prevents the stress-induced cell death in vivo by promotin IKK/NF-kB activation and induces a cell survival effect (Chun et al. 2010). The peritumoral microenvironment contains innate immune cells (macrophages, dendritic cells, natural killer cells) that communicate directly or through the production of cytokines, with T and B lymphocytes to control cancer cell proliferation. The interaction with TLR-4 also participates in the activation of the NF-κB pathway, forming, in this way, a link between chronic inflammation and immune surveillance in cancer (Chow et al. 2012). The same cancer cells, activating TLR-4 by secreted Hsp60, constantly edit and modulate the host anti-tumor immune response and the host immune response modifies tumor immunogenicity and clonal selection. During this process, the balance between anti-tumor and tumor-promoting immunity may be skewed in favour of tumor proliferation and cancer cell vitality (Grivennikov et al. 2010). Hsp60 released by cancer cells, can induce secretion of cytokines and chemokines from inflammatory cells to regulate and induce the inhibition of the host-protective anti-tumor response (Ostrand-Rosenberg and Sinha 2009). Hsp60, localized in the cell membrane, also might be recognized by the immune system which, could potentially attack and eliminate cancer cells. However, whilst this anti-tumoral action on the one hand could be a natural anti-tumoral action, on the other hand, it could simultaneously result in the selection of more aggressive cancer clones that are able to escape the immune surveillance of the host (Cappello et al. 2011).

10.1.2 Hsp60 Level Expression in Cancer

Hsp60 is expressed, at higher levels, in a wide range of human tumor specimens, compared to the normal tissue of the same organ, and elevated levels of this molecular chaperone in tumor cells have been correlated with increased tumor growth and progression (Ciocca and Calderwood 2005; Ghosh et al. 2008; Castilla et al. 2010; Li et al. 2014). A gradual increase of human Hsp60 occurs along the steps in the 'adenoma-to-carcinoma sequence' of the large bowel (Cappello et al. 2003a, b). In a recent work, we evaluated, the comparative immunohistochemical expression levels of Hsp10, 60, 70 and 90 both in the epithelium and lamina propria in biopsies of normal human large bowel mucosa, tubular adenoma with moderate dysplasia, as an example of pre-neoplastic lesion, and invasive colorectal adenocarcinoma (Rappa et al. 2016). The results obtained have shown that only immunopositivity of Hsp10 and Hsp60 increased gradually throughout the all steps from normal mucosa through dysplastic mucosa (adenoma samples) until invasive adenocarcinoma. Hsp60 levels were higher in both the pre-neoplastic lesion and the neoplastic lesion as compared to normal mucosa "Fig. 10.1". The presence of elevated levels of Hsp 60, associated also with high levels of Hsp10, in adenoma indicates that these molecules appear to be involved in the very early steps of large bowel carcinogenesis (Rappa et al. 2016). If confirmed by molecular biology, this could help the pathologist make predictions of disease progression. Hsp60 upregulation in tumor cells of large bowel disease is also directly correlated with higher tumor grade and, thereby, with tumor progression and poor prognosis (Cappello et al. 2005a). Increased levels of Hsp60 have been detected, also, in the carcinogenic steps of other types of tumor. In the dysplasia to carcinoma, sequence of the exocervix, the levels of Hsp60, as measured by immunohistochemistry and Western blotting, were found increased from low-grade squamous intraepithelial lesions (SIL) through high-grade SIL to invasive carcinomas (Cappello et al. 2002) and were correlated with the tumor grade. It is suggested by some authors that the development of cervical cancers and of ovarian cancer has been associated with chronic Chlamydia trachomatis

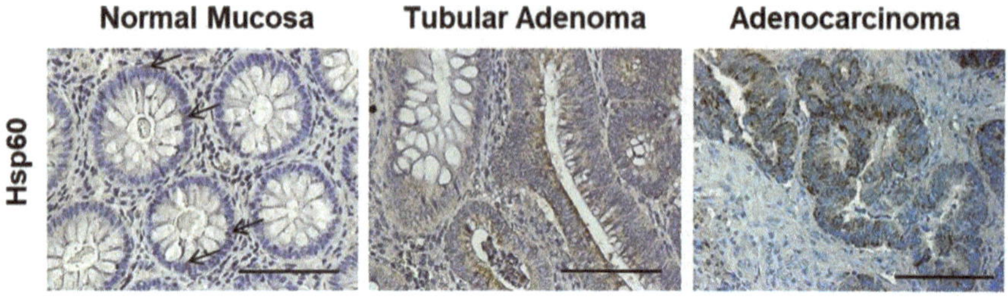

Fig. 10.1 Representative images of immunohistochemical results for Hsp60 in human large bowel biopsies of normal mucosa, tubular adenoma with moderate grade of dysplasia and adenocarcinoma with moderate grade of differentiation. Magnification 400×. Scale bar 100-μm. Arrows shown Hsp60 positivity in the epithelial cells of normal mucosa. Note the slight positivity for Hsp60 in normal epithelial cells. (Modified by Rappa et al. 2016)

infection (Di Felice et al. 2005). They hypothesize that the Hsp60 of Clamydia might induce an anti-apoptotic effect on cancer cells by inhibiting, for example, caspase-3 activation (Di Felice et al. 2005).

The expression levels of Hsp60 have also been studied in samples that highlight the natural course of progression from pre-cancerous lesions to invasive cancer in the human prostate gland. Indeed, increasing levels of Hsp60 and also of its co-chaperone Hsp10, have been observed in low-grade prostatic intraepithelial lesions (L-PIN), in high-grade prostatic intraepithelial lesions (H-PIN) and in prostatic carcinoma of an intermediate grade of differentation (Cappello et al. 2003c). Furthermore, it is shown that Hsp60 is also overexpressed in poorly differentiated prostate carcinomas and its levels are correlated with clinical indicators of a poor prognosis (Castilla et al. 2010). These data demonstrate that the overexpression of Hsp60 appears early on during prostatic tumorigenesis and is even correlated to cancer progression. In the dysplasia-carcinoma sequence of oral mucosa it has been observed that Hsp60 expression was higher in leukoplakia than normal mucosa and higher still in squamous cell carcinoma (Fan et al. 2006). The overexpression of Hsp60 is implicated in yet further types of cancer. In gastric carcinoma, high levels of Hsp60 has been associated with invasion, metastasis and cancer staging (Li et al. 2014) and even in ovarian cancer high-level expression of Hsp60 is indicative of poor prognosis, possibly by inducing drug resistance (Kimura et al. 1993; Schneider et al. 1999). Hsp60 immunopositivity has also been evaluated in a series of brain tumors where it was found to be significantly higher in neuroepithelial tumors compared with healthy control tissues, but not significantly higher or different in meningeal neoplasm compared with normal meninges (Rappa et al. 2013). On the other hand, in some types of tumor subnormal levels of Hsp60 were detected. For example HSP60 expression is inversely correlated with bladder tumor progression in accordance with Lebret (Lebret et al. 2003; Cappello et al. 2006a) and the loss of Hsp60 is related to the development and progression of bronchial cancer in smokers with COPD (Cappello et al. 2005b; 2006b). We have observed, indeed, a loss of expression levels of Hsp60, and of CD1a cells, in a series of squamous cell carcinomas (SCC) of the skin when compared to keratoacantomas (KA) (Cabibi et al. 2015). The reduction of Hsp60 expression was gradual from KA to SSC G1G2 (good and intermediate grade of differentiation) to SCC G3 (low grade of differentiation). Based on these features, one can hypothesize, in KA cases, a synergic action between Hsp60 and CD1a cells in activating the innate immune response. This synergism would produce an antitumor immune effect through activation of dendritic cells, which upon failing, reduces immunosurveillance and allows cancer progression (Coventry and Heinzel 2004; Corrao et al. 2008). These data could help to understand why some keratoacantomas progress into squamous carcinomas more than others.

10.2 Conclusions

The role of Hsp60 in human tumor development is very interesting and its functions, both inside and outside cancer cells, are multiple. The results obtained from the different studies on this molecule have been highlighted and tend to confirm its central position in the homeostasis of the cancer cell. Hsp60 could be a promising candidate for a prognostic biomarker and should help the pathologist to make useful predictions of cancer progression, as in, for example, large bowel carcinogenesis. Indeed, Hsp60 should be take into consideration in design of antitumoral strategies.

Acknowledgements F.C. and F.R. were partially supported by UniPA.

References

Barazi HO, Zhou L, Templeton NS, Krutzsch HC, Roberts DD (2002) Identification of heat shock protein 60 as molecular mediator of alpha 3 beta 1integrin activation. Cancer Res 62:1541–1548

Cabibi D, Conway de Macario E, Ingrao S et al (2015) CD1a-positive cells and Hsp60 (HSPD1) levels in keratoacantomas and squamous cell carcinoma. Cell Stress Chaperones 21:131–137

Campanella C, Bucchieri F, Ardizzone NM et al (2008) Upon oxidative stress, the antiapoptotic Hsp60/procaspase-3 complex persist in mucoepidermoid carcinoma cells. Eur J Histochem 52:221–228

Campanella C, Bucchieri F, Merendino AM et al (2012) The odyssey of Hsp60 from tumor cells to other destinations includes plasma membrane-associated stages and Golgi and exosomal proteintrafficking modalities. PLoS One 7:e42008

Campanella C, Rappa F, Sciumè C et al (2015) Heat shock protein 60 levels in tissue and circulating exosomes in human large bowel cancer before and after ablative surgery. Cancer 121:3230–3239

Cappello F, Zummo G (2005) HSP60 expression during carcinogenesis: a molecular "Proteus" of carcinogenesis? Cell Stress Chaperones 10:263–264

Cappello F, Bellafiore M, Palma A et al (2002) Expression of 60-kD heat shock protein increases during carcinogenesis in the uterine exocervix. Pathobiology 70:83–88

Cappello F, Bellafiore M, David S, Anzalone R, Zummo G (2003a) Ten kilodalton heat shock protein (HSP10) is overexpressed during carcinogenesis of large bowel and uterine exocervix. Cancer Lett 196:35–41
Cappello F, Bellafiore M, Palma A et al (2003b) 60kDa chaperonin (HSP60) is overexpressed during colorectal carcinogenesis. Eur J Histochem 47:105–110
Cappello F, Rappa F, David S, Anzalone R, Zummo G (2003c) Immunohistochemical evaluation of PCNA, p53, HSP60, HSP10 and MUC-2 presence and expression in prostate carcinogenesis. Anticancer Res 23:1325–1331
Cappello F, David S, Rappa F et al (2005a) The expression of HSP60 and HSP10 in large bowel carcinomas with lymph node metastase. BMC Cancer 5:139
Cappello F, Di Stefano A, D'Anna SE, Donner CF, Zummo G (2005b) Immunopositivity of heat shock protein 60 as a biomarker of bronchial carcinogenesis. Lancet Oncol 6:816
Cappello F, David S, Ardizzone N et al (2006a) Expression of heat-shock proteins Hsp10, Hsp27, Hsp60, Hsp70 and Hsp90 in urothelial carcinoma of urinary bladder. J Cancer Mol 2:7
Cappello F, Di Stefano A, David S et al (2006b) Hsp60 and Hsp10 down-regulation predicts bronchial epithelial carcinogenesis in smokers with chronic obstructive pulmonary disease. Cancer 107:2417–2424.3-77
Cappello F, David S, Peri G et al (2011) Hsp60: molecular anatomy and role in colorectal cancer diagnosis and treatment. Front Biosci S3:341–351
Caruso Bavisotto C, Nikolic D, Marino Gammazza A et al (2017) The dissociation of the Hsp60/pro-Caspase-3 complex by bis(pyridyl)oxadiazole copper complex (CubipyOXA) leads to cell death in NCI-H292 cancer cells. J Inorg Biochem 170:8–16
Castilla C, Congregado B, Conde JM et al (2010) Immunohistochemical expression of Hsp60 correlates with tumor progression and hormone resistance in prostate Cancer. Urology 76:1017.e1–1017.e6
Chandra D, Choy G, Tang DG (2007) Cytosolic accumulation of HSP60 during apoptosis with or without apparent mitochondrial release: evidence that its pro-apoptotic or pro-survival functions involve differential interactions with caspase-3. J Biol Chem 282:31289–31301
Chandrasekaran L, He CZ, Al-Barzai H, Krutzsch HC, Irulea-Arispe ML, Robets DD (2000) Cell contact-dependent activation of alpha 3 beta 1integrin modulates endothelial cell responses to thrombospondin-1. Mol Biol Cell 11:2885–2900
Chow MT, Möller A, Smyth MJ (2012) Inflammation and immune surveillance in cancer. Semin Cancer Biol 22:23–32
Chun JN, Choi B, Lee KW et al (2010) Cytosolic Hsp60 in involved in the NF-kB dependent survival of cancer cell via IKK regulation. PLoS One 5:e9422
Ciocca DR, Calderwood SK (2005) Heat shock proteins in cancer: diagnostic, prognostic, predictive, and treatment implications. Cell Stress Chaperones 10:86–103
Cohen-Sfady M, Nussbaum G, Pevsner-Fischer M et al (2005) Control of B-cell responses by toll-like receptors. Nature 438:364–368
Corrao S, La Rocca G, Anzalone R et al (2008) Role of CD1a and Hsp60 in the antitumoral response of oesophageal cancer. Oncol Rev 1:225–232
Coventry BJ, Heinzel S (2004) CD1a in human cancers: a new role for an old molecule. Trends Immunol 25:242–248
Di Felice V, David S, Cappello F, Farina F, Zummo G (2005) Is chlamydial heat shock protein 60 a risk factor for oncogenesis? Cell Mol Life Sci 62:4–9
Fan GK, Chen F, Gen Y (2006) Immunohistochemical analysis of P57(kip2), p53 and Hsp60 expression in premalignant and malignant oral tissue. Oral Oncol 42:147–153
Flohé SB, Brüggermann J, Lendemans S et al (2003) Human heat shock protein 60 induces maturation of dendritic cells versus a Th1-promoting phenotype. J Immunol 170:2340–2348
Ghosh JC, Dohi T, Kang BH, Altieri DC (2008) Hsp60 regulation of tumor cell apoptosis. J Biol Chem 283:5188–5194

Gorska M, Marino Gammazza A, Zmijewski MA et al (2013) Geldanamycin-induced osteosarcoma cell death is associated with hyperacetylation and loss of mitochondrial pool of heat shock protein 60 (hsp60). PLoS One 8:e71135
Grivennikov SI, Greten FR, Karin M (2010) Immunity, inflammation, and cancer. Cell 140:883–899
Kao TY, Chiu YC, Fang WC et al (2015) Mitochondrial Lon regulates apoptosis through the association with Hsp60-mtHsp70 complex. Cell Death Dis 6:e1642
Kimura E, Enns RE, Thiebaut F, Howell SB (1993) Regulation of HSP60 mRNA expression in a human ovarian carcinoma cell line. Cancer Chemother Pharmacol 32:279–285
Lanneau D, Brunet M, Frisan E, Solary E, Fontenay M, Garrido C (2008) Heat shock proteins: essential proteins for apoptosis regulation. J Cell Mol Med 12:743–761
Lebret T, Watson RW, Molinie V (2003) Heat shock proteins HSP27, HSP60, HSP70, and HSP90: expression in bladder carcinoma. Cancer 98:970–977
Li X-s, Xu Q, Fu X-y, Luo W-s (2014) Heat shock protein 60 overexpression is associated with the progression and prognosis in gastric Cancer. PLoS One 9:e107507
Macario AJL, Conway de Macario E (2007) Chaperonopathies by defect, excess, or mistake. Ann N Y Acad Sci 1113:178–191
Merendino AM, Bucchieri F, Campanella C et al (2010) Hsp60 is actively secreted by human tumor cells. PLoS One 5:e9247
Osterloh A, Kalinke U, Weiss S, Fleisher B, Breloer M (2007) Synergistic and differential modulation of immune responses byHSP60 and LPS. J Biol Chem 282:4669–4680
Ostrand-Rosenberg S, Sinha P (2009) Myeloid-derived suppressor cells: linking inflammation and cancer. J Immunol 182:4499–4506
Pasare C, Medzhitov R (2005) Control of B-cell responses by Toll-like receptors. Nature 438:364–368
Rappa F, Farina F, Zummo G et al (2012) HSP-molecular chaperones in cancer biogenesis and tumor therapy: an overview. Anticancer Res 32:5139–5150
Rappa F, Unti E, Baiamonte P, Cappello F, Scibetta N (2013) Different immunohistochemical levels of Hsp60 and Hsp70 in a subset of brain tumors and putative role of Hsp60 in neuroepithelial tumorigenesis. Eur J Histochem 57:e20
Rappa F, Pitruzzella A, Marino Gammazza A et al (2016) Quantitative patterns of Hsps in tubular adenoma compared with normal and tumoral tissue, reveal the value of Hsp10 and Hsp60 in early diagnosis of large bowel cancer. Cell Stress Chaperones 21:927–933
Samali A, Cai J, Zhivotovsky B, Jones DP, Orrenius S (1999) Presence of a pre-apoptotic complex of pro-caspase-3, Hsp60 and Hsp10 in the mitochondrial fraction of Jurkat cells. EMBO J 18:2040–2048
Schneider J, Jiménez E, Marenbach K, Romero H, Marx D, Meden H (1999) Immunohistochemical detection of HSP60 expression in human ovarian cancer. Correlation with survival in a series of 247 patients. Anticancer Res 19:2141–2146
Tsai YP, Yang MH, Huang CH et al (2009) Interaction between HSP60 and βcatenin promotes metastasis. Carcinogenesis 30:1049–1057
Xanthoudakis S, Roy S, Rasper D et al (1999) Hsp60 accelerates the maturation of pro-caspase-3 by upstream activator proteases during apoptosis. EMBO J 18:2049–2056
Zanin-Zhorov A, Nussbaum G, Franitza S, Cohen IR, Lider O (2003) T cells respond to heat shock protein 60 via TRL2: activation of adhesion and inhibition of chemokine receptors. FASEB J 17:1567–1569
Zanin-Zhorov A, Tal G, Shivtiel S et al (2005) Heat shock protein 60 activates cytokineassociated negative regulator suppressor of cytokine signaling 3 in T cells: effects on signaling, chemotaxis, and inflammation. J Immunol 175:276–285
Zhang D, Fei F, Li S et al (2017) The role of βcatenin in the initiation and metastasis of TA2 mice spontaneous breast cancer. J Cancer 8:2114–2123

Part III
HSP60 and Inflammatory Diseases and Disorders

Chapter 11
Anti-human Hsp60 Autoantibodies in Autoimmune and Inflammatory Rheumatic Diseases

Georgios Efthymiou, Lazaros I. Sakkas, and Dimitrios P. Bogdanos

Abstract Autoantibodies against human heat shock protein 60 (Hsp60) have been implicated in the pathogenesis of autoimmune and inflammatory rheumatic diseases (AIRDs), including rheumatoid arthritis (RA), systemic lupus erythematosus (SLE), systemic sclerosis, Sjogren's syndrome and various idiopathic vasculitides. Anti-human Hsp60 autoantibodies can be detected in the sera of patients with AIRDs in various frequencies and levels; therefore, their diagnostic, clinical and pathogenic importance remains enigmatic. Regarding their clinical significance, several studies have implicated them in AIRDs with vascular manifestations, including SLE and primary vasculitides. For other diseases, however, clinical associations have not been comprehensively investigated. Their pathogenetic role is also questionable, although several studies have suggested an apoptotic capacity of anti-Hsp60 autoantibodies on osteoblasts and endothelial cells, linking them with joint inflammation and bone erosion in RA and vascular inflammation in SLE and vasculitides, respectively. In addition, molecular mimicry based on amino acid similarities between human Hsp60 and bacterial Hsp60 or other bacterial antigens has been implicated in the production of cross-reactive autoantibodies in AIRDS, connecting infection with autoimmunity and autoimmune disease. In search of the connection between anti-human Hsp60 autoantibodies and AIRDS, the current chapter reviews research advances in the field and discusses prospective investigations.

Keywords Autoantibody · Autoimmunity · Hsp60 · Inflammation · Rheumatic diseases

G. Efthymiou · L. I. Sakkas · D. P. Bogdanos (✉)
Department of Rheumatology and Clinical Immunology, University General Hospital of Larissa, Faculty of Medicine, School of Health Sciences, University of Thessaly, Larissa, Greece
e-mail: bogdanos@med.uth.gr

A. A. A. Asea, P. Kaur (eds.), *Heat Shock Protein 60 in Human Diseases and Disorders*, Heat Shock Proteins 18,
https://doi.org/10.1007/978-3-030-23154-5_11

Abbreviations

ACPAs	Anti-citrullinated peptide antibodies
AECAs	Anti-endothelial cell antibodies
AIRDs	Autoimmune and inflammatory rheumatic diseases
autoAbs	Autoantibodies
BD	Behçet's disease
CCT	Chaperonin containing t-complex polypeptide 1
E. coli	*Escherichia coli*
EGPA	Eosinophilic granulomatosis with polyangiitis
GCA	Giant cell arteritis
GPA	Granulomatosis with polyangiitis
H. pylori	*Helicobacter pylori*
Hsp	Heat shock protein
IL	Interleukin
JIA	Juvenile idiopathic arthritis
MPO-ANCA	Myeloperoxidase-antineutrophil cytoplasmic antibodies
PR3-ANCA	Peroxidase 3- antineutrophil cytoplasmic antibodies
RA	Rheumatoid arthritis
SjS	Sjogren's syndrome
SLE	Systemic lupus erythematosus
SpA	Spondyloarthritis
SSc	Systemic sclerosis
TA	Takayasu's arteritis
TLR	Toll-like receptor
UCTD	Undifferentiated connective tissue disease

11.1 Introduction

Autoimmune and inflammatory rheumatic diseases (AIRDs), such as rheumatoid arthritis (RA), systemic lupus erythematosus (SLE), Sjogren's syndrome (SjS), systemic sclerosis (SSc), vasculitides and others are characterized by the breakdown of immunological self–tolerance, aberrant activation of immune cells and production of autoantibodies (autoAbs). AutoAbs may result from insufficient clearance of apoptotic or necrotic cells, modification of self-antigens that generates neoepitopes that are perceived as foreign by B cells or from cross-reactivity between foreign and self-antigens (Anders et al. 2005; Bogdanos et al. 2001; Ehser et al. 2013; Getts et al. 2013; Kivity et al. 2009; Polymeros et al. 2006; Saeki and Ishihara 2014; Suurmond and Diamond 2015). The cumulative evidence in support of infectious-triggered autoimmunity has led investigators to introduce the concept of the infectome and autoinfectome, which describes the totality of specific pathogens that are responsible for the development and maintenance of autoreactive immune responses in susceptible individuals (Bogdanos and Sakkas 2017; Bogdanos et al. 2013a, b, 2015). Among the autoAbs with a potential diagnostic, clinical and pathogenic importance are those against human heat shock protein 60 (Hsp60), particularly in atherosclerosis, but equally importantly in AIRDs (Alard et al. 2007, 2008; Mandal et al. 2005). In a similar vein, although unrelated to the topic of the present chapter, several studies have investigated the role of bacterial antibodies against other Hsp antigens as triggers of autoimmunity and organ-specific autoimmune diseases (Bogdanos et al. 2013a, 2015; Dubaniewicz 2010; Kaufmann et al. 1991; Res et al. 1991; Schultz and Arnold 1993; van der Zee et al. 1998).

Hsp60 is a phylogenetically and functionally conserved molecular chaperone that assists in the folding of nascent and denatured proteins (Horvath et al. 2002). The bacterial homologues are GroEL in *Escherichia coli*, Hsp60 in *Chlamydia pneumoniae* and Hsp65 in *Mycobacterium tuberculosis*, all sharing a high sequence similarity with the human Hsp60 (Gupta 1990; Jones et al. 1993). Bacterial Hsp60 is very immunogenic and acknowledged as the "common antigen" of gram-negative bacteria (van Eden et al. 2017). Hence, anti-bacterial Hsp60 antibodies are detected in many infectious diseases (Alard et al. 2007). As we are going to discuss later, infections are commonly found in patients with AIRDS, secondary to immune-suppressive treatment, which weakens immune system and makes the affected individuals prone to infections. Thus, anti-bacterial Hsp60 antibodies may be merely epiphenomena rather than potentially relevant to disease pathogenesis in AIRDs. On the other hand, bacterial anti-Hsp60 antibodies can be found in treatment-naive patients, raising concerns as to whether they bear a pathogenic potential (Kaufmann et al. 1991; Leung and Gershwin 1991; Res et al. 1991; Schultz and Arnold 1993).

In human, under physiological conditions Hsp60 has a housekeeping role and is located in the mitochondrial matrix (Alard et al. 2007). However, under stress conditions Hsp60 can be over-expressed and translocated on the cell membrane (Jamin et al. 2005; Pfister et al. 2005), or can be detected in serum (Davies et al. 2006) and in cell culture supernatant (Basu et al. 2000). In addition, elevated levels of Hsp60 are detected in sera of patients of inflammatory diseases, including Behçet's disease (BD) (Shaker et al. 2007). However, in some autoimmune diseases, such as immune thrombocytopenia, serum Hsp60 levels are decreased (Dolasik et al. 2015), but whether this is due to its decreased expression or its binding and concomitant precipitation by anti-Hsp antibodies remains elusive (Alard et al. 2011; Dolasik et al. 2015; Rai et al. 2015).

Cellular immune responses to human Hsp60 have been detected in almost all chronic, inflammatory diseases. There is evidence that cellular responses to heat shock proteins, including Hsp60, are associated with anti-inflammatory regulation and theses are extensively reported in well-informative reviews (van Eden et al. 2005, 2017).

Humoral immune responses to Hsp60 have been detected in autoimmune and non-autoimmune diseases, including multiple sclerosis (Chiba et al. 2006; Efthymiou et al. 2016), type 1 diabetes mellitus (Horvath et al. 2002), atherosclerosis (Kilic and Mandal 2012; Wick et al. 2004) and cystic fibrosis (de Graeff-Meeder et al. 1993). Antibodies against both bacterial (Efthymiou et al. 2016) and human

Hsp60 (Alard et al. 2007) can also be detected in the sera of healthy individuals. In this case, they are mainly considered as part of physiological, pathogen-induced immune responses (Zugel and Kaufmann 1999), rather than triggers of disease induction.

Several investigators have suggested that the production of anti-human Hsp60 antibodies may be triggered by bacterial Hsp60 via antibody cross-reaction (Fig. 11.1). Not all of them have been able to demonstrate the presence of cross-reactive humoral (and/or cellular) responses (Alard et al. 2008; Efthymiou et al. 2016; Richter et al. 1994). In addition, endogenous Hsp60 can also act as a target of autoAbs, possibly through mechanisms induced by changes in the structure of the protein, post-translational modifications or the formation of immunogenic complexes with other foreign or self-antigens. Several regions of the human Hsp60 molecule have been recognized as antigenic epitopes (Fig. 11.2), such as epitopes located within $Hsp60_{383-447}$ (Boog et al. 1992), three overlapping epitopes spanning $Hsp60_{394-460}$ (Horvath et al. 2002) and an epitope spanning $Hsp60_{286-315}$ corresponding to apical I helix of human Hsp60 (Elfaitouri et al. 2013).

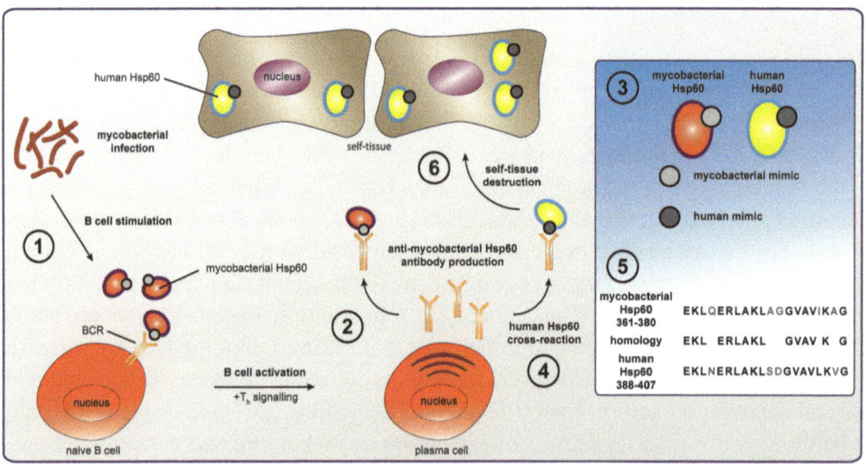

Fig. 11.1 Molecular mimicry and antibody cross-reaction as potential mechanism of infection-induced anti-human Hsp60 antibodies production. Bacterial infection (1) stimulates the production of antigen-specific anti-bacterial antigen antibodies, such as those against bacterial Hsp60 (2). Molecular mimicry (3) between a bacterial Hsp60 epitope and a human Hsp60 epitope leads to the induction of cross-reactive antibodies (4) that recognize both the bacterial and human mimics. To this end, Boog et al. and Horvath et al. have reported on a human Hsp60 highly sequence-homologous epitope who is recognized by autoimmune disease-specific antibodies that cross-react with mycobacterial Hsp60 as noted in the insert (5). Eventually, cross-reactive antibodies (and in a similar vein cross-reactive T-cell responses) mediate self-tissue inflammation and subsequent tissue destruction (6)

Score	Expect	Method	Identities	Positives	Gaps
469 bits(1206)	6e-165	Compositional matrix adjust.	248/528(47%)	364/528(68%)	3/528(0%)

```
Mycobacterial Hsp60  2    AKTIAYDEEARRGLERGLNALADAVKVTLGPKGRNVVLEKKWGAPTITNDGVSIAKEIEL  61
                          AK + +  +AR  + +G++ LADAV VT+GPKGR V++E+ WG+P +T DGV++AK I+L
        Human Hsp60  27   AKDVKFGADARALMLQGVDLLADAVAVTMGPKGRTVIIEQSWGSPKVTKDGVTVAKSIDL  86

Mycobacterial Hsp60  62   EDPYEKIGAELVKEVAKKTDDVAGDGTTTATVLAQALVREGLRNVAAGANPLGLKRGIEK  121
                          +D Y+ IGA+LV++VA  T++ AGDGTTTATVLA+++ +EG   ++ GANP+ ++RG+
        Human Hsp60  87   KDKYKNIGAKLVQDVANNTNEEAGDGTTTATVLARSIAKEGFEKISKGANPVEIRRGVML  146

Mycobacterial Hsp60  122  AVEKVTETLLKGAKEVETKEQIAATAAISA-GDQSIGDLIAEAMDKVGNEGVITVEESNT  180
                          AV+ V   L K +K V T E+IA  A ISA GD+ IG++I++AM KVG +GVITV++  T
        Human Hsp60  147  AVDAVIAELKKQSKPVTTPEEIAQVATISANGDKEIGNIISDAMKKVGRKGVITVKDGKT  206

Mycobacterial Hsp60  181  FGLQLELTEGMRFDKGYISGYFVTDPERQEAVLEDPYILLVSSKVSTVKDLLPLLEKVIG  240
                             +LE+ EGM+FD+GYIS YF+   + Q+   +D Y+LL   K+S+++ ++P LE
        Human Hsp60  207  LNDELEIIEGMKFDRGYISPYFINTSKGQKCEFQDAYVLLSEKKISSIQSIVPALEIANA  266

Mycobacterial Hsp60  241  AGKPLLIIAEDVEGEALSTLVVNKIRGTFKSVAVKAPGFGDRRKAMLQDMAILTGGQVIS  300
                            KPL+IIAEDV+GEALSTLV+N+++   + VAVKAPGFGD RK  L+DMAI TGG V
        Human Hsp60  267  HRKPLVIIAEDVDGEALSTLVLNRLKVGLQVVAVKAPGFGDNRKNQLKDMAIATGGAVFG  326

Mycobacterial Hsp60  301  EE-VGLTLENADLSLLGKARKVVVTKDETTIVEGAGDTDAIAGRVAQIRQEIENSDSDYD  359
                          EE + L LE+     LGK  +V+VTKD+  +++G GD   I  R+ +I ++++ + S+Y+
        Human Hsp60  327  EEGLTLNLEDVQPHDLGKVGEVIVTKDDAMLLKGKGDKAQIEKRIQEIIEQLDVTTSEYE  386

Mycobacterial Hsp60  360  REKLQERLAKLAGGVAVIKAGAATEVELKERKHRIEDAVRNAKAAVEEGIVAGGGVTLLQ  419
                          +EKL ERLAKL+ GVAV+K G  ++VE+ E+K R+ DA+   +AAVEEGIV GGG  LL+
        Human Hsp60  387  KEKLNERLAKLSDGVAVLKVGGTSDVEVNEKKDRVTDALNATRAAVEEGIVLGGGCALLR  446

Mycobacterial Hsp60  420  AAPTLDELKLEG-DEATGANIVKVALEAPLKQIAFNSGLEPGVVAEKVRNLPAGHGLNAQ  478
                            P LD L     D+  G  I+K  L+ P   IA N+G+E  ++ EK+     +  G +A
        Human Hsp60  447  CIPALDSLTPANEDQKIGIEIIKRTLKIPAMTIAKNAGVEGSLIVEKIMQSSSEVGYDAM  506

Mycobacterial Hsp60  479  TGVYEDLLAAGVADPVKVTRSALQNAASIAGLFLTTEAVVADKPEKEK  526
                           G + +++  G+ DP KV R+AL +AA +A L  T E VV + P++EK
        Human Hsp60  507  AGDFVNMVEKGIIDPTKVVRTALLDAAGVASLLTTAEVVVTEIPKEEK  554
```

Fig. 11.2 Alignment using BLASTp2 of human Hsp60 and mycobacterial Hsp60 shows a significant degree of amino acid similarity at various sequences. Double underlined: the sequence of the human $Hsp60_{383-419}$ epitope recognized by antibodies in patients with juvenile chronic arthritis (Boog et al. 1992), overlaps with the epitope recognized by antibodies in patients with type 1 diabetes mellitus (Horvath et al. 2002). Single underlined is the sequence of the human $Hsp60_{286-315}$ epitope recognized by IgG and IgM antibodies in patients with myalgic encephalomyelitis (Elfaitouri et al. 2013)

11.1.1 Rheumatoid Arthritis

Rheumatoid arthritis (RA) is an autoimmune disease characterized mainly by chronic destructive polyarthritis, but may also affect extra-articular organs/tissues (Choy and Panayi 2001; McInnes and Schett 2011). The presence of anti-citrullinated peptide antibodies (ACPAs) is a highly specific biomarker for this disease (Sakkas et al. 2014). ACPAs are directed against various citrullinated antigens, such as fibrinogen, vimentin, α-enolase, collagen type II, histone and others (Sakkas et al. 2014).

Anti-human Hsp60 antibodies are commonly found at low frequency in RA (Table 11.1). The prevalence of anti-human Hsp60 autoAbs was found to be 10% by immunoblotting and 20% by ELISA (Jamin et al. 2005). Jarjour et al. using immunoblotting found that IgM and IgG autoAbs were uncommon in RA and not different from healthy controls (Jarjour et al. 1991). Similarly, several other groups did not detect increased levels of anti-human Hsp60 autoAbs in RA (Dieude et al. 2004; Horvath et al. 2001; Lu et al. 2016). However, elevated titres of IgG autoAbs against two members of the Hsp60 chaperonin family were reported, namely Hsp60 itself

Table 11.1 Anti-human Hsp60 immune responses in autoimmune rheumatic diseases

Disease	References	Technique	Prevalence	HC prevalence	Other major findings
Rheumatoid arthritis	Jarjour et al. (1991)	WB IgG/IgM using cell culture whole extract	5% IgG 5% IgM	2.8% IgG 2.8% IgM	
	Jamin et al. (2005)	WB IgG using recombinant Hsp60;	10% by WB	ND	No difference in titres compared to HC[a]
		ELISA IgG using recombinant Hsp60	20% by ELISA		
	Hirata et al. (1997)	ELISA IgG using recombinant human Hsp60	ND	ND	Higher titres compared to HC
	Yokota et al. (2000)	ELISA IgG using human mitochondrial Hsp60	ND	ND	Higher titres compared to HC
	Horvath et al. (2001)	ELISA IgG using recombinant human Hsp60	ND	ND	No difference in titres compared to HC
	Dieudè et al. (2004)	ELISA IgG using Hsp60	ND	ND	No difference in titres compared to HC
	van Halm et al. (2006)	ELISA IgG using recombinant human Hsp60	ND	ND	Higher titres compared to HC
	Lu et al. (2016)	ELISA using recombinant Hsp60 and purified citHsp60	ND	ND	Higher titres of anti-Hsp60 and anti-citHsp60 compared to HC
Juvenile rheumatoid arthritis	Graeff-Meeder et al. (1993)	ELISA IgG using human Hsp60	100%	ND	Higher titres compared to HC
	Nguyen et al. (2006)	ELISA IgG using recombinant human Hsp60	23/23 by ELISA	ND	
		SDS- and native WB IgG using heat shock proteins	23/23 by SDS-PAGE 23/23 by native PAGE		
	Zlacka et al. (2006)	ELISA IgG, IgM, IgA using recombinant human Hsp60	ND	ND	Lower titres compared to HC
	Wu et al. (2011)	ELISA IgG, IgM, IgA using Hsp60	ND	ND	Higher titres compared to HC

(continued)

Table 11.1 (continued)

Disease	References	Technique	Prevalence	HC prevalence	Other major findings
Spondyloarthritis	Handley et al. (1996)	ELISA IgG using human Hsp60	ND	ND	Higher titres compared to HC
	Carlsen et al. (2013)	ELISA IgG1, IgG2, IgG3, IgG4 using full length, recombinant human Hsp60	ND	ND	Higher titres of IgG3 and IgG1 compared to HC
	Hjelholt et al. (2013)	ELISA IgG1, IgG2, IgG3, IgG4 using full length, recombinant human Hsp60	ND	ND	Higher titres of IgG3 and IgG1 compared to HC
Systematic lupus erythematosus	Jarjour et al. (1991)	WB IgG/IgM using cell culture whole extract	4.2% IgG	2.8% IgG	
			12.5% IgM	2.8% IgM	
	Jamin et al. (2005)	WB IgG using recombinant Hsp60;	76% by WB	ND	Higher titres compared to HC[a]
		ELISA IgG using recombinant Hsp60	39% by ELISA		
	Yokota et al. (2000)	ELISA IgG using human mitochondrial Hsp60	ND	ND	Higher titres compared to HC
	Horvath et al. (2001)	ELISA IgG using recombinant human Hsp60	ND	ND	No difference in titres compared to HC
	Dieudè et al. (2004)	ELISA IgG using Hsp60	ND	ND	No difference in titres compared to HC
Systemic sclerosis	Jarjour et al. (1991)	WB IgG/IgM using cell culture whole extract	6.3% IgG	2.8% IgG	
			12.5% IgM	2.8% IgM	
	Horvath et al. (2001)	ELISA IgG using recombinant human Hsp60	ND	ND	No difference in titres compared to HC
	Jamin et al. (2005)	WB IgG using recombinant Hsp60;	10% by WB	ND	No difference in titres compared to HC[a]
		ELISA IgG using recombinant Hsp60	10% by ELISA		

(continued)

Table 11.1 (continued)

Disease	References	Technique	Prevalence	HC prevalence	Other major findings
Sjogren's syndrome	Yokota et al. (2000)	ELISA IgG using human mitochondrial Hsp60	ND	ND	No difference in titres compared to HC
		ELISA IgG/IgM using human Hsp60	ND	ND	Lower titres compared to HC
	Jamin et al. (2005)	WB IgG using recombinant Hsp60;	20% by WB	ND	No difference in titres compared to HC[a]
		ELISA IgG using recombinant Hsp60	10% by ELISA		
Wegener's granulomatosis	Jamin et al. (2005)	WB IgG using recombinant Hsp60;	56% by WB	ND	Higher titres compared to HC[a]
		ELISA IgG using recombinant Hsp60	44% by ELISA		
Polyarteritis nodosa	Jamin et al. (2005)	WB IgG using recombinant Hsp60;	79% by WB	ND	Higher titres compared to HC[a]
		ELISA IgG using recombinant Hsp60	42% by ELISA		
Microscopic polyangiitis	Jamin et al. (2005)	WB IgG using recombinant Hsp60;	62% by WB	ND	Higher titres compared to HC[a]
		ELISA IgG using recombinant Hsp60	25% by ELISA		
Churg-Strauss syndrome	Jamin et al. (2005)	WB IgG using recombinant Hsp60;	20% by WB	ND	No difference in titres compared to HC[a]
		ELISA IgG using recombinant Hsp60	0% by ELISA		
MPO-ANCA vasculitis	Slot et al. (2006)	ELISA IgG using recombinant human Hsp60	ND	ND	Higher titres compared to PR3-ANCA vasculitis and HC
Takayasu's arteritis	Kumar Chauhan et al. (2004)	ELISA IgG/IgM/IgA using human Hsp60	84% IgG	22% IgG	Higher IgG titres compared to HC
			15% IgM	11% IgM	
			15% IgA	11% IgA	

(continued)

Table 11.1 (continued)

Disease	References	Technique	Prevalence	HC prevalence	Other major findings
Behçet's disease	Hirata et al. (1997)	ELISA IgG using recombinant human Hsp60	ND	ND	No difference in titres compared to HC[a]
	Jamin et al. (2005)	WB IgG using recombinant Hsp60;	44% by WB	ND	Higher titres compared to HC[a]
		ELISA IgG using recombinant Hsp60	22% by ELISA		
	Doino et al. (2017)	ELISA IgG using Hsp60	ND	ND	Higher titres in patients with moderate compared to patients with mild or severe BD symptoms[b]
Polymyositis	Jarjour et al. (1991)	WB IgG/IgM using cell culture whole extract	0% IgG	2.8% IgG	
			0% IgM	2.8% IgM	
	Horvath et al. (2001)	ELISA IgG using recombinant human Hsp60	ND	ND	Lower titres compared to HC, but not statistically significant

[a]Difference is not explicitly indicated, the estimation was extrapolated by figures
[b]In resting saliva. *citHsp60* citrullinated Hsp60, *HC* healthy controls, *ND* not defined, *WB* western blot

and chaperonin containing t-complex polypeptide 1 (CCT), in RA (Yokota et al. 2000). The elevated levels of anti-Hsp60 autoAbs were also reported in two other studies (Hirata et al. 1997; van Halm et al. 2006). Anti-human Hsp60 autoAbs might be of clinical significance for cardiovascular disease in RA (van Halm et al. 2006). Humoral immune responses against Hsp60 have been linked with cardiovascular disease (Zhu et al. 2001) and implicated in the development of atherosclerosis (Kilic and Mandal 2012; Wick et al. 2004). However, anti-Hsp60 autoAbs levels were comparable between RA patients with or without cardiovascular disease (van Halm et al. 2006). Therefore, anti-human Hsp60 autoAbs has not been considered as a marker for cardiovascular risk in RA.

The pathogenetic potential of anti-citrullinated Hsp60 antibodies was elegantly demonstrated by Lu et al. (2016). ACPAs isolated from RA patients can bind to citrullinated Hsp60 on the surface of human mature osteoblasts in vitro, a finding that had been previously reported by mass spectrometry (Goeb et al. 2009). In addition, ACPAs can mediate osteoblast apoptosis by binding to cell surface-expressed citrullinated Hsp60 through Toll-like receptor (TLR) 4 signaling and stimulate IL-6 and IL-8 gene expression (Lu et al. 2016). Both interleukins promote osteoclast proliferation and could contribute to bone erosion in RA (Pathak et al. 2015). Serum levels of anti-citrullinated Hsp60 autoAbs were elevated in RA and correlated with joint damage, but there was no difference in the levels of anti-Hsp60 autoAbs between RA patients and healthy controls (Lu et al. 2016). It would be interesting to determine whether citrullinated Hsp60 is localized on the cell membrane in arthritic joints in RA. The homologue mycobacterial Hsp65 has been detected in the synovial membrane of arthritic joints both in RA and animal models of arthritis (de Graeff-Meeder et al. 1990; Karlsson-Parra et al. 1990). However, anti-Hsp60 antibody levels were found to be lower in synovial fluids than in sera, which argues against local production of antibodies in arthritic joints (Hirata et al. 1997).

The mechanism of molecular mimicry between bacterial and endogenous Hsp60 has been explored in the study of Yokota et al. who demonstrated by inhibition studies that antibodies recognizing CCT, human Hsp60, *E. coli* GroEL and mycobacterial Hsp65 were cross-reactive (Yokota et al. 2000), indicating that these antibodies recognize common epitopes on all four proteins. The observed anti-Hsp60 and anti-CCT antibody titres correlated strongly in sera of patients with rheumatic autoimmune diseases, a correlation that was not affected by age, disease duration or treatment, suggesting a common mechanism of production. Immunoblotting experiments showed that the epitopes recognized by the anti-Hsp60 and anti-CCT autoAbs were conformational and not sequence-specific. Of relevance, when sera of RA patients were absorbed with *E. coli* GroEL the reactivity to human Hsp60 was lost, but when sera were absorbed with human Hsp60 reactivity to *E. coli* GroEL remained (Hirata et al. 1997). Furthermore, in RA the levels of anti-human Hsp60 autoAbs and anti-*E. coli* GroEL abs were elevated in sera but lower in synovial fluid. These results suggest that molecular mimicry is the mechanism of anti-human Hsp60 antibodies in RA, rather than a synovium-localized, RA-specific generation (Hirata et al. 1997). Similar cross-reactivity was detected between mycobacterial $Hsp70_{287-306}$ and human binding immunoglobulin protein $(BiP)_{336-355}$. These data strengthen the hypothesis that molecular mimicry and immunological cross-reactivity, involving bacterial and human Hsp, is a mechanism for anti-human Hsp60 antibody production in RA (Shoda et al. 2016).

11.1.2 Juvenile Idiopathic Arthritis

Juvenile idiopathic arthritis (JIA), the most common pediatric chronic rheumatic disease, represents a group of disorders characterized by chronic joint inflammation (Eisenstein and Berkun 2014). The etiology and pathogenesis of JIA are not yet fully elucidated, although a few genes have been implicated (Hinks et al. 2009, 2010a, b; Yanagimachi et al. 2011). Several studies have investigated humoral responses against human Hsp60 in JIA (Table 11.1). Two studies have reported a universal presence of anti-human Hsp60 autoAbs in the sera of JIA patients (de Graeff-Meeder et al. 1993; Nguyen et al. 2006). Serum levels of anti-human Hsp60 autoAbs were elevated in JIA patients, particularly with polyarticular onset (de Graeff-Meeder et al. 1993; Nguyen et al. 2006). These results were also reported in patients with active, polyarticular JIA (Wu et al. 2011). In contrast, one study reported decreased serum antibody levels in JIA patients compared to healthy controls (Zlacka et al. 2006). Interestingly, levels of anti-human Hsp60 antibodies were much higher in the synovial fluids than matched sera, suggesting local antibody production by plasma cells in arthritic joints (de Graeff-Meeder et al. 1993).

11.1.3 Spondyloarthritis

Spondyloarthritis (SpA) encompasses a group of immune-mediated inflammatory diseases, including ankylosing spondylitis, psoriatic arthritis, reactive arthritis, inflammatory bowel disease-associated arthritis and undifferentiated spondyloarthritis (Terenzi et al. 2018). These diseases share clinical manifestations, such as inflammation of the spine, peripheral arthritis, enthesitis, and anterior uveitis. An early study reported elevated serum levels of IgG anti-human Hsp60 autoAbs in reactive arthritis (Handley et al. 1996) (Table 11.1). Another study examined serum IgG subclass antibodies (IgG1, IgG2, IgG3, and IgG4) against human Hsp60 and Hsp60 from *Chlamydia trachomatis*, *Salmonella enteritidis* and *Campylobacter jejuni* (Hjelholt et al. 2013). IgG1 and IgG3 abs against human Hsp60, but not against bacterial Hsp60, were elevated in SpA and there was only a weak association between antibodies to human and bacterial Hsp60. These findings suggest that cross-reactivity between human and bacterial Hsp60 is unlikely. Another study supported the notion reporting that different IgG subclass IgG3 and IgG1 antibodies against human and bacterial Hsp60, respectively, of the same serum sample argues against cross-reactivity (Carlsen et al. 2013).

11.1.4 Systemic Lupus Erythematosus

Systemic lupus erythematosus (SLE) is a systemic autoimmune disease that can affect almost any organ system, and is characterized by a plethora of autoAbs mostly targeting nuclear antigens (Kaul et al. 2016).

In SLE patients, findings on the frequency of anti-human Hsp60 antibodies are conflicting with one study reporting increased frequency (Jamin et al. 2005), but another not (Jarjour et al. 1991) (Table 11.1). With regard to anti-human Hsp60 autoAbs serum levels, results again are conflicting, with few studies reporting elevated serum levels (Jamin et al. 2005; Yokota et al. 2000), while other do not (Dieude et al. 2004; Horvath et al. 2001; van Paassen et al. 2007).

A potential role for anti-human Hsp60 autoAbs in SLE has arisen for the vascular manifestations of the disease. SLE carries an increased cardiovascular risk, including premature or accelerated atherosclerosis (Zampieri et al. 2005). Anti-Hsp60 autoAbs have been linked to pathogenesis and prognosis of coronary heart disease, vascular disease and atherosclerosis (Kilic and Mandal 2012; Wick et al. 2004; Zhu et al. 2001), possibly through endothelial cytotoxic events (Mayr et al. 1999). For instance, anti-Hsp60 autoAbs bind to endothelial cells and induce apoptosis, an event that facilitates the generation of anti-phospholipid autoAbs and thrombosis (Dieude et al. 2004). This was likely the first study to identify autoAbs against Hsp60 as anti-endothelial cell autoAbs, a heterogeneous class of antibodies that have been detected under autoimmune conditions connected with vascular involvement (D'Cruz et al. 1991; Renaudineau et al. 2002; Salojin et al. 1996). Similarly, by using immunoblotting and immunoabsorption techniques, the complex group of anti-endothelial cell autoAbs were confirmed to contain Hsp60 autoAbs and that their binding to endothelial cells induces apoptosis (Jamin et al. 2005). The presence of anti-human Hsp60 autoAbs and their ability to promote vascular endothelial cell injury and impairment of microcirculation has also been implicated in the development of neuropsychiatric SLE (Kimura et al. 2008). However, the role of anti-human Hsp60 autoAbs in apoptosis induction was questioned, since anti-epithelial cell antibodies not directed against Hsp60 were detected to be responsible for the induction of epithelial cell apoptosis in SLE patients with glomerular vasculopathy (van Paassen et al. 2007).

11.1.5 Systemic Sclerosis

Systemic sclerosis (SSc) is a complex disease of the connective tissue with extensive fibrosis, microvasculopathy and production of a range of disease-related and other autoAbs (Sakkas and Bogdanos 2016). To date, no direct evidence has linked anti-human Hsp60 antibodies with SSc (Table 11.1). The frequency of IgG and IgM anti-human Hsp60 antibodies in SSc patients was "not demonstrable" (Jarjour et al. 1991), and the difference in their levels between patients and controls was not statistically significant (Horvath et al. 2001; Jamin et al. 2005). These data argue against the involvement of anti-Hsp60 autoAbs in SSc. However, elevated serum levels of anti-human-Hsp60 autoAbs were found in in patients with undifferentiated connective tissue disease (UCTD) (Horvath et al. 2001). UCTD is a systemic disorder with a mixture of clinical traits of autoimmune connective tissue diseases, such as SSc, SLE, RA or SjS (Mosca et al. 2014). Although UCTD often differentiates to one of these well-characterized diseases, most patients will maintain the diagnosis of UCTD based on their clinical and laboratory features. Anti-human Hsp60 antibodies in UCTD patients could be further investigated to determine whether they define patients who will later develop one of the connective tissue diseases or not.

Similarly, in a study of autoimmune connective tissue disease that included undifferentiated connective tissue disease, SSc and SLE, serum levels of anti-Hsp60 autoAbs were normal and not correlated with anti-*H. pylori* antibodies (Kalabay et al. 2002). Unfortunately, the authors did not analyze each disease group separately, so any potential disease-specific anti-Hsp60 autoAbs differences were masked.

Nevertheless, the normal levels of anti-human Hsp60 antibodies in autoimmune diseases supported the author's suggestion that these are natural autoAbs (Prohaszka et al. 2001), a class of antibodies protecting the host from both invading antigens and endogenous neo-antigens that are constantly produced (Lobo 2016).

11.1.6 Sjogren's Syndrome

Sjogren's syndrome (SjS) is an autoimmune disease characterized by inflammation of exocrine glands, mainly salivary and lacrimal glands (Tong et al. 2017). A plethora of autoAbs can be detected in SjS patients, with anti-Ro/SSA autoantibodies being the most prevalent ones. Yokota et al. were the first to examine immune responses against human Hsp60 and chaperonin containing t-complex polypeptide 1 (CCT), a member of the Hsp60 family, in SjS (Yokota et al. 2000) (Table 11.1). Serum levels of anti-CCT IgG autoAbs but not anti-Hsp60 autoAbs were elevated in SjS. Similarly, other studies reported no difference in prevalence or levels of anti-Hsp60 autoAbs between SjS patients and healthy controls (Jamin et al. 2005) or even lower levels of anti-Hsp60 IgG and IgM autoAbs (Shovman et al. 2005). Although the relevance of anti-Hsp60 antibodies to SjS pathogenesis is questionable, their lower levels could imply a protective role against the disease. Hence, larger prospective studies are necessary to define the association between particular infections and autoimmunity.

11.1.7 Vasculitis

Vasculitis is defined as an inflammation of blood vessel walls. Although vasculitis can be secondary to infection, here we refer to idiopathic vasculitis. Vasculitis leads to aneurysm and vessel rupture or to vessel wall thickening, stenosis and tissue ischemia and can be primary or secondary to an underlying autoimmune disease (Savage et al. 2000). Primary vasculitis can involve any type of blood vessel and can affect any organ. The clinical manifestations vary according to the size and distribution of the inflamed blood vessels.

Jamin et al. investigated on the role of anti-human Hsp60 antibodies in vasculitis including granulomatosis with polyangiitis (GPA, formerly Wegener's granulomatosis), eosinophilic granulomatosis with polyangiitis (EGPA, formerly Churg-Strauss syndrome), microscopic polyangiitis, polyarteritis nodosa and in vasculitis-associated autoimmune diseases (SLE) (Jamin et al. 2005) (Table 11.1). The frequency of anti-human Hsp60 autoAbs was higher in vasculitic (vasculitis and SLE) compared to non-vasculitic autoimmune rheumatic diseases (RA, SjS), particularly in anti-endothelial cell antibody (AECA)-positive patients. Furthermore, Hsp60 was an important target of AECAs in their apoptotic effect on endothelial cells and appeared to have the capacity to induce apoptosis of endothelial cells

(Jamin et al. 2005). In vitro experiments have demonstrated that this apoptotic capacity occurs in endothelial cells under stress, a condition that involves Hsp60 on the cell surface, requires the interaction of Hsp60 and Hsp70 and is propagated through the chemokine receptor CCR5 (Alard et al. 2009). Of interest, a specific interaction between Hsp60 and ATP synthase has been recognized in vitro on the surface of endothelial cells and has been shown to prevent the pathogenic effect of anti-ATP synthase autoAbs on endothelial cells (Alard et al. 2011), suggesting that the diverse surface interactions of Hsp60 can affect endothelial physiology in various ways. It is also intriguing to speculate that these interactions of surface Hsp60 could form immunogenic complexes that are recognized as foreign by the immune system and may be important in the pathogenesis of vasculitis (Anders et al. 2005; Suurmond and Diamond 2015). The presence of this type of antibodies could render an individual susceptible to endothelial inflammation in the context of a primary vasculitic entity or a secondary vasculitic component of another rheumatic disease (Alard et al. 2008).

The production of anti-human Hsp60 antibodies was reported to be elevated in patients with MPO-ANCA- compared to PR3-ANCA-associated vasculitis and healthy controls (Slot et al. 2006) (Table 11.1). However, inhibition assays did not support a role for infection-triggered pathogenesis of MPO-ANCA-associated vasculitis through molecular mimicry between MPO, human Hsp60 and mycobacterial Hsp65 (Slot et al. 2006).

A link between anti-human Hsp60 autoAbs and vasculitic inflammation has been reported in Takayasu's arteritis (TA). Both the frequency and titres of IgG, but not IgM or IgA, anti-human Hsp60 autoAbs were increased in TA (Kumar Chauhan et al. 2004) (Table 11.1). Of importance, humoral and cellular immune response against both human Hsp60 and mycobacterial Hsp65 were significantly correlated, suggesting cross-reactivity and molecular mimicry as the triggering mechanisms of autoimmunity in TA. Furthermore, anti-aortic endothelial cell antibodies comprise of anti-human Hsp60 autoAbs in the vast majority of TA patients and these autoAbs can induce the expression of adhesion molecules and pro-inflammatory cytokines, as well as apoptosis (Chauhan et al. 2006). In giant cell arteritis (GCA) a study reported elevated levels of anti-human Hsp60 autoAbs but they did not correlate with anti-*Chlamydia pneumoniae* Hsp60 antibodies (Lopez-Hoyos et al. 2008).

11.1.8 Behçet's Disease

Behçet's disease (BD) (also known as Adamantiades-Behcet's disease) is a systemic, inflammatory disorder characterized by recurrent inflammation that manifests as oral apthous ulcers, genital ulcers, uveitis and acne-like skin lesions (Zeidan et al. 2016). No study thus far has investigated on the prevalence of Hsp60 autoAbs in BD. There are conflicting reports regarding serum anti-Hsp60 autoAbs levels with one study reporting elevated levels and another study reporting normal levels (Hirata et al. 1997; Jamin et al. 2005). Higher levels of anti-Hsp60 autoAbs were detected in resting saliva of BD patients with moderate disease compared to patients with mild or severe disease and higher levels were associated with stomatitis for more than 2 weeks and with gingival inflammation (Doino et al. 2017).

11.1.9 Idiopathic Inflammatory Myopathies

Polymyositis is an idiopathic, inflammatory rheumatic disease of the muscles with typical symptoms including symmetrical, proximal muscle weakness, difficulty arising from a seated position, dysphagia and aspiration, arthralgia Raynaud phenomenon and fever (Dalakas 2015). Of the two studies that have measured immune responses against human Hsp60 in patients with polymyositis, none of them managed to detect differences either in prevalence or serum titres of Hsp60 autoAbs (Horvath et al. 2001; Jarjour et al. 1991).

11.1.10 Further Prospectives

A consensus on the role of anti-human Hsp60 antibodies in autoimmune rheumatic diseases has not emerged. The discrepancies that are observed between studies on Hsp60 autoantibodies could be attributed to experimental limitations. Most studies do not measure the prevalence of anti-human Hsp60 antibodies but rather their titres on patients' sera. In addition, almost all studies measure immune responses against human Hsp60 in numerous autoimmune rheumatic diseases, instead of focusing in a single one. This is not a de facto limitation, but it becomes one in cases where small cohorts of each disease were studied. Patient selection and numbers is a major consideration regarding interpretation of the results. Therefore, a more organized investigation of human Hsp60 humoral immune responses in larger patient groups is necessary.

11.2 Conclusions

In conclusion, anti-hsp60 autoantibodies are prevalent in various autoimmune rheumatic and other autoimmune diseases, as well as other chronic inflammatory diseases. Whether are of pathogenic significance or not remains to be seen. Detailed, well-designed studies regarding their prevalence, epitope specificity and IgG subclass distribution is missing and should be performed in large multi-centre studies.

Molecular mimicry and immunological cross-reactivity involving human Hsp60 and bacterial Hsp60 mimics has been implicated in the production of cross-reactive autoantibodies but solid data are still absent, and is premature to connect initiation of infectious triggered hsp60 autoreactivity infection with autoimmunity and autoimmune disease.

Acknowledgements G.E. is supported by a PhD scholarship funded by the Hellenic State Scholarship Foundation (IKY, Greece).

References

Alard JE, Dueymes M, Youinou P, Jamin C (2007) Modulation of endothelial cell damages by anti-Hsp60 autoantibodies in systemic autoimmune diseases. Autoimmun Rev 6:438–443

Alard JE, Dueymes M, Youinou P, Jamin C (2008) HSP60 and anti-HSP60 antibodies in vasculitis: they are two of a kind. Clin Rev Allergy Immunol 35:66–71

Alard JE, Dueymes M, Mageed RA, Saraux A, Youinou P, Jamin C (2009) Mitochondrial heat shock protein (HSP) 70 synergizes with HSP60 in transducing endothelial cell apoptosis induced by anti-HSP60 autoantibody. FASEB J 23:2772–2779

Alard JE, Hillion S, Guillevin L et al (2011) Autoantibodies to endothelial cell surface ATP synthase, the endogenous receptor for hsp60, might play a pathogenic role in vasculatides. PLoS One 6:e14654

Anders HJ, Zecher D, Pawar RD, Patole PS (2005) Molecular mechanisms of autoimmunity triggered by microbial infection. Arthritis Res Ther 7:215–224

Basu S, Binder RJ, Suto R, Anderson KM, Srivastava PK (2000) Necrotic but not apoptotic cell death releases heat shock proteins, which deliver a partial maturation signal to dendritic cells and activate the NF-kappa B pathway. Int Immunol 12:1539–1546

Bogdanos DP, Sakkas LI (2017) From microbiome to infectome in autoimmunity. Curr Opin Rheumatol 29:369–373

Bogdanos DP, Choudhuri K, Vergani D (2001) Molecular mimicry and autoimmune liver disease: virtuous intentions, malign consequences. Liver 21:225–232

Bogdanos DP, Smyk DS, Invernizzi P et al (2013a) Infectome: a platform to trace infectious triggers of autoimmunity. Autoimmun Rev 12:726–740

Bogdanos DP, Smyk DS, Invernizzi P et al (2013b) Tracing environmental markers of autoimmunity: introducing the infectome. Immunol Res 56:220–240

Bogdanos DP, Smyk DS, Rigopoulou EI, Sakkas LI, Shoenfeld Y (2015) Infectomics and autoinfectomics: a tool to study infectious-induced autoimmunity. Lupus 24:364–373

Boog CJ, de Graeff-Meeder ER, Lucassen MA et al (1992) Two monoclonal antibodies generated against human hsp60 show reactivity with synovial membranes of patients with juvenile chronic arthritis. J Exp Med 175:1805–1810

Carlsen TG, Hjelholt A, Jurik AG et al (2013) IgG subclass antibodies to human and bacterial HSP60 are not associated with disease activity and progression over time in axial spondyloarthritis. Arthritis Res Ther 15:R61

Chauhan SK, Tripathy NK, Nityanand S (2006) Antigenic targets and pathogenicity of anti-aortic endothelial cell antibodies in Takayasu arteritis. Arthritis Rheum 54:2326–2333

Chiba S, Yokota S, Yonekura K et al (2006) Autoantibodies against HSP70 family proteins were detected in the cerebrospinal fluid from patients with multiple sclerosis. J Neurol Sci 241:39–43

Choy EH, Panayi GS (2001) Cytokine pathways and joint inflammation in rheumatoid arthritis. N Engl J Med 344:907–916

D'Cruz DP, Houssiau FA, Ramirez G et al (1991) Antibodies to endothelial cells in systemic lupus erythematosus: a potential marker for nephritis and vasculitis. Clin Exp Immunol 85:254–261

Dalakas MC (2015) Inflammatory muscle diseases. N Engl J Med 373:393–394

Davies EL, Bacelar MM, Marshall MJ et al (2006) Heat shock proteins form part of a danger signal cascade in response to lipopolysaccharide and GroEL. Clin Exp Immunol 145:183–189

de Graeff-Meeder ER, Voorhorst M, van Eden W et al (1990) Antibodies to the mycobacterial 65-kd heat-shock protein are reactive with synovial tissue of adjuvant arthritic rats and patients with rheumatoid arthritis and osteoarthritis. Am J Pathol 137:1013–1017 de Graeff-Meeder ER, Rijkers GT, Voorhorst-Ogink MM et al (1993) Antibodies to human HSP60 in patients with juvenile chronic arthritis, diabetes mellitus, and cystic fibrosis. Pediatr Res 34:424–428

Dieude M, Senecal JL, Raymond Y (2004) Induction of endothelial cell apoptosis by heat-shock protein 60-reactive antibodies from anti-endothelial cell autoantibody-positive systemic lupus erythematosus patients. Arthritis Rheum 50:3221–3231

Doino M, Yokoyama M, Sasaki Y, Kondo K, Yasuda Y, Arakawa S (2017) Evaluation of the relationship between salivary concentration of anti-heat shock protein immunoglobulin and clinical manifestations of Behcet's disease. Scand J Rheumatol 46:381–387

Dolasik I, Birtas Atesoglu E, Tarkun P et al (2015) Decreased serum heat shock protein 60 levels in newly diagnosed immune thrombocytopenia patients. Platelets 26:220–223

Dubaniewicz A (2010) *Mycobacterium tuberculosis* heat shock proteins and autoimmunity in sarcoidosis. Autoimmun Rev 9:419–424

Efthymiou G, Dardiotis E, Liaskos C et al (2016) Anti-hsp60 antibody responses based on Helicobacter pylori in patients with multiple sclerosis: (ir)relevance to disease pathogenesis. J Neuroimmunol 298:19–23

Ehser J, Holdener M, Christen S et al (2013) Molecular mimicry rather than identity breaks T-cell tolerance in the CYP2D6 mouse model for human autoimmune hepatitis. J Autoimmun 42:39–49

Eisenstein EM, Berkun Y (2014) Diagnosis and classification of juvenile idiopathic arthritis. J Autoimmun 48–49:31–33

Elfaitouri A, Herrmann B, Bolin-Wiener A et al (2013) Epitopes of microbial and human heat shock protein 60 and their recognition in myalgic encephalomyelitis. PLoS One 8:e81155

Getts DR, Chastain EM, Terry RL, Miller SD (2013) Virus infection, antiviral immunity, and autoimmunity. Immunol Rev 255:197–209

Goeb V, Thomas-L'Otellier M, Daveau R et al (2009) Candidate autoantigens identified by mass spectrometry in early rheumatoid arthritis are chaperones and citrullinated glycolytic enzymes. Arthritis Res Ther 11:R38

Gupta RS (1990) Sequence and structural homology between a mouse T-complex protein TCP-1 and the 'chaperonin' family of bacterial (GroEL, 60–65 kDa heat shock antigen) and eukaryotic proteins. Biochem Int 20:833–841

Handley HH, Yu J, Yu DT, Singh B, Gupta RS, Vaughan JH (1996) Autoantibodies to human heat shock protein (hsp)60 may be induced by *Escherichia coli* groEL. Clin Exp Immunol 103:429–435

Hinks A, Ke X, Barton A et al (2009) Association of the IL2RA/CD25 gene with juvenile idiopathic arthritis. Arthritis Rheum 60:251–257

Hinks A, Eyre S, Ke X et al (2010a) Association of the AFF3 gene and IL2/IL21 gene region with juvenile idiopathic arthritis. Genes Immun 11:194–198

Hinks A, Martin P, Flynn E et al (2010b) Association of the CCR5 gene with juvenile idiopathic arthritis. Genes Immun 11:584–589

Hirata D, Hirai I, Iwamoto M et al (1997) Preferential binding with *Escherichia coli* hsp60 of antibodies prevalent in sera from patients with rheumatoid arthritis. Clin Immunol Immunopathol 82:141–148

Hjelholt A, Carlsen T, Deleuran B et al (2013) Increased levels of IgG antibodies against human HSP60 in patients with spondyloarthritis. PLoS One 8:e56210

Horvath L, Czirjak L, Fekete B et al (2001) Levels of antibodies against C1q and 60 kDa family of heat shock proteins in the sera of patients with various autoimmune diseases. Immunol Lett 75:103–109

Horvath L, Cervenak L, Oroszlan M et al (2002) Antibodies against different epitopes of heat-shock protein 60 in children with type 1 diabetes mellitus. Immunol Lett 80:155–162

Jamin C, Dugue C, Alard JE et al (2005) Induction of endothelial cell apoptosis by the binding of anti-endothelial cell antibodies to Hsp60 in vasculitis-associated systemic autoimmune diseases. Arthritis Rheum 52:4028–4038

Jarjour WN, Jeffries BD, Davis JS, Welch WJ, Mimura T, Winfield JB (1991) Autoantibodies to human stress proteins. A survey of various rheumatic and other inflammatory diseases. Arthritis Rheum 34:1133–1138

Jones DB, Coulson AF, Duff GW (1993) Sequence homologies between hsp60 and autoantigens. Immunol Today 14:115–118

Kalabay L, Fekete B, Czirjak L et al (2002) Helicobacter pylori infection in connective tissue disorders is associated with high levels of antibodies to mycobacterial hsp65 but not to human hsp60. Helicobacter 7:250–256

Karlsson-Parra A, Soderstrom K, Ferm M, Ivanyi J, Kiessling R, Klareskog L (1990) Presence of human 65 kD heat shock protein (hsp) in inflamed joints and subcutaneous nodules of RA patients. Scand J Immunol 31:283–288

Kaufmann SH, Schoel B, van Embden JD et al (1991) Heat-shock protein 60: implications for pathogenesis of and protection against bacterial infections. Immunol Rev 121:67–90

Kaul A, Gordon C, Crow MK et al (2016) Systemic lupus erythematosus. Nat Rev Dis Primers 2:16039

Kilic A, Mandal K (2012) Heat shock proteins: pathogenic role in atherosclerosis and potential therapeutic implications. Autoimmune Dis 2012:502813

Kimura A, Sakurai T, Tanaka Y et al (2008) Proteomic analysis of autoantibodies in neuropsychiatric systemic lupus erythematosus patient with white matter hyperintensities on brain MRI. Lupus 17:16–20

Kivity S, Agmon-Levin N, Blank M, Shoenfeld Y (2009) Infections and autoimmunity – friends or foes? Trends Immunol 30:409–414

Kumar Chauhan S, Kumar Tripathy N, Sinha N, Singh M, Nityanand S (2004) Cellular and humoral immune responses to mycobacterial heat shock protein-65 and its human homologue in Takayasu's arteritis. Clin Exp Immunol 138:547–553

Leung PS, Gershwin ME (1991) The immunobiology of heat shock proteins. J Investig Allergol Clin Immunol 1:23–30

Lobo PI (2016) Role of natural autoantibodies and natural IgM anti-leucocyte autoantibodies in health and disease. Front Immunol 7:198

Lopez-Hoyos M, Alvarez L, Ruiz Soto M, Blanco R, Jose Bartolome M, Martinez-Taboada VM (2008) Serum levels of antibodies to Chlamydia pneumoniae and human HSP60 in giant cell arteritis patients. Clin Exp Rheumatol 26:1107–1110

Lu MC, Yu CL, Yu HC, Huang HB, Koo M, Lai NS (2016) Anti-citrullinated protein antibodies promote apoptosis of mature human Saos-2 osteoblasts via cell-surface binding to citrullinated heat shock protein 60. Immunobiology 221:76–83

Mandal K, Foteinos G, Jahangiri M, Xu Q (2005) Role of antiheat shock protein 60 autoantibodies in atherosclerosis. Lupus 14:742–746

Mayr M, Metzler B, Kiechl S et al (1999) Endothelial cytotoxicity mediated by serum antibodies to heat shock proteins of Escherichia coli and Chlamydia pneumoniae: immune reactions to heat shock proteins as a possible link between infection and atherosclerosis. Circulation 99:1560–1566

McInnes IB, Schett G (2011) The pathogenesis of rheumatoid arthritis. N Engl J Med 365:2205–2219

Mosca M, Tani C, Vagnani S, Carli L, Bombardieri S (2014) The diagnosis and classification of undifferentiated connective tissue diseases. J Autoimmun 48–49:50–52

Nguyen TT, Zlacka D, Vavrincova P, Sedlacek P, Hromadnikova I (2006) Detection of antibodies against 60-, 65- and 70-kDa heat shock proteins in paediatric patients with various disorders using Western blotting and ELISA. Clin Chem Lab Med 44:442–449

Pathak JL, Bakker AD, Verschueren P et al (2015) CXCL8 and CCL20 enhance osteoclastogenesis via modulation of cytokine production by human primary osteoblasts. PLoS One 10:e0131041

Pfister G, Stroh CM, Perschinka H et al (2005) Detection of HSP60 on the membrane surface of stressed human endothelial cells by atomic force and confocal microscopy. J Cell Sci 118:1587–1594

Polymeros D, Bogdanos DP, Day R, Arioli D, Vergani D, Forbes A (2006) Does cross-reactivity between *mycobacterium avium* paratuberculosis and human intestinal antigens characterize Crohn's disease? Gastroenterology 131:85–96

Prohaszka Z, Duba J, Horvath L et al (2001) Comparative study on antibodies to human and bacterial 60 kDa heat shock proteins in a large cohort of patients with coronary heart disease and healthy subjects. Eur J Clin Investig 31:285–292

Rai R, Chauhan SK, Singh VV, Rai M, Rai G (2015) Heat shock protein 27 and its regulatory molecules express differentially in SLE patients with distinct autoantibody profiles. Immunol Lett 164:25–32

Renaudineau Y, Dugue C, Dueymes M, Youinou P (2002) Antiendothelial cell antibodies in systemic lupus erythematosus. Autoimmun Rev 1:365–372

Res P, Thole J, de Vries R (1991) Heat-shock proteins and autoimmunity in humans. Springer Semin Immunopathol 13:81–98

Richter W, Mertens T, Schoel B et al (1994) Sequence homology of the diabetes-associated autoantigen glutamate decarboxylase with coxsackie B4-2C protein and heat shock protein 60 mediates no molecular mimicry of autoantibodies. J Exp Med 180:721–726

Saeki Y, Ishihara K (2014) Infection-immunity liaison: pathogen-driven autoimmune-mimicry (PDAIM). Autoimmun Rev 13:1064–1069

Sakkas LI, Bogdanos DP (2016) Systemic sclerosis: new evidence re-enforces the role of B cells. Autoimmun Rev 15:155–161

Sakkas LI, Bogdanos DP, Katsiari C, Platsoucas CD (2014) Anti-citrullinated peptides as autoantigens in rheumatoid arthritis-relevance to treatment. Autoimmun Rev 13:1114–1120

Salojin KV, Le Tonqueze M, Nassovov EL et al (1996) Anti-endothelial cell antibodies in patients with various forms of vasculitis. Clin Exp Rheumatol 14:163–169

Savage CO, Harper L, Cockwell P, Adu D, Howie AJ (2000) ABC of arterial and vascular disease: vasculitis. BMJ 320:1325–1328

Schultz DR, Arnold PI (1993) Heat shock (stress) proteins and autoimmunity in rheumatic diseases. Semin Arthritis Rheum 22:357–374

Shaker O, Ay El-Deen MA, El Hadidi H, Grace BD, El Sherif H, Abdel Halim A (2007) The role of heat shock protein 60, vascular endothelial growth factor and antiphospholipid antibodies in Behcet disease. Br J Dermatol 156:32–37

Shoda H, Hanata N, Sumitomo S, Okamura T, Fujio K, Yamamoto K (2016) Immune responses to mycobacterial heat shock protein 70 accompany self-reactivity to human BiP in rheumatoid arthritis. Sci Rep 6:22486

Shovman O, Sherer Y, Gilbourd B et al (2005) Low levels of heat shock proteins-60 and -65 autoantibodies in Sjogren's syndrome. Isr Med Assoc J: IMAJ 7:778–780

Slot MC, Theunissen R, van Paassen P, Damoiseaux JG, Tervaert JW (2006) Evaluation of antibodies against human HSP60 in patients with MPO-ANCA associated glomerulonephritis: a cohort study. J Autoimmune Dis 3:4

Suurmond J, Diamond B (2015) Autoantibodies in systemic autoimmune diseases: specificity and pathogenicity. J Clin Invest 125:2194–2202

Terenzi R, Monti S, Tesei G, Carli L (2018) One year in review 2017: spondyloarthritis. Clin Exp Rheumatol 36:1–14

Tong L, Koh V, Thong BY (2017) Review of autoantigens in Sjogren's syndrome: an update. J Inflamm Res 10:97–105

van der Zee R, Anderton SM, Prakken AB, Liesbeth Paul AG, van Eden W (1998) T cell responses to conserved bacterial heat-shock-protein epitopes induce resistance in experimental autoimmunity. Semin Immunol 10:35–41

van Eden W, van der Zee R, Prakken B (2005) Heat-shock proteins induce T-cell regulation of chronic inflammation. Nat Rev Immunol 5:318–330

van Eden W, Jansen MAA, Ludwig I, van Kooten P, van der Zee R, Broere F (2017) The enigma of heat shock proteins in immune tolerance. Front Immunol 8:1599 van Halm VP, Slot MC, Nurmohamed MT, Tervaert JW, Dijkmans BA, Voskuyl AE (2006) Antibodies against human 60 kDa heat shock protein are not associated with cardiovascular disease in patients with rheumatoid arthritis. Ann Rheum Dis 65:590–594

van Paassen P, Duijvestijn A, Debrus-Palmans L, Damoiseaux J, Vroomen M, Tervaert JW (2007) Induction of endothelial cell apoptosis by IgG antibodies from SLE patients with nephropathy: a potential role for anti-endothelial cell antibodies. Ann N Y Acad Sci 1108:147–156

Wick G, Knoflach M, Xu Q (2004) Autoimmune and inflammatory mechanisms in atherosclerosis. Annu Rev Immunol 22:361–403

Wu CT, Ou LS, Yeh KW, Lee WI, Huang JL (2011) Serum heat shock protein 60 can predict remission of flare-up in juvenile idiopathic arthritis. Clin Rheumatol 30:959–965

Yanagimachi M, Miyamae T, Naruto T et al (2011) Association of HLA-A∗02:06 and HLA-DRB1∗04:05 with clinical subtypes of juvenile idiopathic arthritis. J Hum Genet 56:196–199

Yokota SI, Hirata D, Minota S et al (2000) Autoantibodies against chaperonin CCT in human sera with rheumatic autoimmune diseases: comparison with antibodies against other Hsp60 family proteins. Cell Stress Chaperones 5:337–346

Zampieri S, Iaccarino L, Ghirardello A et al (2005) Systemic lupus erythematosus, atherosclerosis, and autoantibodies. Ann N Y Acad Sci 1051:351–361

Zeidan MJ, Saadoun D, Garrido M, Klatzmann D, Six A, Cacoub P (2016) Behcet's disease physiopathology: a contemporary review. Auto Immun Highlights 7:4

Zhu J, Quyyumi AA, Rott D et al (2001) Antibodies to human heat-shock protein 60 are associated with the presence and severity of coronary artery disease: evidence for an autoimmune component of atherogenesis. Circulation 103:1071–1075

Zlacka D, Vavrincova P, Hien Nguyen TT, Hromadnikova I (2006) Frequency of anti-hsp60, −65 and −70 antibodies in sera of patients with juvenile idiopathic arthritis. J Autoimmun 27:81–88

Zugel U, Kaufmann SH (1999) Role of heat shock proteins in protection from and pathogenesis of infectious diseases. Clin Microbiol Rev 12:19–39

Chapter 12
Hsp60 in Inflammatory Disorders

Antonella Marino Gammazza, Giovanni Tomasello, Angelo Leone, and Abdo Jurjus

Abstract Heat shock proteins (HSP) including HSP60 are immunogenic proteins shared by particular microbial agents and mammals. HSP60 has been implicated in multiple inflammatory disorders and autoimmune diseases mostly through its interactions with the immune system. Such diseases include inflammatory bowel disease, chronic obstructive pulmonary disease, Hashimoto's thyroiditis, myasthenia gravis, multiple sclerosis and even atherosclerosis plaques among others. It is present in the cytosol, cell membrane, cell surface as well as in the extracellular space and in the circulation. As a super antigen, HSP60 has the dual role as an immunomodulator and as a biomarker, a node molecule in balance between health and disease. Deciphering the mechanisms of HSP60 interactions with the immune system could lead to the development of new therapeutic strategies.

Keywords Autoimmunity · Homeostasis · HSP60 · Immune system · Immunoregulation · Inflammatory disorders

A. Marino Gammazza (✉)
Department of Biomedicine, Neuroscience and Advanced Diagnostics, University of Palermo, Palermo, Italy

Euro-Mediterranean Institute of Science and Technology, Palermo, Italy

G. Tomasello · A. Leone
Department of Biomedicine, Neuroscience and Advanced Diagnostics, University of Palermo, Palermo, Italy

A. Jurjus
Department of Anatomy, Cell Biology and Physiological Sciences, American University of Beirut, Beirut, Lebanon

Department of Anatomy and Regenerative Biology, George Washington University, Washington, DC, USA

A. A. A. Asea, P. Kaur (eds.), *Heat Shock Protein 60 in Human Diseases and Disorders*, Heat Shock Proteins 18,
https://doi.org/10.1007/978-3-030-23154-5_12

Abbreviations

AChR	Muscle acetylcholine receptor
APC	Antigen-presenting cells
CD	Crohn's disease
COPD	Chronic obstructive pulmonary disease
CR	Chemokine receptor
CREB	cAMP response element-binding protein
CSF	Cerebral spinal fluid
ERK	Extracellular signal–regulated kinases
HSP	Heat shock proteins
HSP60	Heat shock protein 60
HT	Hashimoto's thyroiditis
IBD	Inflammatory bowel disease
IFNɣ	Interferon gamma
IL	Interleukin
LPS	Lipopolysaccharides
MG	Myasthenia gravis
MHC	Major histocompatibility complex
MS	Multiple sclerosis
NFkB	Nuclear factor kappa-light-chain-enhancer of activated B cells
PBMC	Peripheral blood mononuclear cells
PKB	Protein kinase B
TG	Thyroglobulin
TLR	Innate toll-like receptor
TPO	Thyroid peroxidase
Tregs	T regulatory cells
UC	Ulcerative colitis

12.1 Introduction

Molecular chaperones constitute a large group of molecules highly conserved during evolution. They play many roles during the cell cycle and organismal growth by being involved primarily with protein homeostasis (Rappa et al. 2012). Evolutionary data suggest that the primitive chaperones evolved from archaea to reach the complex forms we detected today in humans, probably as a consequence of the constant exposure of living creatures to environmental stressors. Throughout evolution, molecular chaperones were physiologically involved as part of the defensive system against aggressors (temperature elevation, tumors, infections, etc.). For instance, the regulation of their levels could play an important role in maintaining cell viability both under normal and stress conditions. In pathologic conditions, these proteins are recognized as potentially useful biomarkers and therapeutic targets (Rappa et al. 2012; Barone et al. 2016). Likewise, the immune system as the role to protect against aggressors, such as infectious agents. In this respect, it is probably that the chaperoning and the immune system interact to ensure organismal homeostasis both in normal and pathological conditions. A considerable amount of data, over the last decades, have shown that exploring the cytoprotective and immunoregulatory characteristics of chaperones/heat shock proteins (HSP) can open a new avenue for drug discovery and treatment of many important human pathologies (Campanella et al. 2016; Ghosh et al. 2010; Czarnecka et al. 2006).

In this chapter, we review the interaction between the immune system and the chaperoning system with special attention to heat shock proteins 60 (HSP60), focusing on some chronic inflammatory disorders such as IBD, COPD, HT, MG and MS. It is well established that, HSP60 has the capacity to act as a self-antigen, foreign antigen, a carrier of other functional molecules, and as a ligand for innate TLR (Quintana and Cohen 2011). Then, HSP60 has the dual role as an immune modulator and a biomarker, thus giving the possibility to modulate immunity for therapeutic purposes, and to monitor the immune response in health and disease.

12.2 Hsp60 a Multifaceted Molecule that Speaks with the Immune System in Many Voice

HSP60 is one of the most studied HSP, especially in its interaction with the immune system. It was first identified as a protein capable to stimulate human monocyte proinflammatory cytokine synthesis without inducing monocyte activation (Tsan and Gao 2009). This led to assign to the chaperonin proinflammatory properties acting via the same receptors as lipopolysaccharides (LPS) (Tsan and Gao 2009). HSP60 is classically described as an intracellular chaperone, typically a mitochondrial protein, assisting the folding of polypeptides into proteins and their transport inside the cell (Campanella et al. 2014). HSP60 functions as a highly connected chaperone with links to most cellular proteins (Borges and Ramos 2005) since it has been found in the cytosol, cell membrane and surface as well as in the extracellular space and in circulation (Marino Gammazza et al. 2016). The chaperonin was described as a dominant antigen recognized during infections (Van Eden et al. 2005) and has been studied in various immunologic mechanisms involved in tumors, transplantation, tissue regeneration, autoimmune and inflammatory diseases (Rappa et al. 2012; Coelho and Faria 2012; Pei et al. 2016). The immunological relevance of HSP60 was recognized also in physiological conditions by having the capacity to induce self-reactive B and T cell clones even in health status (Coelho and Faria 2012). Although HSP60 represents a fundamental molecule in the intracellular chaperone network, evidence is lacking for the immune-specific function of the chaperonin inside the cell. Besides, the participation of other molecular chaperones, such as HSP70 and HSP90, in the cell biology of antigen processing and presentation (Bendz et al. 2007; Kunisawa and Shastri 2006), and

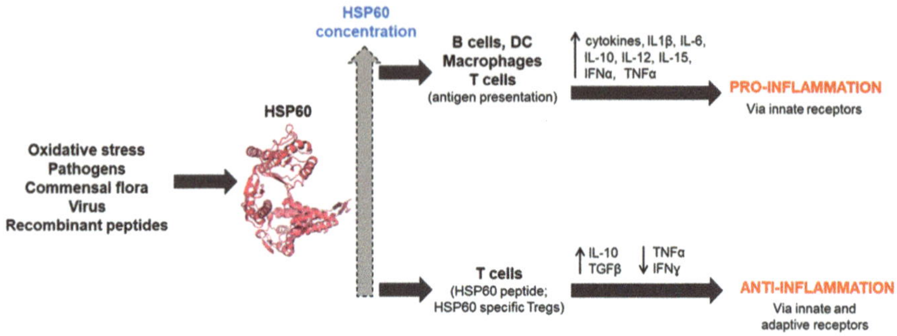

Fig. 12.1 HSP60 can interact with the innate ad the adaptive immune system. HSP60 can act as immunogen-antigen or as auto-antigen. HSP60 can also act as a sort of hormonal factor on cells very near its origin (paracrine effect) or far away (endocrine-like effect). Immune effects of HSP60 are mediated by innate TLR signaling and adaptive receptors. The local concentration of HSP60 is an important factor that determines the choice of pro-inflammatory or anti-inflammatory pathways. An increase in the HSP60 levels caused by stress factors, for instance, can lead to a pro-inflammatory effect via cytokine cascade or via B-cell activation, but it can also lead to an anti-inflammatory effect by increasing IL-10 and TGFβ and decreasing of TNFα by T-cells (Source: modified from Quintana and Cohen 2011). In hot pink a tridimensional model of an HSP60 monomer created using SWISS-MODEL (https://swissmodel.expasy.org/)

in T cell polarization, through their interactions with transcription factors have been published (Bohen et al. 1995; Pratt 1997). HSP90, for example, controls the activity of the ligand-activated transcription factor aryl hydrocarbon receptor (Tsuji et al. 2014), which has been proven recently to play an important role in the differentiation of FoxP3+ T regulatory cells (Treg), Th17 and Tr1 cells (Quintana et al. 2008; Gandhi et al. 2010). At the cell surface, HSP60 as the capacity to interact with TLR2, TLR4 and major histocompatibility complex (MHC) molecules triggering innate and adaptive immune responses (Coelho and Faria 2012) (Fig. 12.1). Increased amounts of HSP60 on the cell's surface was considered to serve as a danger signal for the immune system leading to the activation and maturation of dendritic cells and the generation of a antitumor T-cell response (Pockley et al. 2008; Quintana and Cohen 2011) (Fig. 12.1). Moreover, Hsp60 was found in the extracellular space (Davies et al. 2006) and may be exported outside cells through vesicles like exosomes (Campanella et al. 2014; Thery et al. 2009) or might be released intact or fragmented from damaged or dead cells (Quintana and Cohen 2011). In particular, the immunological activities of exosomes affect immunoregulation mechanisms including modulating antigen presentation, immune activation, immune suppression, immune surveillance, and intercellular communication (Greening et al. 2015; Cappello et al. 2006). The presence of HSP60 in circulation has been linked with various inflammatory conditions (Quintana and Cohen 2011; Henderson and Pockley 2010) even if the exact mechanisms by which HSP60 is secreted into the extracellular space are not well understood. However, it is clear that extracellular HSP60 is a link between body tissue and the immune system acting in paracrine and endocrine fashion (Henderson and Pockley 2010;

Campanella et al. 2012). Tsan and collaborators (Tsan and Gao 2009) suggested that the immunomodulatory role of the chaperonin on the innate immune system results from the presence of bacterial contaminants in preparations of recombinant mammalian HSP60 (Tsan and Gao 2009). However, it has been demonstrated that HSP60 on its own can trigger the activation of innate immune receptors (Henderson et al. 2010). In a review of Quintana and Cohen, they reported that TLR4 signaling, in macrophages and dendritic cells, is activated in response to four sources of HSP60: (1) bacterial HSP60, (2) bacterial or self-HSP60 molecules that bear LPS or other bacterial ligands (3) self-HSP60 molecules produced by infected, transformed, damaged or stressed cells, and (4) peptides of HSP60 (Quintana and Cohen 2011). Moreover, in macrophages the chaperonin when released extracellularly, can interact also with other cell-surface receptors, such as CD14, CD40, causing in turn either pro- or anti-inflammatory effects (reviewed in Quintana and Cohen 2011 and Henderson and Pockley 2010) (Fig. 12.1).

A number of studies reported that HSP60 can induce secretion of cytokines from professional antigen-presenting cells (APC), with consequent activation of T cells (Quintana et al. 2008; Osterloh et al. 2008) (Fig. 12.1). For example, it has been demonstrated that HSP60 can regulate T-cell behavior in inflammation via TLR2 and the down-regulation of chemokine receptor expression (CXCR4 and CCR7) (Zanin-Zhorov et al. 2005). Self-HSP60 can be recognized from T cells as specific antigens both in health and in autoimmune disease and the effect of the chaperonin on these cells is varied and unexpected (Nussbaum et al. 2006). The chaperonin was recognized as a co-stimulator of CD4+CD25+ Tregs via innate TLR2 signaling, with specific changes in Protein kinase B (PKB), Pyk2, p38, extracellular signal–regulated kinases (ERK) and T-bet signaling related to Tregs and nuclear factor kappa-light-chain-enhancer of activated B cells (NFkB) signaling related to T effector cells (Zanin-Zhorov et al. 2006).

Finally, HSP60 can activate, via TLR4 and MyD88 signaling, naive mouse B cells to proliferate, to secrete interleukin (IL)-10 and IL-6, and to upregulate the expression of MHCII and other activation molecules (Cohen-Sfady et al. 2005). Activated B cells present antigens to allogeneic T cells to enhance T cell secretion of both IL-10 and IFNɣ (Cohen-Sfady et al. 2005) (Fig. 12.1). Furthermore, HSP60 can trigger the production of polyclonal IgG3 antibodies (Cohen-Sfady et al. 2005), and have the capacity to play as a second signal to activate specific IgG3 antibodies to LPS (Cohen-Sfady et al. 2005; Quintana and Cohen 2011). Thus, extracellular HSP60 impacts B cell function in terms of cytokine expression, antigen presentation and antibody secretion. These interactions probably reflect the complementarity of two biological mechanisms, the chaperoning and the immune system, that have evolved to defend the cell and the organism, as mentioned earlier in this chapter and other publications. Probably, there is an evolutionary advantage to having an immune system that recognizes and responds to HSP60 in different ways from birth, since upregulation of HSP60 represents a sign of cell stress, the chaperonin might allow the body to respond and to restore tissue homeostasis (Cohen 2007).

12.3 HSP60 in Chronic Inflammatory and Autoimmune Diseases

As mentioned earlier in this chapter, HSP60 is a linking molecule in intercellular immune networks. It has the capacity to interact with both the innate and the adaptive immune systems in mammals (Quintana and Cohen 2011; Marino Gammazza et al. 2017). Many publications reported that some immunological properties of HSP60 arise from its high degree of structural similarity with the prokaryotic counterpart of the chaperonin, triggering the failure of the mechanism of self-non-self-discrimination leads to induction of autoimmunity (Marino Gammazza et al. 2014) and inflammation (Tomasello et al. 2011) and, consequently, to chronic inflammatory disorders (Cappello et al. 2011). Because of these physiological characteristics, HSP60 can be involved in the pathogenesis of a variety of human diseases. Here, we report our work together with the findings of other laboratories, to better illustrate the protein involvement in some chronic inflammatory and autoimmune disorders.

12.3.1 Inflammatory Bowel Disease (IBD)

IBD is a complex of continuum pathologies that result from the interaction of environmental and genetic factors leading to immunological responses and inflammation. Crohn's disease (CD) and ulcerative colitis (UC) are the most common types of IBD of the colon and small intestine. CD can also affect the mouth, oesophagus, stomach and the anus, whereas UC primarily affects the colon and the rectum (Baumgart and Carding 2007). IBD is considered a high-risk condition predisposing to cancer development and the chaperonin could be implicated in the pathogenesis of UC and CD by triggering and/or maintaining inflammation (Tomasello et al. 2011). However, the role of HSP60 is still controversial. In fact, comparative proteomics analysis showed colonocyte mitochondrial dysfunction due to low levels of HSP60 observed in biopsy specimens from both CD and UC (Peetermans et al. 1995; Rodolico et al. 2010). On the contrary, it has been reported that control and CD tissues showed similar quantitative patterns of the protein (Baca-Estrada et al. 1994). Interestingly, published data indicate a cause-effect relationship between bacterial infections and IBD due to the high conservation of the HSP60 sequence between humans and bacteria. For example, IBD-specific T cell epitopes were found in many regions of HSP60 and of Hsp65 sequences (Bene et al. 2002). An old work in 1992, reported increased levels of anti HSP60 IgA antibody in serum from CD and UC patients, after stimulation with the *Mycobacterium tuberculosis* homolog Hsp65, probably as a result of HSP60 release from damaged gut epithelium, or as a result of increased intestinal permeability that facilitates mucosal access of luminal antigens leading to the production of cross-reactive anti-bacterial HSP60 antibodies (Stevens et al. 1992). Moreover, administration of Yersinia enterocolitica HSP60 induced UC-like lesions and autoimmune responses in mice (Sukegawa et al. 2000; Yagita et al. 1999). Recently, it has been demonstrated that prozumab, a monoclonal antibody against the human chaperonin, developed from an antibody against HPS65, suppressed murine colitis by inducing IL-10 secretion from human peripheral blood mononuclear cells (PBMC) (Ulmansky et al. 2015). Moreover, low levels of antibodies against *Escherichia coli* and mycobacterial HSP65 were detected in patients with CD and in both active and inactive UC, whereas no difference was found in the levels of anti-HSP60 antibodies (Huszti et al. 2004; Bene et al. 2002). Other studies showed that pediatric CD was associated with an autoimmune response to HSP60-derived T-cell epitopes after stimulation of biopsy samples with an HSP60/65-derived peptide (Puga et al. 2009). These data suggested that an abnormal immune response to bacterial HSP65 can contribute to a dysregulation of host defences against certain component of the intestinal flora (Bene et al. 2002), and it is reasonable to hypothesized that the use of specific probiotics can counteract gut microbiota imbalance and HSP malfunction and deregulation in IBD (Bellavia et al. 2013).

12.3.2 Chronic Obstructive Pulmonary Disease (COPD)

COPD is a chronic inflammatory disease of the central and peripheral airways as well as the lung parenchyma. It is characterized by an increased number of inflammatory cells such as tissue lymphocytes, macrophages, and neutrophils (Di Stefano et al. 2009). Tobacco smoking is the most common cause of COPD, together with a number of other factors such as air pollution and genetics (Decamer et al. 2012). Long-term exposure to these irritants causes an inflammatory response in the lungs resulting in the narrowing of the small airways and breakdown of lung tissue (Gamble et al. 2007). However, the chaperonin can be involved in maintaining the inflammatory status, since in severe COPD there was a positive correlation between the number of neutrophils and elevated HSP60 levels (Cappello et al. 2011). Recently, it has been proved that in human bronchial epithelial cells, stimulation with HSP60 showed pro-inflammatory properties by the up-regulation of IL-8, IL-10, and CREB (Sangiorgi et al. 2017). Regarding the sequence homologies between bacterial and human HSP60, Chlamydia pneumoniae has been established as a common cause of acute exacerbations of COPD producing HSP60 as a critical proinflammatory factor (Rupp et al. 2004).

12.3.3 Hashimoto's Thyroiditis (HT)

HT is characterized by a prolonged autoimmune response against thyroid tissue that alters significantly the morphology of the gland (Ahmed et al. 2000) and causes primary hypothyroidism in humans (Vanderpump and Tunbridge 2002). The development and progression of the disease include increased levels of antibodies to thyroglobulin (TG) and thyroid peroxidase (TPO), two proteins localized within the thyroid gland cells (Lorini et al. 2003). Because of the interaction of the antibodies with TG and TPO inflammation develops, the gland is destroyed, and the patient develops hypothyroidism (Ahmed et al. 2000). A bioinformatics analysis, conducted in our laboratories, showed that there are regions in the HSP60 sequences with a high degree of similarity with portions of the TG and TPO molecules, supporting the idea that autoantibodies against TG and TPO are likely to recognize HSP60 exposed on the plasma membrane of oncocytes (Marino Gammazza et al. 2014). Moreover, peripheral blood mononuclear cells PBMC from HT patients after stimulation with recombinant HSP60 produce IL-2 and IFN-γ, suggesting that circulating HSP60 levels might be considered as good candidates forrr biomarkers in HT (Tonello et al. 2015).

12.3.4 Myasthenia Gravis (MG)

MG is a T cell-dependent, B cell-mediated autoimmune disease in which autoantibodies against the muscle acetylcholine receptor (AChR) attack the receptor at the neuromuscular junction (Astarloa and Martinez Castrillo 1996). The humoral immune response to HSP60 in MG is still in need of more scrutiny. Seroreactivity to HSP60 was detected in MG patients thus suggesting the involvement of the chaperonin in the development of the disease (Astarloa and Martinez Castrillo 1996). Moreover, a bioinformatics analysis conducted in our laboratory showed that HSP60 proteins from humans, Chlamydia trachomatis and Chlamydia pneumoniae, share sequence segments of high similarity with AChR subunit α1 (Marino Gammazza et al. 2012; Cappello et al. 2010) indicating that AChRa1 autoantibodies production could be elicited and/or maintained by self- and/or bacterial HSP60 (Marino Gammazza et al. 2012; Cappello et al. 2010).

12.3.5 Multiple Sclerosis (MS)

MS is a chronic inflammatory demyelinating disease of the central nervous system with unknown etiology and pathogenesis (Ruiz-Vazquez and de Castro 2003). A common structural motif ("2-6-11" motif) of the chaperonin is able to elicit the immune response of PBMC from MS patients by the release of pro-inflammatory cytokines consistent with a Th1-like pattern (Ruiz-Vazquez and de Castro 2003). Antigen arrays conducted on cerebral spinal fluid (CSF) and serum samples of patients with untreated relapsing-remitting MS showed the presence of different antibody signatures targeting epitopes of different proteins including HSP60 (PMID: Quintana et al. 2012). Moreover, there are experimental evidences suggesting that Helicobacter pylori is a trigger of MS and that the anti-HSP60 seropositivity correlated with age of disease onset (Efthymiou et al. 2016).

12.4 Conclusions

In physiological conditions, HSP60 has the capacity to function as a homeostatic molecule participating in the fine-tuning of inflammation allowing the body to respond and to restore tissue homeostasis. In pathological conditions, as in chronic inflammatory or autoimmune disease, HSP60 can function eliciting autoantibodies production or stimulating the immune cells to produce pro-inflammatory factors and thereby perpetuating inflammation. On this basis, the regulation of HSP60 levels or effects, for example through appropriate peptides or in combination with other modulatory agents, can be considered as new strategies for autoimmune and chronic inflammatory disease treatment. Since it is still unclear what determines the immune regulatory role or the pro-inflammatory activities of the chaperonin, further studies and scientific efforts are necessary to better elucidate the molecular pathways involved and their physiological significance. However, there are no doubts that HSP60 represent a node molecule in the balance between health and disease.

Acknowledgements Part of this work was funded by the Italian National Operational Programme (PON) «Imprese e Competitività» 2014-2020 FESR, grant awarded by the Italian Ministry of Economic Development to the project titled «Gestione di un servizio integrato multicentrico di diagnostica e terapia personalizzata in oncologia» (Project code: F/090012/01-02/X36).

References

Ahmed R, Al-Shaikh S, Akhtar M (2000) Hashimoto thyroiditis: a century later. Adv Anat Pathol 19:181–186

Astarloa R, Martinez Castrillo JC (1996) Humoral response to the human heat shock 60 kDa protein in myasthenia gravis. J Neurol Sci 135:182–183

Baca-Estrada ME, Gupta RS, Stead RH et al (1994) Intestinal expression and cellular immune responses to human heat-shock protein 60 in Crohn's disease. Dig Dis Sci 39:498–506

Barone R, Rappa F, Macaluso F et al (2016) Alcoholic liver disease: a mouse model reveals protection by *Lactobacillus fermentum*. Clin Transl Gastroenterol 7:e138

Baumgart DC, Carding SR (2007) Inflammatory bowel disease: cause and immunobiology. Lancet 369:1627–1640

Bellavia M, Tomasello G, Romeo M et al (2013) Gut microbiota imbalance and chaperoning system malfunction are central to ulcerative colitis pathogenesis and can be counteracted with specifically designed probiotics: a working hypothesis. Med Microbiol Immunol 202:393–406

Bendz H, Ruhland SC, Pandya MJ et al (2007) Human heat shock protein 70 enhances tumor antigen presentation through complex formation and intracellular antigen delivery without innate immune signaling. J Biol Chem 282:31688–31702

Bene L, Fust G, Huszti Z et al (2002) Impaired humoral immune response against mycobacterial 65-kDa heat shock protein (HSP65) in patients with inflammatory bowel disease. Dig Dis Sci 47:1432–1437

Bohen SP, Kralli A, Yamamoto KR (1995) Hold 'em and fold 'em: chaperones and signal transduction. Science 268:1303–1304

Borges JC, Ramos CH (2005) Protein folding assisted by chaperones. Protein Pept Lett 12:257–261

Campanella C, Bucchieri F, Merendino AM et al (2012) The odyssey of Hsp60 from tumor cells to other destinations includes plasma membrane-associated stages and Golgi and exosomal protein-trafficking modalities. PLoS One 7:e42008
Campanella C, Caruso Bavisotto C, Marino Gammazza A et al (2014) Exosomal heat shock proteins as a new players in tumor cell-to-cell communication. J Circ Biomark 3:10
Campanella C, D'Anneo A, Marino Gammazza A et al (2016) The histone deacetylase inhibitor SAHA induces HSP60 nitration and its extracellular release by exosomal vesicles in human lung-derived carcinoma cells. Oncotarget 7:28849–28867
Cappello F, Di Stefano A, David S et al (2006) Hsp60 and Hsp10 down-regulation predicts bronchial epithelial carcinogenesis in smokers with chronic obstructive pulmonary disease. Cancer 107:2417–2424
Cappello F, Marino Gammazza A, Zummo L et al (2010) Hsp60 and AChR cross-reactivity in myasthenia gravis: an update. J Neurol Sci 292:117–118
Cappello F, Caramori G, Campanella C et al (2011) Convergent sets of data from in vivo and in vitro methods point to an active role of Hsp60 in chronic obstructive pulmonary disease pathogenesis. PLoS One 6:e28200
Coelho V, Faria AM (2012) HSP60: issues and insights on its therapeutic use as an immunoregulatory agent. Front Immunol 12:97
Cohen IR (2007) Real and artificial immune systems: computing thestate of the body. Nat Rev Immunol 7:569–574
Cohen-Sfady M, Nussbaum G, Pevsner-Fischer M, Mor F, Carmi P, Zanin-Zhorov A, Lider O, Cohen IR (2005) Heat shock protein 60 activates B cells via the TLR4-MyD88 pathway. J Immunol 175:3594–3602
Czarnecka AM, Campanella C, Zummo G et al (2006) Mitochondrial chaperones in cancer: from molecular biology to clinical diagnostics. Cancer Biol Ther 5:714–720
Davies EL, Bacelar MM, Marshall MJ et al (2006) Heat shock proteins form part of a danger signal cascade in response to lipopolysaccharide and GroEL. Clin Exp Immunol 145:183–189
Decramer M, Janssens W, Miravitlles M (2012) Chronic obstructive pulmonary disease. Lancet 379:1341–1351
Di Stefano A, Caramori G, Gnemmi I et al (2009) Association of increased CCL5 and CXCL7 chemokine expression with neutrophil activation in severe stable COPD. Thorax 64:968–975
Efthymiou G, Dardiotis E, Liaskos C et al (2016) Anti-hsp60 antibody responses based on helicobacter pylori in patients with multiple sclerosis: (ir)relevance to disease pathogenesis. J Neuroimmunol 298:19–23
Gamble E, Grootendorst DC, Hattotuwa K et al (2007) Airway mucosal inflammation in COPD is similar in smokers and ex-smokers: a pooled analysis. Eur Respir J 30:467–471
Gandhi R, Kumar D, Burns EJ et al (2010) Activation of the aryl hydrocarbon receptor induces human type 1 regulatory T cell-like and Foxp3(+) regulatory T cells. Nat Immunol 11:846–853
Ghosh JC, Siegelin MD, Dohi T et al (2010) Heat shock protein 60 regulation of the mitochondrial permeability transition pore in tumor cells. Cancer Res 70:8988–8993
Greening DW, Gopal SK, Xu R et al (2015) Exosomes and their roles in immune regulation and cancer. Semin Cell Dev Biol 40:72–81
Henderson B, Pockley AG (2010) Molecular chaperones and protein-folding catalysts as intercellular signaling regulators in immunity and inflammation. J Leukoc Biol 88:445–462
Henderson B, Calderwood SK, Coates AR et al (2010) Caught with their PAMPs down? The extracellular signaling actions of molecular chaperones are not due to microbial contaminants. Cell Stress Chaperones 15:123–141
Huszti Z, Bene L, Kovacs A et al (2004) Low levels of antibodies against E. coli and mycobacterial 65kDa heat shock proteins in patients with inflammatory bowel disease. Inflamm Res 53:551–555
Kunisawa J, Shastri N (2006) Hsp90alpha chaperones large C-terminally extended proteolytic intermediates in the MHC class I antigen processing pathway. Immunity 24:523–534
Lorini R, Gastaldi R, Traggiai C et al (2003) Hashimoto's thyroiditis. Pediatr Endocrinol Rev 1:205–211

Marino Gammazza A, Bucchieri F, Grimaldi LM et al (2012) The molecular anatomy of human Hsp60 and its similarity with that of bacterial orthologs and acetylcholine receptor reveal a potential pathogenetic role of anti-chaperonin immunity in myasthenia gravis. Cell Mol Neurobiol 32:943–947

Marino Gammazza A, Rizzo M, Citarrella R et al (2014) Elevated blood Hsp60, its structural similarities and cross-reactivity with thyroid molecules, and its presence on the plasma membrane of oncocytes point to the chaperonin as an immunopathogenic factor in Hashimoto's thyroiditis. Cell Stress Chaperones 19:343–353

Marino Gammazza A, Caruso Bavisotto C, David S (2017) HSP60 is a ubiquitous player in the physiological and pathogenic interactions between the chaperoning and the immune systems. Curr Immunol Rev 13:44–55

Marino GA, Caruso Bavisotto C, Barone R, de Macario EC, Macario AJ (2016) Alzheimer's disease and molecular chaperones: current knowledge and the future of chaperonotherapy. CurrPharm Des 22:4040–4049

Nussbaum G, Zanin-Zhorov A, Quintana F, Lider O, Cohen IR (2006 Oct) Peptide p277 of HSP60 signals T cells: inhibition of inflammatory chemotaxis. Int Immunol 18(10):1413–1419

Osterloh A, Veit A, Gessner A, Fleischer B, Breloer M (2008) Hsp60-mediated T cell stimulation is independent of TLR4 and IL-12. Int Immunol 20:433–443

Peetermans WE, D'Haens GR, Ceuppens JL et al (1995) Mucosal expression by B7-positive cells of the 60-kilodalton heat-shock protein in inflammatory bowel disease. Gastroenterology 108:75–82

Pei W, Tanaka K, Huang SC et al (2016) Extracellular HSP60 triggers tissue regeneration and wound healing by regulating inflammation and cell proliferation. NPJ Regen Med 1:16013

Pockley AG, Muthana M, Calderwood SK (2008) The dual immunoregulatory roles of stress proteins. Trends Biochem Sci 33:71–79

Pratt WB (1997) The role of the hsp90-based chaperone system in signal transduction by nuclear receptors and receptors signaling via MAP kinase. Annu Rev Pharmacol Toxicol 37:297–326

Puga Yung GL, Fidler M, Albani E et al (2009) Heat shock protein-derived T-cell epitopes contribute to autoimmune inflammation in pediatric Crohn's disease. PLoS One 4:e7714

Quintana FJ, Cohen IR (2011) The HSP60 immune system network. Trends Immunol 32:89–95

Quintana FJ, Basso AS, Iglesias AH et al (2008) Control of T(reg) and T(H)17 cell differentiation by the aryl hydrocarbon receptor. Nature 23:23

Quintana FJ, Farez MF, Izquierdo G, Lucas M, Cohen IR, Weiner HL (2012) Antigen microarrays identify CNS-produced autoantibodies in RRMS. Neurology 78:532–539

Rappa F, Farina F, Zummo G et al (2012) HSP-molecular chaperones in cancer biogenesis and tumor therapy: an overview. Anticancer Res 32:5139–5150

Rodolico V, Tomasello G, Zerilli M et al (2010) Hsp60 and Hsp10 increase in colon mucosa of Crohn's disease and ulcerative colitis. Cell Stress Chaperones 15:877–884

Ruiz-Vazquez E, de Castro P (2003) "2-6-11" motif in heat shock protein 60 and central nervous system antigens: a preliminary study in multiple sclerosis patients. J Physiol Biochem 59:1–9

Rupp J, Droemann D, Goldmann T et al (2004) Alveolar epithelial cells type II are major target cells for *C. pneumoniae* in chronic but not in acute respiratory infection. FEMS Immunol Med Microbiol 41:197–203

Sangiorgi C, Vallese D, Gnemmi I et al (2017) HSP60 activity on human bronchial epithelial cells. Int J Immunopathol Pharmacol 30:333–340

Stevens TR, Winrow VR, Blake DR et al (1992) Circulating antibodies to heat-shock protein 60 in Crohn's disease and ulcerative colitis. Clin Exp Immunol 90:271–274

Sukegawa Y, Kamiya S, Yagita A et al (2000) Induction of autoimmune colitis by Yersinia enterocolitica 60-kilodalton heat-shock protein. Scand J Gastroenterol 35:1188–1193

Thery C, Ostrowski M, Segura E (2009) Membrane vesicles as conveyors of immune responses. Nat Rev Immunol 9:581–593

Tomasello G, Rodolico V, Zerilli M et al (2011) Changes in immunohistochemical levels and subcellular localization after therapy and correlation and colocalization with CD68 suggest a pathogenetic role of Hsp60 in ulcerative colitis. Appl Immunohistochem Mol Morphol 19:552–561

Tonello L, Conway de Macario E, Marino Gammazza A et al (2015) Data mining-based statistical analysis of biological data uncovers hidden significance: clustering Hashimoto's thyroiditis patients based on the response of their PBMC with IL-2 and IFN-gamma secretion to stimulation with Hsp60. Cell Stress Chaperones 20:391–395

Tsan MF, Gao B (2009) Heat shock proteins and immune system. J Leukoc Biol 85:905–910

Tsuji N, Fukuda K, Nagata Y, Okada H, Haga A, Hatakeyama S, Yoshida S, Okamoto T, Hosaka M, Sekine K, Ohtaka K, Yamamoto S, Otaka M, Grave E, Itoh H (2014) The activation mechanism of the aryl hydrocarbon receptor (AhR) by molecular chaperone HSP90. FEBS Open Bio 4:796–803

Ulmansky R, Landstein D, Moallem E et al (2015) A humanized monoclonal antibody against heat shock protein 60 suppresses murine arthritis and colitis and skews the cytokine balance toward an anti-inflammatory response. J Immunol 194:5103–5109

Van Eden W, van der Zee R, Prakken B (2005) Heat-shock proteins induce T-cell regulation of chronic inflammation. Nat Rev Immunol 5:318–330

Vanderpump MP, Tunbridge WM (2002) Epidemiology and prevention of clinical and subclinical hypothyroidism. Thyroid 12:839–347

Yagita A, Sukegawa Y, Maruyama S et al (1999) Mouse colitis induced by *Escherichia coli* producing Yersinia enterocolitica 60-kilodalton heat-shock protein: light and electron microscope study. Dig Dis Sci 44:445–451

Zanin-Zhorov A, Tal G, Shivtiel S, Cohen M, Lapidot T, Nussbaum G, Margalit R, Cohen IR, Lider O (2005) Heat shock protein 60 activates cytokine-associated negative regulator suppressor of cytokine signaling 3 in T cells: effects on signaling, chemotaxis, and inflammation. J Immunol 175:276–285

Zanin-Zhorov A, Cahalon L, Tal G, Margalit R, Lider O (2006 Jul) Cohen IR. Heat shock protein 60 enhances CD4+ CD25+ regulatory T cell function via innate TLR2 signaling. J Clin Invest 116(7):2022–2032. Epub 2006 Jun 8. Retraction in: J Clin Invest. 2018 Jun 1;128(6):2651

Part IV
HSP60 and Cardiovascular Diseases and Disorders

Chapter 13
Hsp60 in Atherosclerosis: Past, Present and Future

Bojana Jakic, Georg Wick, and Giuseppe Cappellano

Abstract Atherosclerosis is a multifactorial inflammatory disease of the arteries that manifests itself with calcified plaque formation within endothelial cells and the smooth muscle cell layer of vessels. T cells that recognize endogenous Hsp60 on endothelial cells initiate the disease. Here, we first describe the initial experiments that led to the discovery of Hsp60 as an autoantigenic driver of atherosclerosis. Then, we address numerous epidemiological and experimental studies performed by our lab and others that have firmly established Hsp60 as an autoantigen. In addition, we describe the pathogenic mechanisms mediated by Hsp60 and list known inducers of ectopic Hsp60 expression. Finally, we discuss the potential of Hsp60-based vaccination against atherosclerosis.

Keywords Atherosclerosis · Autoimmunity · Hsp60 · Immunoregulation · T cell · Vaccination

B. Jakic (✉)
Laboratory of Autoimmunity, Division of Experimental Pathophysiology and Immunology, Biocenter, Medical University of Innsbruck, Innsbruck, Austria

Division of Translational Cell Genetics, Department for Pharmacology and Genetics, Medical University of Innsbruck, Innsbruck, Austria
e-mail: bojana.jakic@i-med.ac.at

G. Wick
Laboratory of Autoimmunity, Division of Experimental Pathophysiology and Immunology, Biocenter, Medical University of Innsbruck, Innsbruck, Austria

G. Cappellano
Laboratory of Autoimmunity, Division of Experimental Pathophysiology and Immunology, Biocenter, Medical University of Innsbruck, Innsbruck, Austria

Interdisciplinary Research Center of Autoimmune Diseases (IRCAD), Department of Health Sciences, University of Piemonte Orientale (UPO), Novara, Italy

A. A. A. Asea, P. Kaur (eds.), *Heat Shock Protein 60 in Human Diseases and Disorders*, Heat Shock Proteins 18,
https://doi.org/10.1007/978-3-030-23154-5_13

Abbreviations

AA	Adjuvant arthritis
ApoE	Apolipoprotein E
ARFY	Atherosclerosis risk factor in female youngsters
ARMY	Atherosclerosis risk factor in male youngsters
CAD	Coronary artery disease
CFA	Complete Freund's adjuvant
cHsp60	Chlamydia pneumoniae-derived Hsp60
CMV	Cytomegalovirus
CRP	C-reactive protein
DC	Dendritic cell(s)
EC	Endothelial cell(s)
ELISA	Enzyme-linked immunosorbent assay
eNOS	Endothelial nitric oxide synthase 3
ERK	Extracellular signal-regulated kinase
HCAEC	Human coronary artery endothelial cell(s)
hHsp60	Human Hsp60
HNE	4-hydroxy-2-nonenal
Hsp	Heat shock protein
Hsp60	Heat shock protein 60
HUVEC	Human umbilical vein endothelial cell(s)
i.p.	Intraperitoneal
i.v.	Intravenous
ICAM-1	Intercellular adhesion molecule 1
IFA	Incomplete Freund's adjuvant
IFN-γ	Interferon γ
IL	Interleukin
LDL	Low-density lipoprotein
LOX-1	Lectin-like oxidized LDL receptor
LPS	Lipopolysaccharide
MALT	Muscosa-associated lymphoid tissue
MAPK	Mitogen-activated protein kinase
mbHsp60	Mycobacterial Hsp60
MDSCs	Myeloid-derived suppressor cells
MHC	Major histocompatibility complex
MLN	Mesenteric lymph node
NZW	New Zealand white rabbits
OVA	Ovalbumin
oxLDL	Oxidized LDL
RA	Rheumatoid arthritis
s.c.	Subcutaneous
sHsp60	Soluble/circulating Hsp60
TGF-β	Transforming growth factor β

Th	T-helper cells
TLR	Toll-like receptor
TNF-α	Tumor necrosis factor α
Treg	Regulatory T cell
VALT	Vascular-associated lymphoid tissue
VCAM-1	Vascular cell adhesion molecule 1
VSMC	Vascular smooth muscle cell(s)

13.1 Introduction

Atherosclerosis, the worldwide killer number one, is a chronic cardiovascular disease characterized by a thickening and hardening (sclerosis) of the arterial wall associated with loss of elasticity due to infiltration of the intima, the innermost layer of the arterial wall, with smooth muscle cells, and increased deposition of extracellular matrix proteins. As outlined below, the earliest stage of atherosclerosis is an inflammatory lesion in the intima dominated by an accumulations of lymphoid cells at certain predilection sites, namely arterial branching points, followed by so-called fatty streaks harboring abundant lipid-laden macrophages (foam cells) that further proceed to severe, complicated, and irreversible late manifestations called atherosclerotic plaques (Ross 1999). In this chapter, we summarize published research findings from our laboratory and others on the role of heat shock protein 60 (Hsp60) in the initiation and progression of atherosclerosis.

Various heat shock proteins (HSP) play important roles during the course of different diseases in animals and humans, both as primary pathogenic factors and as secondary phenomena. This also holds true for autoimmune diseases in general (Van Eden et al. 2007) and cardiovascular diseases in particular. Thus, serum levels of Hsp10 and Hsp60 are increased in type I diabetes and Hsp10 is overexpressed in B-cell lymphomas; however, no specific correlation of Hsp10 expression with the development of cardiovascular diseases has been found so far. Hsp27, another small HSP, triggers anti-apoptotic mechanisms and seems to exert a protective role in atherogenesis. Hsp40 associates with Hsp70 and promotes cardiovascular diseases. Hsp70 is an important chaperone the expression of which increases in pro- and eukaryotic cells when subjected to any kind of stress. We did not, however, find evidence for a primary atherosclerosis-promoting pathophysiological role of Hsp70. Hsp90 also acts as a potent chaperone, but again has no proven atherogenic or – protective properties. It is, however, abundantly expressed in late atherosclerotic lesions, predominantly in the shoulder region of instable plaques. In ApoE$^{-/-}$ mice, Hsp90 has been shown to curtail the development of atherosclerosis (summarized in Table 1, Wick et al. 2014). In conclusion, none of these HSP plays a primary atherogenic role. Our own work during the last decades has focused on the possible atherogenic potential of Hsp60, and the interpretation of data from our own and other labs culminated in the formulation of the Autoimmune Concept of Atherosclerosis (Wick et al. 1992, 1995, 2004). In the present chapter, we will briefly summarize the major steps in the past in this area of research, address recent work that corroborates this concept and give an outline on possible future lines of research.

13.1.1 The Past: How HSP60 Became an Atherogenic Molecule

In the 1980s, members of our group were studying the role of changes in lipid metabolism on the phenotypic and functional age-dependent alterations of lymphoid cells (Traill et al. 1990). In the course of this work, we observed an increased expression of low-density lipoprotein (LDL)-receptors on the surface of lymphocytes from elderly donors and an increased plasma membrane viscosity due to an elevated molar ratio of free cholesterol to phospholipids (Huber et al. 1991). For determining cell membrane viscosity, we used a modified method employed for measurements of single cells by flow cytometry (Böck et al. 1989). We then finally decoded the defects of the intracellular lipid metabolism responsible for the age-dependent alterations of lymphoid cell function (Stulnig et al. 1995). In 1988, we presented these data at a Meeting on Aging that we organized in Seefeld, Tyrol. There, we also briefly alluded to the possible role of our observations in the relative loss of immunological self-recognition and thus increased autoimmunity – our major field of expertise – in older age. Later at the bar, some of our colleagues teased us saying that we looked at every immunological finding from the perspective of autoimmunologists and that we would perhaps end up claiming that even atherosclerosis was an autoimmune disease. Interestingly, we had already earlier floated thoughts along these lines, albeit without even planning or performing any experimental work.

Back in the lab after the conference, we seriously discussed applying our expertise in autoimmunology to tackling this problem. This decision was further promoted by our knowledge as immunopathologists that hallmarks of inflammation had already been described more than a century earlier (Mayerl et al. 2006). As a matter of fact, this phenomenon had recently been "rediscovered" by several groups (Ross 1999; Ridker et al. 2002; Hansson and Libby 2006; Jongstra-Bilen et al. 2006). However, most of these latter studies were performed on patients and experimental animals with fully developed, late-stage atherosclerosis, i.e. a situation that was not of prime interest to us. Pathohistological or proteomic analyses of atherosclerotic plaques certainly revealed a panoply of cellular elements and expressed proteins that reflect a situation far from the stage of the initiation of the lesion (Danesh et al. 1997; Daugherty et al. 1997). In contrast to these studies, our own work on organ-specific and systemic autoimmune diseases always focused on two specific issues, viz. (a) what are the first cells/molecules that initiate the disease, and (b) how can our data be interpreted from an evolutionary viewpoint?

At the beginning of our project on The Immunology of Atherosclerosis, we therefore first performed pathohistological studies comparing early to late human atherosclerotic lesions (Xu et al. 1990). In contrast to the then prevalent dogma, these studies showed that T cells preceded the appearance of macrophages/foam cells in early lesions, while later the number of foam cells by far exceeded that of lymphoid cells. These observations were later confirmed in further studies on specimens of early lesions (Millonig et al. 2002). Encouraged by these results, we started to perform appropriate animal experiments using the methods and criteria for induction and definition of autoimmune diseases established by E. Witebsky and N. R. Rose (Rose and Witebsky 1956; Witebsky and Rose 1956). We reasoned that should atherosclerosis be an autoimmune disease, autoantigens should be present within atherosclerotic lesions. We therefore first immunized three groups of normocholesterolemic young New Zealand White (NZW) rabbits with either delipidated proteins from human plaques or plaques from hypercholesterolemic LDLr$^{-/-}$ rabbits (Watanabe rabbits) and a control group with ovalbumin (OVA) (Xu et al. 1992). The antigenic protein solutions were emulsified with complete Freund's adjuvant (CFA) consisting of mineral oil, heat-killed mycobacteria and an emulsifier. If our hypothesis were true, plaque protein-immunized groups should develop atherosclerosis and the OVA-immunized rabbits should remain unaffected. To our surprise, all three groups were afflicted with atherosclerosis. Since CFA was the only common denominator, we then immunized NZW rabbits with CFA alone. The results were what we had expected – the NZW rabbits got atherosclerosis while controls receiving incomplete Freund's adjuvant (IFA) without mycobacteria remained unaffected. Other labs working on the pathogenesis of rheumatoid arthritis (RA) had shown earlier that immunization with a constituent of CFA, namely mycobacterial Hsp60 (designated as mbHsp65 due to a slightly higher molecular weight than Hsp60 of other pro- and eukaryotics) induced an RA-like disease (adjuvant arthritis – AA) in certain strains of rats and mice (Anderton et al. 1994). We therefore immunized NZW rabbits with pyrogen-free recombinant mbHsp65 and were again able to induce atherosclerosis (Xu et al. 1992) (Fig. 13.1). Early lesions of rabbits harbored Hsp60-reactive T cells (Xu et al. 1993). Later, it was shown that this approach also worked in mice, although much better in hypercholesterolemic knock- out (LDLr$^{-/-}$ or ApoE$^{-/-}$) than in wild-type strains (George et al. 1999). Interestingly, our rabbits did not simultaneously develop AA. As demonstrated later, T cells of rats with AA recognize other (arthritogenic) mbHsp65 epitopes than mice with atherosclerosis (atherogenic epitopes) (CORDIS- Final Report Summary of TOLERAGE: Normalization of immune reactivity in old age – from basic mechanisms to clinical application; http://www.tolerage.eu).

13.1.1.1 Epidemiology

In mice, mbHsp65-induced atherosclerosis could be transferred by T cells to unimmunized histocompatible recipients (George et al. 2001). In rabbits, mbHsp65-induced atherosclerosis could be prevented by T-cell depletion (Metzler et al. 1999). We then switched back to human patients and showed a correlation of anti-mbHsp65 antibodies (Xu et al. 1999) and soluble human Hsp60 (shHsp60) (Xu et al. 2000;

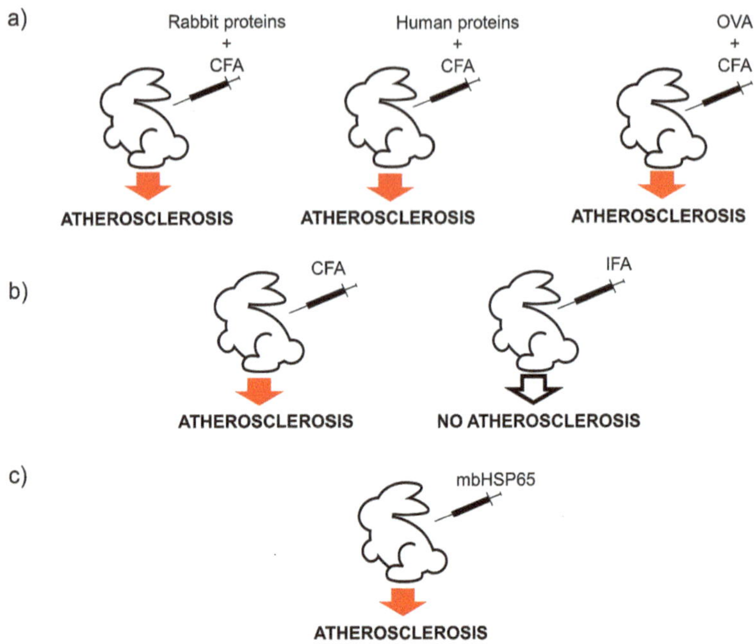

Fig. 13.1 Initial experiments in rabbits showing that HSP60 is a driver of atherosclerosis (**a**) Rabbits were immunized with proteins derived wither from rabbit or human atherosclerotic plaques, and OVA was used as a control, all emulsified in CFA. All groups developed atherosclerosis. (**b**) CFA, the common denominator in the first line of experiments, induced atherosclerosis, whereas IFA did not. (**c**) mbHSP65, a common constituent of CFA, was alone enough to replicate experiments with CFA immunization. Figure made in Powerpoint with free online images from clkr.com

Xiao et al. 2005) in the serum of patients with advanced atherosclerosis. The sera for these and many follow-up studies were obtained from participants in the BRUNECK-study, a long-running, population-based atherosclerosis prevention project started in 1990 and still ongoing (Willeit and Kiechl 1993). Serum titers of anti-mbHsp65 or cross reactive anti-hHsp60 (auto)antibodies not only represented a marker for atherosclerosis in the BRUNECK cohort, but follow-up studies showed that they also have prognostic value (Xu et al. 1999). Later, we had the opportunity to analyze male (atherosclerosis risk factor in male youngsters – ARMY study) (Knoflach et al. 2003) and female (atherosclerosis risk factor in female youngsters – ARFY study) (Knoflach et al. 2009) young subjects, applying the BRUNECK study methodology. It became evident in the these two cohorts of young subjects that the number of Hsp60-reactive T cells (but not antibodies) showed a significant correlation with early atherosclerosis demonstrable with ultrasound, while the opposite was true for the elderly participants of the BRUNECK study (Knoflach et al. 2007). From a possible clinical epidemiological and therapeutic viewpoint, our demonstration that the first inflammatory stage of atherosclerosis is still reversible, but persistence of risk factors, such as high serum cholesterol levels, turn it into an advanced, irreversible stage, is of special importance (Xu et al. 1996).

13.1.1.2 Experimental Work

An important next step in our work was the identification of possible atherogenic Hsp60 epitopes. HSP60 of different bacterial species display more than 95% sequence homology at the DNA and protein levels and greater than 55% homology still exists between bacterial and mammalian, e.g. human or murine, Hsp60. At certain molecular domains this latter homology even extends to more than 70% (Craig et al. 1993; Karlin and Brocchieri 2000). Therefore, every human being harbors antibodies and T cells reacting with bacterial HSP60 acquired either by prior bacterial or parasite infections or vaccinations. Incidentally, virus genomes do not code for viral HSP60, but viruses carry HSP60 from their target cells in their envelope, e.g. human Hsp60 derived from human immunodeficiency virus infected CD4+ T cells (Bartz et al. 1994). Antibodies against HSP60 from a given bacterial species extensively cross-react with Hsp60 from other species, e.g. anti-mbHsp65 with GroEL, the HSP60 of *E. coli* (Mayr et al. 1999), *H. pylori* (Mayr et al. 2000) or *C. pneumoniae* (Mayr et al. 1999) and so do T cells (Rossmann et al. 2008). Importantly, both anti-bacterial HSP60-reactive antibodies and T cells cross-react with eukaryotic Hsp60, e.g. appearing on human vascular endothelial cells (EC) stressed in vitro with classical atherosclerosis risk factors (Wick et al. 1995). Also, stressed but not unstressed vascular ECs can be lysed in a complement-dependent fashion or via antibody-dependent cellular cytotoxicity (ADCC) by cytotoxic anti-Hsp60 human antibodies (Schett et al. 1995). Using phage display and other biochemical methods, we first identified potentially atherogenic mbHSP65 B cell epitopes (Metzler et al. 1997; Perschinka et al. 2003, 2007). Interestingly, we also identified a B-cell epitope that had T-cell-stimulating potential, i.e. displayed linear sequence homology. Experiments successfully identifying Hsp60 T cell epitopes will be discussed further in detail below.

In mammalian cells, Hsp60 is encoded in the nucleus, but expressed in the mitochondria. From there, it is translocated to the cytoplasm and finally to the cell surface (Pfister et al. 2005), where it acts as a danger signal recognized by pre-existing Hsp60-reactive (auto)antibodies and T cells. Hsp60 released from damaged or dying cells, e.g. EC, is biochemically modified and recognized via "physiological" autoimmunity, thus functioning as a "garbage removal" mechanism. In elaborate studies, we have scrutinized the stressor and thus Hsp60-inducing potential of various classical atherosclerosis risk factors on human vascular EC in vitro. The results of these experiments at both RNA and protein levels showed that all risk factors, from infections over smoking and mechanical stress (hypertension) to high oxidized LDL (oxLDL) levels, advanced glycation end products (AGE) (Grundtman et al. 2011) and high salt concentration (Jakic et al. 2017) lead to Hsp60 expression demonstrated within as well as on the surface of endothelial target cells for pre-existing anti-Hsp60 humoral and cellular immunity. Using appropriate imaging techniques, we also were able to visualize the expression of Hsp60 on the arterial surface in vivo in rats with experimentally increased unilateral blood pressure (Hochleitner et al. 2000), and rats (Seitz et al. 1996) and rabbits stressed with bacterial lipopolysaccharide (LPS) as a surrogate for infection (Wick et al. 2008). Arterial EC turned out to be more susceptible than venous EC to stress-induced Hsp60 expression due to being pre-stressed by lifelong higher arterial blood pressure and pulsatile flow conditions. This principle was also further proven in vivo in rats and in vitro by immunohistological studies of fragments of human venous cardiac bypass conduits (Wick et al. 2012a). Importantly, the different stressors, including certain drugs, not only lead to the expression of Hsp60 on vascular ECs, but simultaneously induce the expression of adhesion molecules, thus providing the prerequisites for interaction of Hsp60-reactive T cells with their targets (Seitz et al. 1996; Amberger et al. 1997).

13.1.1.3 Vascular-Associated Lymphoid Tissue

In the course of these studies, several other observations were made that have functional impact in the context of atherogenesis. Among these, the discovery of a Langerhans cell-like network of vascular-associated dendritic cells in the arterial intima with increased density at the known predilection sites for atherosclerotic lesions at the arterial branching points was especially important regarding its potential for local antigen presentation (Millonig et al. 2001a, b, 2002). In analogy to another surface-bound branch of the immune system, the mucosa-associated lymphoid tissue (MALT), we designated pre-existing accumulations of mononuclear cells in the healthy arterial intima of healthy humans as the vascular-associated lymphoid tissue (VALT) (Wick et al. 1997). We hypothesized that these foci of tertiary lymphoid tissue might exert a similar local patrolling and protective role as the MALT, in this case, the inner, i.e. the vascular, surface (Waltner-Romen et al. 1998). The VALT together with blood-borne lymphoid cells – might contribute to the initial process of atherogenesis. Recently, other groups have emphasized a possible atherogenic role of the periarterial aggregates of mononuclear cells (Hu et al. 2015) and we have accordingly modified our VALT structure by dividing it into an inner and outer VALT. In summary, the Autoimmune Concept of Atherosclerosis, as depicted in Fig. 13.2, is built on a large body of data, originating in our lab and further supported by data from other groups. Without underestimating the pathophysiological role of Hsp60, these data support our theory that (a) atherosclerosis starts as an autoimmune reaction of T cells against Hsp60 expressed on arterial EC stressed by classical risk factors and is further accelerated and perpetuated by anti-Hsp60 (auto)antibodies, and (b) this concept provides a good evolutionary example of pleiotropic antagonism, namely the effect of genes the action of which is beneficial in youth – such as protective anti-Hsp60 immunity – may turn detrimental if expressed by an inappropriate target cell later in life (Wick et al. 2012a).

13.1.2 Hsp60 as an Atherosclerotic Autoantigen

It is now firmly established that the autoimmune recognition of Hsp60 by T cells is an initiator and driver of cardiovascular-associated pathologies. In this section, we summarize the earliest publications and recent research that support the pathogenic role of HSP60 in atherosclerosis.

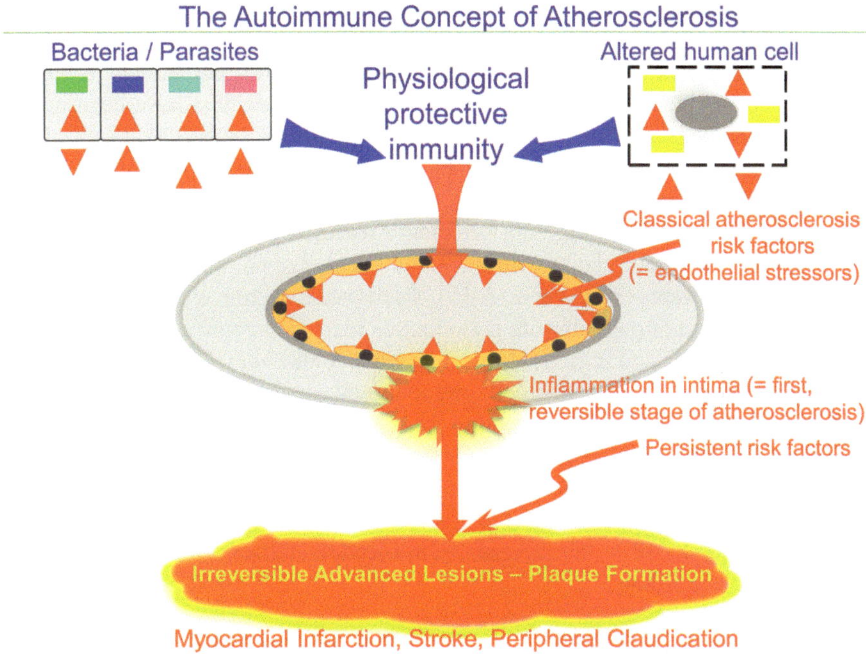

Fig. 13.2 The autoimmune concept of atherosclerosis
Microbiotic derived proteins (depicted as green, blue, turquoise or pink rectangles) and microbiotic HSP60 (red triangles) lead to protective immunity upon infection. In addition, endogenously derived HSP60 (red triangles) and other endogenous proteins (yellow rectangles) also lead to protective immunity under physiological conditions. However, should one expose the arteries to stress, with for example with a high fat diet or smoking, ECs will express HSP60 in the mitochondria, the cytoplasm and the surface, together with adhesion molecules. This sets the stage for (auto) recognition of pre-formed (during infection of vaccination) T cells that react towards HSP60 and enter the intima with the help of EC adhesion molecules. This early stage of disease is still reversible, should one stop to stress the ECs. If on the other hand, EC stressors, e.g. smoking, persist, a fully developed, irreversible, calcified plaque forms, potentially leading to deadly outcomes. Figure made in Powerpoint

13.1.2.1 Circulating Endogenous Hsp60

There are several reports that circulating Hsp60 in humans is associated with cardiovascular pathologies, among them atrial fibrillation (Maan et al. 2016), coronary heart disease (Zhang et al. 2008), carotid stiffness (Ellins et al. 2008), atherosclerosis (Xu et al. 2000) and others (Grundtman et al. 2011). Interestingly, our and other groups have reported circulating endogenous Hsp60 in healthy and asymptotic individuals as well (Pockley et al. 1999; Xiao et al. 2005) suggesting that Hsp60 is released even before the onset of disease. Moreover, we have demonstrated that T-cell immunity toward endogenous and mycobacterial Hsp60 manifests itself already at a young age (Knoflach et al. 2003, 2009).

13.1.2.2 Microbial Hsp60

Due to the high homology to prokaryotic HSP60, infectious burden has also been correlated to incidence of atherosclerosis. Micro-organisms that have been implicated in diseases include *C. pneumoniae*, P. gingivalis, *H. pylori*, M. tuberculosis and cytomegalovirus, among others (Campbell and Rosenfeld 2015). Immunohistochemical analysis of arterial specimens from patients with coronary artery disease that underwent endarterectomy revealed co-localization of cHsp60 and hHsp60, which correlated with increased C-reactive protein (CRP) and major adverse cardiac events (Tuleta et al. 2015). T cells isolated from either atheromatous plaques or gingiva of patients with periodontitis reacted to peptide19, derived from P. gingivalis Hsp60, and to hHsp60 in vitro (Choi et al. 2011). Hsp60 derived from F. nucleatum transformed THP-1 cells into foam cells in vitro, and induced increased expression of atherosclerosis promoting factors such as LDL, CRP and interleukin (IL)-6 in $ApoE^{-/-}$ mice when injected intraperitoneally (i.p.) (Lee et al. 2012). Ayada et al. demonstrated that orally administered *H. pylori* to $Ldlr^{+/-}ApoE^{+/-}$ double heterozygous mice induced a *H. pylori*-derived Hsp60-specific T helper 1 (Th1) response accompanied by an increase in transendothelial migration (Schett et al. 1997; Ayada et al. 2009).

13.1.2.3 Induction of Atherosclerosis by Administration of Hsp60

That Hsp60 can induce atherosclerosis was first established in rabbits injected with mbHSP65 emulsified in IFA (Xu et al. 1992) as mentioned above. Subsequent experiments in which bacterial Hsp60 and peptides thereof were administered subcutaneously (s.c.), intravenously (i.v.) and i.p. helped confirm that Hsp60 induces and exacerbates atherosclerosis (George et al. 1999; Lin et al. 2011; Grundtman et al. 2015; Hu et al. 2018). However, oral or nasal administration of Hsp60 induced regulatory T cells (Tregs) and anti-inflammatory cytokines, as discussed in more detail further on (Hu et al. 2018). To start with, it is possible to induce fatty streak formation, a pre-stage for atherosclerosis development, even in athero-resistant strains such as C57BL/6 (wild-type) mice. As reported above, s.c. administration with either recombinant HSP65 or mbHsp65 induced atherosclerosis (George et al. 1999). Furthermore, we could demonstrate in $ApoE^{-/-}$ mice that s.c. injection of either whole mbHsp65 protein or peptides thereof either enhanced or ameliorated atherosclerosis (Grundtman et al. 2015). Interestingly, i.p. injection of Hsp60 derived from F. nucleatum aggravated disease in $ApoE^{-/-}$ mice (Lee et al. 2012).

13.1.2.4 Cellular Immunity to Hsp60

Hsp60 has been shown to directly induce inflammatory phenotypes on ECs, vascular smooth muscle cells (VSMCs), monocytes, dendritic cells (DCs) and T cells, all components involved in the pathogenesis of atherosclerosis. When monocytes derived from peripheral blood of healthy individuals were stimulated with hHsp60 in vitro, they produced more IL-12p70 and interferon (IFN)-γ, suggesting they could induce Th1-polarization in situ (Justo-Junior et al. 2018). Deniset et al. used an in vitro system with rabbit VSMCs that were transfected with an adenovirus to overexpress either mitochondrial or cytosolic hHsp60 and showed in both settings that overexpression led to VSMC proliferation (Deniset et al. 2018). Choi et al. reported in vivo balloon-induced hyperplasia to be dependent on Hsp60 expression in VSMC, as a selective deletion of cytosolic Hsp60 in VSMCs resulted in reduction in intima-media thickening and a less inflammatory phenotype (Choi et al. 2015). Human DCs derived from peripheral blood monocytes have been shown to activate T cells in a major histocompatibility class (MHC)-II restricted manner after treatment with Hsp60 in vitro. These MHC-II-Hsp60-activated T cells produced IL-17 and IFN-γ (Rahman et al. 2017). Work from our laboratory identified T cells isolated from early lesions, classification based on (Stary 1992), and found that early infiltrating auto-reactive T cells recognize a distinct set of hHsp60-derived peptides, as opposed to late lesion- derived T cells (Almanzar et al. 2012). This is the first ever functional report on T cells isolated from human lesions in early-stage disease (Almanzar et al. 2012), and will be described in more detail below.

13.1.2.5 Hsp60 Binding and Signaling Pathways

Several groups have shown that Hsp60 binds to endothelial cells, T cells or monocytes through Toll-like receptors (TLR)2 and 4 (Ohashi et al. 2000; Brea et al. 2011; Lin et al. 2013; Zhao et al. 2015; Huang et al. 2016). Specifically, Ohashi et al. demonstrated for the first time that endogenous hHsp60 activated mouse bone marrow-derived macrophages in vitro through TLR4 (Ohashi et al. 2000). It was also shown that cHsp60 induced the expression of TLR4 in vitro on human coronary artery endothelial cells (HCAECs) accompanied by phosphorylation of MAPKs p38, ERK1/2 and c-Jun N-terminal kinases (Lin et al. 2011). The authors then went on to show that Hsp60 impairs vascular endothelial cell function, through TLR4-dependent mechanisms, using a TLR4 mutant mouse model for induced hind limb ischemia (Lin et al. 2011). Brea et al. incubated human umbilical vein ECs (HUVECs) with serum from ischemic stroke patients and showed an inflammatory response, in the form of IL-6 and tumor necrosis factor (TNF)-α production, while E-selectin was reduced when TLR2/4 or Hsp60 was blocked (Brea et al. 2011).

HCAEC treated in vitro with cHsp60 displayed a reduction in endothelial nitric oxide synthase 3 (eNOS) production and eNOS-related gene expression, which was accompanied by activation of the MAPK signaling pathway, i.e. ERK phosphorylation. eNOS is considered to have a protective function on the cardiovascular system. By inhibiting MAPK signaling molecules, the authors could reverse cHsp60-induced eNOS reduction (Chen et al. 2009), thus showing that Hsp60 has a direct influence on EC fitness. Similarly, it was shown that Hsp60 could induce VSMC migration in vitro, which was dependent on TLR4 and induced MAPK/ERK signaling (Zhao et al. 2015). Furthermore, Chanine et al. showed that rabbit VSMC incubated with *C. pneumoniae* and oxLDL in vitro displayed upregulation of Hsp60 and phosphorylation of ERK1/2. The upregulation of Hsp60 could subsequently be reversed by inhibition of ERK1/2 pathway (Chahine et al. 2011). Kol et al. showed, on the other hand, that treatment of ECs with either cHsp60 or hHsp60 induced nuclear factor-κB signaling (Kol et al. 1999). The inflammatory effects of Hsp60 have in part been inhibited by Annexin 5A, possibly by direct binding to Hsp60 and thus hindering activation of DCs in vitro (Rahman et al. 2017). In contrast, Annexin A2 seems to play a pathogenic role in Behçet's disease, and its expression is correlated to that of Hsp60 (Chen et al. 2015).

CD4 Th1 cells are the most prominent T cells in atherosclerotic lesions, and they are primed by antigen-presenting cells with cognate antigen loaded in MHC-II, such as DCs or ECs. Both, mouse and human endogenous and bacteria-derived Hsp60 peptides have been found in MHC-II context and were able to prime autologous T cells in vitro (Anderton et al. 1993; Zugel et al. 1995; Michaelsson et al. 2002). We have also demonstrated an increase of γδ-T cells in the early lesions compared to peripheral blood (Kleindienst et al. 1993), which indicates that Hsp60 recognition and activation by T cells may occur in the absence of MHC presentation, since γδ-T cells do not require cross-presentation (Fisch et al. 1990). In addition, Th17 are also important drivers of disease (Gao et al. 2010).

13.1.2.6 Hsp60-Induced Adhesion Molecules and Cytokines

Isolated HCAECs treated with cHsp60 in vitro displayed expression of vascular cell adhesion molecule (VCAM)-1 and intercellular adhesion molecule (ICAM-1), as demonstrated by western blotting (Huang et al. 2016). Kol et al. demonstrated that both, cHsp60 and hHsp60 induced VCAM-1 and ICAM-1 on ECs (Kol et al. 1999). In addition, i.v. administered cHsp60 to rabbits induced VCAM-1 expression on abdominal aorta ECs (Huang et al. 2012). Moreover, P. gingivalis-derived Hsp60 reduced vascular endothelial-cadherin and eNOS in HUVECs, whereas apoptosis is induced (Wu et al. 2016). Hsp60 from another periodontitis associated microbe, F. nucleatum, also induced ICAM-1, VCAM-1, E-selectin and IL-8 and monocyte chemoattractant protein in vitro in human microvascular ECs (Lee et al. 2012). Furthermore, hHsp60 activated mouse macrophages in vitro and induced TNF-α production, measured with enzyme-linked immunosorbent assay (Ohashi et al. 2000). In early studies, our lab showed that heat shock (42 °C) induced Hsp60 in human venous and arterial ECs. In addition, Hsp60 expression together with VCAM-1 and ICAM-1 was induced by TNF-α (Amberger et al. 1997). Together, these data show that soluble Hsp60 can induce many factors that contribute to activating ECs and enhancing lymphocyte adhesion and transmigration. Important to consider, as mentioned above, is that arterial ECs are more sensitive to stressors than venous ECs, because they have undergone mechanical pre-stress by higher arterial blood pressure and pulsations (Wick et al. 2012a).

13.1.2.7 Hsp60 and LDL: Partners in Crime

Interestingly, Hsp60 and oxLDL share one common scavenger receptor on ECs and foam cells, namely lectin-like oxLDL receptor (LOX-1) (Yoshimoto et al. 2011). 4-hydroxy-2-noneal (HNE), a by-product of lipids during oxidative stress, can form adducts with Hsp60 in phorbol 12-myristate 13-acetate-activated THP-1 monocytes that were stimulated with HNE and oxLDL (Arcaro et al. 2015), suggesting that oxLDL and Hsp60 can act in unison on foam cells. Moreover, s.c. administration of recombinant HSP65 into wild-type or ApoE$^{-/-}$ mice aggravated atherosclerosis, disrupted cholesterol efflux, increased serum LDL and simultaneously led to a decrease in high-density lipoprotein, which is generally considered to be atheroprotective (Sun et al. 2014). Human coronary artery endothelial cells treated with P. gingivalis-derived Hsp60 or cHsp60 showed an upregulation of LOX-1. These cells also displayed an increase in oxLDL uptake and conversion of THP-1 cells to foam cells (Lin et al. 2011; Huang et al. 2016). In addition, when the cHsp60 was injected i.v. into high-fat-fed rabbits, atherosclerosis was induced and expression of LOX-1 was enhanced (Lin et al. 2011). These data indicate that, in some cases, Hsp60 and oxLDL can act in synergy and thus accelerate disease progression.

13.1.3 Inducers of Hsp60

An understanding of the inducers of ectopic expression and translocation of endogenous Hsp60 on the cell surface is of crucial importance. Under physiological conditions, Hsp60 is expressed in the mitochondria (although encoded in the nucleus) and then translocated to the cytoplasm where it functions as a chaperone (Soltys and Gupta 1997). Under pathological conditions, Hsp60 is also translocated to the cell surface. As such, finding the risk factors for abnormal induction of Hsp60 means to find risk factors for atherosclerosis development. Since circulating T cells recognize Hsp60 that is on the surface of endothelial cells lining the arteries, this section will focus primarily on mediators that induce Hsp60, on ECs in particular and on other cells in general, involved in atherogenesis. Figure 13.3 summarizes the most important findings for the Autoimmune Concept of Atherosclerosis on ECs.

13.1.3.1 High-Fat Diet (LDL)

A high-fat diet and subsequent obesity, in particular visceral adipose tissue, is one of the strongest risk factors for atherosclerosis (Writing Group et al. 2016). DCs derived from human peripheral blood treated with oxLDL upregulated Hsp60 expression as shown by qPCR. This effect could be reversed upon addition of atorvastatin (Frostegard et al. 2016). In contrast, in early studies, we could not show induction of Hsp60 on human endothelial cells subjected to oxLDL (Amberger et al. 1997).

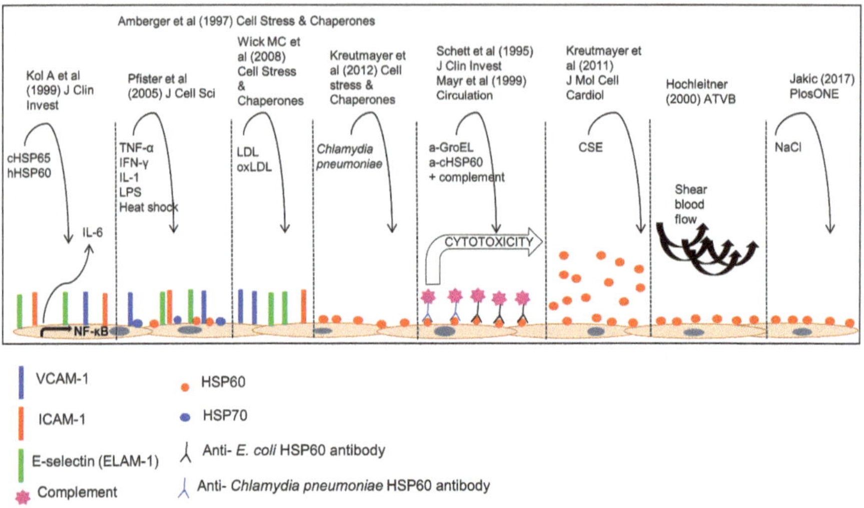

Fig. 13.3 Endothelial cell stressors
Stressors such as HSPs, cytokines, bacteria, bacterial products, antibodies, heat, shear blood flow, chemicals (CSE and salt) and lipids can all induce various components of either adhesion molecules or endogenous HSP expression on the surface of ECs, thus making them a target for pre-existing HSP reactive T cells. References are mentioned in the text. Figure made in Powerpoint

13.1.3.2 Infections

As mentioned above, infections with several pathogens have been correlated to the prevalence of atherosclerosis (Campbell and Rosenfeld 2015). Our lab has provided evidence that infection of HUVECs with C. pneunomiae in vitro can lead to surface expression of hHsp60, as demonstrated by confocal microscopy (Kreutmayer et al. 2013). In vitro infection of porcine coronary arteries with *C. pneumoniae* revealed co-localization of hHsp60 (eukaryotic), cHsp60 and TLR4 on the surface of ECs (Deniset et al. 2012). Specifically, i.v. administration of LPS, a common bacterial endotoxin, into rabbits, together with radiolabeled anti-Hsp60 antibodies, resulted in LPS-induced Hsp60 expression on the surface of ECs, as shown by PET-CT scanning (Wick et al. 2008). Lastly, a correlation was found between cytotoxic CagA seropositive individuals and intima-media thickness of carotid arteries and to anti-mbHSP65 antibodies (Mayr et al. 2000, 2003).

13.1.3.3 Smoking

Research from our lab has shown by fluorescence microscopy of live cells in cell culture that cigarette smoke extract (CSE) can induce Hsp60 expression on the surface of HUVECs. In addition, these HUVECs released more Hsp60 into the supernatant (Kreutmayer et al. 2011).

13.1.3.4 Sodium and Cadmium

A diet rich in salt (NaCl) is a known additional risk factor associated with increased cardiovascular events (Powe and Bibbins-Domingo 2016). Physiologically, high plasma sodium content can increase resting blood pressure, and thus disturb sheer blood flow, a risk factor discussed further below (Suckling et al. 2012). We have demonstrated that incubation of HUVECs with increasing concentrations of sodium chloride correlated to increased intracellular and surface expression of hHsp60, as shown by microscopy, reflecting plasma sodium levels in normonatremic to hypernatremic individuals. Moreover, the salt-induced Hsp60 expression correlated to an increase in apoptosis (Jakic et al. 2017). An interesting finding from an evolutionary point of view was that a clam species (Ruditapes philippinarum) upregulated Hsp60 expression under low-salinity-induced stress (Ding et al. 2018). Cadmium is an important constituent of urban pollution and it has been demonstrated that it can act as an independent risk factor for early atherosclerosis (Messner et al. 2009). In vivo, cadmium-fed ApoE$^{-/-}$ mice developed atherosclerosis, with lesion sites expressing VCAM-1 and Hsp60 on the surface of ECs (Knoflach et al. 2011). Furthermore, cadmium is a major cigarette smoke constituent, and in young smokers with higher serum cadmium there was stress-associated Hsp60 gene expression on ECs (Bernhard et al. 2006).

13.1.3.5 Shear Stress

Turbulent shear stress rather than laminar shear stress is active at branching points in the vascular system. Turbulent stress exposes endothelial cells of the arteries to higher relative blood pressure, and increases the time for contact with potentially detrimental agents (Wick et al. 2012b). Using an in vitro culture system with a rotational viscometer, HUVECs were exposed to shear stress resembling that of arterial blood flow. The endothelial cells expressed Hsp60 in a time-dependent manner. These findings were confirmed in the right carotid artery of Lewis rats that are physiologically subjected to higher shear stress (Hochleitner et al. 2000). These same areas, namely the intima of carotid arteries, were populated already at a young age (children <10 years old) by CD3+ T cells, CD68+ macrophages and CD1a+ DCs (Waltner-Romen et al. 1998). As mentioned above, we designated these cell accumulations as VALT. Further research has been carried out on VALT by us and others (Bobryshev and Lord 2001; Millonig et al. 2001a; Langohr et al. 2008).

13.1.3.6 Pro-inflammatory Mediators

Platelet-derived growth factor BB and IL-8 added to HUVECs induced Hsp60 secretion, as shown by western blot of cell supernatant (Zhao et al. 2015). Moreover, treatment of HUVECs with TNF-α entailed increased expression of Hsp60 and ICAM-1, as shown by flow cytometry and western blot (Amberger et al. 1997; Wu et al. 2012).

Homocysteine, another CVD risk factor, particularly in hemodialysis patients, also increased the expression of Hsp60 in HUVECs, as shown by qPCR and western blotting (Capasso et al. 2012).

13.1.4 Humoral Immunity to Hsp60

Generally, T cells recognizing autologous Hsp60 are considered to be the initiators of atherosclerosis, whereas B cells and circulating antibodies produced by plasma cells are secondary and accelerating agents.

In addition to pathogen-specific antibodies, antibodies against bacterial Hsp60 develop after infection or vaccination. Due to the high homology between prokaryotic and eukaryotic Hsp60, cross-reactions occur when endogenous Hsp60 is expressed ectopically. For instance, there are several reports that anti-Hsp60 antibodies are associated with cerebrovascular events, such as stroke (Banecka-Majkutewicz et al. 2014; Galovic et al. 2016). Data from the Framingham study showed correlation between anti-Hsp60 antibodies, serum IL-2 and TNF-α with coronary calcium score in asymptomatic patients (Damluji et al. 2015). The presence of both, cHsp60 and hHsp60 in situ in patients with unstable angina, has been associated with circulating levels of anti-Hsp60 autoantibodies (Andrie et al. 2011). Immunohistochemical analysis of cHsp60 and hHsp60 of human coronary arteries correlated with circulating anti-Hsp60 autoantibodies and increased the predisposition for major adverse cardiac events (Tuleta et al. 2015). Our laboratory has linked circulating anti-Hsp65 antibodies to carotid atherosclerosis (Xu et al. 1999). More recently, we also showed circulating anti-Hsp60 antibodies in the serum of patients with late lesions that correlate with disease, but not in those with early lesions. However, Hsp60 T cell-reactive cells showed correlation to both early and late lesions (Almanzar et al. 2012). Furthermore, patients that receive hemodialysis are at an increased risk for developing CVDs, and have increased circulating anti-Hsp60 antibodies, compared to healthy controls (Musial et al. 2009; Esposito et al. 2011). In Balb/c mice with experimentally induced thrombi, there was enhanced thrombus formation and with larger thrombi (Dieude et al. 2009).

Interestingly, Wang et al. found a natural IgM antibody in mouse that selectively recognized epitopes on both oxLDL and Hsp60 of Aggerigatibacter actinomycetemcomitans (Wang et al. 2016). Natural antibodies, which are germline encoded, of the IgM class are generally considered to be protective (Binder et al. 2005). Finally, it has been reported that natural anti-Hsp60 might be an inherited trait, as they have been detected in the fetus-maternal interface, indicating that a certain risk for atherosclerosis might already be passed on at birth (Varbiro et al. 2010).

13.1.5 Hsp60 Vaccination for Prevention and Treatment of Atherosclerosis

In general, the treatment for atherosclerosis comprises drugs to lower lipid levels together with anti-inflammatory drugs to inhibit the development of atherosclerotic lesions. Despite their initial success, we know that the events leading to plaque formation involve complex inflammatory/immune processes that are poorly treated by available drugs in the market (Bittencourt and Cerci 2015). Taking for granted that atherosclerosis is a Thl immune-mediated disease, efforts are being directed to designing specific strategies to modulate inflammation by shifting the immune response toward Th2 (Schulte et al. 2008) and/or inducing regulatory T-cell (Treg) responses, strategies shown to be beneficial in treatment of atherosclerosis (Foks et al. 2015). However, in our experimental work, we first focused on prevention of atherosclerosis, and only later proceeded to the therapeutic/intervention. While all possible atherosclerosis-specific antigens are not yet completely characterized, there is a bulk of evidence to show that Hsp60 is a key antigen in driving initiation of the disease, and thus makes it a promising prevention and treatment target. This paragraph describes the state of art of vaccination strategies targeting the Hsp60 full molecule or using Hsp60-derived peptides thereof applied via different routes, namely oral, nasal and s.c. to induce tolerance in animal models of atherosclerosis.

Peripheral tolerance is achieved by mucosal antigen delivery without an adjuvant to shut down the immune response. So far, one of the classical approaches to induce tolerance is the oral delivery of the antigen. In mammals, the gut is continuously exposed to vast amounts of foreign antigenic material derived from food and commensal organisms (also termed microbiota) that colonize the mucosa of the small and large intestine. Oral tolerance is referred to as the non-responsiveness to intestinal antigens and it is mediated by Tregs (Cebula et al. 2013). In fact, in vivo studies showed that adoptive transfer of orally induced Foxp3+ Tregs can transfer oral tolerance to naïve animals, a process which is diminished by the depletion of CD25+ cells, suggesting that Tregs are crucial in regulating this process (Dubois et al. 2003). CD103+ DCs are also involved in establishment of oral tolerance. In particular, CD103+ DCs from the intestinal lamina propria and mesenteric lymph nodes (MLNs) have the unique ability to induce the expression of gut-homing molecules on T cells and are potent in the induction of Foxp3+ Tregs (Jaensson et al. 2008). Finally, another cell type involved in oral tolerance are myeloid-derived suppressor cells (MDSCs) designated as CD11b+Gr-1+ (Veglia et al. 2018). Interestingly, it was shown that in vivo transfer of suppressor CD11b+Gr-1+ myeloid-derived suppressor cells in LDLr$^{-/-}$ mice protected them from atherosclerosis (Foks et al. 2016).

Hu et al. showed that preventive oral feeding with full-length Hsp60 reduced atherosclerotic plaque size at the aortic root in ApoE$^{-/-}$ mice fed with western-type diet. This protective effect was MDSCs-mediated, since their percentages were increased in both peripheral blood and spleen of Hsp60-orally treated mice and not in the PBS control group. Moreover, Hsp60-treated mice secreted higher plasma levels of IL-10 and lower levels of the pro-inflammatory cytokines such as IFN-γ and IL-17A (Hu et al. 2018). Finally, at mRNA levels, an increase of Foxp3 expression and other Tregs-related transcriptional factors were found (Hu et al. 2018).

In our recent studies, we also investigated the effect of oral Hsp60 on atherosclerosis in both young and aged $ApoE^{-/-}$ mice (Grundtman et al. 2015; Wick et al. 2018). We used prokaryotic Hsp60, taking mbHSP65 as a potent representative, to induce experimental atherosclerosis after s.c. immunization of $ApoE^{-/-}$ mice with recombinant mbHSP65 protein emulsified in IFA. Interestingly, feeding with mbHS65 protein resulted in amelioration of mbHSP65-induced atherosclerosis in both young and aged mice. In agreement with previous studies (Foks et al. 2016; Hu et al. 2018), we also found that the atheroprotective effect, upon oral treatment with prokaryotic Hsp60, was sustained by the increase in the proportion of Foxp3+ Tregs in both young and aged mice (Wick et al. 2018) and by the secretion of higher IL-10 plasma levels in young mice (Grundtman et al. 2015). All animals and humans develop protective innate and adaptive immunity against Hsp60 (due to vaccinations or infections). On the one hand, Hsp60 should be recognized and reacted against to fight infection, but on the other, autologous Hsp60 should still be tolerated to avoid autoimmunity (Wick et al. 2004). Since immunity against Hsp60 seems to represent an important protective basic defense mechanism, tolerization against the whole Hsp60 molecule cannot be an optimal preventive or curative approach that could be translated to the human system. To navigate this complex situation, new approaches using bioinformatic tools are aiming to design novel Hsp60 vaccines targeting specific epitopes of the molecule. In line with these observations, we identified for the first time potentially atherogenic/atheroprotective peptides, both for mice (Grundtman et al. 2015) and humans (Almanzar et al. 2012) that could be optimal future candidates for the development of an oral vaccine against atherosclerosis, targeting specific epitopes of Hsp60 while leaving the immune response to the rest of molecule intact. With a systematic screening of mbHsp65-derived peptides, we found in $ApoE^{-/-}$ mice that seven peptides gave the highest proliferative response (Grundtman et al. 2015). We then assessed the in vivo potential atherogenic or atheroprotective effects of mbHsp65 peptides. Interestingly, we found that s.c. immunization with in vitro stimulatory peptides either promoted or decreased atherosclerosis or had no effect. In silico analysis showed that these peptides bound MHC class II molecules with high to intermediate affinity, inducing a CD4+ T cell response (Grundtman et al. 2015). Moreover, in mice, four peptides cover the same peptide regions as recognized by human T cells (Almanzar et al. 2012). In particular, the amino acids in mbHSP65-derived peptide 33 were conserved in hHsp60, mouse Hsp60 and mbHsp65 proteins (Almanzar et al. 2012). Interestingly, it was reported earlier that a hHsp60 peptide, which overlaps with our peptide 33, was recognized by a CD4+ T cell clone obtained from atherosclerotic plaques of C. pneumonia seropositive patients (Benagiano et al. 2005). Another study showed that s.c. immunization with two immunodominant Hsp60 peptides derived from P. gingivalis had an anti- or pro-atherogenic role in an ApoE$^{-/-}$ mouse model of infection-triggered atherosclerosis (Jeong et al. 2015).

In humans, our group described the reactivity of T cells isolated from early, clinically still inapparent atherosclerotic lesions to potentially atherogenic peptides of human Hsp60. We found that CD4+ T cells from early lesions secreted higher amount of IFN-γ and IL-17 cytokines. Moreover, T cells proliferate against the whole human Hsp60 and, similar to our mice data, we found that a specific peptide pool gave the highest rate of proliferation, suggesting the presence of human atherogenic peptides involved in the early phase of the disease (Almanzar et al. 2012). Targeted multi-epitope therapeutic vaccines have been shown to be successful in cancer treatment (Donaldson et al. 2017) and recently it was suggested that they could be highly effective in atherosclerosis, too. In fact, it was shown that oral administration of a recombinant antigen expressing epitopes from Hsp60, Apolipoprotein B (ApoB) and *C. pneumoniae* outer membrane protein (Cpn) resulted in amelioration of atherosclerosis in the mouse model (Mundkur et al. 2014). The percentages of CD11c+CD103+ from MLNs and Peyer's patches were significantly increased in Hsp60/ApoB/Cpn-treated mice. In addition, the frequencies of CD4+ IFN-γ and IL-17A+ T cells were also significantly reduced. By contrast, a significant increase in the frequencies of Foxp3+ and CTLA-4+ Tregs was observed (Mundkur et al. 2014). The authors proposed a mechanism where the peptides might activate tolerogenic DC in the gut that in turn induce Treg-mediated responses. Moreover, Tregs can also induce the shift of macrophages toward to M2 phenotype, leading to a reduction in plaque inflammation (Thota et al. 2017).

Using an immunoinformatics based-approach, Karkhah et al. designed a multi-epitope vaccine from Hsp60 and calreticulin proteins (Karkhah et al. 2017). Calreticulin was chosen since it has Th2-skewing property. Furthermore, it interacts with the mammalian scavenger receptor and at the same time induces a Th2 response (Rzepecka et al. 2009). Using in silico analysis, four epitopes from Hsp60 and two epitopes from calreticulin proteins were identified and selected as IL-4-inducing epitopes. As expected, at least one of those predicted Hsp60 IL-4-inducing epitopes (aa sequence: GEEGLTLNLEDVQPH) was identical to peptide 66 in human early lesions identified by our lab (Almanzar et al. 2012). Computational analysis showed that the predicted epitopes had a strong potential to induce B-cell mediated immune responses and to shift T-cell mediated immune responses toward protective Th2-type (Karkhah et al. 2017). The same group performed an in silico analysis which showed the potential for atherosclerosis treatment of a chimeric vaccine consisting of a different combination of three atherosclerotic autoantigens, namely hHsp60, ApoB-100, and β2 glycoprotein I (Karkhah and Amani 2016). Another in vivo study demonstrated that s.c. immunization with a chimeric vaccine combining both ApoB-100 and hHsp60 peptides resulted in a significant reduction of early atherosclerotic lesions in ApoE$^{-/-}$ mice (Li et al. 2011). Intriguingly, the same protective effect was achieved even when mice were treated with hHsp60 alone. Moreover, it was shown in vitro that, upon stimulation with hHsp60, CD4+ T cells secreted significantly higher levels of TGF- β and lower levels of IFN- γ in comparison with ApoB-100 peptide alone. By contrast, DCs produced more IL-10 upon stimulation with both peptides. These results suggested that hHsp60 peptide was the predominant component of the vaccine being the most effective in ameliorating atherosclerosis (Li et al. 2011). The shift toward Th2 response was also observed upon oral administration of L. lactis delivering mbHsp65 in $LDLr^{-/-}$ mice (Jing et al. 2011).

Nasal delivery of antigen represents another mucosal route to the immune system. In contrast to oral feeding, upon nasal vaccination the antigen remains intact, – since enzymatic cleavage occurs in the gut – and thus is directly absorbed in the upper respiratory tract. Most importantly, to be effective, it requires a smaller amount of desirable antigen. In fact, low doses of antigen would preferentially induce Tregs, suppressing disease by releasing antigen-non-specific cytokines (bystander suppression). In contrast, high doses of antigens induce T-cell anergy (Friedman and Weiner 1994). In rabbits, nasal administration of mbHsp65 protein reduced atherosclerotic lesions in high-cholesterol diet-driven atherosclerosis. mbHsp65 protein and multifaceted vaccines targeting simultaneously HSP65 and lipids were more effective in controlling atherosclerotic lesions (Jun et al. 2012; Long et al. 2012). Moreover, in both studies, HSP65-treated rabbits displayed high serum levels of IL-10 and lower levels of IFN- γ (Jun et al. 2012; Long et al. 2012). In mice, nasal Hsp60 vaccination resulted in a significant reduction of atherosclerosis in the early stage of disease. An increase of Foxp3+ Tregs was observed in Hsp60-treated mice. Staining of T cells using a known activation marker for Tregs, namely glycoprotein Glycoprotein A repetitions predominant (GARP), revealed that Tregs were also activated in the spleen and cervical lymph nodes. More abundant Foxp3 mRNA was found in atherosclerotic plaques of Hsp60-treated mice. In contrast, no differences in GARP mRNA were observed between untreated and treated mice. Moreover, nasal application of Hsp60 increased the production of TGF-β and IL-10 and downregulated Th1 and Th17 responses. Similar to the studies in rabbits, the atheroprotective effect was only partly abrogated by IL-10 neutralization, which could be explained by the increase of TGF-β1 that can also mediate the anti-atherosclerotic effect in addition to its pro-fibrotic effect leading to arterial stiffness (Zhong et al. 2016). The pro-fibrotic activity of TGF-β1 could lead to increased cross-linking of ECM proteins, and thereby contribute to vascular stiffness (Foote et al. 2016). In agreement with this observation, Li et al. showed that the increased production of TGF-β in the Hsp60-treated mice ameliorated disease, and this effect was mediated by the increased percentages of CD4+LAP+ Tregs (Li et al. 2012).

S.c. immunization with mbHsp65 in alum adjuvant alone or when combined with anti-CD45RB antibody administered i.p. (an isoform of CD45 which is important for the regulation of T-cell activation) protected $ApoE^{-/-}$ mice against progression of early atherosclerosis (Klingenberg et al. 2012). However, the major reduction in atherosclerotic plaques was not observed when mice were treated with the antibody alone, which suggests that to be effective, synergy with mbHSP65 was required. Indeed, it did not enhance the proportion of Foxp3+ Tregs and it might lead to a transient reduction in peripheral T cells (Klingenberg et al. 2012). In contrast to this study, Hu et al. showed that s.c. vaccination with Hsp60 emulsified in IFA exacerbated atherosclerosis in $ApoE^{-/-}$ mice (Hu et al. 2018). Finally,

Xiong et al. reported that s.c. immunization with a peptide of human Hsp60, also emulsified in IFA, aggravated atherosclerosis in rabbits fed a high cholesterol diet (Xiong et al. 2016). The discrepancies between these studies might be due to the different vaccine formulations used that could have an intrinsic effect on atherosclerosis (Khallou-Laschet et al. 2006). There is accumulating experimental evidence on the successful use of Hsp60 whole molecule and Hsp60-derived peptides as immunomodulatory agents in atherosclerosis. Hopefully, in the near future, these preclinical studies might lead to translational approaches in humans. This is not far from becoming a reality, since vaccines targeting HSP (namely HSPE7, which is composed of HSP65 and human papilloma viral protein E7) in cervical cancer are already tested in clinical trials (source: http://clinicaltrials.gov/).

13.2 Conclusions

In summary, the pathogenic role of Hsp60 in atherosclerosis has been firmly established over the past two and a half decades, using various experimental approaches and epidemiological longitudinal studies. Currently, numerous groups are working towards establishing an Hsp60-based vaccine and deciphering its functional mechanisms, both in vitro and in vivo. It is rewarding and satisfying to see that our original discovery of the initiating antigenic role of Hsp60 in atherogenesis and the subsequent formulation of the Autoimmune Concept of Atherosclerosis have triggered such a broad new field of cardiovascular research, advanced by so many groups, extending from basic work to new preventive and therapeutic clinical approaches.

Acknowledgments We would like to thank past and present members of the Laboratory of Autoimmunity for their contributions to the work on Hsp60. We would also like to thank funding agencies that have supported our research; FP-7HEALTH EU funded TOLERAGE, Austrian Science Fund (FWF), Tyrolean Science Fund (Tiroler Wissenschaft Fonds, TWF), the Jubiläumsfonds of the Austrian National Bank (Österreichische National- bank, ÖNB) and the Lore and Udo Saldow Donation. Lastly, we would like to thank Rajam Csordas for critical evaluation of the English language.

References

Almanzar G, Ollinger R, Leuenberger J et al (2012) Autoreactive Hsp60 epitope-specific T-cells in early human atherosclerotic lesions. J Autoimmun 39(4):441–450

Amberger A, Maczek C, Jurgens G et al (1997) Co-expression of ICAM-1, VCAM-1, ELAM-1 and Hsp60 in human arterial and venous endothelial cells in response to cytokines and oxidized low-density lipoproteins. Cell Stress Chaperones 2(2):94–103

Anderton SM, van der Zee R, Goodacre JA (1993) Inflammation activates self Hsp60-specific T cells. Eur J Immunol 23(1):33–38

Anderton SM, van der Zee R, Noordzij A, van Eden W (1994) Differential mycobacterial 65-kDa heat shock protein T cell epitope recognition after adjuvant arthritis-inducing or protective immunization protocols. J Immunol 152(7):3656–3664

Andrie RP, Bauriedel G, Braun P et al (2011) Prevalence of intimal heat shock protein 60 homologues in unstable angina and correlation with anti-heat shock protein antibody titers. Basic Res Cardiol 106(4):657–665

Arcaro A, Daga M, Cetrangolo GP et al (2015) Generation of adducts of 4-Hydroxy-2-nonenal with heat shock 60 kDa protein 1 in human promyelocytic HL-60 and Monocytic THP-1 cell lines. Oxidative Med Cell Longev 2015:296146

Ayada K, Yokota K, Hirai K et al (2009) Regulation of cellular immunity prevents Helicobacter pylori-induced atherosclerosis. Lupus 18(13):1154–1168

Banecka-Majkutewicz Z, Grabowski M, Kadzinski L et al (2014) Increased levels of antibodies against heat shock proteins in stroke patients. Acta Biochim Pol 61(2):379–383

Bartz SR, Pauza CD, Ivanyi J et al (1994) An Hsp60 related protein is associated with purified HIV and SIV. J Med Primatol 23(2–3):151–154

Benagiano M, D'Elios MM, Amedei A et al (2005) Human 60-kDa heat shock protein is a target autoantigen of T cells derived from atherosclerotic plaques. J Immunol 174(10):6509–6517

Bernhard D, Rossmann A, Henderson B et al (2006) Increased serum cadmium and strontium levels in young smokers: effects on arterial endothelial cell gene transcription. Arterioscler Thromb Vasc Biol 26(4):833–838

Binder CJ, Shaw PX, Chang MK et al (2005) The role of natural antibodies in atherogenesis. J Lipid Res 46(7):1353–1363

Bittencourt MS, Cerci RJ (2015) Statin effects on atherosclerotic plaques: regression or healing? BMC Med 13:260

Bobryshev YV, Lord RS (2001) Vascular-associated lymphoid tissue (VALT) involvement in aortic aneurysm. Atherosclerosis 154(1):15–21

Böck G, Huber LA, Wick G, Traill KN (1989) Use of a FACS III for fluorescence depolarization with DPH. J Histochem Cytochem 37(11):1653–1658

Brea D, Blanco M, Ramos-Cabrer P et al (2011) Toll-like receptors 2 and 4 in ischemic stroke: outcome and therapeutic values. J Cereb Blood Flow Metab 31(6):1424–1431

Campbell LA, Rosenfeld ME (2015) Infection and atherosclerosis development. Arch Med Res 46(5):339–350

Capasso R, Sambri I, Cimmino A et al (2012) Homocysteinylated albumin promotes increased monocyte-endothelial cell adhesion and up-regulation of MCP1, Hsp60 and ADAM17. PLoS One 7(2):e31388

Cebula A, Seweryn M, Rempala GA et al (2013) Thymus-derived regulatory T cells contribute to tolerance to commensal microbiota. Nature 497(7448):258–262

Chahine MN, Deniset J, Dibrov E et al (2011) Oxidized LDL promotes the mitogenic actions of Chlamydia pneumoniae in vascular smooth muscle cells. Cardiovasc Res 92(3):476–483

Chen C, Chai H, Wang X, Lin PH, Yao Q (2009) Chlamydia heat shock protein 60 decreases expression of endothelial nitric oxide synthase in human and porcine coronary artery endothelial cells. Cardiovasc Res 83(4):768–777

Chen P, Yan H, Tian Y et al (2015) Annexin A2 as a target endothelial cell membrane autoantigen in Behcet's disease. Sci Rep 5:8162

Choi J, Lee SY, Kim K, Choi BK (2011) Identification of immunoreactive epitopes of the Porphyromonas gingivalis heat shock protein in periodontitis and atherosclerosis. J Periodontal Res 46(2):240–245

Choi B, Choi M, Park C et al (2015) Cytosolic Hsp60 orchestrates the survival and inflammatory responses of vascular smooth muscle cells in injured aortic vessels. Cardiovasc Res 106(3):498–508

Craig EA, Gambill BD, Nelson RJ (1993) Heat shock proteins: molecular chaperones of protein biogenesis. Microbiol Rev 57(2):402–414

Damluji AA, Ramireddy A, Al-Damluji MS et al (2015) Association between anti-human heat shock protein-60 and interleukin-2 with coronary artery calcium score. Heart 101(6):436–441

Danesh J, Collins R, Peto R (1997) Chronic infections and coronary heart disease: is there a link? Lancet 350(9075):430–436

Daugherty A, Pure E, Delfel-Butteiger D et al (1997) The effects of total lymphocyte deficiency on the extent of atherosclerosis in apolipoprotein $E^{-/-}$ mice. J Clin Invest 100(6):1575–1580

Deniset JF, Hedley TE, Dibrov E, Pierce GN (2012) Chlamydophila pneumoniae infection induces alterations in vascular contractile responses. Am J Pathol 180(3):1264–1272

Deniset JF, Hedley TE, Hlavackova M et al (2018) Heat shock protein 60 involvement in vascular smooth muscle cell proliferation. Cell Signal 47:44–51

Dieude M, Gillis MA, Theoret JF et al (2009) Autoantibodies to heat shock protein 60 promote thrombus formation in a murine model of arterial thrombosis. J Thromb Haemost 7(4):710–719

Ding J, Li J, Yang D et al (2018) Molecular characteristics of a novel Hsp60 gene and its differential expression in Manila clams (Ruditapes philippinarum) under thermal and hypotonic stress. Cell Stress Chaperones 23(2):179–187

Donaldson B, Al-Barwani F, Pelham SJ et al (2017) Multi-target chimaeric VLP as a therapeutic vaccine in a model of colorectal cancer. J Immunother Cancer 5(1):69

Dubois B, Chapat L, Goubier A et al (2003) Innate CD4+CD25+ regulatory T cells are required for oral tolerance and inhibition of CD8+ T cells mediating skin inflammation. Blood 102(9):3295–3301

Ellins E, Shamaei-Tousi A, Steptoe A et al (2008) The relationship between carotid stiffness and circulating levels of heat shock protein 60 in middle-aged men and women. J Hypertens 26(12):2389–2392

Esposito P, Tinelli C, Libetta C et al (2011) Impact of seropositivity to Chlamydia pneumoniae and anti-hHsp60 on cardiovascular events in hemodialysis patients. Cell Stress Chaperones 16(2):219–224

Fisch P, Malkovsky M, Braakman E et al (1990) Gamma/delta T cell clones and natural killer cell clones mediate distinct patterns of non-major histocompatibility complex-restricted cytolysis. J Exp Med 171(5):1567–1579

Foks AC, Lichtman AH, Kuiper J (2015) Treating atherosclerosis with regulatory T cells. Arterioscler Thromb Vasc Biol 35(2):280–287

Foks AC, Van Puijvelde GH, Wolbert J et al (2016) CD11b+Gr-1+ myeloid-derived suppressor cells reduce atherosclerotic lesion development in LDLr deficient mice. Cardiovasc Res 111(3):252–261

Foote CA, Castorena-Gonzalez JA, Ramirez-Perez FI et al (2016) Arterial stiffening in western diet-fed mice is associated with increased vascular elastin, transforming growth factor-beta, and plasma neuraminidase. Front Physiol 7:285

Friedman A, Weiner HL (1994) Induction of anergy or active suppression following oral tolerance is determined by antigen dosage. Proc Natl Acad Sci U S A 91(14):6688–6692

Frostegard J, Zhang Y, Sun J, Yan K, Liu A (2016) Oxidized low-density lipoprotein (OxLDL)-treated dendritic cells promote activation of T cells in human atherosclerotic plaque and blood, which is repressed by statins: microRNA let-7c is integral to the effect. J Am Heart Assoc 5(9):e003976

Galovic R, Flegar-Mestric Z, Vidjak V, Matokanovic M, Barisic K (2016) Heat shock protein 70 and antibodies to heat shock protein 60 are associated with cerebrovascular atherosclerosis. Clin Biochem 49(1–2):66–69

Gao Q, Jiang Y, Ma T et al (2010) A critical function of Th17 proinflammatory cells in the development of atherosclerotic plaque in mice. J Immunol 185(10):5820–5827

George J, Shoenfeld Y, Afek A et al (1999) Enhanced fatty streak formation in C57BL/6J mice by immunization with heat shock protein-65. Arterioscler Thromb Vasc Biol 19(3):505–510

George J, Afek A, Gilburd B, Shoenfeld Y, Harats D (2001) Cellular and humoral immune responses to heat shock protein 65 are both involved in promoting fatty-streak formation in LDL-receptor deficient mice. J Am Coll Cardiol 38(3):900–905

Grundtman C, Kreutmayer SB, Almanzar G, Wick MC, Wick G (2011) Heat shock protein 60 and immune inflammatory responses in atherosclerosis. Arterioscler Thromb Vasc Biol 31(5):960–968

Grundtman C, Jakic B, Buszko M et al (2015) Mycobacterial heat shock protein 65 (mbHSP65)-induced atherosclerosis: preventive oral tolerization and definition of atheroprotective and atherogenic mbHSP65 peptides. Atherosclerosis 242(1):303–310

Hansson GK, Libby P (2006) The immune response in atherosclerosis: a double-edged sword. Nat Rev Immunol 6(7):508–519

Hochleitner BW, Hochleitner EO, Obrist P et al (2000) Fluid shear stress induces heat shock protein 60 expression in endothelial cells in vitro and in vivo. Arterioscler Thromb Vasc Biol 20(3):617–623

Hu D, Mohanta SK, Yin C et al (2015) Artery tertiary lymphoid organs control aorta immunity and protect against atherosclerosis via vascular smooth muscle cell lymphotoxin beta receptors. Immunity 42(6):1100–1115

Hu Y, Chen Z, Jiang L et al (2018) Effects of oral and subcutaneous administration of Hsp60 on myeloid-derived suppressor cells and atherosclerosis in ApoE−/− mice. Biochem Biophys Res Commun 498(4):701–706

Huang CY, Shih CM, Tsao NW et al (2012) GroEL1, from Chlamydia pneumoniae, induces vascular adhesion molecule 1 expression by p37(AUF1) in endothelial cells and hypercholesterolemic rabbit. PLoS One 7(8):e42808

Huang CY, Shih CM, Tsao NW et al (2016) The GroEL protein of Porphyromonas gingivalis regulates atherogenic phenomena in endothelial cells mediated by upregulating toll-like receptor 4 expression. Am J Transl Res 8(2):384–404

Huber LA, Xu QB, Jurgens G et al (1991) Correlation of lymphocyte lipid composition membrane microviscosity and mitogen response in the aged. Eur J Immunol 21(11):2761–2765

Jaensson E, Uronen-Hansson H, Pabst O et al (2008) Small intestinal CD103+ dendritic cells display unique functional properties that are conserved between mice and humans. J Exp Med 205(9):2139–2149

Jakic B, Buszko M, Cappellano G, Wick G (2017) Elevated sodium leads to the increased expression of Hsp60 and induces apoptosis in HUVECs. PLoS One 12(6):e0179383

Jeong E, Kim K, Kim JH et al (2015) Porphyromonas gingivalis Hsp60 peptides have distinct roles in the development of atherosclerosis. Mol Immunol 63(2):489–496

Jing H, Yong L, Haiyan L et al (2011) Oral administration of Lactococcus lactis delivered heat shock protein 65 attenuates atherosclerosis in low-density lipoprotein receptor-deficient mice. Vaccine 29(24):4102–4109

Jongstra-Bilen J, Haidari M, Zhu SN et al (2006) Low-grade chronic inflammation in regions of the normal mouse arterial intima predisposed to atherosclerosis. J Exp Med 203(9):2073–2083

Jun L, Jie L, Dongping Y et al (2012) Effects of nasal immunization of multi-target preventive vaccines on atherosclerosis. Vaccine 30(6):1029–1037

Justo-Junior AS, Villarejos LM, Lima XTV et al (2018) Monocytes of patients with unstable angina express high levels of chemokine and pattern-recognition receptors. Cytokine 113:61–67

Karkhah A, Amani J (2016) A potent multivalent vaccine for modulation of immune system in atherosclerosis: an in silico approach. Clin Exp Vaccine Res 5(1):50–59

Karkhah A, Saadi M, Nouri HR (2017) In silico analyses of heat shock protein 60 and calreticulin to designing a novel vaccine shifting immune response toward T helper 2 in atherosclerosis. Comput Biol Chem 67:244–254

Karlin S, Brocchieri L (2000) Heat shock protein 60 sequence comparisons: duplications, lateral transfer, and mitochondrial evolution. Proc Natl Acad Sci U S A 97(21):11348–11353

Khallou-Laschet J, Tupin E, Caligiuri G et al (2006) Atheroprotective effect of adjuvants in apolipoprotein E knockout mice. Atherosclerosis 184(2):330–341

Kleindienst R, Xu Q, Willeit J et al (1993) Immunology of atherosclerosis. Demonstration of heat shock protein 60 expression and T lymphocytes bearing alpha/beta or gamma/delta receptor in human atherosclerotic lesions. Am J Pathol 142(6):1927–1937

Klingenberg R, Ketelhuth DF, Strodthoff D, Gregori S, Hansson GK (2012) Subcutaneous immunization with heat shock protein-65 reduces atherosclerosis in Apoe(−)/(−) mice. Immunobiology 217(5):540–547

Knoflach M, Kiechl S, Kind M et al (2003) Cardiovascular risk factors and atherosclerosis in young males: ARMY study (atherosclerosis risk-factors in male youngsters). Circulation 108(9):1064–1069

Knoflach M, Kiechl S, Mayrl B et al (2007) T-cell reactivity against Hsp60 relates to early but not advanced atherosclerosis. Atherosclerosis 195(2):333–338

Knoflach M, Kiechl S, Penz D et al (2009) Cardiovascular risk factors and atherosclerosis in young women: atherosclerosis risk factors in female youngsters (ARFY study). Stroke 40(4):1063–1069

Knoflach M, Messner B, Shen YH et al (2011) Non-toxic cadmium concentrations induce vascular inflammation and promote atherosclerosis. Circ J 75(10):2491–2495

Kol A, Bourcier T, Lichtman AH, Libby P (1999) Chlamydial and human heat shock protein 60s activate human vascular endothelium, smooth muscle cells, and macrophages. J Clin Invest 103(4):571–577

Kreutmayer SB, Messner B, Knoflach M et al (2011) Dynamics of heat shock protein 60 in endothelial cells exposed to cigarette smoke extract. J Mol Cell Cardiol 51(5):777–780

Kreutmayer S, Csordas A, Kern J et al (2013) Chlamydia pneumoniae infection acts as an endothelial stressor with the potential to initiate the earliest heat shock protein 60-dependent inflammatory stage of atherosclerosis. Cell Stress Chaperones 18(3):259–268

Langohr IM, HogenEsch H, Stevenson GW, Sturek M (2008) Vascular-associated lymphoid tissue in swine (*Sus scrofa*). Comp Med 58(2):168–173

Lee HR, Jun HK, Kim HD, Lee SH, Choi BK (2012) Fusobacterium nucleatum GroEL induces risk factors of atherosclerosis in human microvascular endothelial cells and ApoE(−/−) mice. Mol Oral Microbiol 27(2):109–123

Li J, Zhao X, Zhang S et al (2011) ApoB-100 and Hsp60 peptides exert a synergetic role in inhibiting early atherosclerosis in immunized ApoE-null mice. Protein Pept Lett 18(7):733–740

Li H, Ding Y, Yi G, Zeng Q, Yang W (2012) Establishment of nasal tolerance to heat shock protein-60 alleviates atherosclerosis by inducing TGF-beta-dependent regulatory T cells. J Huazhong Univ Sci Technolog Med Sci 32(1):24–30

Lin FY, Lin YW, Huang CY et al (2011) GroEL1, a heat shock protein 60 of Chlamydia pneumoniae, induces lectin-like oxidized low-density lipoprotein receptor 1 expression in endothelial cells and enhances atherogenesis in hypercholesterolemic rabbits. J Immunol 186(7):4405–4414

Lin YW, Huang CY, Chen YH et al (2013) GroEL1, a heat shock protein 60 of Chlamydia pneumoniae, impairs neovascularization by decreasing endothelial progenitor cell function. PLoS One 8(12):e84731

Long J, Lin J, Yang X et al (2012) Nasal immunization with different forms of heat shock protein-65 reduced high-cholesterol-diet-driven rabbit atherosclerosis. Int Immunopharmacol 13(1):82–87

Maan A, Jorgensen NW, Mansour M et al (2016) Association between heat shock protein-60 and development of atrial fibrillation: results from the Multi-Ethnic Study of Atherosclerosis (MESA). Pacing Clin Electrophysiol 39(12):1373–1378

Mayerl C, Lukasser M, Sedivy R et al (2006) Atherosclerosis research from past to present – on the track of two pathologists with opposing views, Carl von Rokitansky and Rudolf Virchow. Virchows Arch 449(1):96–103

Mayr M, Metzler B, Kiechl S et al (1999) Endothelial cytotoxicity mediated by serum antibodies to heat shock proteins of Escherichia coli and Chlamydia pneumoniae: immune reactions to heat shock proteins as a possible link between infection and atherosclerosis. Circulation 99(12):1560–1566

Mayr M, Kiechl S, Willeit J, Wick G, Xu Q (2000) Infections, immunity, and atherosclerosis: associations of antibodies to Chlamydia pneumoniae, Helicobacter pylori, and cytomegalovirus with immune reactions to heat-shock protein 60 and carotid or femoral atherosclerosis. Circulation 102(8):833–839

Mayr M, Kiechl S, Mendall MA et al (2003) Increased risk of atherosclerosis is confined to CagA-positive Helicobacter pylori strains: prospective results from the Bruneck study. Stroke 34(3):610–615

Messner B, Knoflach M, Seubert A et al (2009) Cadmium is a novel and independent risk factor for early atherosclerosis mechanisms and in vivo relevance. Arterioscler Thromb Vasc Biol 29(9):1392–1398
Metzler B, Schett G, Kleindienst R et al (1997) Epitope specificity of anti-heat shock protein 65/60 serum antibodies in atherosclerosis. Arterioscler Thromb Vasc Biol 17(3):536–541
Metzler B, Mayr M, Dietrich H et al (1999) Inhibition of arteriosclerosis by T-cell depletion in normocholesterolemic rabbits immunized with heat shock protein 65. Arterioscler Thromb Vasc Biol 19(8):1905–1911
Michaelsson J, Teixeira de Matos C, Achour A et al (2002) A signal peptide derived from Hsp60 binds HLA-E and interferes with CD94/NKG2A recognition. J Exp Med 196(11):1403–1414
Millonig G, Niederegger H, Rabl W et al (2001a) Network of vascular-associated dendritic cells in intima of healthy young individuals. Arterioscler Thromb Vasc Biol 21(4):503–508
Millonig G, Niederegger H, Wick G (2001b) Analysis of the cellular composition of the arterial intima with modified en face techniques. Lab Investig 81(4):639–641
Millonig G, Malcom GT, Wick G (2002) Early inflammatory-immunological lesions in juvenile atherosclerosis from the Pathobiological Determinants of Atherosclerosis in Youth (PDAY)-study. Atherosclerosis 160(2):441–448
Mundkur L, Ponnusamy T, Philip S et al (2014) Oral dosing with multi-antigenic construct induces atheroprotective immune tolerance to individual peptides in mice. Int J Cardiol 175(2):340–351
Musial K, Szprynger K, Szczepanska M, Zwolinska D (2009) Heat shock proteins in children and young adults on chronic hemodialysis. Pediatr Nephrol 24(10):2029–2034
Ohashi K, Burkart V, Flohe S, Kolb H (2000) Cutting edge: heat shock protein 60 is a putative endogenous ligand of the toll-like receptor-4 complex. J Immunol 164(2):558–561
Perschinka H, Mayr M, Millonig G et al (2003) Cross-reactive B-cell epitopes of microbial and human heat shock protein 60/65 in atherosclerosis. Arterioscler Thromb Vasc Biol 23(6):1060–1065
Perschinka H, Wellenzohn B, Parson W et al (2007) Identification of atherosclerosis-associated conformational heat shock protein 60 epitopes by phage display and structural alignment. Atherosclerosis 194(1):79–87
Pfister G, Stroh CM, Perschinka H et al (2005) Detection of Hsp60 on the membrane surface of stressed human endothelial cells by atomic force and confocal microscopy. J Cell Sci 118(Pt 8):1587–1594
Pockley AG, Bulmer J, Hanks BM, Wright BH (1999) Identification of human heat shock protein 60 (Hsp60) and anti-Hsp60 antibodies in the peripheral circulation of normal individuals. Cell Stress Chaperones 4(1):29–35
Powe NR, Bibbins-Domingo K (2016) Dietary salt, kidney disease, and cardiovascular health. JAMA 315(20):2173–2174
Rahman M, Steuer J, Gillgren P et al (2017) Induction of dendritic cell-mediated activation of T cells from atherosclerotic plaques by human heat shock protein 60. J Am Heart Assoc 6(11):e006778
Ridker PM, Rifai N, Rose L, Buring JE, Cook NR (2002) Comparison of C-reactive protein and low-density lipoprotein cholesterol levels in the prediction of first cardiovascular events. N Engl J Med 347(20):1557–1565
Rose NR, Witebsky E (1956) Studies on organ specificity. V. Changes in the thyroid glands of rabbits following active immunization with rabbit thyroid extracts. J Immunol 76(6):417–427
Ross R (1999) Atherosclerosis – an inflammatory disease. N Engl J Med 340(2):115–126
Rossmann A, Henderson B, Heidecker B et al (2008) T-cells from advanced atherosclerotic lesions recognize hHsp60 and have a restricted T-cell receptor repertoire. Exp Gerontol 43(3):229–237
Rzepecka J, Rausch S, Klotz C et al (2009) Calreticulin from the intestinal nematode Heligmosomoides polygyrus is a Th2-skewing protein and interacts with murine scavenger receptor-A. Mol Immunol 46(6):1109–1119
Schett G, Xu Q, Amberger A et al (1995) Autoantibodies against heat shock protein 60 mediate endothelial cytotoxicity. J Clin Invest 96(6):2569–2577

Schett G, Metzler B, Kleindienst R et al (1997) Salivary anti-hsp65 antibodies as a diagnostic marker for gingivitis and a possible link to atherosclerosis. Int Arch Allergy Immunol 114(3):246–250
Schulte S, Sukhova GK, Libby P (2008) Genetically programmed biases in Th1 and Th2 immune responses modulate atherogenesis. Am J Pathol 172(6):1500–1508
Seitz CS, Kleindienst R, Xu Q, Wick G (1996) Coexpression of heat-shock protein 60 and intercellular-adhesion molecule-1 is related to increased adhesion of monocytes and T cells to aortic endothelium of rats in response to endotoxin. Lab Investig 74(1):241–252
Soltys BJ, Gupta RS (1997) Cell surface localization of the 60 kDa heat shock chaperonin protein (Hsp60) in mammalian cells. Cell Biol Int 21(5):315–320
Stary HC (1992) Composition and classification of human atherosclerotic lesions. Virchows Arch A Pathol Anat Histopathol 421(4):277–290
Stulnig TM, Klocker H, Harwood HJ Jr et al (1995) In vivo LDL receptor and HMG-CoA reductase regulation in human lymphocytes and its alterations during aging. Arterioscler Thromb Vasc Biol 15(7):872–878
Suckling RJ, He FJ, Markandu ND, MacGregor GA (2012) Dietary salt influences postprandial plasma sodium concentration and systolic blood pressure. Kidney Int 81(4):407–411
Sun H, Shen J, Liu T et al (2014) Heat shock protein 65 promotes atherosclerosis through impairing the properties of high density lipoprotein. Atherosclerosis 237(2):853–861
Thota LN, Ponnusamy T, Philip S, Lu X, Mundkur L (2017) Immune regulation by oral tolerance induces alternate activation of macrophages and reduces markers of plaque destabilization in Apob(tm2Sgy)/Ldlr(tm1Her/J) mice. Sci Rep 7(1):3997
Traill KN, Huber LA, Wick G, Jurgens G (1990) Lipoprotein interactions with T cells: an update. Immunol Today 11(11):411–417
Tuleta I, Reek D, Braun P et al (2015) Influence of intimal Chlamydophila pneumoniae persistence on cardiovascular complications after coronary intervention. Infection 43(1):51–57
Van Eden W, Wick G, Albani S, Cohen I (2007) Stress, heat shock proteins, and autoimmunity: how immune responses to heat shock proteins are to be used for the control of chronic inflammatory diseases. Ann N Y Acad Sci 1113:217–237
Varbiro S, Biro A, Cervenak J et al (2010) Human anti-60 kD heat shock protein autoantibodies are characterized by basic features of natural autoantibodies. Acta Physiol Hung 97(1):1–10
Veglia F, Perego M, Gabrilovich D (2018) Myeloid-derived suppressor cells coming of age. Nat Immunol 19(2):108–119
Waltner-Romen M, Falkensammer G, Rabl W, Wick G (1998) A previously unrecognized site of local accumulation of mononuclear cells. The vascular-associated lymphoid tissue. J Histochem Cytochem 46(12):1347–1350
Wang C, Kankaanpaa J, Kummu O et al (2016) Characterization of a natural mouse monoclonal antibody recognizing epitopes shared by oxidized low-density lipoprotein and chaperonin 60 of Aggregatibacter actinomycetemcomitans. Immunol Res 64(3):699–710
Wick G, Kleindienst R, Dietrich H, Xu Q (1992) Is atherosclerosis an autoimmune disease? Trends Food Sci Technol 3:114–119
Wick G, Schett G, Amberger A, Kleindienst R, Xu Q (1995) Is atherosclerosis an immunologically mediated disease? Immunol Today 16(1):27–33
Wick G, Romen M, Amberger A et al (1997) Atherosclerosis, autoimmunity, and vascular-associated lymphoid tissue. FASEB J 11(13):1199–1207
Wick G, Knoflach M, Xu Q (2004) Autoimmune and inflammatory mechanisms in atherosclerosis. Annu Rev Immunol 22:361–403
Wick MC, Mayerl C, Backovic A et al (2008) In vivo imaging of the effect of LPS on arterial endothelial cells: molecular imaging of heat shock protein 60 expression. Cell Stress Chaperones 13(3):275–285
Wick G, Buhr N, Fraedrich G, Grundtman CA (2012a) Darwinian-evolutionary concept for atherogenesis: the role of immunity to Hsp60. In: Inflammation and Atherosclerosiseds. Springer, Wien, pp 171–196

Wick G, Oellinger R, Almanzar G (2012b) The Vascular-Associated Lymphoid Tissue (VALT). In: Wick G, Grundtman C (eds) Infammation and atherosclerosis. Springer, Wien, pp 77–86
Wick G, Jakic B, Buszko M, Wick MC, Grundtman C (2014) The role of heat shock proteins in atherosclerosis. Nat Rev Cardiol 11(9):516–529
Wick C, Onestingel E, Demetz E, Dietrich H, Wick G (2018) Oral tolerization with mycobacterial heat shock protein 65 reduces chronic experimental atherosclerosis in aged mice. Gerontology 64(1):36–48
Willeit J, Kiechl S (1993) Prevalence and risk factors of asymptomatic extracranial carotid artery atherosclerosis. A population-based study. Arterioscler Thromb 13(5):661–668
Witebsky E, Rose NR (1956) Studies on organ specificity. IV. Production of rabbit thyroid antibodies in the rabbit. J Immunol 76(6):408–416
Writing Group, M, Mozaffarian D, Benjamin EJ et al (2016) Heart disease and stroke statistics-2016 update: a report from the American Heart Association. Circulation 133(4):e38–e360
Wu HJ, Hao J, Wang SQ, Jin BL, Chen XB (2012) Protective effects of ligustrazine on TNF-alpha-induced endothelial dysfunction. Eur J Pharmacol 674(2–3):365–369
Wu C, Guo S, Niu Y et al (2016) Heat-shock protein 60 of Porphyromonas gingivalis may induce dysfunction of human umbilical endothelial cells via regulation of endothelial-nitric oxide synthase and vascular endothelial-cadherin. Biomed Rep 5(2):243–247
Xiao Q, Mandal K, Schett G et al (2005) Association of serum-soluble heat shock protein 60 with carotid atherosclerosis: clinical significance determined in a follow-up study. Stroke 36(12):2571–2576
Xiong Q, Feng J, Zhang Y et al (2016) Promotion of atherosclerosis in high cholesterol diet-fed rabbits by immunization with the P277 peptide. Immunol Lett 170:80–87
Xu QB, Oberhuber G, Gruschwitz M, Wick G (1990) Immunology of atherosclerosis: cellular composition and major histocompatibility complex class II antigen expression in aortic intima, fatty streaks, and atherosclerotic plaques in young and aged human specimens. Clin Immunol Immunopathol 56(3):344–359
Xu Q, Dietrich H, Steiner HJ et al (1992) Induction of arteriosclerosis in normocholesterolemic rabbits by immunization with heat shock protein 65. Arterioscler Thromb 12(7):789–799
Xu Q, Kleindienst R, Waitz W, Dietrich H, Wick G (1993) Increased expression of heat shock protein 65 coincides with a population of infiltrating T lymphocytes in atherosclerotic lesions of rabbits specifically responding to heat shock protein 65. J Clin Invest 91(6):2693–2702
Xu Q, Kleindienst R, Schett G et al (1996) Regression of arteriosclerotic lesions induced by immunization with heat shock protein 65-containing material in normocholesterolemic, but not hypercholesterolemic, rabbits. Atherosclerosis 123(1–2):145–155
Xu Q, Kiechl S, Mayr M et al (1999) Association of serum antibodies to heat-shock protein 65 with carotid atherosclerosis : clinical significance determined in a follow-up study. Circulation 100(11):1169–1174
Xu Q, Schett G, Perschinka H et al (2000) Serum soluble heat shock protein 60 is elevated in subjects with atherosclerosis in a general population. Circulation 102(1):14–20
Yoshimoto R, Fujita Y, Kakino A et al (2011) The discovery of LOX-1, its ligands and clinical significance. Cardiovasc Drugs Ther 25(5):379–391
Zhang X, He M, Cheng L et al (2008) Elevated heat shock protein 60 levels are associated with higher risk of coronary heart disease in Chinese. Circulation 118(25):2687–2693
Zhao Y, Zhang C, Wei X et al (2015) Heat shock protein 60 stimulates the migration of vascular smooth muscle cells via Toll-like receptor 4 and ERK MAPK activation. Sci Rep 5:15352
Zhong Y, Tang H, Wang X et al (2016) Intranasal immunization with heat shock protein 60 induces CD4(+) CD25(+) GARP(+) and type 1 regulatory T cells and inhibits early atherosclerosis. Clin Exp Immunol 183(3):452–468
Zugel U, Schoel B, Yamamoto S et al (1995) Crossrecognition by CD8 T cell receptor alpha beta cytotoxic T lymphocytes of peptides in the self and the mycobacterial Hsp60 which share intermediate sequence homology. Eur J Immunol 25(2):451–458

Chapter 14
Cardiac Myopathy in Conditional Hsp60 Transgenic Mice

Tsung-Hsien Chen, Tzu-Ying Chen, Shan-Wen Liu, Mei-Ru Chen, Yen-Ling Chen, Gin-Yen Lin, Arunim Guchait, Ching-Han Hsu, and Kurt Ming-Chao Lin

Abstract The mitochondrial chaperonin Hsp60 (*Hspd1*) plays important roles in sustaining cellular viability, regulate cellular functions and maintain homeostasis. Mutations in the Hsp60 gene or erratic expression has been frequently observed in wide-ranging human diseases. Targeting Hsp60 to ameliorate the prognosis of mitochondrial dysfunction-related diseases were proposed in the past. Genetically engineered mice provide a compelling tool to investigate the aetiology and pathogenesis of these diseases. Eventually, this will benefit the development of therapeutics towards these physiological complications. Conventional Hsp60 transgenic mice are often neonatally lethal. We've generated a unique conditional Hsp60 transgenic (Tg) mouse model to investigate the mitochondrial activities and demonstrated that

T.-H. Chen
Institute of Biomedical Engineering and Nanomedicine, National Health Research Institutes, Miaoli, Taiwan

Chiayi Christian Hospital, Chiayi, Taiwan

T.-Y. Chen · M.-R. Chen · Y.-L. Chen · G.-Y. Lin · K. M.-C. Lin (✉)
Institute of Biomedical Engineering and Nanomedicine, National Health Research Institutes, Miaoli, Taiwan
e-mail: klin@nhri.org.tw

S.-W. Liu
Institute of Biomedical Engineering and Nanomedicine, National Health Research Institutes, Miaoli, Taiwan

Institute of Biomedical Engineering and Environmental Sciences, National Tsing Hua University, Hsinchu, Taiwan

A. Guchait
Institute of Biomedical Engineering and Nanomedicine, National Health Research Institutes, Miaoli, Taiwan

Department of Life Sciences, National Central University, Taoyuan, Taiwan

C.-H. Hsu
Institute of Biomedical Engineering and Environmental Sciences, National Tsing Hua University, Hsinchu, Taiwan

A. A. A. Asea, P. Kaur (eds.), *Heat Shock Protein 60 in Human Diseases and Disorders*, Heat Shock Proteins 18,
https://doi.org/10.1007/978-3-030-23154-5_14

ubiquitous expression of human Hsp60 protein in mice leads to neonatal death due to septum defect (ASD) and cardiac myopathy. This chapter concisely reviews recent advances regarding manipulating Hsp60 levels in cells and mouse models along with depicts our quest to develop transgenic mice to study Hsp60-related human diseases.

Keywords Cardiac myopathy · Chaperonopathies · Conditional transgenic mice · Heart failure · Hsp60 · Mitochondrial molecular chaperone

Abbreviations

ASD	Atrial septum defect
ATP	Adenosine triphosphate
CAGGS	CMV enhancer/chicken β-actin
CLU	Clusterin
CMC	Cardiomyocytes
CNS	Central nervous system
H&E	Hematoxylin and eosin
HIIEα	Histocompatibility complex class II-Eα
HSP	Heat shock proteins
HSR	Heat shock response
IKK	Inhibitor of κB kinase
KO	Knockout
LVDP	Left ventricular developed pressure
NF-κB	Nuclear factor-κB
NOD	Nonobese diabetic
ROS	Reactive oxygen species
SPG13	Spastic paraplegia 13
Tg	Transgenic
UPR^{mt}	Mitochondrial unfolded protein response

14.1 Introduction

In all organisms, a distinct group of proteins termed as the heat shock proteins (HSP) are specially synthesized under heat stress and by a diverse range of external impetus including increased temperature (Currie et al. 1988), pressure overload (Katayose et al. 1993), ischemia (Richard et al. 1996), hypoxia (Heads et al. 1995), or changes in chemical environment. Commonly, HSP are molecular chaperones important for physiological and protective roles in cells, as they facilitate crucial activities, such as protein folding, material transport, and cellular signalling. Triggered by stress or protein denaturation, HSP facilitate to preserve the natural metabolic, structural and functional stability of the cell, as a protective feedback responses by preserving in their native forms and refolding denatured proteins (Benjamin and McMillan 1998; Liu et al. 2012; Macario and Conway de Macario 2007). This brisk induction of HSP in response to stress, attributed to the heat shock response (HSR) (Shamovsky and Nudler 2008).

14.1.1 The Curious Case of Hsp60

In general, HSP have been categorized based on their molecular weights. Mitochondrial chaperonin Hsp60 (Hspd1) together with Hsp10 (Hspe1) are constitutively expressed in normal condition as a folding machine for refolding imported proteins into the mitochondrial matrix (Ciocca and Calderwood 2005). Thus, Hsp60 is crucial in cell survival and to maintain mitochondrial functions, including the TCA cycle, respiration, and synthesis of ATP (Hartl 1996; Horwich et al. 2007). Hsp60 is concurrently induced under stress conditions by various stressors like heat shock, DNA damage, oxidative stress, and the unfolded protein response in mitochondria (Gupta and Knowlton 2002; Habich and Burkart 2007; Ohashi et al. 2000; Wick 2000). Hsp60 induction can result in pro-survival or pro-death consequences depending on the tissue type and stressors.

Upregulation of Hsp60 is an indicator of mitochondrial stress. This is well demonstrated in mitochondrial unfolded protein response (UPR^{mt}), where under stress condition nucleus-encoded mitochondrial chaperones (Hsp60, Hsp10 and mtHsp70) are induced by still poorly defined mitochondria-to-nucleus communication (Juwono and Martinus 2016). UPR^{mt} has been identified in worm, flies, and mouse, but the signalling pathways responsible for sensing the mitochondrial stress and activating nuclear gene transcription are only identified in flies and worms mediating transcriptional activation of Hsp60, including ATFS-1, DVE-1, UBL-5, and chromatin remodelling factor. Other forms of mitochondrial stress including oxidative stress (low concertation of hydrogen peroxide), hyperglycemic condition (100 mM glucose), and respiration stress (sodium azide 50 mM), were shown to result in ROS production, inhibition of mitochondrial dehydrogenase, and the induction of Hsp60 and mtHsp70 (Hall and Martinus 2013; Pellegrino et al. 2013). UPR^{mt} was shown to influence longevity, innate immunity, and diseases affecting the central nervous system (CNS) (Jovaisaite et al. 2014).

In human, *Hsp60* gene is situated on chromosome 2 and it shares a bidirectional promoter with *Hsp10* gene (Wu et al. 2017). Three major domains of Hsp60 are: the apical, intermediate, and equatorial domains (Sigler et al. 1998). The mechanistic integral biology of Hsp60, regarding substrate folding, has been investigated extensively. However, in the past few years, there has been an upsurge of interest about Hsp60, as roles of mitochondrial, cytosolic, and extracellular Hsp60 have been widely documented in numerous diseases. Although most of the Hsp60 proteins are transported and stayed in the mitochondrial matrix, they also appear in the cytoplasm. The elevated level of extracellular Hsp60, at least in part due to enhanced Hsp60 secretion, have been associated with type 2 diabetes, cancer, cardiovascular, and immunity-related diseases (Cappello et al. 2014; Caruso Bavisotto et al. 2017; Deocaris et al. 2006; Hohfeld and Hartl 1994).

The over-expression of Hsp60 has been reported to be linked with various cancers including colorectal cancer (Hamelin et al. 2011), hepatocellular carcinoma (Abdalla and Haj-Ahmad 2012), gastric cancer (Giaginis et al. 2009; Li et al. 2014), large bowel cancer (Campanella et al. 2015), prostate cancer (Skvortsov et al. 2011), head and neck cancer (Tsai et al. 2009), breast cancer (Desmetz et al. 2008), ovarian cancer (Hjerpe et al. 2013) and cervical cancer (Hwang et al. 2009). There are several reports that Hsp60 promotes cancer cell survival, such as in neuroblastoma cells by binding and inhibiting the intracellular CLU (clusterin) (Chaiwatanasirikul and Sala 2011). In another report, cytosolic Hsp60 interacts and regulates the inhibitor of κB kinase (IκB kinase or IKK) in human cervical cancer HeLa cells, which lead to the survival of cancer cells via nuclear factor-κB (NF-κB) (Chun et al. 2010). Inhibition of Hsp60 leads to caspase-dependent apoptosis and suppress tumour growth (Ghosh et al. 2010). By interacting with β-catenin, over-expression of Hsp60 promotes metastatic phenotypes in cancer cells (Tsai et al. 2009). In a murine model of ovarian cancer, treatment of tumour cells with a proteasome inhibitor, bortezomib, ensues into the upregulation of Hsp60 and Hsp90 on the surface of cancer cells and promotes phagocytosis by dendritic cells (Chang et al. 2012). Moreover, an anti-leukemic agent, azacytidine, has been reported to induce over-expression of Hsp60 in tumour cells (Tian et al. 2013). The pro-apoptotic role of Hsp60 in HeLa and Jurkat cells was also reported two decades ago (Samali et al. 1999; Xanthoudakis et al. 1999). Loss of Hsp60 expression has been documented, in the case of esophageal squamous cell carcinoma (Faried et al. 2004), ovarian cancer (Schneider et al. 1999) and bladder carcinoma (Lebret et al. 2003).

The role of Hsp60 in metabolic diseases has not been explored enough. Increased level of Hsp60 was observed in metabolic diseases, such as type 2 diabetes mellitus patients (Juwono and Martinus 2016; Yuan et al. 2011). Hsp60 has been identified as a mediator of adipose tissue inflammation and circulating Hsp60 levels were found elevated in obese individuals compared to lean controls (Märker et al. 2012). Moreover, it has been reported that obese mice develop an autoimmune response to Hsp60, which partially responsible for metabolic anomalies (Şelli et al. 2017). A recent study showed that elevated Hsp60 secretion as the response to IL-1β increases the phosphorylation of ERK, JNK, and p38 MAPK, and further augment the inflammation primarily via TLR4-p38 MAPK axis (Swaroop et al. 2016). Endurance exercise training increases Hsp60 expression in skeletal muscle, particularly in the type I muscle fibres and in the blood (Barone et al. 2016). This is the first report demonstrating differential responses to exercise, in various muscle types, by varying Hsp60 induction. Also, exercise increases Hsp60 expression level in the subcutaneous adipose tissue of diabetic and obese individuals, concomitantly alleviates inflammation (Khadir et al. 2018).

Interestingly, with age and in the case of metabolic diseases, Hsp60 expression has been reduced in the heart. According to the study, caloric restriction increases lifespan, improve cardiovascular activities and restore ageing-related abatement of Hsp60 expression in the heart (Colotti et al. 2005). Hsp60 is involved in protecting cardiac myopathy by preserving mitochondrial function, ATP synthesis and by suppressing cardiac myocyte apoptosis (Rizzo et al. 2011). In vitro studies have shown Hsp60 over-expression can result in cell survival or in cell death, depending on the cell type and models of the study. In neonatal rat cardiomyocytes (CMC), cells infected with an adenoviral construct by concomitantly overexpressing Hsp60 and Hsp10 were reported to be protected against simulated ischemia, whereas, cells infected with adenoviral constructs by overexpression of only Hsp60 or Hsp10 was less effective to ischemic injury (Lau et al. 1997). A follow-up study showed that combined or individual overexpression of Hsp60 and Hsp10 protect myocytes against apoptosis, preserve mitochondrial integrity and capability for ATP generation after simulated ischaemia and reperfusion (Lin et al. 2001). In the case of heart failure, cardiomyocytes secrete Hsp60 and its presence in the serum related to the severity of heart failure and cardiovascular risk (Bonanad et al. 2013; Nahas et al. 2014). Hsp60 is released via exosomes by adult cardiomyocytes and ectopic trafficking of Hsp60 to the cell surface may lead to the loss of myocyte and heart failure progression (Gupta and Knowlton 2007; Lin et al. 2007). Moreover, in cardiac myocytes, cytosolic Hsp60 interacts with apoptotic molecules Bax and Bak (Gupta and Knowlton 2005; Kirchhoff et al. 2002). Also, extracellular Hsp60 (exHsp60) binds to cardiac myocytes and involves in apoptosis (Kim et al. 2009). Even though, the involvement of Hsp60 in apoptosis of CMC was demonstrated in various in-vitro studies, its role and underlying molecular mechanisms for resulting in mitochondrial dysfunction and apoptosis in CMC remain elusive in vivo. Thus, genetically engineered mice models are needed to examine the underlying mechanisms of Hsp60 on the pathogenesis of cardiovascular risk and its possibilities in prognosis. Amid the inconsistency and complexity of in-vitro study reports, the role of Hsp60 as a potential biomarker and therapeutic target for the diagnosis and prognosis will remain clouded without the development of in vivo Hsp60 expressing transgenic mouse models.

14.1.2 Transgenic Hsp60 Mouse Models: A Brief History and New Possibilities

Though HSP can be induced by a variety of stimulants, yet in cell culture studies HSR is primarily induced by increasing temperature. Protocols used to induce the synthesis of HSP by exposing the cells to 40–45 °C heating for 15–20 min. Apparently, this temperature is standard in terms of thermotolerance. However, for cells, this temperature is extreme and may lead to disturbance to the cell's cytoskeleton and cytotoxicity. The heat-associated impairment includes the disintegration of the organization of keratin filaments (Shyy et al. 1989), actin filaments (Glass et al. 1985; van Bergen en Henegouwen and Linnemans 1987) and other undesired alterations in cellular metabolism. Because of the significance of Hsp60 in numerous diseases, transgenic animal models with inducible and tissue-specific Hsp60 expression will be beneficial to understand the pathogenesis and prognosis of these diseases. Moreover, by the development and introduction of genetically engineered animals which overexpress Hsp60 at any desired level, most of the problems related to thermal/stress induction of Hsp60 can be avoided. The significant benefit of using transgenic mouse models is, it's achievable to induce and intensify the level of Hsp60 in a tissue-specific manner, without the introduction of other metabolic alterations. The genetically engineered Hsp60 mouse models in human diseases are summarized in Table 14.1.

The story begins with the significance of Hsp60 in autoimmune diabetes. Hsp60 and Hsp70 of both prokaryotic and eukaryotic origins were identified as antigens of human diseases involving innate immunity (Dieude et al. 2004; Quintana and

Table 14.1 Hsp60 mouse models in human diseases and disorders

Model	Strain	Hsp60	Promoter[a]	Disease	References
Transgenic mice	NOD	MuHsp60	*H-2Ea*^{K}	Reducing autoimmune diabetes	Birk et al. (1996)
	C57BL/6	HuHsp60 lacks a MTS (AA 1–26)	CAGGS	Controlling mitochondrial-derived ROS through *NF-κB* target gene expression	Chun et al. (2010)
	FVB	HuHsp60	PGK	Chondrocyte proliferation and articular cartilage thickening	Ko et al. (2016)
Conditional transgenic mice	FVB	HuHsp60	CAGGS/CAGGS	Neonatal death, atrial septal defects	Chen et al. (2015)
	FVB/B6 hybrid	HuHsp60	CAGGS/*Myh6*	Cardiovascular disorders	Unpublished
Knockdown mice	Heterozygous			Loss of the *Hspd1* gene is lethal, disproportionately large number of male offspring	Christensen et al. (2010)
	C57BL/6			Late onset motor neuron disorder	Magnoni et al. (2013)
	C57BL/6			Hypothalamic insulin resistance, mitochondrial dysfunction	Kleinridders et al. (2013)
Conditional knockout mice	C57BL/6		-/*Villin*	Activates the UPR^{mt}, mitochondrial dysfunction	Berger et al. (2016)

PGK phosphoglycerate kinase promoter, *UPR*^{mt} mitochondrial unfolded protein response
[a]Transgenic promoter/Cre promoter

Cohen 2011; Quintana et al. 2004; Tanaka et al. 1999; van Eden et al. 2005; Zugel and Kaufmann 1999). By using nonobese diabetic (NOD) mice as a spontaneous mouse model of type I diabetes, murine Hsp60 transgene induced by the major histocompatibility complex class II-Eα (HIIEα) promoter was generated in NOD strain. The researchers achieved to express Hsp60 distinctly in the thymus and bone marrow and also have shown a significantly restrained propensity to autoimmunity induced diabetes mellitus in this nonobese diabetic (NOD) HIIEα-HSP60 Tg mice (Birk et al. 1996).

The existence of cytosolic Hsp60 involved in cellular signalling has been shown in certain cell types, such as cardiac myocytes and hepatocytes (Gupta and Knowlton 2002; Lai et al. 2007; Park et al. 2003). The researchers have expressed human Hsp60, lacking mitochondrial targeting sequence (MTS; amino acids 1–26 according to human sequence) into CAGGS transgenic vector in C57BL/6j mice. This transgenic mouse study, expressing truncated Hsp60 instead of the complete Hsp60, demonstrated that the resultant cytosolic Hsp60 Tg mice were impervious to hepatic stress with increased cell survival (Chun et al. 2010). As this study reported, Hsp60 directly interacts and influence the activation of the inhibitor of κB kinase (IκB kinase or IKK) and regulate mitochondrial-derived reactive oxygen species (ROS) via nuclear factor-κB (NF-κB) target gene expression, and this mechanism consequently leads to cell survival. A previous study also showed that Hsp60 interacts with the IKK (Cappello et al. 2008). As mitochondrial ROS has been related with human diseases like cancer, degenerative diseases, therefore, further research on the pro-survival role of cytosolic Hsp60 which fails to enter mitochondria, but regulates the ROS production through cytosolic pathways can shed a light on new therapeutics for these maladies (Coelho and Faria 2012).

In human, Hsp60 is encoded by *Hspd1* gene located within Chromosome 2. Its dysfunction is associated with some hereditary diseases such as autosomal dominantly inherited hereditary spastic paraplegia 13 (SPG13) and autosomal recessively inherited hypomyelinating leukodystrophy termed MitCHAP-60, caused by mutations in the *Hspd1* gene at equatorial domain of Hsp60 protein (Bross et al. 2008; Hansen et al. 2007; Hansen et al. 2003; Hansen et al. 2002; Magen et al. 2008), with a functional consequence affecting only the central nervous system. Hsp60 knockout (KO) mice were not successfully produced until recently. It was shown that Hsp60 homozygous KO mice which lack both functional *Hspd1* alleles, are lethal at early embryonical stage (at 7.5 dpc); by contrast, the heterozygous *Hspd1*$^{+/-}$ mouse, in which Hsp60 expression had been reduced by 50% in most organs was postnatally viable up to a few weeks (Christensen et al. 2010). The *Hspd1*$^{+/-}$ mice developed a late-onset, gradual dysfunction in motor functions due to *Hspd1* haploinsufficiency ensues in the hereditary spastic paraplegia-like features in mice, suggests a role for Hsp60 in late-onset motor neuron disorder (Magnoni et al. 2013). This heterozygous *Hspd1*$^{+/-}$ mice in combination with tissue-specific cre mouse have the possibility to serve as valuable mouse models and shed light on mechanistic details for diseases related to mitochondrial functional deficiencies and neurodegenerative disorders such as, Parkinson's disease, Alzheimer's disease, Huntington's disease, and multiple sclerosis (Dutta et al. 2006; Kwong et al. 2006).

Hspd1$^{+/-}$ mice present swollen mitochondria and deficient complex III activity in spinal cord and brain cortex with an increase of protein carbonylation (oxidation of protein side chains), indicative of increased ROS generation in these tissues. In the affected tissue, the decreased level of complex III subunit ubiquinone cytochrome c core protein1 (Uqcrc1), and the increase of ROS levels may be due to increased turnover of matrix superoxide dismutase (SOD2) as a result of impaired protein folding (Magnoni et al. 2014).

A prevalent characteristic of human obesity is leptin resistance and it's linked with insulin resistance and mitochondrial dysfunction (Myers et al. 2008). A recent study reported obesity is linked to mitochondrial dysfunction in the hypothalamus due to the reduction of Hsp60 and demonstrated Hsp60 as a leptin-induced mitochondrial chaperone. This study investigated a new perspective of Hsp60 in obesity and type 2 diabetes by documenting decreased Hsp60 in the brain of diabetic mice and humans. Mitochondrial dysfunction and lowered Hsp60 expression lead to weakened hypothalamic insulin signalling in mice. Heterozygous obliteration of Hsp60 in the hypothalamus results in mitochondrial dysfunction, elevated ROS, and insulin resistance. Hsp60 downregulation in the hypothalamus was also achieved by bilaterally injecting lentiviral vector enclosing shRNA against Hsp60 into the ventral hypothalamus, and resulted in insulin resistance in the mice. Thus, by using knockdown mouse model, Hsp60 has been found as a novel mediator correlates leptin/insulin crosstalk in the brain (Kleinridders et al. 2013).

Control of intestinal epithelial stemness is important for tissue homeostasis and disturbances in epithelial function can lead to gastrointestinal tract diseases (Sartor 2006). To understand how Hsp60 regulates the epithelial cell homeostasis in the intestine, the researchers have established the epithelial-specific knockout mice. In intestinal epithelial cell (IEC)-specific mouse model, intestinal epithelial-specific Hsp60 deletion resulted in defected UPR^{mt} and leads to mitochondrial dysfunction, impedes epithelial stem cell homeostasis (Berger et al. 2016). This finding may suggest that Hsp60 induction can potentially be beneficial by aggravating or simulating local UPR^{mt} in targeted lesions.

A newly reported study confirmed the pathological role of Hsp60 in osteoarthritis. Transgenic mice that overexpress human Hsp60 driven by phosphoglycerate kinase promoter were generated, which had higher chondrocyte proliferation along with thicker articular cartilage compared to wild-type mice. These findings suggest a therapeutic potential of targeting Hsp60 for osteoarthritis (Ko et al. 2016).

14.1.3 Conditional Hsp60 Transgenic Mouse Models to Study Cardiovascular Disorders

Protective roles of Hsp60, together with Hsp10, in the cardiovascular system by maintaining mitochondrial function and protecting from ischemia/reperfusion injury has been shown previously through a mechanism involving collaborative folding by Hsp60 and Hsp10 (Lau et al. 1997; Lin et al. 2001, 2004). Reduction in

Hsp60 expression and subsequent decline of insulin-like growth factor-1 receptor (IGF-1R) signalling in cardiac muscle cells have been implicated in the development of diabetic cardiomyopathy (Shan et al. 2003). We've generated a unique Hsp60 Tg mouse model (G-lox-HSP60) in FVB strain driven by a ubiquitous CMV early enhancer/chicken β-actin promoter (CAGGS) to investigate the mitochondrial function and generation of ROS in tissues isolated from our Hsp60 Tg mouse (Chen et al. 2015). This Hsp60 Tg vector allows tissue-specific induction of human Hsp60 in Tg mice. We reported that ubiquitous expression of human Hsp60 protein in mice results in neonatal death due to septum defect (ASD) and cardiac myopathy (Fig. 14.1). This conditional Hsp60 transgenic mice model surmounted the early lethality of the conventional transgenic method.

In our another project, we've generated conditional Hsp60 transgenic mice for heart-specific Hsp60 expression involving the G-Lox-HSP60 Tg vector and *Myh6*-creER^{T2} Tg vector driven by the *Myh6* promoter. A strong induction of human Hsp60 expression in the heart of double-Tg mice by 2-week tamoxifen feeding, was validated by western blotting. In the double Tg mice, we've observed the rapid induction of cardiac hypertrophy and dilated heart failure within 6–8 weeks after the tamoxifen treatment, which supports the hypothesis that perturbation of the Hsp60 level in

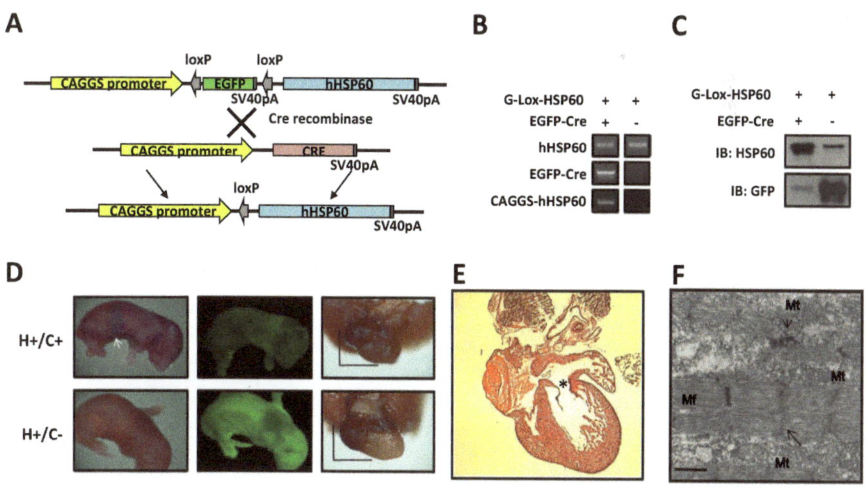

Fig. 14.1 (**a**) G-Lox-HSP60 and EGFP-Cre Tg vectors. The CAGGS promoter was used to drive both Tg vectors. After the LoxP sites were rejoined using the Cre DNA recombinase, the Hsp60 transcript was expressed. (**b**) Analysis of possible littermate genotypes using PCR on tailDNA. PCR amplification using the primer pair complementary to human Hsp60 (top row), amplification of EGFP-Cre (middle row), and the abridged sequence in the recombined vector from CAGGS promoter to human Hsp60 (bottom row). (**c**) Western blotting for Hsp60 and EGFP proteins in samples of B. (**d**) Pictures of neonatal Tg litters. The white arrow indicates cyanosis and abdominal bleeding in H$^+$/C$^+$ neonates; middle, fluorescent images of the same mice; right, the lungs and heart of neonatal mice. Scale bar = 3 mm. (**e**) Atrial septal defect in H$^+$/C$^+$ neonatal mice. (**f**) Transmission electron microscopy showing ultrastructure of myofibril defect in H$^+$/C$^+$ neonatal heart. This research was originally published in *Biomed Res Int.* (Chen et al. 2015)

cardiac myocytes can result in mitochondrial and calcium dysregulation, and in certain circumstances, precipitate cardiomyopathy. By using Langendorff isolated heart perfusion model and simulated ischemia/reperfusion protocol, the left ventricular developed pressure (LVDP) of hearts isolated from double-Tg mice were recorded. Strikingly, 4 weeks after Hsp60 induction, both LVDP at baseline and after reperfusion were significantly higher than uninduced double Tg mice or wild-type mice. Baseline LVDP as well as after reperfusion LVDP have been diminished at 6–7 weeks since the induction of Hsp60 (Fig. 14.2). The results suggest an opportunity of employing Hsp60 induction for treating diseases involving a reduced Hsp60 level, such as in brains and muscles during ageing and diabetes. This hypothesis has been further supported by a recent report showing benefit from enhanced Hsp60 expression during endurance training (Barone et al. 2016). Optimal Hsp60 induction in cardiac myocytes can be beneficial for cardiac function and ameliorate cardiac ischemic injuries. Whereas, prolonged Hsp60 induction may result in pathological consequences in the heart such as pathological remodelling and hypertrophy. Thus, this will be a potential conditional transgenic mice model to study cardiac myopathy.

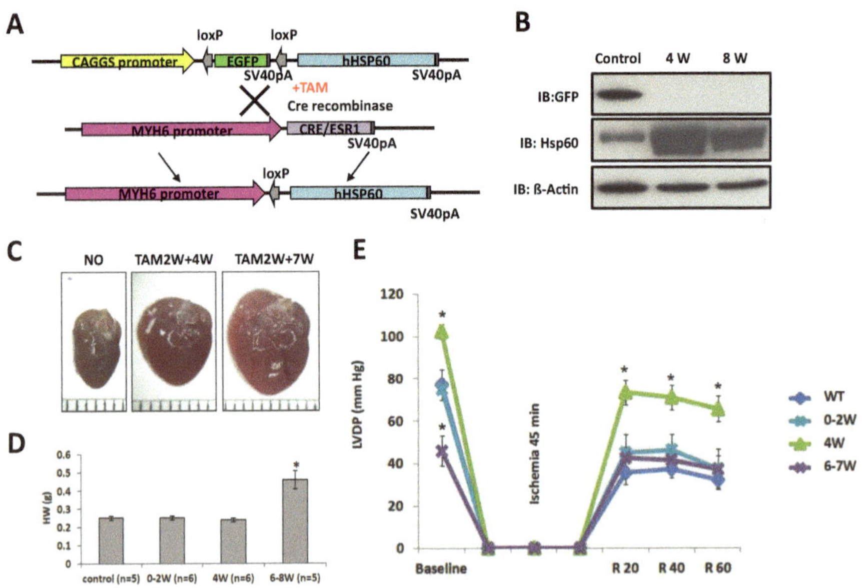

Fig. 14.2 (**a**) G-Lox-HSP60 and *Myh6*-creER^{T2} Tg vectors. (**b**) Western blotting for Hsp60 and GFP proteins in hearts at 4th week or 8th week after the period of tamoxifen feeding. (**c**) Bright-field images of hearts from double-Tg mice, which haven't received tamoxifen and sacrificed at 4 or 7 weeks after receiving tamoxifen. (**d**) Heart weight of double-Tg mice received control chow and mice sacrificed at 0–2, 4, 6–8 weeks after receiving tamoxifen. (**e**) Left ventricular developed pressure (LVDP) in wild-type and double-Tg mice hearts, that were uninduced or induced for 0–2, 4, or 6–7 weeks and subjected to Langendorff preparation. Simulated ischemia/reperfusion was induced by stopping the flow for 45 mins followed by reperfusion. Both LVDP at baseline and after reperfusion in hearts of double Tg mice at 4 weeks after Hsp60 induction, were significantly higher than other groups. Baseline LVDP as well as after reperfusion LVDP have been diminished at 6–7 weeks after the induction of Hsp60 ($*p < 0.05$)

14.2 Conclusions

Lately, the involvement of Hsp60 with a wide array of human diseases has gained increasing interests and focus on Hsp60. Though, Hsp60 has been explored extensively for more than three decades from the molecular, genetic, or protein aspects, its involvement in complex biological pathways are not yet completely explored. In vitro studies on Hsp60 have come a long way to discover the enormous amount of information on fundamental mechanisms and biological roles. As we've seen in this chapter that there are many contrasting results regarding the impact of Hsp60 in cells, in vivo, or in diseases. With the advances of molecular biology and cell analysis techniques such as cellular imaging, cryo EM, and due to the rapid developments in metabolic research, the unsolved puzzles of mitochondrial Hsp60 can be revisited, particularly about the molecular mechanisms of Hsp60 in the context of myopathy and protection against noxious stress. Moreover, the presence of extramitochondrial Hsp60 adds a new dimension to the research of Hsp60. The development of genetically engineered mouse models will enable to decode this highly complex mitochondrial chaperonin in in vivo settings. Knowledge to be gained will be beneficial to the better diagnosis and prognosis of diseases such as Hsp60 chaperonopathies, cancer, cardiovascular disorders, type 2 diabetes mellitus and other metabolic diseases.

Acknowledgements The research works were supported by intramural grants from the National Health Research Institutes of Taiwan [BNPP02-014]. We would like to thank Dr. Vivia Kao of Chia Nan University of Pharmacy and Science, Tainan, Taiwan for generating conditional constructs and Dr. Chun-Hong Chen of NHRI for valuable discussions.

References

Abdalla MA, Haj-Ahmad Y (2012) Promising urinary protein biomarkers for the early detection of hepatocellular carcinoma among high-risk hepatitis C virus Egyptian patients. J Cancer 3:390–403

Barone R, Macaluso F, Sangiorgi C et al (2016) Skeletal muscle heat shock protein 60 increases after endurance training and induces peroxisome proliferator-activated receptor gamma coactivator 1 alpha1 expression. Sci Rep 6:19781

Benjamin IJ, McMillan DR (1998) Stress (heat shock) proteins: molecular chaperones in cardiovascular biology and disease. Circ Res 83:117–132

Berger E, Rath E, Yuan D et al (2016) Mitochondrial function controls intestinal epithelial stemness and proliferation. Nat Commun 7:13171

Birk OS, Douek DC, Elias D et al (1996) A role of Hsp60 in autoimmune diabetes: analysis in a transgenic model. Proc Natl Acad Sci 93:1032

Bonanad C, Nunez J, Sanchis J et al (2013) Serum heat shock protein 60 in acute heart failure: a new biomarker? Congest Heart Fail 19:6–10

Bross P, Naundrup S, Hansen J et al (2008) The Hsp60-(p.V98I) mutation associated with hereditary spastic paraplegia SPG13 compromises chaperonin function both in vitro and in vivo. J Biol Chem 283:15694–15700

Campanella C, Rappa F, Sciume C et al (2015) Heat shock protein 60 levels in tissue and circulating exosomes in human large bowel cancer before and after ablative surgery. Cancer 121:3230–3239
Cappello F, Conway de Macario E, Marasa L, Zummo G, Macario AJ (2008) Hsp60 expression, new locations, functions and perspectives for cancer diagnosis and therapy. Cancer Biol Ther 7:801–809
Cappello F, Marino Gammazza A, Palumbo Piccionello A et al (2014) Hsp60 chaperonopathies and chaperonotherapy: targets and agents. Expert Opin Ther Targets 18:185–208
Caruso Bavisotto C, Cappello F, Macario AJL et al (2017) Exosomal HSP60: a potentially useful biomarker for diagnosis, assessing prognosis, and monitoring response to treatment. Expert Rev Mol Diagn 17:815–822
Chaiwatanasirikul KA, Sala A (2011) The tumour-suppressive function of CLU is explained by its localisation and interaction with HSP60. Cell Death Dis 2:e219
Chang C-L, Hsu Y-T, Wu C-C et al (2012) Immune mechanism of the antitumor effects generated by bortezomib. J Immunol 189:3209–3220
Chen TH, Liu SW, Chen MR et al (2015) Neonatal death and heart failure in mouse with transgenic HSP60 expression. Biomed Res Int 2015:539805
Christensen JH, Nielsen MN, Hansen J et al (2010) Inactivation of the hereditary spastic paraplegia-associated Hspd1 gene encoding the Hsp60 chaperone results in early embryonic lethality in mice. Cell Stress Chaperones 15:851–863
Chun JN, Choi B, Lee KW et al (2010) Cytosolic Hsp60 is involved in the NF-kappaB-dependent survival of cancer cells via IKK regulation. PLoS One 5:e9422
Ciocca DR, Calderwood SK (2005) Heat shock proteins in cancer: diagnostic, prognostic, predictive, and treatment implications. Cell Stress Chaperones 10:86–103
Coelho V, Faria AM (2012) HSP60: issues and insights on its therapeutic use as an immunoregulatory agent. Front Immunol 2:97
Colotti C, Cavallini G, Vitale RL et al (2005) Effects of aging and anti-aging caloric restrictions on carbonyl and heat shock protein levels and expression. Biogerontology 6:397–406
Currie RW, Karmazyn M, Kloc M, Mailer K (1988) Heat-shock response is associated with enhanced postischemic ventricular recovery. Circ Res 63:543–549
Deocaris CC, Kaul SC, Wadhwa R (2006) On the brotherhood of the mitochondrial chaperones mortalin and heat shock protein 60. Cell Stress Chaperones 11:116–128
Desmetz C, Bibeau F, Boissiere F et al (2008) Proteomics-based identification of HSP60 as a tumor-associated antigen in early stage breast cancer and ductal carcinoma in situ. J Proteome Res 7:3830–3837
Dieude M, Senecal JL, Raymond Y (2004) Induction of endothelial cell apoptosis by heat-shock protein 60-reactive antibodies from anti-endothelial cell autoantibody-positive systemic lupus erythematosus patients. Arthritis Rheum 50:3221–3231
Dutta R, McDonough J, Yin X et al (2006) Mitochondrial dysfunction as a cause of axonal degeneration in multiple sclerosis patients. Ann Neurol 59:478–489
Faried A, Sohda M, Nakajima M, Miyazaki T, Kato H, Kuwano H (2004) Expression of heat-shock protein Hsp60 correlated with the apoptotic index and patient prognosis in human oesophageal squamous cell carcinoma. Eur J Cancer 40:2804–2811
Ghosh JC, Siegelin MD, Dohi T, Altieri DC (2010) Heat shock protein 60 regulation of the mitochondrial permeability transition pore in tumor cells. Cancer Res 70:8988–8993
Giaginis C, Daskalopoulou SS, Vgenopoulou S, Sfiniadakis I, Kouraklis G, Theocharis SE (2009) Heat shock Protein-27, −60 and −90 expression in gastric cancer: association with clinicopathological variables and patient survival. BMC Gastroenterol 9:14
Glass JR, DeWitt RG, Cress AE (1985) Rapid loss of stress fibers in Chinese hamster ovary cells after hyperthermia. Cancer Res 45:258–262
Gupta S, Knowlton AA (2002) Cytosolic heat shock protein 60, hypoxia, and apoptosis. Circulation 106:2727–2733
Gupta S, Knowlton AA (2005) HSP60, Bax, apoptosis and the heart. J Cell Mol Med 9:51–58

Gupta S, Knowlton AA (2007) HSP60 trafficking in adult cardiac myocytes: role of the exosomal pathway. Am J Phys Heart Circ Phys 292:H3052–H3056
Habich C, Burkart V (2007) Heat shock protein 60: regulatory role on innate immune cells. Cell Mol Life Sci 64:742–751
Hall L, Martinus RD (2013) Hyperglycaemia and oxidative stress upregulate HSP60 & HSP70 expression in HeLa cells. Springerplus 2:431
Hamelin C, Cornut E, Poirier F et al (2011) Identification and verification of heat shock protein 60 as a potential serum marker for colorectal cancer. FEBS J 278:4845–4859
Hansen JJ, Durr A, Cournu-Rebeix I et al (2002) Hereditary spastic paraplegia SPG13 is associated with a mutation in the gene encoding the mitochondrial chaperonin Hsp60. Am J Hum Genet 70:1328–1332
Hansen JJ, Bross P, Westergaard M et al (2003) Genomic structure of the human mitochondrial chaperonin genes: HSP60 and HSP10 are localised head to head on chromosome 2 separated by a bidirectional promoter. Hum Genet 112:71–77
Hansen J, Svenstrup K, Ang D et al (2007) A novel mutation in the HSPD1 gene in a patient with hereditary spastic paraplegia. J Neurol 254:897–900
Hartl FU (1996) Molecular chaperones in cellular protein folding. Nature 381:571–579
Heads RJ, Yellon DM, Latchman DS (1995) Differential cytoprotection against heat stress or hypoxia following expression of specific stress protein genes in myogenic cells. J Mol Cell Cardiol 27:1669–1678
Hjerpe E, Egyhazi S, Carlson J et al (2013) HSP60 predicts survival in advanced serous ovarian cancer. Int J Gynecol Cancer 23:448–455
Hohfeld J, Hartl FU (1994) Role of the chaperonin cofactor Hsp10 in protein folding and sorting in yeast mitochondria. J Cell Biol 126:305–315
Horwich AL, Fenton WA, Chapman E, Farr GW (2007) Two families of chaperonin: physiology and mechanism. Annu Rev Cell Dev Biol 23:115–145
Hwang YJ, Lee SP, Kim SY et al (2009) Expression of heat shock protein 60 kDa is upregulated in cervical cancer. Yonsei Med J 50:399–406
Jovaisaite V, Mouchiroud L, Auwerx J (2014) The mitochondrial unfolded protein response, a conserved stress response pathway with implications in health and disease. J Exp Biol 217:137–143
Juwono J, Martinus RD (2016) Does Hsp60 provide a link between mitochondrial stress and inflammation in diabetes mellitus? J Diabetes Res 2016:6
Katayose D, Isoyama S, Fujita H, Shibahara S (1993) Separate regulation of heme oxygenase and heat shock protein 70 mRNA expression in the rat heart by hemodynamic stress. Biochem Biophys Res Commun 191:587–594
Khadir A, Kavalakatt S, Cherian P et al (2018) Physical exercise enhanced heat shock protein 60 expression and attenuated inflammation in the adipose tissue of human diabetic obese. Front Endocrinol (Lausanne) 9:16
Kim SC, Stice JP, Chen L et al (2009) Extracellular heat shock protein 60, cardiac myocytes, and apoptosis. Circ Res 105:1186–1195
Kirchhoff SR, Gupta S, Knowlton AA (2002) Cytosolic heat shock protein 60, apoptosis, and myocardial injury. Circulation 105:2899–2904
Kleinridders A, Lauritzen HP, Ussar S et al (2013) Leptin regulation of Hsp60 impacts hypothalamic insulin signaling. J Clin Invest 123:4667–4680
Ko JY, Sun YC, Li WC, Wang FS (2016) Chaperonin 60 regulation of SOX9 ubiquitination mitigates the development of knee osteoarthritis. J Mol Med (Berl) 94:755–769
Kwong JQ, Beal MF, Manfredi G (2006) The role of mitochondria in inherited neurodegenerative diseases. J Neurochem 97:1659–1675
Lai HC, Liu TJ, Ting CT et al (2007) Regulation of IGF-I receptor signaling in diabetic cardiac muscle: dysregulation of cytosolic and mitochondria HSP60. Am J Phys Endocrinol Metab 292:E292–E297

Lau S, Patnaik N, Sayen MR, Mestril R (1997) Simultaneous overexpression of two stress proteins in rat cardiomyocytes and myogenic cells confers protection against ischemia-induced injury. Circulation 96:2287–2294

Lebret T, Watson RW, Molinie V et al (2003) Heat shock proteins HSP27, HSP60, HSP70, and HSP90: expression in bladder carcinoma. Cancer 98:970–977

Li X-s, Xu Q, Fu X-y, Luo W-s (2014) Heat shock protein 60 overexpression is associated with the progression and prognosis in gastric cancer. PLoS One 9:e107507

Lin KM, Lin B, Lian IY, Mestril R, Scheffler IE, Dillmann WH (2001) Combined and individual mitochondrial HSP60 and HSP10 expression in cardiac myocytes protects mitochondrial function and prevents apoptotic cell deaths induced by simulated ischemia-reoxygenation. Circulation 103:1787–1792

Lin KM, Hollander JM, Kao VY, Lin B, Macpherson L, Dillmann WH (2004) Myocyte protection by 10 kD heat shock protein (Hsp10) involves the mobile loop and attenuation of the Ras GTPase pathway. FASEB J 18:1004–1006

Lin L, Kim SC, Wang Y et al (2007) HSP60 in heart failure: abnormal distribution and role in cardiac myocyte apoptosis. Am J Phys Heart Circ Phys 293:H2238–H2247

Liu T, Daniels CK, Cao S (2012) Comprehensive review on the HSC70 functions, interactions with related molecules and involvement in clinical diseases and therapeutic potential. Pharmacol Ther 136:354–374

Macario AJ, Conway de Macario E (2007) Molecular chaperones: multiple functions, pathologies, and potential applications. Front Biosci 12:2588–2600

Magen D, Georgopoulos C, Bross P et al (2008) Mitochondrial hsp60 chaperonopathy causes an autosomal-recessive neurodegenerative disorder linked to brain hypomyelination and leukodystrophy. Am J Hum Genet 83:30–42

Magnoni R, Palmfeldt J, Christensen JH et al (2013) Late onset motoneuron disorder caused by mitochondrial Hsp60 chaperone deficiency in mice. Neurobiol Dis 54:12–23

Magnoni R, Palmfeldt J, Hansen J, Christensen JH, Corydon TJ, Bross P (2014) The Hsp60 folding machinery is crucial for manganese superoxide dismutase folding and function. Free Radic Res 48:168–179

Märker T, Sell H, Zilleßen P et al (2012) Heat shock protein 60 as a mediator of adipose tissue inflammation and insulin resistance. Diabetes 61:615–625

Myers MG, Cowley MA, Munzberg H (2008) Mechanisms of leptin action and leptin resistance. Annu Rev Physiol 70:537–556

Nahas EA, Nahas-Neto J, Orsatti CL et al (2014) The 60- and 70-kDa heat-shock proteins and their correlation with cardiovascular risk factors in postmenopausal women with metabolic syndrome. Cell Stress Chaperones 19:559–568

Ohashi K, Burkart V, Flohe S, Kolb H (2000) Cutting edge: heat shock protein 60 is a putative endogenous ligand of the toll-like receptor-4 complex. J Immunol 164:558–561

Park SG, Lee SM, Jung G (2003) Antisense oligodeoxynucleotides targeted against molecular chaperonin Hsp60 block human hepatitis B virus replication. J Biol Chem 278:39851–39857

Pellegrino MW, Nargund AM, Haynes CM (2013) Signaling the mitochondrial unfolded protein response. Biochim Biophys Acta 1833:410–416

Quintana FJ, Cohen IR (2011) The HSP60 immune system network. Trends Immunol 32:89–95

Quintana FJ, Hagedorn PH, Elizur G, Merbl Y, Domany E, Cohen IR (2004) Functional immunomics: microarray analysis of IgG autoantibody repertoires predicts the future response of mice to induced diabetes. Proc Natl Acad Sci U S A 101(Suppl 2):14615–14621

Richard V, Kaeffer N, Thuillez C (1996) Delayed protection of the ischemic heart – from pathophysiology to therapeutic applications. Fundam Clin Pharmacol 10:409–415

Rizzo M, Macario AJ, de Macario EC et al (2011) Heat shock protein-60 and risk for cardiovascular disease. Curr Pharm Des 17:3662–3668

Samali A, Cai J, Zhivotovsky B, Jones DP, Orrenius S (1999) Presence of a pre-apoptotic complex of pro-caspase-3, Hsp60 and Hsp10 in the mitochondrial fraction of jurkat cells. EMBO J 18:2040–2048

Sartor RB (2006) Mechanisms of disease: pathogenesis of Crohn's disease and ulcerative colitis. Nat Clin Pract Gastroenterol Hepatol 3:390–407
Schneider J, Jimenez E, Marenbach K, Romero H, Marx D, Meden H (1999) Immunohistochemical detection of HSP60-expression in human ovarian cancer. Correlation with survival in a series of 247 patients. Anticancer Res 19:2141–2146
Şelli ME, Wick G, Wraith DC, Newby AC (2017) Autoimmunity to HSP60 during diet induced obesity in mice. Int J Obes (2005) 41:348–351
Shamovsky I, Nudler E (2008) New insights into the mechanism of heat shock response activation. Cell Mol Life Sci 65:855–861
Shan YX, Yang TL, Mestril R, Wang PH (2003) Hsp10 and Hsp60 suppress ubiquitination of insulin-like growth factor-1 receptor and augment insulin-like growth factor-1 receptor signaling in cardiac muscle: implications on decreased myocardial protection in diabetic cardiomyopathy. J Biol Chem 278:45492–45498
Shyy TT, Asch BB, Asch HL (1989) Concurrent collapse of keratin filaments, aggregation of organelles, and inhibition of protein synthesis during the heat shock response in mammary epithelial cells. J Cell Biol 108:997–1008
Sigler PB, Xu Z, Rye HS, Burston SG, Fenton WA, Horwich AL (1998) Structure and function in GroEL-mediated protein folding. Annu Rev Biochem 67:581–608
Skvortsov S, Schafer G, Stasyk T et al (2011) Proteomics profiling of microdissected low- and high-grade prostate tumors identifies Lamin A as a discriminatory biomarker. J Proteome Res 10:259–268
Swaroop S, Sengupta N, Suryawanshi AR, Adlakha YK, Basu A (2016) HSP60 plays a regulatory role in IL-1β-induced microglial inflammation via TLR4-p38 MAPK axis. J Neuroinflammation 13:27
Tanaka T, Yamakawa N, Koike N, Suzuki J, Mizuno F, Usui M (1999) Behcet's disease and antibody titers to various heat-shock protein 60s. Ocul Immunol Inflamm 7:69–74
Tian E, Tang H, Xu R, Liu C, Deng H, Wang Q (2013) Azacytidine induces necrosis of multiple myeloma cells through oxidative stress. Proteome Sci 11:24
Tsai YP, Yang MH, Huang CH et al (2009) Interaction between HSP60 and beta-catenin promotes metastasis. Carcinogenesis 30:1049–1057
van Bergen en Henegouwen PM, Linnemans AM (1987) Heat shock gene expression and cytoskeletal alterations in mouse neuroblastoma cells. Exp Cell Res 171:367–375
van Eden W, van der Zee R, Prakken B (2005) Heat-shock proteins induce T-cell regulation of chronic inflammation. Nat Rev Immunol 5:318
Wick G (2000) Atherosclerosis – an autoimmune disease due to an immune reaction against heat-shock protein 60. Herz 25:87–90
Wu J, Liu T, Rios Z, Mei Q, Lin X, Cao S (2017) Heat shock proteins and cancer. Trends Pharmacol Sci 38:226–256
Xanthoudakis S, Roy S, Rasper D et al (1999) Hsp60 accelerates the maturation of pro-caspase-3 by upstream activator proteases during apoptosis. EMBO J 18:2049–2056
Yuan J, Dunn P, Martinus RD (2011) Detection of Hsp60 in saliva and serum from type 2 diabetic and non-diabetic control subjects. Cell Stress Chaperones 16:689–693
Zugel U, Kaufmann SH (1999) Role of heat shock proteins in protection from and pathogenesis of infectious diseases. Clin Microbiol Rev 12:19–39

Part V
HSP60 and Neurological and Neurosciences

Chapter 15
Heat Shock Protein 60: An Effective Target Candidate in Neurological Diseases Treatment

Babita Sharma, Madhu Smita, Ishu Khangwal, Rajat Maheshwari, and Arun Kumar Dangi

Abstract The accumulation of intra- and extracellular misfolded proteins is found to play an imperative role in the progression of several neurological disorders including epilepsy, Alzheimer's disease, brain tumors, etc. which ultimately causes death worldwide. Heat shock protein (Hsp60) was found to be an important biomolecule that plays an essential role in the removal or degradation of these misfolded proteins and also act as a biomarker in disease prognosis. In neurological diseases, these systems are compromised due to deregulation or mutation of Hsp60 resulting in a large amount of aggregated proteins accumulation. Therefore, the development of novel and more efficient Hsp60 modulators is essential that can modulate the Hsp60 and involved pathways for the treatment of neurological diseases. For this propose, several In silico computational tools are developed, and many tools are under development. In this chapter, we discuss the role of Hsp60 in some more prominent neurological diseases and modulators developed in this direction. Further, we also highlight advanced computational tools that could be used for designing more Hsp60 modulators.

Keywords Alzheimer disease · Brain tumor · Epilepsy · Heat shock protein 60 · Hsp60 modulators neurological disorders

B. Sharma · I. Khangwal · A. K. Dangi (✉)
Enzyme Technology and Protein Bioinformatics Laboratory, Department of Microbiology, Maharshi Dayanand University, Rohtak, Haryana, India

M. Smita
Rahul Chaudhry and Partners, Gurugram, Haryana, India

R. Maheshwari
Department of Microbiology, Maharshi Dayanand University, Rohtak, Haryana, India

A. A. A. Asea, P. Kaur (eds.), *Heat Shock Protein 60 in Human Diseases and Disorders*, Heat Shock Proteins 18,
https://doi.org/10.1007/978-3-030-23154-5_15

Abbreviations

AD	Alzheimer's disease
FLD	Frontotemporal lobar degeneration
Hsp	Heat shock protein
INDD	Integrated neurodegenerative disease database
PDT	Photodynamic therapy
SRSs	Spontaneous recurrent seizures
TLE	Temporal lobe epilepsy

15.1 Introduction

Heat shock proteins (Hsp) are a special kind of protein generally found in all three kingdom prokaryotes, archaea and eukaryotes including human beings (Kim et al. 2013). The primary roles of Hsp are to help in the folding of nascent polypeptides to their native and accurate confirmation, prevent the aggregation of other proteins, degrade severely damaged proteins, and to regulate the apoptosis (Akerfelt et al. 2010). The failure of this cellular process leads to several neurological disorders. Although, different kinds of Hsp, like Hsp40, Hsp60, Hsp70, Hsp90, and Hsp100, among them, Hsp60 has been reported a critical molecule involved in several neurological disorders (Bross et al. 2012).

Hsp60 involved in translocation, folding and assembly of different proteins. It acts as a molecular chaperone which cooperates with other chaperones such as Hsp10/GroES, Hsp70 to fulfill their specific functions (Hartl 1996). Hsp60 (HspD1) is most commonly present in mitochondria, help in the transportation of nascent polypeptide from the cytoplasm to the mitochondrial matrix. It mainly works in combination with Hsp10 (Cpn10), resides in the mitochondria (Veereshwarayya et al. 2006). In addition to its classical chaperone function, Hsp60 is also involved in the replication and transmission of mitochondrial DNA molecule. It has been reported that mutations in Hsp60 molecules, unregulated the mtDNA levels that cause transmission defects (Samluk et al. 2018).

There are mainly two functions perform by Hsp60 concerning mitochondrial protein transport. First is to catalyze the folding of proteins destined for the matrix and then maintains protein in an unfolded state for transportation across the inner membrane of the mitochondria. Hsp60 binds to incoming proteins and induces conformational and structural changes, the hydrophobic portion of Hsp60 is responsible for maintaining the protein in unfolded conformation during transmembrane transport (Hartl and Hayer-Hartl 2002; Natalello et al. 2013). Further, consequent changes in the concentration of ATP hydrolyze the bonds between the protein and Hsp60 which signals the protein to exit from the mitochondria. Hsp60 is also capable of distinguishing between proteins designated for export and proteins destined to remain in the mitochondrial matrix by looking for an amphiphilic alpha-helix of 15–20 residues (Priya et al. 2013).

Additionally, some evidence indicates that Hsp60 also represents a regulator of an enormous and diverse range of cellular activities as far beyond to identify roles in protein folding and chaperoning (M van Noort et al. 2017). Now, it is well understood that Hsp60 to be involved in several processes such as synaptic transmission, autophagy, ER stress response, protein kinase and cell death signaling (Bie et al. 2016). Further, dysfunction of Hsp60 has robust effects on the fate of cells in neurological injury and disease states that leads to a variety of severe neurodegenerative disorders including Alzheimer' and Parkinson's disease, familial amyloidotic polyneuropathy and bovine spongiform encephalopathy as well as Jacob-Creutzfeldt disease (Cardinale et al. 2014). Deficiency in Hsp60 has recently been identified as being associated with atypical mitochondrial diseases in combination with multisystem failure (Amor et al. 2014). Further, proteomic analysis revealed that Hsp60 acts as an interactor of Parkin, PINK1, α-synuclein, and DJ-1. Among all these proteins PINK1 has different mitochondrial localization, but all these are involved in Parkinson's disease (Campanella et al. 2012). Hsp60 is among the most potent suppressors of neurodegeneration in animal models. Thus, Hsp60 provides a potential target for protective pharmacotherapy in many neurologic disorders (Benarroch 2011). This chapter presents the information about the role of Hsp60 in different neurological disorder with their multiple functions and dynamics of Hsp modulators with different computational tools designed to predict the conformational changes occurs during binding of Hsp60 with different proteins.

15.1.1 HSP60 in Neurological Diseases

Neuron dysfunction is caused by abnormal aggregation of misfolded or mutant proteins which is a vague and significant medical challenging condition (Cardinale et al. 2014). The plagued brain region and disruption of daily activities including sensory and motor functions are major clinical manifestations. These include a problem in moving, speaking, swallowing, breathing and cognitive dysfunctions. In different neurodegenerative diseases, misfolding and accumulation of proteins are known to be a significant cause of neuronal death and loss of synapses (Bross et al. 2012). There are various cellular processes which exacerbate or attenuate the diseases process (Wyttenbach et al. 2000). There are several mechanisms involved to clear off these protein aggregates which include molecular chaperones, the autophagy pathway, and the ubiquitin-proteasome system. Among all these mechanisms chaperone-mediated autophagy is known to be the most effective and promising approach to remove misfolded protein aggregation in the cells (Maiti et al. 2014).

Protein quality control mechanism of the mitochondrial matrix has been taken up by two protein system mitochondrial *m*-AAA protease system paraplegin and Afg312 (Neuspiel 2008). Defects of genes encoding in these systems lead to neurological disorders like hereditary spastic paraplegia SPG7, spinocerebellar ataxia SCA and early onset spatic ataxia-neuropathy syndrome (Rugarli and Langer 2012). Along with these findings mutation in the gene encoding, Hsp60 is associated with another form of hereditary spastic paraplegia SPG13 (Hansen et al. 2002). This disease is a subgroup of heterogeneous neurological disorders characterized by primary motor neuron degeneration, and the disease affects motor and sensory neurons with the longest axons. This degeneration is known as dying back degeneration (Laser et al. 2003).

A further role of Hsp60 was supported by findings that homozygosity for a missense mutation in the HSPD1 gene which is found to be associated with fatal hypomyelinatingleukodystrophy (MitCHAP60) (Magen et al. 2008). MRI revealed severe hypomyelination in the central nervous system and characterized by roatary nystagmus, early onset and progressive spastic paraplegia which is followed by neurological decline and worsening situations with severe mental retardation (Hansen et al. 2007). MitCHAP60 mutations for heterozygous siblings are entirely asymptomatic which recommends that a single mutant allele does not considerably impair the overall activity of Hsp60 in mitochondria (Bross et al. 2012).

15.1.1.1 Epilepsy

It is a central nervous system disorder affecting approximately 1% of the world population (Sendrowski and Sobaniec 2013). It is characterized by spontaneous recurrent seizures (SRSs) with a high frequency of neuronal discharges, unusual behavior, sensations and sometimes loss of consciousness (Gu and Daltone 2017). Temporal lobe epilepsy (TLE) is the most common one in adult humans which is characterized by the advanced development of SRSs from temporal lobe foci and morphological fluctuations in the hippocampus (Liu et al. 2008). Generally, TLE is initiated by head injury or stroke, infection in brain or delirious seizures which induces a status epilepticus (SE) and epileptogenesis is the period between initial injury and seizure. This period duration is 5–10 years in which many neurobiological events occur. At the site of injury, inflammation develops which involves glial and endothelial cells (Van Liefferinge et al. 2013). Clinically for diagnostic and prognosis purposes, biochemical measurements of inflammatory mediators in blood serum serve as a powerful tool. Hsp serves as promising candidates can be used as biomarkers among neuroinflammatory mediators including TLE (Chang et al. 2012). Hsp60, a constitutively expressed mitochondrial protein is expressed endogenously in astrocytes, neurons, oligodendrocytes, and microglia, ependymal cells of the brain. Thus, Hsp60 distribution in many of the cells of the brain suggests its active participation in normal and pathological conditions (Gu and Daltone 2017).

Gammazza et al. 2015 studied the expression and distribution of Hsp60 in the hippocampus of rats. They induce partial seizures in anesthetized rats which were based on the phenomenon of maximal dentate activation (MDA) which was recorded in the dentate gyrus (DG). This was induced by repetitive electrical stimulation of the perforant path in these rats and conducted analysis using western blot and immunohistochemistry of hippocampal tissues. Epileptic rats were assessed for Hsp60 by using ELISA. A similar technique was used to assay Hsp60 levels in the bloodstream of patients suffering from temporal lobe epilepsy. It was found that immunoreactivity of Hsp60 in rat models of TLE increased. The epileptic rats and patients with seizures showed a high level of circulating Hsp60 in the bloodstream. Their results demonstrate that Hsp60 can be potentially used as a biomarker and diagnostic for patients with these types of seizures.

15.1.1.2 Alzheimer's Disease

Alzheimer's disease (AD) is a type of dementia which is irreversible progressive and slowly destroys memory and thinking skills (Sonkusare et al. 2005). It is currently sent as the 6th leading cause of death in the US after heart disease and cancer (Rodgers 2002). In the AD, the expression of an Hsp60 is associated with the deposition of Aβ tangles of neurofibrils (Venegas and Heneka 2017). Current studies on Hsp showed that they prevent the accumulation of Aβ tangles. Papuc et al. 2015 studied the humoral response against Hsp60 in the AD and found out its potential as biomarkers for early diagnosis of disease. It has been known that anti-60 KD Hsp60 antibodies are present in the serum of healthy humans and the serum of patients with inflammatory and autoimmune disorders. A cellular signaling pathway involved in Hsp60-mediated NLRP3 inflammasome activation and subsequent IL-1β production is described in Fig. 15.1. So, it might be hypothesized that AD patients may be accompanied by the presence of anti-HSP antibodies due to the inflammatory process present in the brains of these patients. They didn't find any significant difference in antibody titer of the healthy individual as compared to AD patients. So, they concluded that anti Hsp60 antibodies that are present in these patients might belongs to a natural human immune system and does not significantly induce any effect on immunoreactivity against Hsp60.

Hsp60 is a mitochondrial chaperone which is required for its homeostasis and there are many human diseases associated with mutation in Hsp60 (also known as *HSPD1* gene) (Meng et al. 2018).V98I mutation in Hsp60 was reported to be associated with hereditary spastic paraplegia SPG 13 which is a rare neurodegenerative disorder characterized by weakness of the lower Limbs (Bross et al. 2008). Another mutation in Hsp60 (D27G) was identified from patients suffering in an autosomal recessive neurodegenerative disorder also called MitCHAP-60 disease (Magen et al. 2008). It is characterized by hypomyelination and leukodystrophy in the brain, and it was found that D27G mutant was less stable in forming hepta and decameric oligomers as compared to wild-type (Meng et al. 2018).

15.1.1.3 Brain Tumors

The frequency of development of brain tumors accounts for 2% of human neoplasms in adults and 20% in pediatric patients (McKeever 1999). These tumors comprise a different group of neoplasms which arises in the brain and its

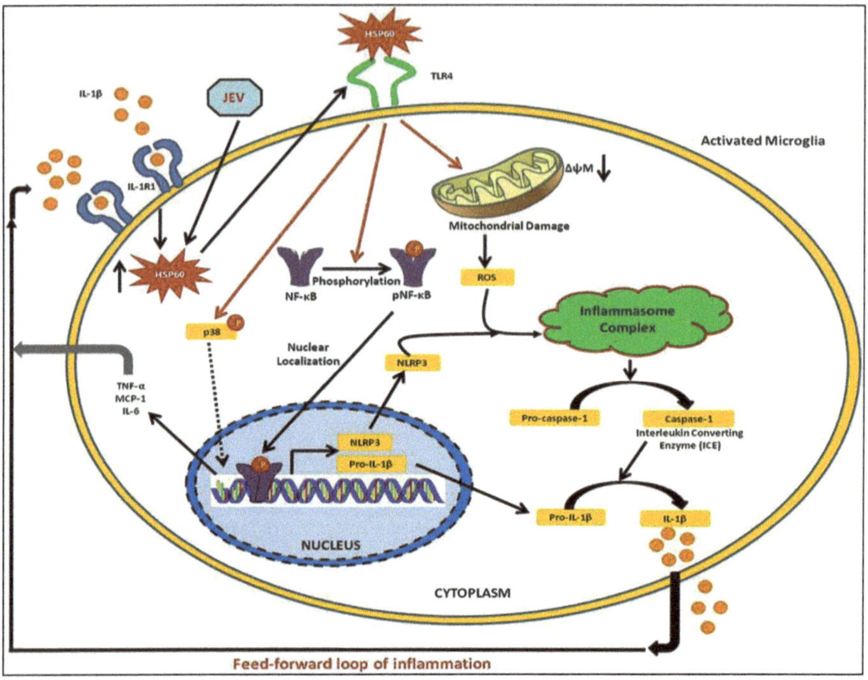

Fig. 15.1 Scheme of the signaling pathway involved in Hsp60-mediated NLRP3 inflammasome activation and subsequent IL-1β production. IL-1β induces its production by the activated microglia in an Hsp60-dependent manner. Hsp60, after being up-regulated by IL-1β, gets secreted outside and binds with TLR4 of the microglia to activate p38 MAPK. Binding of Hsp60 with TLR4 facilitates NF-κB phosphorylation, mitochondrial damage, and ROS generation and finally activates NLRP3 inflammasome leading to IL-1β production. JEV also augments Hsp60 production and thus influences inflammasome complex to induce a consecutive expression of IL-1β and, in turn, induces an exaggerated immune response

surrounding structures as shown in Fig. 15.2. Hsp had a promising role for cell survival functions and believed to be involved in the development of some human diseases known as chaperonopathies (Macario and Macario 2005). According to some studies, chaperones favors the tumor by inhibiting tumor cell apoptosis and presence of high levels of various Hsp in human brain tumors might be useful in early detection of disease (Campanella et al. 2012). Rappa et al. 2013 studied the level and cellular distribution of Hsp60 and Hsp70 in a series of brain tumors. They found the significantly higher level of Hsp60 in neuroepithelial tumors as compared to meningeal neoplasms, and Hsp60 was mainly present in the cytoplasmic level. They hypothesized that increment of level of Hsp60 is not by passive phenomenon but might be implicated in tumor progression. In another study by Kato et al. 2001, demonstrated expression and immunopositivity of Hsp60 in a series of 158 human brain tumors as compared to normal brain tissues.

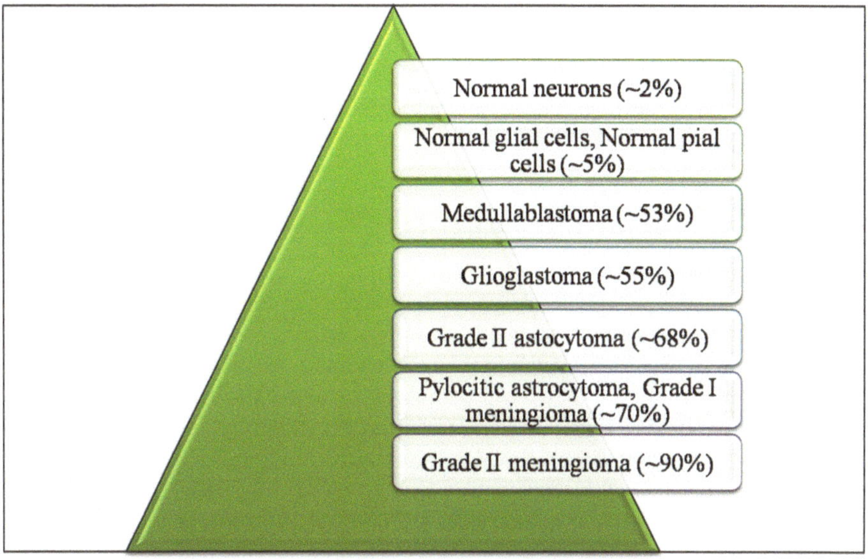

Fig. 15.2 An analysis of the immunopositivity of Hsp60 in brain tumors

15.1.2 HSP60 Modulators

There is a need to develop Hsp60 modulators that can target Hsp60 and useful in therapeutics of these disease conditions (Nakamura and Minegishi 2013). Already numerous compounds have been designed to focus on Hsp90, but few have been developed up till now for Hsp60 (Capello et al. 2014). Most of the Hsp60 modulators that are developed to date are bioactive compounds from studies of chemo proteomics. These modulators are classified into two types Type I and Type II (Pace et al. 2013). Type 1 inhibitors are work through inhibiting ATP binding and its hydrolysis. These inhibitors inhibit Hsp60 refolding activity as ATP-dependent conformation changes are affected (Meng et al. 2018). Type II inhibitors covalently react with specific cysteine residues in Hsp60. There is some natural product based Hsp-inhibitors and other from synthetic source (Radons 2017). Mizoribine is the first smallest organic molecule to be known as an Hsp60 inhibitor (Mizuno et al. 1974). This compound act as imidazole nucleoside antibiotic isolated from *Eupenicillium brefeldianum.* It does not contain any antimicrobial activity, but it is a potential immunosuppressor (Itoh et al. 1999).

Another natural compound which inhibits Hsp60 is epolactaene isolated from fungus *Penicillium* sp. Recently, MyrtucommulaneA (MC) is a non prenylatedacylphloroglucinol which contains antibacterial, antioxidant, anti-inflammatory, antitumor properties and isolated from human leukemia cells (Nagumo et al. 2004). Experiments have shown that Hsp60 is a direct target of MC. Indeed there is an excellent opportunity to develop potential therapeutic by targeting Hsp60 because the expression level of Hsp60 is increased in almost all neurodegenerative disorders

(Nagumo et al. 2005). Few small modulators of Hsp60 have been identified which includes both natural products and synthetic molecules but these all molecules does not share any common structural motif or pharmacophore, so it will be crucial to understanding how these inhibitors can interact with Hsp60.

15.1.3 Computational Tools Used in Drug Discovery

Hsp60 is a complex molecule having a large number of therapeutic application in drug discovery, by stimulation of different genes and protein effectively (Abramovitch 2018). The pro-tumoral strategies in Hsp60 include stimulation of pro-apoptotic pathways, pro-survival pathways, surface expression, and antitumor immune responses. The target of Hsp60 inhibitors is ATP binding sites and hydrolysis sites whose change in conformation can lead to the tumor (Nakamoto et al. 2018). A variety of docking databases can help to identify these conformational changes such as PDB bind database, some commercial programs such as LigandFit, Glide, GOLD, MOE Dock, and Surflex-Dock and some academic programs such as AutoDock, AutoDock Vina, LeDock, rDock, and UCSF DOCK also helpful in this direction (Padmadas et al. 2018). These programs can help in finding the accuracies of binding pose prediction (sampling power) and binding affinity (scoring power). Docking can assist not only in identifying these ligands with their specific sites also they can contribute to predicting the interactions between them (de Ruyck et al. 2016). Another place for Hsp60 attack is oxidation sites and binding sites with specific ligands. Improvement in computational tools such as target/ligand databases, homology modeling, ligand fingerprint methods, etc. can sort out the difficulties of predicting ligand activity, based on only similarities or dissimilarities. Various computer-aided drug designing can act as a companion for efficient drugs development (Hauser et al. 2017). Tumor-derived Hsp60 act as a loyal candidate for anti-tumor vaccines; derived peptides shows high immunomodulatory effect such as in arthritis. In addition to this, levels of Hsp60 influence by the levels of flavonoids and other proteins (Carlson 2002). Certain web tools such as SwissADME can detect these physicochemical properties, pharmacokinetics, drug-likeness. Some methods in cheminformatics such as BOILED-Egg and iLOGP are used to support drug discovery endeavors (Daina et al. 2017). They function collaboratory and interacts with others also; some proteins inhibit the expression of other Hsp such as Hsp60, Hsp70, Hsp27, and Hsp47. Photodynamic therapy (PDT) can help in identifying treatments for different oncologic and nononcologic lesions; it can act as a powerful tool for modulating *HSP60/HSPD1* gene expression tumor cell resistance to anti-tumor agents induced by PDT might be related to Hsp60 overexpression. For example, *E. coli* Hsp90 and the DnaK system work synergistically to remodel of the client protein. *E. coli* Hsp90 and DnaK interact both in vivo and in vitro, additional evidence justify that *E. coli* Hsp90 and the DnaK system function together (Genest et al. 2011).

Some DNA vaccines encoded with Hsp60 directly targets different disease such as arthritis, diabetes, obesity, etc. so, they can be modified and predicted by using various computational strategies. Moreover, various computational studies can be used to predict the epitopes design for particular diseases especially mycobacterium Hsp60which is responsible for T-cell activations. Here, Docking of peptides to the binding groove of MHC I proteins can be used to study the binding interactions of 15 antigenic CTL epitopes with three class I major histocompatibility complex (MHC I). The self-reactive T-cell population can be protected by the cross-reactivity between non-self- and self-Hsp60. Exosome-based tumor vaccines represent an exciting approach in inducing strong anti-tumor immune responses CD8(+) T cells have been recognized as the significant T-cell subset responsible for the anti-tumor effect of Hsp60-containing exosomes interpretation of anti-tumor vaccines based on Hsp60-containing exosomes act as novel anti-cancer therapies. In addition to this, chemotherapy possibly impacts Hsp60 expression has been the subject of numerous investigation (Leelananda and Lindert 2016).

In neurological disorders especially that targets the genes and protein expression, Hsp60 plays a significant role. Hsp60 plays a crucial role in protecting the brain, as it consists of proteins consisting of microglial triggering receptor expressed in myeloid (TREM). They bind on the surface of neuroblastoma cells and astrocytes when exposed to the surface of Hsp60. Under nonpathological conditions Hsp60-complexed TREM-2 synchronize functions of brain cells, these types of correlations can be studied by using various computational strategies such as Brain-coX which provide a wide range of transcriptomics studies based on brain and gene expressions. In microglia, it was found that different interactions of extracellular Hsp60 with microglial LOX-1 boost the production of pro-inflammatory factors (IL-1β, NO and ROS) and propagate neuronal damage. For a better analysis of gene networks for hsp60 in different neurological disorders, NeuroDNet is a web server using Mysql - 5.0.18 - Win32 and PHP - 5.2.0 (Bendl et al. 2016; Jorgensen 2004). For the comparative analysis, an Integrated Neurodegenerative Disease Database (INDD) has been made of frontotemporal lobar degeneration (FLD) (Sliwoski et al. 2014). Extracellular Hsp60 induces an inflammatory response and soluble neuronal injury signal having a direct link with neuroinflammation and neurotoxicity. More informative studies can be done by Alzforum, by which diagnostics and treatment related to Alzheimer disease and related disorders can be retrieved quickly. Alzforum has developed other ten open-access databases- AlzBiomarker, AlzGene, AlzPedia, AlzRisk, Antibodies, Brain Banks, Mutations, Research Models, Protocols, Therapeutics which directly links it to all aspects of AD and PD (Weinberg et al. 2017). In addition to this other database such as CIDeR and CombiROC can be used for integrative studies of metabolic and neurological disorders, it can help in determining best possible biomarkers for specific diseases (Scott et al. 2016).

15.2 Conclusions

In recent years, there has been a spectacular augment in cases where individual suffering from neurological disorders worldwide. It was observed from different studies that Hsp60 deregulation or malfunction play a significant role in the progression of these diseases. In a healthy individual, Hsp60 involves in refolding, or degradation of misfolded proteins thus prevents the development of the disease. So, the development of Hsp60 modulators could be a better and effective treatment option for several neurological diseases. In this direction, several modulators such as Mizoribine and Myrtucommulane A were discovered from natural sources that can inhibit or modulate the Hsp60 activity and pathways where Hsp60 takes part. Further, several computer-based advanced docking and simulation models have been developed that can be used for designing and screening of more potent modulators for disease treatment.

Acknowledgements The author acknowledges Maharshi Dayanand University, Rohtak for providing infrastructure and lab facility for the compilation of this interesting and informative book chapter.

References

Abramovitch RB (2018) *Mycobacterium tuberculosis* reporter strains as tools for drug discovery and development. IUBMB Life 70:818–825

Åkerfelt M, Morimoto RI, Sistonen L (2010) Heat shock factors: integrators of cell stress, development and lifespan. Nat Rev Mol Cell Biol 11:545

Amor S, Peferoen LA, Vogel DY, Breur M, van der Valk P, Baker D, van Noort JM (2014) Inflammation in neurodegenerative diseases–an update. Immunology 142:151–166

Benarroch EE (2011) Heat shock proteins multiple neuroprotective functions and implications for neurologic disease. Neurology 76:660–667

Bendl J, Stourac J, Sebestova E, Vavra O, Musil M, Brezovsky J et al (2016) HotSpot Wizard 2.0: automated design of site-specific mutations and smart libraries in protein engineering. Nucleic Acids Res 44:W479–W487

Bie AS, Fernandez-Guerra P, Birkler RI, Nisemblat S, Pelnena D, Lu X, Deignan JL, Lee H, Dorrani N, Corydon TJ, Palmfeldt J (2016) Effects of a mutation in the HSPE1 gene encoding the mitochondrial co-chaperonin HSP10 and its potential association with a neurological and developmental disorder. Front Mol Biosci 3:65

Bross P, Naundrup S, Hansen J, Nielsen MN, Christensen JH, Kruhøffer M, Palmfeldt J, Corydon TJ, Gregersen N, Ang D, Georgopoulos C (2008) The Hsp60-(p. V98I) mutation associated with hereditary spastic paraplegia SPG13 compromises chaperonin function both in vitro and in vivo. J Biol Chem 283(23):15694–15700

Bross P, Magnoni R, SigaardBie A (2012) Molecular chaperone disorders: defective Hsp60 in neurodegeneration. Curr Top Med Chem 12(22):2491–2503

Campanella C, Bucchieri F, Merendino AM, Fucarino A, Burgio G, Corona DF, de Macario EC (2012) The odyssey of Hsp60 from tumor cells to other destinations includes plasma membrane-associated stages and Golgi and exosomal protein-trafficking modalities. PLoS One 7(7):e42008

Cappello F, Marino Gammazza A, Palumbo Piccionello A, Campanella C, Pace A, Conway de Macario E, Macario AJ (2014) Hsp60 chaperonopathies and chaperonotherapy: targets and agents. Expert Opin Ther Targets 18(2):185–208

Cardinale A, Chiesa R, Sierks M (2014) Protein misfolding and neurodegenerative diseases. Int J Biochem Cell Biol 2014:2014

Carlson HA (2002) Protein flexibility and drug design: how to hit a moving target. Curr Opin Chem Biol 6:447–452

Chang CC, Lui CC, Lee CC, Chen SD, Chang WN, Lu CH, Chuang YC (2012) Clinical significance of serological biomarkers and neuropsychological performances in patients with temporal lobe epilepsy. BMC Neurol 12:15

Daina A, Michielin O, Zoete V (2017) SwissADME: a free web tool to evaluate pharmacokinetics, drug-likeness and medicinal chemistry friendliness of small molecules. Sci Rep 7:42717

de Ruyck J, Brysbaert G, Blossey R, Lensink MF (2016) Molecular docking as a popular tool in drug design, an in silico travel. Adv Appl Bioinforma Chem: AABC 9:1

Gammazza AM, Colangeli R, Orban G, Pierucci M, Di Gennaro G, Bello ML, Muscat R (2015) Hsp60 response in experimental and human temporal lobe epilepsy. Sci Rep 5:9434

Genest O, Hoskins JR, Camberg JL, Doyle SM, Wickner S (2011) Heat shock protein 90 from *Escherichia coli* collaborates with the DnaK chaperone system in client protein remodeling. Proc Natl Acad Sci 108:8206–8211

Gu B, Daltone KA (2017) Models and detection of spontaneous recurrent seizures in laboratory rodents. Zool Res 38(4):171–179

Hansen JJ, Dürr A, Cournu-Rebeix I, Georgopoulos C, Ang D, Nielsen MN, Davoine CS, Brice A, Fontaine B, Gregersen N, Bross P (2002) Hereditary spastic paraplegia SPG13 is associated with a mutation in the gene encoding the mitochondrial chaperonin Hsp60. Am J Hum Genet 70(5):1328–1332

Hansen J, Svenstrup K, Ang D, Nielsen MN, Christensen JH, Gregersen N, Bross P (2007) A novel mutation in the HSPD1 gene in a patient with hereditary spastic paraplegia. J Neurol 254(7):897–900

Hartl FU (1996) Molecular chaperones in cellular protein folding. Nature 381:571–579

Hartl FU, Hayer-Hartl M (2002) Molecular chaperones in the cytosol: from nascent chain to folded protein. Science 295:1852–1858

Hauser AS, Attwood MM, Rask-Andersen M, Schiöth HB, Gloriam DE (2017) Trends in GPCR drug discovery: new agents, targets and indications. Nat Rev Drug Discov 16:829

Itoh H, Komatsuda A, Wakui H, Miura AB, Tashima Y (1999) Mammalian HSP60 is a major target for an immunosuppressant mizoribine. J Biol Chem 274(49):35147–35151

Jorgensen WL (2004) The many roles of computation in drug discovery. Science (80-) 303:1813–1818

Kato S, Kato M, Hirano A, Takikawa M, Ohama E (2001) The immunohistochemical expression of stress-response protein (srp) 60 in human brain tumours: relationship of srp 60 to the other five srps, proliferating cell nuclear antigen and p53 protein. Histol Histopathol 16(3):809–820

Kim YE, Hipp MS, Bracher A, Hayer-Hartl M, Ulrich Hartl F (2013) Molecular chaperone functions in protein folding and proteostasis. Annu Rev Biochem 82:323–355

Laser H, Mack TG, Wagner D, Coleman MP (2003) Proteasome inhibition arrests neurite outgrowth and causes “dying-back” degeneration in primary culture. J Neurosci Res 74(6):906–916

Leelananda SP, Lindert S (2016) Computational methods in drug discovery. Beilstein J Org Chem 12:2694

Liu XY, Yang JL, Chen LJ, Zhang Y, Yang ML, Wu YY, Wei YQ (2008) Comparative proteomics and correlated signaling network of rat hippocampus in the pilocarpine model of temporal lobe epilepsy. Proteomics 8(3):582–603

M van Noort J, Bugiani M, Amor S (2017) Heat shock proteins: old and novel roles in neurodegenerative diseases in the central nervous system. CNS Neurol Disord-Drug Targets 16:244–256

Macario AJ, de Macario EC (2005) Sick chaperones, cellular stress, and disease. N Engl J Med 353(14):1489–1501

Magen D, Georgopoulos C, Bross P, Ang D, Segev Y, Goldsher D, Heno B (2008) Mitochondrial hsp60 chaperonopathy causes an autosomal-recessive neurodegenerative disorder linked to brain hypomyelination and leukodystrophy. Am J Hum Genet 83(1):30–42
Maiti P, Manna J, Veleri S, Frautschy S (2014) Molecular chaperone dysfunction in neurodegenerative diseases and effects of curcumin. Biomed Res Int 2014:2014
McKeever PE (1999) The brain, spinal cord and meninges. Diagn Surg Pathol 1:408–409
Meng Q, Li BX, Xiao X (2018) Toward developing chemical modulators of Hsp60 as potential therapeutics. Front Mol Biosci 5:35
Mizuno K, Tsujino M, Takada M, Hayashi M, Atsumi K, Asano K, Matsuda T (1974) Studies on bredinin. J Antibiot 27(10):775–782
Nagumo Y, Kakeya H, Yamaguchi J, Uno T, Shoji M, Hayashi Y, Osada H (2004) Structure–activity relationships of epolactaene derivatives: structural requirements for inhibition of Hsp60 chaperone activity. Bioorg Med Chem Lett 14(17):4425–4429
Nagumo Y, Kakeya H, Shoji M, Hayashi Y, Dohmae N, Osada H (2005) Epolactaene binds human Hsp60 Cys442 resulting in the inhibition of chaperone activity. Biochem J 387(3):835–840
Nakamoto H, Amaya Y, Komatsu T, Suzuki T, Dohmae N, Nakamura Y et al (2018) Stimulation of the ATPase activity of Hsp90 by zerumbone modification of its cysteine residues destabilizes its clients and causes cytotoxicity. Biochem J 475:2559–2576. BCJ20180230
Nakamura H, Minegishi H (2013) HSP60 as a drug target. Curr Pharm Des 19(3):441–451
Natalello A et al (2013) Biophysical characterization of two different stable misfolded monomeric polypeptides that are chaperone-amenable substrates. J Mol Biol 425:1158–1171
Neuspiel M (2008) Mitochondrial dynamics, mechanisms and meaning: Investigations into fusion and fission proteins Mitofusin 2 and MAPL (Doctoral dissertation, University of Ottawa (Canada)
Pace A, Barone G, Lauria A, Martorana A, Palumbo Piccionello A, Pierro P, Angileri F (2013) Hsp60, a novel target for antitumor therapy: structure-function features and prospective drugs design. Curr Pharm Des 19(15):2757–2764
Padmadas N, Panda PK, Durairaj S (2018) Binding patterns associated Aß-HSP60 p458 conjugate to HLA-DR-DRB allele of human in Alzheimer's disease: an in silico approach. Interdiscip Sci Comput Life Sci 10:93–104
Papuć E, Kurys-Denis E, Krupski W, Tatara M, Rejdak K (2015) Can antibodies against glial derived antigens be early biomarkers of hippocampal demyelination and memory loss in Alzheimer's disease? J Alzheimers Dis 48:115–121
Priya S, Sharma SK, Goloubinoff P (2013) Molecular chaperones as enzymes that catalytically unfold misfolded polypeptides. FEBS Lett 587:1981–1987
Radons J (2017) The ATP-driven Hsp60 machinery: biological and clinical implications. Curr Immunol Rev 13(1):19–43
Rappa F, Unti E, Baiamonte P, Cappello F, Scibetta N (2013) Different immunohistochemical levels of Hsp60 and Hsp70 in a subset of brain tumors and putative role of Hsp60 in neuroepithelial tumorigenesis. Eur J Histochem 57(2):e20
Rodgers AB (2002) Alzheimer's disease: unraveling the mystery, vol 1. National Institutes of Health
Rugarli EI, Langer T (2012) Mitochondrial quality control: a matter of life and death for neurons. EMBO J 31(6):1336–1349
Samluk L, Chroscicki P, Chacinska A (2018) Mitochondrial protein import stress and signaling. Curr Opin Physiol 3:41–48
Scott DE, Bayly AR, Abell C, Skidmore J (2016) Small molecules, big targets: drug discovery faces the protein–protein interaction challenge. Nat Rev Drug Discov 15:533
Sendrowski K, Sobaniec W (2013) Hippocampus, hippocampal sclerosis and epilepsy. Pharmacol Rep 65(3):555–565
Sliwoski G, Kothiwale S, Meiler J, Lowe EW (2014) Computational methods in drug discovery. Pharmacol Rev 66:334–395

Sonkusare SK, Kaul CL, Ramarao P (2005) Dementia of Alzheimer's disease and other neurodegenerative disorders—memantine, a new hope. Pharmacol Res 51(1):1–7

Van Liefferinge J, Massie A, Portelli J, Di Giovanni G, Smolders I (2013) Are vesicular neurotransmitter transporters potential treatment targets for temporal lobe epilepsy? Front Cell Neurosci 7:139

Veereshwarayya V, Kumar P, Rosen KM, Mestril R, Querfurth HW (2006) Differential effects of mitochondrial heat shock protein 60 and related molecular chaperones to prevent intracellular β-amyloid-induced inhibition of complex IV and limit apoptosis. J Biol Chem 281:29468–29478

Venegas C, Heneka MT (2017) Danger-associated molecular patterns in Alzheimer's disease. J Leukoc Biol 101(1):87–98

Weinberg BA, Yu H, Rosen R, Staudt J, Ross MB, Marschke G (2017) Identifying potentially transformative research: a case study of Alzheimer's disease research. Alzheimers Dement 13(7):860

Wyttenbach A, Carmichael J, Swartz J, Furlong RA, Narain Y, Rankin J, Rubinsztein DC (2000) Effects of heat shock, heat shock protein 40 (HDJ-2), and proteasome inhibition on protein aggregation in cellular models of Huntington's disease. Proc Natl Acad Sci 97(6):2898–2903

Chapter 16
Hsp60 in Modifications of Nervous System Homeostasis and Neurodegeneration

Celeste Caruso Bavisotto, Federica Scalia, Alessandro Pitruzzella, Magdalena Górska-Ponikowska, Claudia Marino, and Giulio Taglialatela

Abstract Hsp60 is a critical chaperonin for its role in preserving cell survival and protecting mitochondria against stress conditions. Indeed, mutations or malfunctions of Hsp60 are involved in several human diseases, either genetic or acquired, some of them affecting also the brain. In this chapter, we present several experimental observations supporting the role of Hsp60 in some neurodegenerative diseases. Further, Hsp60, as multifunctional protein, contributes to the protein folding system, to protect mitochondria and is involved in several other cellular pathways that are known to be affected in these diseases. Furthermore, due to its role outside of the mitochondria and in the extracellular fluids, it has also been suggested that Hsp60 has a role in triggering neuroinflammation. Taken together, these considerations strongly suggest the important role for Hsp60 in neurodegenerative diseases and might propose Hp60 as an attractive target for developing future therapies.

C. Caruso Bavisotto (✉)
Department of Biomedicine, Neuroscience and Advanced Diagnostics (BIND), Section of Human Anatomy, University of Palermo, Palermo, Italy

Euro-Mediterranean Institute of Science and Technology (IEMEST), Palermo, Italy

Institute of Biophysics, National Research Council, Palermo, Italy
e-mail: celeste.carusobavisotto@unipa.it; celeste.carusobavisotto@pa.ibf.cnr.it

F. Scalia · A. Pitruzzella
Department of Biomedicine, Neuroscience and Advanced Diagnostics (BIND), Section of Human Anatomy, University of Palermo, Palermo, Italy

Euro-Mediterranean Institute of Science and Technology (IEMEST), Palermo, Italy

M. Górska-Ponikowska
Department of Medical Chemistry, Medical University of Gdansk, Gdańsk, Poland

Institute of Biomaterials and Biomolecular Systems, Department of Biophysics, University of Stuttgart, Stuttgart, Germany

C. Marino · G. Taglialatela
Mitchell Center for Neurodegenerative Diseases, Department of Neurology, University of Texas Medical Branch, Galveston, TX, USA

A. A. A. Asea, P. Kaur (eds.), *Heat Shock Protein 60 in Human Diseases and Disorders*, Heat Shock Proteins 18,
https://doi.org/10.1007/978-3-030-23154-5_16

Keywords Exosomes · Heat shock proteins (Hsp) · Hsp60 · Molecular chaperones · Neurodegenerative disease · Theranostic tool

Abbreviations

ACADS	Acyl-CoA dehydrogenase gene
AD	Alzheimer's disease
APP	Amyloid precursor protein
ATPase	Adenosine triphosphatase
Aβ	Amyloid-β peptide
BBB	Blood brain barrier
CNS	Central nervous system
CS	Chaperone system
HD	Huntington's disease
Hsp60	Heat shock protein 60 kDa
HSPD1	Heat shock protein family D
HTT	Huntingtin gene
NDDs	Neurodegenerative disorders
NFTs	Neurofibrillary tangles
PD	Parkinson's disease
sHsp	Small heat shock proteins
TLR	Toll-like receptor

16.1 Introduction

Neurodegenerative disorders (NDDs) are age-related pathologies in which genetic determinants and environmental stressors play a complex role. These pathologies have been often classified as "conformational diseases" or "protein misfolding diseases" (Soto 2003; Butterfiel et al. 2012; Kayed et al. 2003), as they are characterized by synaptic dysfunction and loss of neurons associated with pathologically altered proteins. These proteins are highly prone to aggregate and to form insoluble deposits both in brain and in peripheral organs (Carrell and Lomas 1997; Korsak and Kozyreva 2015).

Despite these neurodegenerative diseases share the common mechanism of protein misfolding, each disease differs from the other for the anatomical onset, and for the primary structure of the misfolded peptide involved (Kayed et al. 2003; Armstrong 2012; Macario and Conway de Macario 2001). Few examples of these disorders are: Alzheimer's disease, which is mainly caused by the misfolding of amyloid beta and tau proteins originating from the hippocampus (Selkoe and Hardy 2016); Parkinson's disease, which is characterized by the misfolding of α-synuclein and neuronal loss originating in the substantia nigra (Poewe et al. 2017); Huntington's disease, which is characterized by poli-repeated hungtintin protein, originating from a mutation in the *HTT* gene originating in the caudate nucleus (Zielonka et al. 2015). Molecular chaperones are critical for the correct function of the stress-response machinery and they are also involved in the maintenance of both proteome integrity and protein homeostasis (proteostasis). However, it has been demonstrated that during aging this capacity can be reduced or compromised, thus leading to the manifestation of various protein-aggregation diseases (Hartl et al. 2011).

Chaperones are ubiquitous proteins highly conserved throughout evolution, from bacteria to humans (Lindquist 1986). They can be classified by molecular weight: small heat shock proteins (sHsp) family from 10 to 30 kDa; Hsp40 family (40 kDa); Hsp60 (or chaperonins) with molecular weight close to 60 kDa; Hsp70 family (70 kDa); Hsp90 family (83–90 kDa); and Hsp100/110 family with molecular weight equal or higher than 100 kDa (Kampinga et al. 2009). Among all components of the Chaperone System (CS), the chaperonin Hsp60 (also called Cpn60 or HspD1) is the focus of this chapter, with particular attention to its role in the Central Nervous System (CNS).

The chaperonin Hsp60, along with its co-chaperonin Hsp10, is essential for cell survival, since assists protein folding, protein trafficking and cellular homeostasis (Macario and Conway de Macario 2005). Even though the typical function of Hsp60 involves stress response, where it is considered mainly as a cytoprotective factor, there is evidence suggesting the presence of various pathological conditions associated with Hsp60 failure or malfunction (Cappello et al. 2013a, b). In mammalian CNS a mild deficiency of Hsp60 primarily affects neuronal and/or glial cells, whereas a more severe deficiency of Hsp60 would affect all tissues and would not be compatible with life (Christensen et al. 2010). The pathological conditions in which Hsp60 promotes these malfunctions are referred as "chaperonopathies" (Macario et al. 2013; Macario and Conway de Macario 2018), and appears that are needed new research approaches, to further understand Hsp60 functions, and therefore propose it as a novel therapeutic tool.

16.2 Central Nervous System Cytology: Neurons and Glia

The brain and the spinal cord constitute the CNS and are protected inside the cranium and the vertebral column respectively. Bundles of axons called nerves connect the CNS to the rest of the body (Peters et al. 1978). Two main types of cells characterizing the CNS are: (1) neurons specialized in impulse conduction and signals exchanging with other neurons and/or specialized cells; (2) neuroglial cells (collectively known as the neuroglia) with many important physiological functions (Table 16.1). Both neurons and neuroglia cells develop from the dorsal ectoderm of the early embryo. Overall, these cells are responsible for most of the functional characteristics of nervous tissue (Peters et al. 1978).

Table 16.1 Cell types and associated neurodegenerative disease of the CNS

Cell type	Characteristics	Functions	Associated neurodegenerative disease	References
Neurons	Polarized cells	Send and receive chemically mediated electrical signals	Alzheimer's disease; Huntington's disease; Parkinson's disease; Lewy body disease; multiple system atrophy; frontotemporal dementia; amyotrophic lateral sclerosis	Ahmed et al. (2012), Beyer et al. (2009), Jellinger (2012), Seelaar et al. (2011), and Wijesekera and Leigh (2009)
Astrocytes	Small cells with extensive and highly branched processes	Guide neuronal development Regulate synaptic communication via regulation of neurotransmitter levels	Alzheimer's disease; Huntington's disease; Parkinson's disease; amyotrophic lateral sclerosis; Alexander's disease	Bellavista et al. (2014), Ferrer et al. (2014), Mizielinska et al. (2013), Pasanen et al. (2014), and Tong et al. (2014)
Oligodendrocytes	Small cells with small processes	Form myelin	Alzheimer's disease; Parkinson's disease; amyotrophic lateral sclerosis; multiple system atrophy	Bellavista et al. (2014), Ferrer et al. (2014), Goedert (2001), Mizielinska et al. (2013), Morell and Norton (1980), and Pasanen et al. (2014)
Microglial cells	Cells with small nucleus and long, thin, very motile processes	Remove dead cells and cell debris	Alzheimer's disease; amyotrophic lateral sclerosis; multiple sclerosis	Bellavista et al. (2014) and Mizielinska et al. (2013)
Ependymal cells	Large cell with distinct round nucleus	Produce and circulate cerebrospinal fluid	–	Jiménez et al. (2014)

16.2.1 Neurons

Neurons are highly specialized cells that send and receive chemical and electrical signals (Peters et al. 1978; Kandel et al. 2000). Neurons exhibit great variability in both size and shape, despite they commonly present a rounded or polygonal area, or soma, that includes the nucleus, the organelles and its cytoplasm. Further, these cells have one or few axons, so send signals to other neurons, and numerous cell processes, or dendrites that receive signals from other neurons. Most neurons of the CNS have several dendrites and are multipolar in shape. By reaching out in various directions, dendrites increase the ability of neurons to receive input from different sources at the same time through the synapses (Peters et al. 1978). A distinguishing characteristic of neurons is their complexity and highly differentiated structure. The number and morphology of their processes (either axonal or dendritic) commonly categorize them. Neuronal cell bodies also contain a prominent Golgi apparatus, similarly to other secretory cells, which is necessary to satisfy the high vesicles synthesis for neurotransmitter release (Illis 1999; Peters et al. 1978).

16.2.2 Glial Cells

Glial cells were first recognized as a distinct cell type in the mid-1800s. The name "neuroglia," assigned by Rudolph Virchow, means "nerve glue," since these cells were thought to support brain function, by keeping neurons in their proper position (Jacobson 1991). Despite glial cells are still thought to be cells of support for neurons, their role is considerably more complex. Indeed, it has been shown that neurons and glia can interact in interdependent ways, influencing each other's development, differentiation, and physiological function.

Astrocytes are thought to play different important roles: (1) they modulate the exchange of materials between capillaries and neuronal cells; (2) they provide nutritive or trophic support to neurons; (3) they regulate the extracellular ionic composition and they are involved in the intake of neurotransmitters that are released by neurons; (4) they remove debris produced after injury, particularly from degenerating synaptic terminals; and (5) they contribute to cellular compartmentation in densely packed neuropil zones by separating neuronal processes (Jäkel and Dimou 2017; Steward 1989).

Further, glia is widespread and heterogeneous, and there are two principal types of glia that are distinguished by size and embryonic origin: macroglia, astroglia and oligodendroglia, the latter being the larger type of glial cells, originating from the neural plate. Microglia are smaller cells and are thought to originate from the mesoderm. Astroglia or astrocytes have small cell bodies and extensive and highly branched processes. They are involved in the induction, development, and regulation of the blood brain barrier (BBB) (Abbott 2002). There are two kinds of astrocytes, which differ in appearance and location: fibrous astrocytes, occurring predominantly in the white matter, and protoplasmic astrocytes, which are found mainly in grey matter (He and Sun 2007).

Oligodendroglia or oligodendrocytes have inferior dimensions than astrocytes and have fewer and shorter processes. The oligodendrocytes can be classified in two types: intrafascicular oligodendroglia, which are found in fiber tracts where they are the myelin-forming elements; and perineuronal oligodendroglia, which are not associated with axon and found near neuronal cell bodies as satellite cells (Morell and Norton 1980).

Microglia are the smallest of the glial elements. They have short, thin processes that contact neurons and capillaries and can migrate even in the mature CNS, and they play an important function in the removal of degeneration debris following injury (Jäkel and Dimou 2017).

Ependymal cells line overlying the ventricles as an epithelial layer. Particularly, these cells are interconnected by gap junctions, and at the ventricular side, ependymal cells have numerous microvilli and cilia (Jiménez et al. 2014).

Glial cells are essential for proper neuronal function. However, the exact nature of their role in the brain function remains to be addressed, even though it is well established that in neurodegenerative diseases they can lose their normal function of supporting cells, thus leading to astrogliosis and microglial proliferation and subsequent damaged brain tissues (Table 16.1) (Bellavista et al. 2014; Ferrer et al. 2014; Goedert 2001; Mizielinska et al. 2013; Pasanen et al. 2014; Perry and Holmes 2014; Tong et al. 2014).

16.3 Hsp60 Canonical and Non-canonical Cell Localization and Functions

Hsp60 is commonly classified as a mitochondrial protein, as it is mainly found in the mitochondrion. However, its coding gene, *HSPD1*, is located in the cell nucleus (Hansen et al. 2003). In fact, after being synthesized in the cytosol as Hsp60 precursor, it migrates into mitochondria, where, once processed in its active form, contributes to the intra-mitochondrial protein homeostasis (Hartl et al. 1989) (Fig. 16.1). Therefore, under normal conditions, there are at least two intracellular populations of Hsp60, one in the cytosol and one in the mitochondria (Chandra 2007; Vilasi et al. 2018). Hsp60 is crucial for life and regulate cell homeostasis. This protein acts as a molecular chaperone, to fulfill its specific functions and it also cooperates with other chaperones (Böttinger et al. 2015; Cappello et al. 2008; Marino Gammazza et al. 2012). In the mitochondria, Hsp60, along with Hsp10, assist the folding of a subset of mitochondrial matrix proteins (Okamoto et al. 2015, 2017), refolds nascent polypeptides and assist their translocation as unfolded linear chains from the cytoplasm to the mitochondrial matrix (Marino Gammazza et al. 2012). The understanding of the structure and function of Hsp60 is mainly based on research done on the bacterial homologue GroEL, as both proteins share approximately the 50% of sequence identity. Hsp60 forms a stable tetradecameric double-ring conformation in which it is possible to identify three domains: (1) apical domain that interacts with the unfolded substrate proteins and serves as binding site for Hsp10; (2) intermediate domain that join the apical to the (3) equatorial domains, which hold the ATPase activity (Horwich et al. 2007; Vilasi et al. 2014). Effectively, the Hsp60/Hsp10 complex has been indicated to be able to function as a single ring (Nielsen et al. 1999; Nielsen and Cowan 1998). However, seem that the human Hsp60 exists in a dynamic

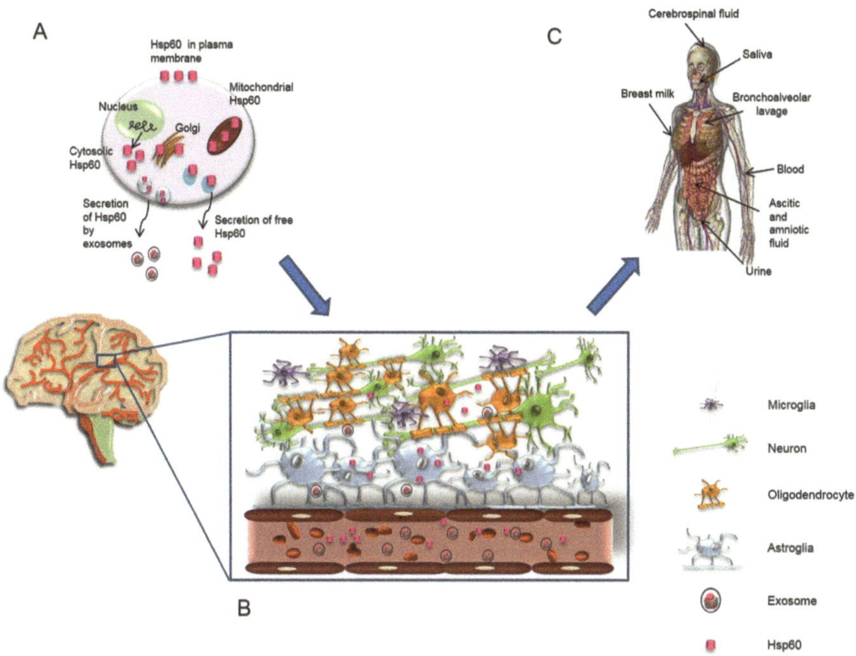

Fig. 16.1 Hsp60: canonical and non-canonical localization. Hsp60 was initially described as a mitochondrial molecular chaperone. During the last years, novel functions have been proposed for Hsp60, relatively to intracellular and extracellular cell compartments, where Hsp60 has been shown to have a multifaceted role, depending on the cell type and its localization. However, these roles are still not completely understood (Caruso Bavisotto et al. 2017). After its synthesis, Hsp60 is translocated and localized in several compartments within the cell, such as mitochondria, Golgi apparatus, plasma membrane, and cytoplasm, but Hsp60 is also secreted outside the cell by canonical secretion pathway as a "free" Hsp60, or by exosomes (**a**). Thus, Hsp60 has also been identified in the extracellular space. According to different studies, the protein can be secreted into the extracellular matrix, for instance, to regulate the tissue regeneration, initiating cell proliferation, regulating inflammation (Campanella et al. 2012, 2014; Caruso Bavisotto et al. 2017; Nakamura and Minegishi 2013; Soltys and Gupta 1997). It mediates the interaction between cells of the immune system and other cell types (**b**). The extracellular Hsp60 can be found as well in human peripheral blood as free or membrane-bound showing immunosuppressive or immunostimulating effects. The exosomes, considered a new mechanism of intercellular exchange of information in normal and pathological conditions, are present in body fluids such as blood, urine, breast milk, saliva, cerebrospinal fluid, bronchoalveolar lavage, ascitic, and amniotic fluids. Therefore despite further studies are needed, the presence of Hsp60 in the exosomes cannot be excluded (**c**). Exosomes can be considered tools for improving diagnostic procedures for diagnosis, patient stratification and/or prognosis of disease outcome. These considerations could contribute to reduce diagnostic mistakes and to improve future therapeutic decisions

equilibrium between monomers, single ring heptamers and double ring dodecamers (Levy-Rimler et al. 2001, Okamoto T. et al. (2015).

Hsp60 is critical for assisting mitochondrial protein folding and is also crucial for the proper function of mitochondria, as it regulates mitochondrial permeability (Ghosh et al. 2010) and favors the folding of mitochondrial matrix enzymes

(Corydon et al. 2005). Moreover, there is a possible link between Hsp60-dependent protein quality control and the oxidative stress scavenging systems that may contribute to increased levels of reactive oxygen species under folding stress (Magnoni et al. 2014).

In addition to its role in protein folding, Hspo60 is implicated in intracellular protein trafficking (Bukau and Horwich 1998) peptide-hormone signaling (Monreal-Flores et al. 2017), and in pro-apoptotic and pro-survival pathways (Caruso Bavisotto et al. 2017; Gorska et al. 2013; Pace et al. 2013). These events are prerogative of both mitochondrial and cytosolic Hsp60 isoforms, and its cell distribution appears to modulate several phenomena in both normal and pathological conditions (Gupta and Knowlton 2005; Gupta and Knowlton 2007; Shan et al. 2003). Further, our studies and various others have shown that Hsp60 also resides and functions outside the mitochondria (Cappello et al. 2008), namely in the cytosol (Chandra et al. 2007), in intracellular vesicles, in the Golgi (Soltys and Gupta 1996, 2000), at the plasma membrane (Soltys and Gupta 1997), and in body fluids (Campanella et al. 2012, 2014, 2015b; Caruso Bavisotto et al. 2017) (Fig. 16.1). Furthermore, Hsp60 also plays non-canonical roles not related to protein homeostasis: innate and adaptive immunities, inflammation, autoimmunity, and cancer (Cappello et al. 2014; Macario and Conway de Macario 2018; Marino Gammazza et al. 2014, 2017b; Vilasi et al. 2018). In the cytosol, Hsp60 upregulates various pathways that can also lead to pathologies, such as facilitating the survival of malignant cells, by protecting them from apoptosis and/or senescence (Marino Gammazza et al. 2017a). Further, it has been demonstrated that Hsp60 facilitates apoptosis through the interaction with mitochondrial Hsp70 (mortalin) (Deocaris et al. 2006; Wadhwa et al. 2005), survivin and p53 (Ghosh et al. 2008; Cappello et al. 2008). Therefore, Hsp60 can be either pro-survival or proapoptotic depending on other cellular factors (Chandra et al. 2007): whether it binds to pro-caspase 3 has a pro-apoptotic role, because allows its activation in normal cells (Samali et al. 1999). Conversely, in tumor cells, as Hsp60 does not activate the caspase pathway, it facilitates the survival of malignant cell, thus, contributing to tumorigenesis (Caruso Bavisotto et al. 2017) by preventing apoptosis (Kirchhoff et al. 2002). In several conditions, Hsp60 it has been shown to be either increased or decreased in its levels and the pathophysiological significance of these quantitative variations seem to be dependent on the environment in which the chaperonin resides (Marino Gammazza et al. 2017b). At the plasma membrane, the extra-mitochondrial Hsp60 may mediate the interaction between cells of the immune system and other cell types. Indeed, receptors for Hsp60 have been observed on various inflammatory cells. Particularly, the most described receptors for HSP60 are toll-like receptor (TLR) 2 and TLR-4 and CD14, CD30, and CD54, expressed predominately on lymphocytes and macrophages (Marino Gammazza et al. 2014; Nakamura and Minegishi 2013).

Interestingly, Hsp60 may be also secreted in the extracellular space and in the blood circulation (Campanella, et al. 2015a, b; Marino Gammazza et al. 2014; Novo et al. 2011; Rizzo et al. 2012), both as a free from or through the secretion of extracellular vesicles, such as exosomes (Fig. 16.1). Extracellular Hsp60 has been postulated to have effects on neutrophils (Cappello et al. 2011; Tomasello et al.

2011) and macrophages. With the latter, Hsp60 has been found to interact with macrophage-specific surface receptors and to modulate either pro- or anti-inflammatory effects (Macario et al. 2010). For example, Hsp60 can induce secretion of cytokines from professional antigen-presenting cells, with consequent activation of T-cells (Gupta et al. 2008; Merendino et al. 2010; Osterloh et al. 2004; Soltys and Gupta 1996).

The secretion of Hsp60 through exosomes is a multi-step process including its (1) accumulation in the cytosol; (2) translocation to the plasma membrane; (3) internalization by lipid rafts into multivesicular bodies; (4) and secretion via exosomes (Campanella et al. 2012; Caruso Bavisotto et al. 2017). Exosomes are extracellular vesicles of 20–150 nm in diameter, considered as one of the most important mediators of cell-to-cell communication regarding their involvement in pathophysiological processes (Caruso Bavisotto et al. 2013). Exosomes may circulate throughout the body and thereby can reach target cells and affect both their properties and functions. It is well established that Hsp60 accumulates in exosomes released by cells and its presence in exosomes may constitute a sort of cell fingerprint. Exosomes circulating throughout the body may reach distinct target cells, thus affecting their properties and functions. Due to the above, exosomal Hsp60 has been proposed as a promising "theranostics" tool and novel biomarker useful for diagnosing and assessing prognosis of a variety of diseases (Caruso Bavisotto et al. 2017b; Gupta and Knowlton 2007; Hayoun et al. 2012; Lv et al. 2012; Malik et al. 2013) (Fig. 16.1).

16.4 Nervous System Diseases in Which Hsp60 Contributes to Disease Pathogenesis

Given the important role of Hsp60 in cellular and extracellular mechanisms, such as protein trafficking and/or degradation, cell differentiation, signal transduction (Caruso Bavisotto et al. 2018; Graziano et al. 2018; Vilasi et al. 2018), it is not surprising that any malfunctioning of this chaperonin leads to pathogenic conditions, as seen in different neurodegenerative diseases. Hsp60 is endogenously expressed in astrocytes, neurons, microglia, oligodendrocytes, and ependymal cells (D'Souza and Brown 1998). This distribution suggests an active participation of this chaperonin in many functions of the brain, in both normal and pathological conditions.

As previously stated, Hsp60 is implicated in the regulation of "proteostasis", by modulating not only protein activity but also by targeting misfolded or aggregated proteins for refolding or for degradation and translocation (Czarnecka et al. 2006). Therefore, any failure of these essential cellular quality control mechanisms contributes to a variety of pathologies, referred as "proteinopathies". These pathologies are characterized by protein dysfunctions, accumulation of specific misfolded or conformationally altered proteins (Bayer 2015). In the CNS, the protective role of Hsp60 is even more important; here in fact, Hsp60 must assist the complex activities and the exclusive morphology of the cells of the nervous system. Much more than the other cells of the human organism, neurons need mitochondria to be functional for energy support and for proper synaptic transmission (Lee et al. 2018). Therefore, it is not surprising that mitochondrial dysfunction contributes to neurological disorders. Despite the phenotypic diversity and the heterogeneous genetic origin, many neurodegenerative disorders share some common features like mitochondrial dysfunction, alteration of cellular quality control mechanisms, oxidative stress, neuroinflammation, and impaired subcellular trafficking (Bross et al. 2012). In the context of this chapter, we distinguish two main groups of neurological disorders that can be linked to Hsp60 functions: (1) mitochondrial dysfunctions diseases and (2) misfolded proteins accumulation diseases.

16.4.1 Mitochondrial Dysfunctions and Related Diseases

The involvement of Hsp60 in the pathogenesis of neurological diseases brought the attention of the scientific community to the identification of mutations in the *HSPD1* gene that might be responsible for neurodegeneration. In fact, m it has been shown that mutations in the *HSPD1* gene are associated with an autosomal dominant form of hereditary spastic paraplegia (designated by its genetic locus as SPG13) and an autosomal recessive hypomyelinating leukodystrophy termed MitCHAP60 disease (Bross et al. 2012).

16.4.1.1 Hereditary Spastic Paraplegia

Hereditary spastic paraplegias (HSPs) represent a clinically and genetically heterogeneous group of neurodegenerative disorders that are characterized by progressive spasticity and weakness of the lower limbs with a typically late-onset (Hansen et al. 2002). In 2000 has been mapped for the first time a new locus associated with SPG13 in a French family (Fontaine et al. 2000). This first analysis gave an idea that the penetrance of autosomal dominant HSP is age-dependent and incomplete: affected patients had a different symptomatology, while individuals who carried the disease haplotype were diagnosed as neurological normal upon clinical examination. The latter individuals were defined as unaffected individuals probably due to their younger age at the onset of the disease, compared to the average of population carrying the disease haplotype (Fontaine et al. 2000).Subsequently, in 2002, the gene encoding for the human mitochondrial Hsp60 and its co-chaperonin Hsp10 was localized on chromosome 2, at the cytogenetic position 2q33.1, which is the same region where the locus for HSP (SPG13 [MIM $*$ 605280]) had been mapped in a French family (Hansen et al. 2002).

After the determination of the Hsp60/Hsp10 genomic structure (*HSPD1* [GenBank accession number AJ250915]), Hansen's group performed the sequencing of the 16 exons encoding for the two genes and the bidirectional promoter region in two affected members of the family with SPG13, and was also detected that both individuals were heterozygous for a variation G → A at position 292 of the cDNA of Hsp60. It results in the replacement of a valine residue at position 72 with isoleucine (V98I), which, as demonstrated by in vitro analyzes, revealed that the Hsp60 V72I mutant did not allow a correct growth of *E. coli* (Hansen et al. 2002). Until these findings, the investigation of the impact of genetic variation in the HSP60 and Hsp10 genes in genetic diseases has been limited by the common knowledge of the tidings of the human chaperonin genes, and the inconsistent nucleotide and mapping databases. Therefore, the Hansen's group characterized the entire nucleotide sequence of the human Hsp60/Hsp10 by experimentally determining the chromosomal localization and characterizing the promoter region. From these analyzes, it was discovered that the fragments analyzed in healthy patients showed a different sequence and different chromosome location when compared to the sequence deposited in the Human Genome Assembly (build 30) (Hansen et al. 2003).

In 2008, the mutation c292 G > C, the V72I, was renamed in p. V98I (Bross et al. 2008). The researchers performed experiments to analyze the properties of mutant protein expression in assisting the folding of an in vitro model protein and have also engineered a flexible *E. coli* system to monitor all different phenotypes. Their results showed that the V98I mutation affected ATPase activity. Incorporation of only one or two ATPase mutant subunits has a dramatic effect; whereas complexes composed of Hsp60-(p.V98I) and wild-type Hsp60 subunits appear to possess reduced but residual activity. Regarding patients affected by SPG13, it has been observed the formation of hetero-complexes of both mutated and wild type proteins, probably leading to a reduction of chaperonin activity and subsequent reduction of protein folding activity and a premature substrate degradation. Interestingly, despite this impairment can be compensated in other tissues thanks to the compensatory up-regulation of encoding genes for other chaperones, in neuronal cells this compensatory mechanism could be limited by the long distances (in some cases more than 1 m) between the axonal mitochondria and the nucleus at the cell body (Bross et al. 2008).

Around the same time, has been identified by others a novel missense mutation (p.Gln461Glu) in the *HSPD1* gene in one out of 23 Danish index patients diagnosed with HSP (Hansen et al. 2007). Functional studies showed that *E. coli* mutant cells deleted in the groES/groEL genes but complemented with Hsp10 and Hsp60 clearly display reduced growth, compared to cells complemented by the wild type Hsp60. Those results suggested that the function of the novel p.Gln461Glu (Q461E) mutant protein is mildly compromised and is consistent with p.Gln461Glu being associated with HSP but having a low penetrance. On the other side, the p.Val98Ile Hsp60 mutant previously associated with a highly penetrant late-onset HSP in a large family was shown to be more severely functionally impaired (Hansen et al. 2007).

To investigate how spastic paraplegia (SPG13)-associated mutant of *HSPD1* affects mitochondrial morphology, Miyamoto et al. carried out in vitro experiments on Cos-7 transfected cells. They evaluate that the length of the mitochondria of cells transfected for the V98I mutation was 30% smaller than the control cells transfected with HPSD1 wildtype; the number of mitochondria of cells harboring the mutation was larger compared to wildtype cells; and V98I *HSPD1* mutant cells incorporated 50% less dye than the cells transfected for *HSPD1* wild type (Miyamoto et al. 2016).

The possibility of interpreting the degree of Hsp60 involvement in pathogenesis was complicated when it was clear that in many monogenic diseases, the phenotype is not solely determined by gene variations in the primary target gene but is subjected to modification by both modifier genes and environmental influences. In complex multi-factorial diseases, it is never a single gene, but rather a set of several gene variants (also termed modifiers that can contribute along with environmental factors to the disease phenotype (Bross et al. 2007). For instance, the p.[Gly563Ala] Hsp60 polymorphism appears to modify the spastin (SPG4) phenotype, causing an earlier onset of the disease symptoms (Hewamadduma et al. 2008); recently, has been reported a rare phenomenon of two siblings, a younger brother with hypomyelinating leukodystrophy and an elder brother with a severe intellectual disability and autistic features. They have independent de novo variants of *HSPD1*c.139T > G (p.Leu47Val) and HIP1c.1393G > A (p.Glu465Lys), respectively. These novel variants were predicted to be pathogenic and both patients also had a known MECP2 variant, c.499C > T (p.Arg167Trp) (Yamamoto et al. 2018) (Table 16.2).

16.4.1.2 MitCHAP60

In 2008 has been identified a disorder termed MitCHAP-60 disease (Magen et al. 2008). This neurodegenerative disorder causes a "complicated" SPG in which patients showed diffuse hypomyelination and leukodystrophy, and myelin is not formed properly. It is allelic to the pure autosomal-dominant SPG13 but is a disease-associated to a recessive missense mutation of Hsp60. In a large inbred Israeli Bedouin kindred has been detected a homozygous missense mutation g.1512A/G in exon 2 at position 86 of the cDNA sequence, causing an aspartic acid/glycine exchange at amino acid 29 of the Hsp60 protein sequence (D29G or D3G in the mature form). D29G mutation in *HSPD1* segregated completely with the disease; all ten affected children were homozygous for the mutation and all parents were heterozygous. The pathogenic effect of the D29G mutation was verified by an *E. coli* Hsp60/Hsp10-GroES/GroEL complementation assay demonstrating a reduced ability of the D29G-Hsp60 mutant protein to support *E. coli* survival at all temperatures, but especially at the higher ones (Magen et al. 2008).

Subsequently, researchers tried to determine the molecular basis of the MitCHAP-60 disease caused by the D3G mutation. The first studies were focused on the entropic destabilization of Hsp60 (Parnas et al. 2009). Particularly, they demonstrated that at high protein concentrations, the D3G mutant can assemble into oligomers (heptamers and tetradecamers), but the D3G oligomers are unstable when compared with wild-type mHsp60 and dissociate into monomers rapidly upon dilution. This destabilization effects on the mHsp60 oligomer ATPase function (Parnas et al. 2009). Afterwards, the focus shifted to the relationship between the disease-associated mutation of *HSPD1* and mitochondrial dynamics. Studying the effect of D29G missense mutation on mitochondrial morphological changes in Cos-7 living

Table 16.2 An overview of the historical literature about Hereditary Spastic Paraplegia

Year of publication	Work of literature	Breakthrough	References
2000	A new locus for autosomal dominant pure spastic paraplegia, on chromosome 2q24-q34	Identified locus for HSP	Fontaine et al. (2000)
2002	Hereditary Spastic Paraplegia SPG13 is associated with a mutation in the gene encoding the mitochondrial chaperonin Hsp60	SPG13 is associated with V72I mutation in Hsp60 gene	Hansen et al. (2002)
2003	Genomic structure of the human mitochondrial chaperonin genes: Hsp60 and Hsp10 are localised head to head on chromosome 2 separated by a bidirectional promoter	Experimental study of the entire Hsp60/Hsp10 nucleotide sequence	Hansen et al. (2003)
2007	A novel mutation in the *HSPD1*gene in a patient with hereditary spastic paraplegia	Identified the missense mutation Q461E and its effects in *E. coli* system	Hansen et al. (2007)
2008	The Hsp60-(p.V98I) mutation associated with hereditary spastic paraplegia SPG13 compromises chaperonin function both in vitro and in vivo	The mutation V72I was renamed V98I. In vivo and in vitro data suggest a dysfunction of the Hsp60-(p.V98I) mutant	Bross et al. (2008)
2008	Hsp60 is a rare cause of hereditary spastic paraparesis, but may act as a genetic modifier	G563A Hsp60 polymorphism appears to modify the spastin (SPG4) phenotype	Hewamadduma et al. (2008)
2016	Data supporting mitochondrial morphological changes by SPG13-associated *HSPD1* mutants	Aberrant mitochondrial morphological changes with decreased activities of Hsp60-(p.V98I) mutants	Miyamoto et al. (2016)
2018	Independent occurrence of de novo *HSPD1* and HIP1 variants in brothers with different neurological disorders–leukodystrophy and autism	L47V *HSPD1* variant was predicted to be pathogenic in association with others mutated genes	Yamamoto et al. (2018)

cells has been demonstrated that the mutation affects mitochondrial fission and fusion cycles and mitochondrial membrane potential (Miyamoto et al. 2015)

More recently, it has been reported an additional case with MitCHAP-60 disease caused by the mutation previously reported in the Israeli Bedouin family (Magen et al. 2008). The patient was homozygous for an ACADS mutation and had the same missense mutation (Asp-29-to-Gly) on *HSPD*1 gene, previously described (Kusk et al. 2016). Therefore, also in this case it is possible to attribute a disturbed interaction between ACADS and *HSPD1*. Because of their multifactorial fashion, to create animal models for these diseases can be challenging, despite it has been reported that transgenic mice expressing HLD4-associated (Asp-29-to-Gly) mutant of *HSPD*1 exhibit a defect in myelination in brain (Miyamoto et al. 2017) (Table 16.3).

Table 16.3 An overview of the historical literature about MitCHAP60 disease

Year of publication	Work of literature	Breakthrough	References
2008	Mitochondrial Hsp60 chaperonopathy causes an autosomal-recessive neurodegenerative disorder linked to brain hypomyelination and leukodystrophy	Identified a homozygous missense mutation (D29G) in *HSPD1* gene	Magen et al. (2008)
2009	TheMitCHAP-60 disease is due to entropic destabilization of the human mitochondrial Hsp60 oligomer	D3G (or D29G) mutation severely impairs the ATPase activity of mHsp60 and to destabilize the oligomeric structure of the protein	Parnas et al. (2009)
2015	Hypomyelinating leukodystrophy-associated missense mutation in *HSPD1* blunts mitochondrial dynamics on chromosome 2 separated by a bidirectional promoter	Effects of D29G missense mutation on mitochondrial morphological changes	Miyamoto et al. (2015)
2016	Hypomyelinating Leukodystrophy due to *HSPD1* mutations: a new patient	Additional case with MitCHAP-60 disease caused by D29G mutation	Kusk et al. (2016)
2017	Defective myelination in mice harboring hypomyelinating leukodystrophy-associated *HSPD1* mutation	Transgenic mice expressing D29G mutation exhibit a defect in myelination in brain	Miyamoto et al. (2017)

16.4.2 Misfolded Protein Accumulation and Related Diseases

Neuronal dysfunction caused by the dysfunctional aggregation of proteins is a key feature shared by several neurodegenerative diseases, in which clinical manifestations depend on the brain region affected, and may involve the disruption of physiological activities including sensory, motor and cognitive functions. Particularly, misfolding and aggregation of proteins are thought to be one of the leading causes of synaptic loss and neuronal death observed in different neurodegenerative diseases. The aggregation of misfolded proteins is highly regulated, and it also depends on genetic and environmental factors (Maiti et al. 2014) (Table 16.4).

16.4.2.1 Alzheimer's Disease

Alzheimer's disease (AD) is a neurodegenerative condition characterized by protein aggregation leading to toxic oligomer formation, synaptic loss and dementia. The protein aggregation mainly involves the amyloid-β peptide (Aβ), which accumulates in extracellular deposits called senile or neuritic plaques, and tau protein, whose hyperphosphorylation and aggregation lead to intraneuronal inclusions

Table 16.4 An overview of the role of Hsp60 in the Misfolded proteins accumulation diseases

Disease	Work of literature	Breakthrough	References
Alzheimer's disease	Hsp60 levels were significantly decreased in hippocampi of APP-transgenic mice then controls	Aβ oligomers might contribute to changing the expression of the chaperonin.	Takano et al. (2013)
	In a rat model of AD, the expression of Hsp60 is significantly decreased in the parietal cortex of AD subjects and in the cerebella	Hsp60 could have protective role against intracellular Aβ stress through the maintenance of mitochondrial functions	Jiang et al. (2013) and Veereshwarayya et al. (2006)
	Hsp60 mediates translocation of APP/Aβ to the mitochondria	Hsp60 influences the Aβ amyloid aggregation process mediating its mislocalization and cell disfunctions	Mangione et al. (2016) and Walls et al. (2012)
	Hsp60 and Aβ peptide can directly interact in several cell compartments and in the extracellular space		
	Hsp60 is highly expressed by activated microglia and it is release in extracellular space	Hsp60 induces neuroinflammation	Cappello et al. (2013a, b), Sun et al. (2018), Wojsiat et al. (2015), and Zhang et al. (2017)
	Lymphocytes from AD patients overexpress Hsp60.		
Parkinson's disease	In mutant yeast for Hsp60, it was observed accumulation of misfolded proteins	Hsp60 has a protective role in PD	Cheng et al. (1989)
	Increased expression of Hsp60 in dopaminergic neurons	Hsp60 favors regeneration and remodelling changes in dopaminergic neurons	Kuter et al. (2016), and Zhao et al. (2016)
	Hsp60 decrease in dopaminergic neuron and located on the surface of activated microglia and in extracellular environment	Hsp60 is involved in the neuronal toxicity and degeneration	Feng et al. (2013)
	Upregulation of Hsp60 in dopaminergic cell line	Role in neurodegeneration	Noelker et al. (2014)
Huntington's disease	Overexpression of Hsp60 family protein suppresses mutant huntingtin aggregation and toxicity in vitro	Supposed positive role in Huntington's disease	Wang et al. (2009)
Creutzfeldt-Jakob disease	Prion protein PrPc interacts with molecular chaperones of the Hsp60 family	The role of Hsp60 in the disease pathogenesis remains unknown	Edenhofer et al. (1996), and Satoh et al. (2005)

called neurofibrillary tangles (NFTs) (Querfurth and La Ferla 2010; Haass et al. 2012; Campanella et al. 2018; Marino Gammazza et al. 2016).The role of Hsp60 in AD is still unclear. Proteomic analysis of hippocampi of APP-transgenic mice shows abundant levels of Aβ oligomers from 8 months of age, but no amyloid plaques is found even at 24 months of age. Further, data revealed that chaperones, including Hsp60, is significantly decreased compared to controls. Therefore, it has been hypothesized that Aβ oligomers might modulate the expression of the chaperons (Takano et al. 2013).However, it has been reported contrasting data about the effect of the expression level of Hsp60: some studies demonstrated the neuroprotective roles of Hsp60 but, conversely, other authors have attributed harmful effects to the elevated expression levels of Hsp60 in AD. In a rat model of AD, the expression of Hsp60 is significantly decreased in the parietal cortex of AD subjects and in the cerebella, suggesting a protective role of the chaperonin (Jiang et al. 2013). In addition, *in vitro* experiments sustain the neuroprotective role of Hsp60. Particularly, Hsp60 would provide protection against the intracellular stress induced by Aβ oligomers through the maintenance of mitochondrial oxidative phosphorylation state and functionality of tricarboxylic acid cycle enzymes (Veereshwarayya et al. 2006). Conversely, it has been demonstrated a negative effect of Hsp60, because Hsp60 bind APP/Aβ and this link is increased in AD mitochondria (Walls et al. 2012). Probably, Hsp60 mediates the translocation of APP to the mitochondria, determining APP/Aβ mislocalization and mitochondrial dysfunction (Walls et al. 2012). In addition, Hsp60 and Aβ peptide can directly interact, not only in mitochondria but also in other cell compartments, as well as in the extracellular space, thus influencing the Aβ amyloid aggregation process (Mangione et al. 2016).

At the extracellular level, Hsp60 is known to can mediate neuroinflammation and therefore might contribute to some possible negative effect. Particularly, Hsp60 is highly expressed in activated microglia and it has been shown that its extracellular release induces a neuro-inflammatory effect, leading to neuronal cell death (Sun et al. 2018). Therefore, the inhibition of Hsp60 expression or its release could be advantageous to prevent neurodegeneration (Zhang et al. 2017). Furthermore, in lymphocytes isolated from AD patients, were detected higher levels of Hsp60 levels when compared to controls (Cappello et al. 2013a, b; Wojsiat et al. 2015) thus suggesting to further study the involvement of Hsp60 in AD, in order to confirm Hsp60 as a possible early marker in the diagnosis of the disease. When a peptide derived from Hsp60 was used as an adjuvant in Aβ immunization in a mouse model of AD (Hsp60 peptide-Aβ), the murine immune system was induced to produce anti-Aβ-specific antibodies, thus resulting in an upregulation of both humoral and cellular immune responses associated with a significant reduction of cerebral amyloid burden (Nemirovsky et al. 2011) (Table 16.4).

16.4.2.2 Parkinson's Disease

Parkinson's disease (PD) is the most common neurodegenerative movement disorder and second most common form of dementia (Poewe et al. 2017). This disease is characterized by progressive degeneration of dopaminergic neurons in the *substantia nigra* and by the accumulation of Lewy bodies in the affected neurons, mainly comprised by aggregated α-synuclein (Dawson and Dawson 2003). For its functions in the mitochondria, Hsp60 is an attractive candidate for PD. It has been shown that in yeast cells carrying a null mutation in *HSPD1* are observed severe defects in folding of mitochondrial proteins and are non-viable. In contrast, yeast cells with conditional mutations in Hsp60 accumulate misfolded proteins that are similar to α-synculein aggregates found in PD (Cheng et al. 1989). Increased expression of Hsp60 was observed during and after dopaminergic neurons degeneration and compensation (Kuter et al. 2016), which is in accordance with its role on tissue regeneration and remodelling changes (Zhao et al. 2016).

In addition, it was demonstrated a role for Hsp60 in microglia activation in a rat model of Parkinson's disease. Particularly, in this study, the authors demonstrated that Hsp60 was decreased in dopaminergic neuron and located on the surface of activated microglia, where could contribute to activating microglia. Furthermore, in vitro experiments showed that Hsp60 was incremented in cell supernatants, thus proposing a mechanism of Hsp60 release and a subsequent role in the induction of neuronal toxicity (Feng et al. 2013). The role of Hsp60 in PD is strongly suggested by other in vivo and in vitro experiments where Hsp60 has been observed to be upregulated and then released by a dopaminergic neurons, and possibly contributing actively to neuronal degeneration (Noelker et al. 2014) (Table 16.4).

16.4.2.3 Huntington's Disease

In literature, there are few studies concerning the role of Hsp60 in Huntington's disease. This autosomal dominant neurodegenerative disease is caused by a pathological expansion of a CAG trinucleotide sequence that encodes for the polyglutamine repeat in the huntingtin (Htt) protein, alteration leading to either a loss of function or gain of toxicity of this protein due to the formation of toxic neuronal inclusions. The anatomical regions affected by HD are mainly the caudate nucleus, the striatal and cortical areas of the brain whose alteration lead to movement disorders and cognitive decline (Imarisio et al. 2008, Kim and Fung (2014). It has been hypothesized that, a protein of the Hsp60 family occurs in polyglutamine aggregate formation and toxicity, suggesting a neuroprotective role (Wang et al. 2009) (Table 16.4).

16.4.2.4 Creutzfeldt-Jakob Disease

Hsp60 is the major interactor for the prion particle in prion disease (Edenhofer et al. 1996; Satoh et al. 2005). However, the mechanism of interaction remains unknown (Table 16.4).

16.5 Conclusions

The results described in the preceding sections strongly suggest that Hsp60 is a key chaperonin with a critical role in the regulation of a variety of cellular and extracellular processes in the central nervous system. As other chaperons, Hsp60 is critical for maintaining cell proteostasis and modulates several metabolic pathways residing in the mitochondria. Therefore, its affected functions certainly can contribute to the onset of fatal neurodegenerative diseases. However, the role of Hsp60 in these disorders is still not fully understood, as there are several controversial findings in this field. Nevertheless, future research is aiming to further investigate the molecular pathways in which Hsp60 is involved. Due to the better health care, hygiene and healthier life styles, the life span is rapidly increasing but, this also correlates with a dramatic increase of age-related diseases, disability, dementia and other related dysfunctions. Considering these aspects, the biomedical research is primarily focusing on elucidating possible molecular mechanisms leading to these diseases to design innovative tools for early diagnostic. To this purpose, Hsp60 constitutes an attractive target for developing specific and sensitive diagnostic tests for preventive therapies or patient follow-ups due to the early detection of altered Hsp60 levels in some neurodegenerative diseases. Further, given its dual activity, Hsp60 can be an attractive candidate for designing new drugs to either inhibit or promote the function of Hsp60 for future personalized medicine.

Acknowledgements A.P., C.C.B. and F.S. were partially supported by UniPA.

References

Abbott NJ (2002) Astrocyte-endothelial interactions and blood-brain barrier permeability. J Anat 200(6):629–638. Retrieved August 28, 2018 from internet: http://www.ncbi.nlm.nih.gov/pubmed/12162730

Ahmed Z, Asi YT, Sailer A, Lees AJ, Houlden H, Revesz T, Holton JL (2012) The neuropathology, pathophysiology and genetics of multiple system atrophy. Neuropathol Appl Neurobiol 38(1):4–24. https://doi.org/10.1111/j.1365-2990.2011.01234.x

Armstrong RA (2012) On the "classification" of neurodegenerative disorders: discrete entities, overlap or continuum? Folia Neuropathol 50(3):201–208. Retrieved August 2, 2018 from internet: http://www.ncbi.nlm.nih.gov/pubmed/23023335

Bayer TA (2015) Proteinopathies, a core concept for understanding and ultimately treating degenerative disorders? Eur Neuropsychopharmacol 25(5):713–724. https://doi.org/10.1016/j.euroneuro.2013.03.007

Bellavista E, Santoro A, Galimberti D, Comi C, Luciani F, Mishto M (2014) Current understanding on the role of standard and immunoproteasomes in inflammatory/immunological pathways of multiple sclerosis. Autoimmune Dis 2014:739705. https://doi.org/10.1155/2014/739705

Beyer K, Domingo-Sàbat M, Ariza A (2009) Molecular pathology of Lewy body diseases. Int J Mol Sci 10(3):724–745. https://doi.org/10.3390/ijms10030724

Böttinger L, Oeljeklaus S, Guiard B, Rospert S, Warscheid B, Becker T (2015) Mitochondrial heat shock protein (Hsp) 70 and Hsp10 cooperate in the formation of Hsp60 complexes. J Biol Chem 290(18):11611–11622. https://doi.org/10.1074/jbc.M115.642017

Bross P, Li Z, Hansen J, Hansen JJ, Nielsen MN, Corydon TJ, Georgopoulos C, Ang D, Lundemose JB, Niezen-Koning K, Eiberg H, Yang H, Kølvraa S, Bolund L, Gregersen N (2007) Single-nucleotide variations in the genes encoding the mitochondrial Hsp60/Hsp10 chaperone system and their disease-causing potential. J Hum Genet 52(1):56–65. https://doi.org/10.1007/s10038-006-0080-7

Bross P, Naundrup S, Hansen J, Nielsen MN, Christensen JH, Kruhøffer M, Palmfeldt J, Corydon TJ, Gregersen N, Ang D, Georgopoulos C, Nielsen KL (2008) The Hsp60-(p.V98I) mutation associated with hereditary spastic paraplegia SPG13 compromises chaperonin function both in vitro and in vivo. J Biol Chem 283(23):15694–15700. https://doi.org/10.1074/jbc.M800548200

Bross P, Magnoni R, Bie AS (2012) Molecular chaperone disorders: defective Hsp60 in neurodegeneration. Curr Top Med Chem 12(22):2491–2503

Bukau B, Horwich AL (1998) The Hsp70 and Hsp60 chaperone machines. Cell 92(3):351–366

Butterfield S, Hejjaoui M, Fauvet B, Awad L, Lashuel HA (2012) Chemical strategies for controlling protein folding and elucidating the molecular mechanisms of amyloid formation and toxicity. JMB 421:204–236

Campanella C, Bucchieri F, Merendino AM, Fucarino A, Burgio G, Corona DFV, Barbieri G, David S, Farina F, Zummo G, de Macario EC, Macario AJL, Cappello F (2012) The odyssey of Hsp60 from tumor cells to other destinations includes plasma membrane-associated stages and Golgi and exosomal protein-trafficking modalities. PLoS One 7(7):e42008. https://doi.org/10.1371/journal.pone.0042008

Campanella C, Caruso Bavisotto C, Marino Gammazza A, Nikolic D, Rappa F, David S, Cappello F, Bucchieri F, Fais S (2014) Exosomal heat shock proteins as new players in tumour cell-to-cell communication. J Circulating Biomark 1. https://doi.org/10.5772/58721

Campanella C, D'Anneo A, Marino Gammazza A, Caruso Bavisotto C, Barone R, Emanuele S, Lo Cascio F, Fais S, Conway de Macario E, Macario AJL (2015a) The histone deacetylase inhibitor SAHA induces HSP60 nitration and its extracellular release by exosomal vesicles in human lung-derived carcinoma cells. Oncotarget 7(20):28849–28867. https://doi.org/10.18632/oncotarget.6680. a

Campanella C, Rappa F, Sciumè C, Marino Gammazza A, Barone R, Bucchieri F, David S, Curcurù G, Caruso Bavisotto C, Pitruzzella A, Geraci G, Modica G, Farina F, Zummo G, Fais S, Conway de Macario E, Macario AJL, Cappello F (2015b) Heat shock protein 60 levels in tissue and circulating exosomes in human large bowel cancer before and after ablative surgery. Cancer 121(18):3230–3239. https://doi.org/10.1002/cncr.29499.b

Campanella C, Pace A, Caruso Bavisotto C, Marzullo P, Marino Gammazza A, Buscemi S, Palumbo Piccionello A (2018) Heat shock proteins in Alzheimer's disease: role and targeting. Int J Mol Sci 19(9):2603. https://doi.org/10.3390/ijms19092603

Cappello F, Conway de Macario E, Marasà L, Zummo G, Macario AJL (2008) Hsp60 expression, new locations, functions and perspectives for cancer diagnosis and therapy. Cancer Biol Ther 7(6):801–809

Cappello F, Caramori G, Campanella C, Vicari C, Gnemmi I, Zanini A, Spanevello A, Capelli A, La Rocca G, Anzalone R, Bucchieri F, D'Anna SE, Ricciardolo FLM, Brun P, Balbi B, Carone M, Zummo G, Conway de Macario E, Macario AJL, Di Stefano A (2011) Convergent sets of data from in vivo and in vitro methods point to an active role of Hsp60 in chronic obstructive pulmonary disease pathogenesis. PLoS One 6(11):e28200. https://doi.org/10.1371/journal.pone.0028200

Cappello F, Angileri F, Conway de Macario E, Macario AJL (2013a) Chaperonopathies and chaperonotherapy. Hsp60 as therapeutic target in cancer: potential benefits and risks. Curr Pharm Des 19(3):452–457

Cappello F, Conway de Macario E, Marino Gammazza A, Bonaventura G, Carini F, Czarnecka AM, Farina F, Zummo G, Macario AJL (2013b) Hsp60 and human aging: les liaisons dangereuses. Front Biosci (Landmark Edition) 18:626–637. Retrieved May 22, 2017 from internet: http://www.ncbi.nlm.nih.gov/pubmed/23276948

Cappello F, Marino Gammazza A, Palumbo Piccionello A, Campanella C, Pace A, Conway de Macario E, Macario AJL (2014) Hsp60 chaperonopathies and chaperonotherapy: targets and agents. Expert Opin Ther Targets 18(2):185–208. https://doi.org/10.1517/14728222.2014.856417

Carrell RW, Lomas DA (1997) Conformational disease. Lancet 350(9071):134–138. https://doi.org/10.1016/S0140-6736(97)02073-4

Caruso Bavisotto C, Marino Gammazza A, Rappa F, Fucarino A, Pitruzzella A, David S, Campanella C (2013) Exosomes: can doctors still ignore their existence? EuroMediterranean Biomed J 8:137–139. https://doi.org/10.3269/1970-5492.2013.8.22

Caruso Bavisotto C, Nikolic D, Marino Gammazza A, Barone R, Lo Cascio F, Mocciaro E, Zummo G, Conway de Macario E, Macario AJ, Cappello F, Giacalone V, Pace A, Barone G, Palumbo Piccionello A, Campanella C (2017a) The dissociation of the Hsp60/pro-Caspase-3 complex by bis(pyridyl)oxadiazole copper complex (CubipyOXA) leads to cell death in NCI-H292 cancer cells. J Inorg Biochem 170:8–16. https://doi.org/10.1016/j.jinorgbio.2017.02.004. b

Caruso Bavisotto C, Graziano F, Rappa F, Marino Gammazza A, Logozzi M, Fais S, Maugeri R, Bucchieri F, Conway de Macario E, Macario A, Cappello F, Iacopino D, Campanella C (2018) Exosomal chaperones and miRNAs in gliomagenesis: state-of-art and theranostics perspectives. Int J Mol Sci 19(9):2626. https://doi.org/10.3390/ijms19092626

Caruso Bavisotto C, Cappello F, Macario AJL, Conway de Macario E, Logozzi M, Fais S, Campanella C (2017b) Exosomal HSP60: a potentially useful biomarker for diagnosis, assessing prognosis, and monitoring response to treatment. Expert Rev Mol Diagn 17(9):815–822. https://doi.org/10.1080/14737159.2017.1356230. a

Chandra D, Choy G, Tang DG (2007) Cytosolic accumulation of HSP60 during apoptosis with or without apparent mitochondrial release: evidence that its pro-apoptotic or pro-survival functions involve differential interactions with caspase-3. J Biol Chem 282(43):31289–31301. https://doi.org/10.1074/jbc.M702777200

Cheng MY, Hartl FU, Martin J, Pollock RA, Kalousek F, Neuper W, Hallberg EM, Hallberg RL, Horwich AL (1989) Mitochondrial heat-shock protein hsp60 is essential for assembly of proteins imported into yeast mitochondria. Nature 337(6208):620–625. https://doi.org/10.1038/337620a0

Christensen JH, Nielsen MN, Hansen J, Füchtbauer A, Füchtbauer E-M, West M, Corydon TJ, Gregersen N, Bross P (2010) Inactivation of the hereditary spastic paraplegia-associated *Hspd1* gene encoding the Hsp60 chaperone results in early embryonic lethality in mice. Cell Stress Chaperones 15(6):851–863. https://doi.org/10.1007/s12192-010-0194-x

Corydon TJ, Hansen J, Bross P, Jensen TG (2005) Down-regulation of Hsp60 expression by RNAi impairs folding of medium-chain acyl-CoA dehydrogenase wild-type and disease-associated proteins. Mol Genet Metab 85(4):260–270. https://doi.org/10.1016/j.ymgme.2005.04.003

Czarnecka AM, Campanella C, Zummo G, Cappello F (2006) Mitochondrial chaperones in cancer: from molecular biology to clinical diagnostics. Cancer Biol Ther 5(7):714–720. https://doi.org/10.4161/cbt.5.7.2975

D'Souza SM, Brown IR (1998) Constitutive expression of heat shock proteins Hsp90, Hsc70, Hsp70 and Hsp60 in neural and non-neural tissues of the rat during postnatal development. Cell Stress Chaperones 3(3):188–199

Dawson TM, Dawson VL (2003) Molecular pathways of neurodegeneration in Parkinson's disease. Science 302(5646):819–822. https://doi.org/10.1126/science.1087753

Deocaris CC, Kaul SC, Wadhwa R (2006) On the brotherhood of the mitochondrial chaperones mortalin and heat shock protein 60. Cell Stress Chaperones 11(2):116–128

Edenhofer F, Rieger R, Famulok M, Wendler W, Weiss S, Winnacker EL (1996) Prion protein PrPc interacts with molecular chaperones of the Hsp60 family. J Virol 70(7):4724–4728

Feng M j, Zhang L, Liu Z, Zhou P, Lu X (2013) The expression and release of Hsp60 in 6-OHDA induced in vivo and in vitro models of Parkinson's disease. Neurochem Res 38(10):2180–2189. https://doi.org/10.1007/s11064-013-1127-8

Ferrer I, López-González I, Carmona M, Arregui L, Dalfó E, Torrejón-Escribano B, Diehl R, Kovacs GG (2014) Glial and neuronal tau pathology in tauopathies: characterization of disease-specific phenotypes and tau pathology progression. J Neuropathol Exp Neurol 73(1):81–97. https://doi.org/10.1097/NEN.0000000000000030

Fontaine B, Davoine CS, Dürr A, Paternotte C, Feki I, Weissenbach J, Hazan J, Brice A (2000) A new locus for autosomal dominant pure spastic paraplegia, on chromosome 2q24-q34. Am J Hum Genet 66(2):702–707. https://doi.org/10.1086/302776

Ghosh JC, Dohi T, Kang BH, Altieri DC (2008) Hsp60 regulation of tumor cell apoptosis. J Biol Chem 283(8):5188–5194. https://doi.org/10.1074/jbc.M705904200

Ghosh JC, Siegelin MD, Dohi T, Altieri DC (2010) Heat shock protein 60 regulation of the mitochondrial permeability transition pore in tumor cells. Cancer Res 70(22):8988–8993. https://doi.org/10.1158/0008-5472.CAN-10-2225

Goedert M (2001) Alpha-synuclein and neurodegenerative diseases. Nat Rev Neurosci 2(7):492–501. https://doi.org/10.1038/35081564

Gorska M, Marino Gammazza A, Zmijewski MA, Campanella C, Cappello F, Wasiewicz T, Kuban-Jankowska A, Daca A, Sielicka A, Popowska U, Knap N, Antoniewicz J, Wakabayashi T, Wozniak M (2013) Geldanamycin-induced osteosarcoma cell death is associated with hyperacetylation and loss of mitochondrial pool of heat shock protein 60 (hsp60). PLoS One 8(8):e71135. https://doi.org/10.1371/journal.pone.0071135

Graziano F, Caruso Bavisotto C, Marino Gammazza A, Rappa F, Coway de Macario E, Macario AJL, Cappello F, Campanella C, Maugeri R, Iacopino DG (2018) Chaperonology: the third eye on brain gliomas. Brain Sci 8(6):110. https://doi.org/10.3390/brainsci8060110

Gupta S, Knowlton AA (2005) HSP60, Bax, apoptosis and the heart. J Cell Mol Med 9(1):51–58

Gupta S, Knowlton AA (2007) HSP60 trafficking in adult cardiac myocytes: role of the exosomal pathway. Am J Physiol Heart Circ Physiol 292(6):3052–3056. https://doi.org/10.1152/ajpheart.01355.2006

Gupta RS, Ramachandra NB, Bowes T, Singh B (2008) Unusual cellular disposition of the mitochondrial molecular chaperones Hsp60, Hsp70 and Hsp10. Novartis Found Symp 291:59–68. Retrieved December 2, 2015 from internet: http://www.ncbi.nlm.nih.gov/pubmed/18575266

Haass C, Kaether C, Thinakaran G, Sisodia S (2012) Trafficking and proteolytic processing of APP. Cold Spring Arb Perspect Med 2:a006270

Hansen J, Dürr A, Cournu-Rebeix I, Georgopoulos C, Ang D, Nielsen MN, Davoine C-S, Brice A, Fontaine B, Gregersen N, Bross P (2002) Hereditary spastic paraplegia SPG13 is associated with a mutation in the gene encoding the mitochondrial chaperonin Hsp60. Am J Hum Genet 70(5):1328–1332. https://doi.org/10.1086/339935

Hansen J, Bross P, Westergaard M, Nielsen M, Eiberg H, Børglum A, Mogensen J, Kristiansen K, Bolund L, Gregersen N (2003) Genomic structure of the human mitochondrial chaperonin genes: HSP60 and HSP10 are localised head to head on chromosome 2 separated by a bidirectional promoter. Hum Genet 112(1):71–77. https://doi.org/10.1007/s00439-002-0837-9

Hansen J, Svenstrup K, Ang D, Nielsen MN, Christensen JH, Gregersen N, Nielsen JE, Georgopoulos C, Bross P (2007) A novel mutation in the *HSPD1* gene in a patient with hereditary spastic paraplegia. J Neurol 254(7):897–900. https://doi.org/10.1007/s00415-006-0470-y

Hartl FU, Pfanner N, Nicholson DW, Neupert W (1989) Mitochondrial protein import. Biochim Biophys Acta 988(1):1–45

Hartl FU, Bracher A, Hayer-Hartl M (2011) Molecular chaperones in protein folding and proteostasis. Nature 475(7356):324–332. https://doi.org/10.1038/nature10317

Hayoun D, Kapp T, Edri-Brami M, Ventura T, Cohen M, Avidan A, Lichtenstein RG (2012) HSP60 is transported through the secretory pathway of 3-MCA-induced fibrosarcoma tumour cells and undergoes N-glycosylation. FEBS J 279(12):2083–2095. https://doi.org/10.1111/j.1742-4658.2012.08594.x

He F, Sun YE (2007) Glial cells more than support cells? Int J Biochem Cell Biol 39(4):661–665. https://doi.org/10.1016/j.biocel.2006.10.022

Hewamadduma CAA, Kirby J, Kershaw C, Martindale J, Dalton A, McDermott CJ, Shaw PJ (2008) HSP60 is a rare cause of hereditary spastic paraparesis, but may act as a genetic modifier. Neurology 70(19):1717–1718. https://doi.org/10.1212/01.wnl.0000311395.31081.70

Horwich AL, Fenton WA, Chapman E, Farr GW (2007) Two families of chaperonin: physiology and mechanism. Annu Rev Cell Dev Biol 23(1):115–145. https://doi.org/10.1146/annurev.cellbio.23.090506.123555

Illis L (1999) The central nervous system structure and function. Spinal Cord 37(4):310–310. https://doi.org/10.1038/sj.sc.3100793

Imarisio S, Carmichael J, Korolchuk V, Chen CW, Saiki S, Rose C, Krishna G, Davies JE, Ttofi E, Underwood BR, Rubinsztein DC (2008) Huntington's disease: from pathology and genetics to potential therapies. Biochem J 412(2):191–209. https://doi.org/10.1042/BJ20071619

Jacobson M (1991) Developmental neurobiology. Springer US, Boston. https://doi.org/10.1007/978-1-4757-4954-0

Jäkel S, Dimou L (2017) Glial cells and their function in the adult brain: a journey through the history of their ablation. Front Cell Neurosci 11:24. https://doi.org/10.3389/fncel.2017.00024

Jellinger KA (2012) Interaction between pathogenic proteins in neurodegenerative disorders. J Cell Mol Med 16(6):1166–1183. https://doi.org/10.1111/j.1582-4934.2011.01507.x

Jiang YQ, Wang XL, Cao XH, Ye ZY, Li L, Cai WQ (2013) Increased heat shock transcription factor 1 in the cerebellum reverses the deficiency of Purkinje cells in Alzheimer's disease. Brain Res 1519:105–111. https://doi.org/10.1016/j.brainres.2013.04.059

Jiménez AJ, Domínguez-Pinos MD, Guerra MM, Fernández-Llebrez P, Pérez-Fígares JM (2014) Structure and function of the ependymal barrier and diseases associated with ependyma disruption. Tissue Barriers 2(1):e28426. https://doi.org/10.4161/tisb.28426

Kampinga HH, Hageman J, Vos MJ, Kubota H, Tanguay RM, Bruford E a, Cheetham ME, Chen B, Hightower LE (2009) Guidelines for the nomenclature of the human heat shock proteins. Cell Stress Chaperones 14(1):105–111. https://doi.org/10.1007/s12192-008-0068-7

Kandel ER, Schwartz JH, Jessell TM, Siegelbaum SA, Hudspet AJ, Mack S et al (2000) Principles of neural science, 4th edn. McGraw-Hill Ed, New York

Kayed R, Head E, Thompson JL, Mclntire TM, Milton SC, Cotman CW, Glabe CG (2003) Common structure of soluble amyloid oligomers implies common mechanism of pathogenesis. Science 300(5618):486–489

Kim SD, Fung VS (2014) An update on Huntington's disease: from the gene to the clinic. Curr Opin Neurol 27(4):477–483. https://doi.org/10.1097/WCO.0000000000000116

Kirchhoff SR, Gupta S, Knowlton AA (2002) Cytosolic heat shock protein 60, apoptosis, and myocardial injury. Circulation 105(24):2899–2904

Korsak M, Kozyreva T (2015) Beta amyloid hallmarks: from intrinsically disordered proteins to Alzheimer's disease. In: Felli I, Pierattelli R (eds) Intrinsically disordered proteins studied by NMR spectroscopy. Advances in experimental medicine and biology, vol 870. Springer, Cham. https://doi.org/10.1007/978-3-319-20164-1_14

Kusk MS, Damgaard B, Risom L, Hansen B, Ostergaard E (2016) Hypomyelinating Leukodystrophy due to *HSPD1* mutations: a new patient. Neuropediatrics 47(5):332–335. https://doi.org/10.1055/s-0036-1584564

Kuter K, Kratochwil M, Marx S-H, Hartwig S, Lehr S, Sugawa MD, Dencher NA (2016) Native DIGE proteomic analysis of mitochondria from substantia nigra and striatum during neuronal degeneration and its compensation in an animal model of early Parkinson's disease. Arch Physiol Biochem 122(5):238–256. https://doi.org/10.1080/13813455.2016.1197948

Lee A, Hirabayashi Y, KyuKwon S, LLewis T Jr, Polleux F (2018) Emerging roles of mitochondria in synaptic transmission and neurodegeneration. Curr Opin Physiol 3:82–93

Levy-Rimler G, Viitanen P, Weiss C, Sharkia R, Greenberg A, Niv A, Lustig A, Delarea Y, Azem A (2001) The effect of nucleotides and mitochondrial chaperonin 10 on the structure and chaperone activity of mitochondrial chaperonin 60. Eur J Biochem 268(12):3465–3472

Lindquist S (1986) The heat-shock response. Annu Rev Biochem 55(1):1151–1191. https://doi.org/10.1146/annurev.bi.55.070186.005443

Lv LH, Wan YL, Lin Y, Zhang W, Yang M, Li G-L, Lin HM, Shang CZ, Chen Y-J, Min J (2012) Anticancer drugs cause release of exosomes with heat shock proteins from human hepatocellular carcinoma cells that elicit effective natural killer cell antitumor responses in vitro. J Biol Chem 287(19):15874–15885. https://doi.org/10.1074/jbc.M112.340588

Macario AJL, Conway De Macario E (2001) Molecular chaperones and age-related degenerative disorders. Adv Cell Aging Gerontol 7:131–162. https://doi.org/10.1016/S1566-3124(01)07018-3

Macario AJL, Conway de Macario E (2005) Sick chaperones, cellular stress, and disease. N Engl J Med 353(14):1489–1501. https://doi.org/10.1056/NEJMra050111

Macario AJ, Conway de Macario E (2018) The Chaperonopathies. Diseases with defective molecular chaperones. In: Fink G (ed) Stress physiology, biochemistry, and pathology. Handbook of stress. Handbook of stress. Elsevier, Edinburgh

Macario AJL, Cappello F, Zummo G, Conway de Macario E (2010) Chaperonopathies of senescence and the scrambling of interactions between the chaperoning and the immune systems. Ann N Y Acad Sci 1197:85–93. https://doi.org/10.1111/j.1749-6632.2010.05187.x

Macario AJL, Conway de Macario E, Cappello F (2013) The chaperonopathies : diseases with defective molecular chaperones. Springer, Dordrecht

Magen D, Georgopoulos C, Bross P, Ang D, Segev Y, Goldsher D, Nemirovski A, Shahar E, Ravid S, Luder A, Heno B, Gershoni-Baruch R, Skorecki K, Mandel H (2008) Mitochondrial hsp60 chaperonopathy causes an autosomal-recessive neurodegenerative disorder linked to brain hypomyelination and leukodystrophy. Am J Hum Genet 83(1):30–42. https://doi.org/10.1016/j.ajhg.2008.05.016

Magnoni R, Palmfeldt J, Hansen J, Christensen JH, Corydon TJ, Bross P (2014) The Hsp60 folding machinery is crucial for manganese superoxide dismutase folding and function. Free Radic Res 48(2):168–179. https://doi.org/10.3109/10715762.2013.858147

Maiti P, Manna J, Veleri S, Frautschy S (2014) Molecular chaperone dysfunction in neurodegenerative diseases and effects of curcumin. Biomed Res Int 2014:1–14. https://doi.org/10.1155/2014/495091

Malik ZA, Kott KS, Poe AJ, Kuo T, Chen L, Ferrara KW, Knowlton AA (2013) Cardiac myocyte exosomes: stability, HSP60, and proteomics. Am J Physiol Heart Circ Physiol 304(7):954–965. https://doi.org/10.1152/ajpheart.00835.2012

Mangione MR, Vilasi S, Marino C, Librizzi F, Canale C, Spigolon D, Bucchieri F, Fucarino A, Passantino R, Cappello F, Bulone D, San Biagio PL (2016) Hsp60, amateur chaperone in amyloid-beta fibrillogenesis. Biochim Biophys Acta 1860(11 Pt A):2474–2483. https://doi.org/10.1016/j.bbagen.2016.07.019

Marino Gammazza A, Bucchieri F, Grimaldi LME, Benigno A, Conway De Macario E, Macario AJL, Zummo G, Cappello F (2012) The molecular anatomy of human Hsp60 and its similarity with that of bacterial orthologs and acetylcholine receptor reveal a potential pathogenetic role of anti-chaperonin immunity in myasthenia gravis. Cell Mol Neurobiol 32(6):943–947. https://doi.org/10.1007/s10571-011-9789-8

Marino Gammazza A, Rizzo M, Citarrella R, Rappa F, Campanella C, Bucchieri F, Patti A, Nikolic D, Cabibi D, Amico G, Conaldi PG, San Biagio PL, Montalto G, Farina F, Zummo G, Conway de Macario E, Macario AJL, Cappello F (2014) Elevated blood Hsp60, its structural similarities and cross-reactivity with thyroid molecules, and its presence on the plasma membrane of oncocytes point to the chaperonin as an immunopathogenic factor in Hashimoto's thyroiditis. Cell Stress Chaperones 19(3):343–353. https://doi.org/10.1007/s12192-013-0460-9

Marino Gammazza A, Caruso Bavisotto C, Barone R, Conway de Macario E, Macario AJL (2016) Alzheimer's disease and molecular chaperones: current knowledge and the future of chaperonotherapy. Curr Pharm Des 22(26):4040–4049. https://doi.org/10.2174/1381612822666160518141437

Marino Gammazza A, Campanella C, Barone R, Caruso Bavisotto C, Gorska M, Wozniak M, Carini F, Cappello F, D'Anneo A, Lauricella M, Zummo G, Conway de Macario E, Macario AJL, Di Felice V (2017a) Doxorubicin anti-tumor mechanisms include Hsp60 post-translational modifications leading to the Hsp60/p53 complex dissociation and instauration of replicative senescence. Cancer Lett 385:75–86. https://doi.org/10.1016/j.canlet.2016.10.045. a

Marino Gammazza A, Caruso Bavisotto C, David S, Barone R, Rappa F, Campanella C, Conway de Macario E, Cappello F, Macario AJL (2017b) HSP60 is a ubiquitous player in the physiological and pathogenic interactions between the chaperoning and the immune systems. Curr Immunol Rev 13(1):44–55. https://doi.org/10.2174/1573395513666170412170540. b

Merendino AM, Bucchieri F, Campanella C, Marcianò V, Ribbene A, David S, Zummo G, Burgio G, Corona DFV, Conway de Macario E, Macario AJL, Cappello F (2010) Hsp60 is actively secreted by human tumor cells. PLoS One 5(2):e9247. https://doi.org/10.1371/journal.pone.0009247

Miyamoto Y, Eguchi T, Kawahara K, Hasegawa N, Nakamura K, Funakoshi-Tago M, Tanoue A, Tamura H, Yamauchi J (2015) Hypomyelinating leukodystrophy-associated missense mutation in *HSPD1* blunts mitochondrial dynamics. Biochem Biophys Res Commun 462(3):275–281. https://doi.org/10.1016/j.bbrc.2015.04.132

Miyamoto Y, Megumi FT, Hasegawa N, Eguchi T, Tanoue A, Tamura H, Yamauchi J (2016) Data supporting mitochondrial morphological changes by SPG13-associated *HSPD1* mutants. Data Brief 6:482–488. https://doi.org/10.1016/j.dib.2015.12.038

Miyamoto Y, Kawahara K, Torii T, Yamauchi J (2017) Defective myelination in mice harboring hypomyelinating leukodystrophy-associated *HSPD1* mutation. Mol Genet Metab Rep 11:6–7. https://doi.org/10.1016/j.ymgmr.2017.03.003

Mizielinska S, Lashley T, Norona FE, Clayton EL, Ridler CE, Fratta P, Isaacs AM (2013) C9orf72 frontotemporal lobar degeneration is characterised by frequent neuronal sense and antisense RNA foci. Acta Neuropathol 126(6):845–857. https://doi.org/10.1007/s00401-013-1200-z

Monreal-Flores J, Espinosa-García MT, García-Regalado A, Arechavaleta-Velasco F, Martínez F (2017) The heat shock protein 60 promotes progesterone synthesis in mitochondria of JEG-3 cells. Reprod Biol 17(2):154–161. https://doi.org/10.1016/j.repbio.2017.04.001

Morell P, Norton WT (1980) Myelin. Sci Am 242(5):88–90

Nakamura H, Minegishi H (2013) HSP60 as a drug target. Curr Pharm Des 19(3):441–451

Nemirovsky A, Fisher Y, Baron R, Cohen IR, Monsonego A (2011) Amyloid beta-HSP60 peptide conjugate vaccine treats a mouse model of Alzheimer's disease. Vaccine 29(23):4043–4050. https://doi.org/10.1016/j.vaccine.2011.03.033

Nielsen KL, Cowan NJ (1998) A single ring is sufficient for productive chaperonin-mediated folding in vivo. Mol Cell 2(1):93–99

Nielsen KL, McLennan N, Masters M, Cowan NJ (1999) A single-ring mitochondrial chaperonin (Hsp60-Hsp10) can substitute for GroEL-GroES in vivo. J Bacteriol 181(18):5871–5875

Noelker C, Morel L, Osterloh A, Alvarez-Fischer D, Lescot T, Breloer M, Gold M, Oertel WH, Henze C, Michel PP, Dodel RC, Lu L, Hirsch EC, Hunot S, Hartmann A (2014) Heat shock protein 60: an endogenous inducer of dopaminergic cell death in Parkinson disease. J Neuroinflammation 11(1):86. https://doi.org/10.1186/1742-2094-11-86

Novo G, Cappello F, Rizzo M, Fazio G, Zambuto S, Tortorici E, Marino Gammazza A, Corrao S, Zummo G, Conway de Macario E, Macario AJL, Assennato P, Novo S, Li Volti G, Volti GL (2011) Hsp60 and heme oxygenase-1 (Hsp32) in acute myocardial infarction. Transl Res: J Lab Clin Med 157(5):285–292. https://doi.org/10.1016/j.trsl.2011.01.003

Okamoto T, Ishida R, Yamamoto H, Tanabe-Ishida M, Haga A, Takahashi H, Takahashi K, Goto D, Grave E, Itoh H (2015) Functional structure and physiological functions of mammalian wild-type HSP60. Arch Biochem Biophys 586:10–19. https://doi.org/10.1016/j.abb.2015.09.022

Okamoto T, Yamamoto H, Kudo I, Matsumoto K, Odaka M, Grave E, Itoh H (2017) HSP60 possesses a GTPase activity and mediates protein folding with HSP10. Sci Rep 7(1):16931. https://doi.org/10.1038/s41598-017-17167-7

Osterloh A, Meier-Stiegen F, Veit A, Fleischer B, von Bonin A, Breloer M (2004) Lipopolysaccharide-free heat shock protein 60 activates T cells. J Biol Chem 279(46):47906–47911. https://doi.org/10.1074/jbc.M408440200

Pace A, Barone G, Lauria A, Martorana A, Piccionello AP, Pierro P, Terenzi A, Almerico AM, Buscemi S, Campanella C, Angileri F, Carini F, Zummo G, Conway de Macario E, Cappello F, Macario AJL (2013) Hsp60, a novel target for antitumor therapy: structure-function features and prospective drugs design. Curr Pharm Des 19(15):2757–2764

Parnas A, Nadler M, Nisemblat S, Horovitz A, Mandel H, Azem A (2009) The MitCHAP-60 disease is due to entropic destabilization of the human mitochondrial Hsp60 oligomer. J Biol Chem 284(41):28198–28203. https://doi.org/10.1074/jbc.M109.031997

Pasanen P, Myllykangas L, Siitonen M, Raunio A, Kaakkola S, Lyytinen J, Tienari PJ, Pöyhönen M, Paetau A (2014) Novel α-synuclein mutation A53E associated with atypical multiple system atrophy and Parkinson's disease-type pathology. Neurobiol Aging 35(9):2180.e1–2180.e5. https://doi.org/10.1016/j.neurobiolaging.2014.03.024

Perry VH, Holmes C (2014) Microglial priming in neurodegenerative disease. Nat Rev Neurol 10(4):217–224. https://doi.org/10.1038/nrneurol.2014.38

Peters A, Palay SL, Webster HD (1978) The fine structure of the nervous system: the neurons and supporting cells. Ann Neurol 4(6):588–588. https://doi.org/10.1002/ana.410040660

Poewe W, Seppi K, Tanner CM, Halliday GM, Brundin P, Volkmann J, Schrag AE, Lang AE (2017) Parkinson disease. Nat Rev Dis Primers 3:17013

Querfurth, La Ferla (2010) Alzheimer's disease. N Ingl J Med 362:329–344

Rizzo M, Cappello F, Marfil R, Nibali L, Marino Gammazza A, Rappa F, Bonaventura G, Galindo-Moreno P, O'Valle F, Zummo G, Conway de Macario E, Macario AJL, Mesa F (2012) Heat-shock protein 60 kDa and atherogenic dyslipidemia in patients with untreated mild periodontitis: a pilot study. Cell Stress Chaperones 17(3):399–407. https://doi.org/10.1007/s12192-011-0315-1

Samali A, Cai J, Zhivotovsky B, Jones DP, Orrenius S (1999) Presence of a pre-apoptotic complex of pro-caspase-3, Hsp60 and Hsp10 in the mitochondrial fraction of jurkat cells. EMBO J 18(8):2040–2048. https://doi.org/10.1093/emboj/18.8.2040

Satoh J, Onoue H, Arima K, Yamamura T (2005) The 14-3-3 protein forms a molecular complex with heat shock protein Hsp60 and cellular prion protein. J Neuropathol Exp Neurol 64(10):858–868. Retrieved September 2, 2018 from internet: http://www.ncbi.nlm.nih.gov/pubmed/16215457

Seelaar H, Rohrer JD, Pijnenburg YAL, Fox NC, van Swieten JC (2011) Clinical, genetic and pathological heterogeneity of frontotemporal dementia: a review. J Neurol Neurosurg Psychiatry 82(5):476–486. https://doi.org/10.1136/jnnp.2010.212225

Selkoe DJ, Hardy J (2016) The amyloid hypothesis of Alzheimer's disease at 25 years. EMBO Mol Med 8:595–608. https://doi.org/10.15252/emmm.201606210

Shan YX, Liu TJ, Su HF, Samsamshariat A, Mestril R, Wang PH (2003) Hsp10 and Hsp60 modulate Bcl-2 family and mitochondria apoptosis signaling induced by doxorubicin in cardiac muscle cells. J Mol Cell Cardiol 35(9):1135–1143

Soltys BJ, Gupta RS (1996) Immunoelectron microscopic localization of the 60-kDa heat shock chaperonin protein (Hsp60) in mammalian cells. Exp Cell Res 222(1):16–27. https://doi.org/10.1006/excr.1996.0003

Soltys BJ, Gupta RS (1997) Cell surface localization of the 60 kDa heat shock chaperonin protein (hsp60) in mammalian cells. Cell Biol Int 21(5):315–320. https://doi.org/10.1006/cbir.1997.0144

Soltys BJ, Gupta RS (2000) Mitochondrial proteins at unexpected cellular locations: export of proteins from mitochondria from an evolutionary perspective. Int Rev Cytol 194:133–196

Soto C (2003) Unfolding the role of protein misfolding in neurodegenerative diseases. Natl Rev 4:49–60

Steward O (1989) Principles of cellular, molecular, and developmental neuroscience. Springer, New York

Sun Y, Zheng J, Xu Y, Zhang X (2018) Paraquat-induced inflammatory response of microglia through HSP60/TLR4 signaling. Hum Exp Toxicol, 960327118758152. https://doi.org/10.1177/0960327118758152

Takano M, Yamashita T, Nagano K, Otani M, Maekura K, Kamada H, Tsunoda S, Tsutsumi Y, Tomiyama T, Mori H, Matsuura K, Matsuyama S (2013) Proteomic analysis of the hippocampus in Alzheimer's disease model mice by using two-dimensional fluorescence difference in gel electrophoresis. Neurosci Lett 534:85–89. https://doi.org/10.1016/j.neulet.2012.11.010

Tomasello G, Rodolico V, Zerilli M, Martorana A, Bucchieri F, Pitruzzella A, Marino Gammazza A, David S, Rappa F, Zummo G, Damiani P, Accomando S, Rizzo M, Conway de Macario E, Macario AJL, Cappello F (2011) Changes in immunohistochemical levels and subcellular localization after therapy and correlation and colocalization with CD68 suggest a pathogenetic role of Hsp60 in ulcerative colitis. Appl Immunohistochem Mol Morphol 19(6):552–561. https://doi.org/10.1097/PAI.0b013e3182118e5f

Tong X, Ao Y, Faas GC, Nwaobi SE, Xu J, Haustein MD, Anderson MA, Mody I, Olsen ML, Sofroniew MV, Khakh BS (2014) Astrocyte Kir4.1 ion channel deficits contribute to neuronal dysfunction in Huntington's disease model mice. Nat Neurosci 17(5):694–703. https://doi.org/10.1038/nn.3691

Veereshwarayya V, Kumar P, Rosen KM, Mestril R, Querfurth HW (2006) Differential effects of mitochondrial heat shock protein 60 and related molecular chaperones to prevent intracellular beta-amyloid-induced inhibition of complex IV and limit apoptosis. J Biol Chem 281(40):29468–29478. https://doi.org/10.1074/jbc.M602533200

Vilasi S, Carrotta R, Mangione MR, Campanella C, Librizzi F, Martorana V, Marino Gammazza A, Ortore MG, Vilasi A, Pocsfalvi G, Burgio G, Corona D, Palumbo Piccionello A, Zummo G, Bulone D, Conway de Macario C, Macario AJL, San Biagio PL, Cappello F (2014) Human Hsp60 with its mitochondrial import signal occurs in solution as heptamers and tetradecamers remarkably stable over a wide range of concentrations. PLoS One 9(5):e97657. https://doi.org/10.1371/journal.pone.0097657

Vilasi S, Bulone D, Caruso Bavisotto C, Campanella C, Marino Gammazza A, San Biagio PL, Cappello F, Conway de Macario E, Macario AJL (2018) Chaperonin of group I: oligomeric spectrum and biochemical and biological implications. Front Mol Biosci 4:99. https://doi.org/10.3389/fmolb.2017.00099

Wadhwa R, Takano S, Kaur K, Aida S, Yaguchi T, Kaul Z, Hirano T, Taira K, Kaul SC (2005) Identification and characterization of molecular interactions between mortalin/mtHsp70 and HSP60. Biochem J 391(Pt 2):185–190. https://doi.org/10.1042/BJ20050861

Walls KC, Coskun P, Gallegos-Perez JL, Zadourian N, Freude K, Rasool S, Blurton-Jones M, Green KN, LaFerla FM (2012) Swedish Alzheimer mutation induces mitochondrial dysfunction mediated by HSP60 mislocalization of amyloid precursor protein (APP) and beta-amyloid. J Biol Chem 287(36):30317–30327. https://doi.org/10.1074/jbc.M112.365890

Wang HQ, Xu YX, Zhao XY, Zhao H, Yan J, Sun XB, Guo JC, Zhu CQ (2009) Overexpression of F0F1-ATP synthase α suppresses mutant huntingtin aggregation and toxicity in vitro. Biochem Biophys Res Commun 390(4):1294–1298. https://doi.org/10.1016/j.bbrc.2009.10.139

Wijesekera LC, Leigh PN (2009) Amyotrophic lateral sclerosis. Orphanet J Rare Dis 4(1):3. https://doi.org/10.1186/1750-1172-4-3

Wojsiat J, Prandelli C, Laskowska-Kaszub K, Martín-Requero A, Wojda U (2015) Oxidative stress and aberrant cell cycle in Alzheimer's disease lymphocytes: diagnostic prospects. J Alzheimer's Dis: JAD 46(2):329–350. https://doi.org/10.3233/JAD-141977

Yamamoto T, Yamamoto-Shimojima K, Ueda Y, Imai K, Takahashi Y, Imagawa E, Miyake N, Matsumoto N (2018) Independent occurrence of de novo *HSPD1* and *HIP1* variants in brothers with different neurological disorders – leukodystrophy and autism. Hum Genome Var 5(1):18. https://doi.org/10.1038/s41439-018-0020-z

Zhang R, Li Y, Hou X, Miao Z, Wang Y (2017) Neuroprotective effect of heat shock protein 60 on matrine-suppressed microglial activation. Exp Ther Med 14(2):1832–1836. https://doi.org/10.3892/etm.2017.4691

Zhao C, Li H, Zhao XJ, Liu ZX, Zhou P, Liu Y, Feng MJ (2016) Heat shock protein 60 affects behavioral improvement in a rat model of Parkinson's disease grafted with human umbilical cord mesenchymal stem cell-derived dopaminergic-like neurons. Neurochem Res 41(6):1238–1249. https://doi.org/10.1007/s11064-015-1816-6

Zielonka D, Mielcarek MG, Landwehrmeyer B (2015) Update on Huntington's disease: advances in care and emerging therapeutic options. Parkinsonism Relat Disord 21(3):169–178

Part VI
HSP60 and Skeletal Muscle Diseases and Disorders

Chapter 17
Hsp60 in Skeletal Muscle: From Molecular Anatomy to Pathophysiology

Rosario Barone, Valentina Di Felice, Dario Coletti, and Alberto J. L. Macario

Abstract The chaperoning system of an organism is composed of the entire set of chaperones, co-chaperones, and chaperone co-factors and their interactors and receptors. Its functions pertain typically to protein homeostasis but also to many other activities inside and outside cells. In the skeletal muscle, with its multi-molecular structures rich in proteins and their continuous rearrangements, the chaperoning system plays a crucial role. However, little is known about the details of the workings of the chaperoning system in skeletal muscle development and during exercise and disease. Molecular chaperones are surely involved in muscle formation and maintenance under physiologic conditions and under stress but if abnormal or involved in a pathogenic pathway can cause disease, a chaperonopathy. There are many genetic and acquired chaperonopathies affecting muscles primarily. For these reasons, we have begun to study chaperones in skeletal muscle, focusing on Hsp60. This chaperone is essential for muscle activity since it maintains the functionality of the respiratory chain inside the mitochondria, among other critical functions such as defense against oxidative stress. In addition, in skeletal muscle, Hsp60 occurs in the cytosol and the extracellular space but its functions in these non-canonical locations remain to be elucidated.

R. Barone (✉) · V. Di Felice
Department of Biomedicine, Neurosciences and Advanced Diagnostics (BiND), University of Palermo, Palermo, Italy

Euro-Mediterranean Institute of Science and Technology (IEMEST), Palermo, Italy

D. Coletti
Department of Anatomical, Histological, Forensic & Orthopaedic Sciences, Section of Histology & Medical Embryology, Sapienza University of Rome, Rome, Italy

Sorbonne University, B2A Biological Adaptation and Aging, Paris, France

A. J. L. Macario
Euro-Mediterranean Institute of Science and Technology (IEMEST), Palermo, Italy

School of Medicine, Department of Microbiology and Immunology, University of Maryland at Baltimore-Institute of Marine and Environmental Technology (IMET), Columbus Center, Baltimore, MD, USA

A. A. A. Asea, P. Kaur (eds.), *Heat Shock Protein 60 in Human Diseases and Disorders*, Heat Shock Proteins 18,
https://doi.org/10.1007/978-3-030-23154-5_17

Keywords Atrophy · Chaperonopathies · Dystrophy · Exercise · Hsp60 · Myosin heavy chain · Skeletal muscle

Abbreviations

EDL	extensor digitorum longus
HSP	heat shock proteins
MHC	myosin heavy chain
PGC-1α	Peroxisome proliferator-activated receptor gamma coactivator 1 alpha
ROS	reactive oxygen species

17.1 Introduction

The human body responds quickly and specifically to the various physiological challenges to maintain a constant internal environment, a process known as homeostatic control. Skeletal muscle, the most abundant tissue in the human body, is highly plastic in response to different stimuli and stress. Physical exercise (Hoppeler 1986), hypoxia (Hoppeler et al. 2008), weightlessness (Desplanches 1997), disuse (Bodine 2013), and nutritional modifications (Vogt et al. 2003) all induce strong responses of the skeletal muscle. Repeated stimuli by mechanical loading or other factors such as hormonal, metabolic, and neuronal inducers regulate both gene expression and protein function affecting muscle homeostasis (Flück and Hoppeler 2003). In particular, the adaptations of skeletal muscle after physical exercise are related to the type of training: for example, strength training determines hypertrophy and hyperplasia, while endurance training increases mitochondria and oxidative capacity.

Skeletal muscle typically contains a heterogeneous population of muscle fibers, which can be classified into several groups based on properties such as contractile speed, myosin heavy chain (MHC) expression, and metabolic capacity (Spangenburg and Booth 2003; Zierath and Hawley 2004). In addition, physiologically each individual muscle fiber can change its properties, such as mechanical (isometric tension) and morphological parameters (cross section area), and molecular patterns (MHC isoforms) to adapt to exogenous stimuli. Other conditions like aging, some diseases, and disuse cause weakening, atrophy and loss of muscle strength, with important consequences, such as compromised mobility. The latter condition is known as muscle wasting and is characterized by a disarrangement of the sarcomeric proteins, a decrease in protein synthesis, and an increase in protein catabolism. The chaperoning system participates in many cellular functions from assisting protein folding and assembling of multimolecular complexes to maintaining the correct structures of enzymes (Czarnecka et al. 2006; Georgopoulos and Welch 1993; Voos 2013; Macario and Conway de Macario 2019). In addition, extracellular chaperones contribute to the intercommunication between different cells, tissues, and organs (Cappello et al. 2008; Henderson et al. 2010). Skeletal muscle is a tissue that relies on the chaperoning system for both, the control of intracellular components and intercellular communications. In response to various types of stress, cells rapidly produce a set of proteins, components of the chaperoning system, the heat-shock proteins (HSP), which help to protect cells against stress, and some are important and ubiquitously functioning as chaperones. In brief, HSP, many of which are molecular chaperones with functions in maintaining protein homeostasis under normal, physiological conditions and as a defense against stress, are classified by molecular weight in various groups: super heavy, 100, 90, 70, 60, 40, and small Hsp (sHsp), including chaperones within the following ranges in kDa: 100 or higher, 81–99, 65–80, 55–64, 35–54, and 34 or lower, respectively (Macario 1995; Macario and Conway de Macario 2019). Members of four of these groups have been detected in skeletal muscle: small Hsp, Hsp60, Hsp70, and Hsp90 (Moresi et al. 2009; Morton et al. 2009). Hsp60 is one of the least studied HSP in skeletal muscle, in particular after exercise, a condition which profoundly affects muscle stress and homeostasis. This protein is primarily localized to the mitochondria, although we currently know that it occurs also in extra-mitochondrial sites, such as cytosol, plasma-cell membrane, inside exosomes, extracellular space, and circulation, both in normal and pathological conditions (Campanella et al. 2012; Cappello et al. 2008). This Chapter focuses on recent advances in the understanding of the role of Hsp60 in skeletal muscle.

17.2 Levels and Distribution of Hsp60 in Skeletal Muscle

Skeletal muscle fibers are a complex and heterogeneous system mostly characterized by the abundance of contractile proteins (myofilaments) organized in myofibrils. Different isoforms of contractile proteins account for the different histochemical and functional characteristics of the various muscle fiber types which as we shall explain also differ in their contents of Hsp60. Skeletal muscle is capable of producing several HSP in response to stress and ATP deficit, in particular Hsp60 (Ornatsky et al. 1995; Bornman et al. 1995). It was known for some time that exercise was accompanied by increased expression of intramitochondrial Hsp60 but its release into the extracellular space and circulation was demonstrated more recently, when it was shown that the chaperonin is secreted by muscle fibers via exosomes (Barone et al. 2016).

Little can be found in the literature about the expression of the Hsp60 during physical exercise or whether it is expressed in the skeletal muscle fibers in proportion to the mitochondrial content and its oxidative capacity. In rats trained with an endurance protocol of 8 weeks, a significant increase in Hsp60 levels in *plantaris* muscle and no difference in *soleus* muscle were shown (Mattson et al. 2000). These findings are consistent with the hypothesis that the adaptive response to running may require augmented expression of Hsp60 to support protein import and folding.

Also, it was shown a significant increase in Hsp60 levels in *soleus* muscle in trained relative to untrained rats, while its levels were unchanged in *lateral gastrocnemius* after training (Samelman 2000). These results indicate that exercise can modify the basal expression of Hsp60 with a pattern that differs between muscles and this observation is consistent with the hypothesis that physical exercise may activate a protective mechanism to stress (Samelman 2000). In contrast, no significant differences in Hsp60 levels were observed by others in rats with different time-point of endurance training in *plantaris* and *gastrocnemius* muscles (Ogata et al. 2009; Moura et al. 2014).

We could find very few reports on studies in humans on Hsp60 levels after training. For instance, the acute effect of a single bout of endurance training was investigated and it was found that the levels of Hsp60 in the *vastus lateralis* were significantly higher in trained subjects than in sedentary ones (Morton et al. 2008); likewise, it was demonstrated that Hsp60 in sedentary and trained subjects was not fiber type specific in the *vastus lateralis* muscle (Folkesson et al. 2013). Our research group measured the levels of Hsp60 in the fibers of the posterior muscles group (*gastrocnemius*, *soleus*, and *plantaris*) of Balb/c mice after they completed a 6-week program of endurance training (Barone et al. 2016). Our data showed a differential quantitative distribution of Hsp60 within a single muscle group and changes in response to exercise. By performing immunohistochemical determinations of Hsp60 levels in muscle cross-sections and by comparing the content of Hsp60 to that of MHCs, it was possible to unveil fiber-type-specific levels of Hsp60 in skeletal muscle. Hsp60 was elevated in type I, IIa, and IIx muscle fibers, while type IIb fibers appeared only slightly positive for Hsp60 (Fig. 17.1) (Barone et al. 2016). Endurance exercise training induced a significant increase in the Hsp60 levels in type I fibers, an increase which may represent a physiological adaptation in response to exercise. Furthermore, we demonstrated a correlation *in vitro* between Hsp60 and PGC-1α (Peroxisome proliferator-activated receptor gamma coactivator 1 alpha) levels; the latter is considered the lead factor of mitochondrial biogenesis (Barone et al. 2016; Barone et al. 2017). In agreement with the above, earlier data exist

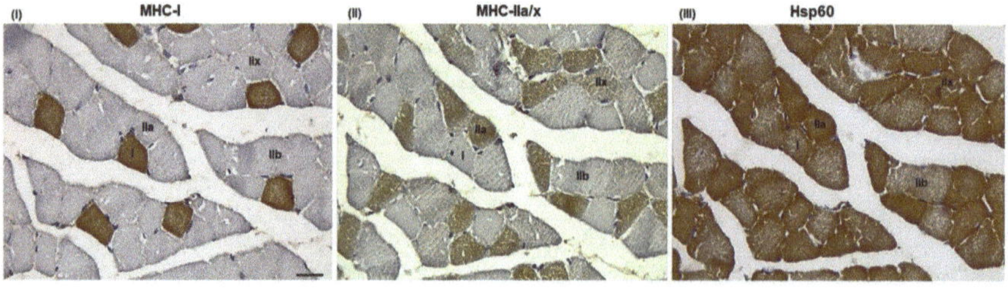

Fig. 17.1 Immunohistochemistry of the posterior group of hind limb muscles. Immunohistochemistry for MHC-I (I), MHC-IIa/x (II) and Hsp60 (III) in serial cross-sections of the posterior group of hind limb muscles (*gastrocnemius*, and *plantaris).* Type IIb fibers are negative to antibodies anti-MHC-I and anti-MHC-IIa/x. Hsp60, heat shock protein 60; MHC-I, myosin heavy chain I; MHC-IIa/x myosin heavy chain IIa/x; Bar 25 μm

showing that the increased expression of Hsp60 in trained mice facilitates protein import and folding, inducing mitochondrial biogenesis (Hood et al. 2000).

17.3 Expression and Role of Hsp60 in Pathological Conditions Involving Skeletal Muscle

Hsp60 plays a crucial role in the reduction of oxidative stress. Silencing of Hsp60 in cell lines increased reactive oxygen species (ROS) production (Tang et al. 2016) and induced mitochondrial dysfunction (Kleinridders et al. 2013). Hsp60 suppresses excessive ROS production and promotes resistance to oxidative stress and the maintenance of mitochondrial function. Hsp60 content was significantly higher in the *extensor digitorum longus* muscle (EDL) of heat-stressed diabetic rats than in non-stressed diabetic rats, and its levels were similar to normal rat skeletal muscle (Nonaka et al. 2018). Heat-stressed diabetic rats maintained Hsp60 content, resulting in the prevention of excessive ROS production and resistance to oxidative stress in the skeletal muscle. The *diaphragm* muscle but not the *quadriceps* and *gastrocnemius* muscles of dystrophic-trained mice showed decreased Hsp60 levels compared to dystrophic-sedentary mice, suggesting exhaustion of potentially protective mechanisms in the *diaphragm* (Morici et al. 2017). Probably, the ability to maintain a high level of Hsp60 contributes to the positive effects of exercise training in limb muscles, while the decreased levels found in the *diaphragm* could contribute to the overall lack of major positive effects of mild exercise in this anatomic district.

17.3.1 Hypothesis on Alternative Roles of Hsp60 in Skeletal Muscle

Well established observations on Hsp60 in muscle tissue are its mitochondrial localization and its increase in response to physical exercise. However, the biological relevance of Hsp60 translocation from the mitochondrion to the cytoplasm and to the plasma-cell membrane, and its secretion with extracellular vesicles, such as exosomes, remain elusive. Similarly, the regulation of Hsp60 by exercise needs more research on our suggestion that it could contribute to PGC-1α upregulation, ultimately leading to improved endurance. A generally well-known function of Hsp60 is its involvement in the folding of mitochondrial proteins, or in the re-folding of these proteins when they are partially denatured by stress. We have shown that Hsp60 increases preferentially in certain muscle fiber types, and that the serum of untrained mice at time zero does not contain abundant exosomes and does not contain much Hsp60 (Barone et al. 2016). We have also shown that: (a) following training, exosomes contain higher levels of Hsp60 than exosomes from untrained animals; (b) during training PGC-1α and its isoforms increase; and (c) Hsp60 in vitro interacts with PGC-1α. We do not know how Hsp60 can enter the target cell and activate the transcription of PGC-1α, but a close correlation between Hsp60 and PGC-1α occurs and, as a consequence, the mitochondrial biogenesis pathway is activated upon training. Another explanation of the increase in Hsp60 in skeletal muscle upon exercise is that in the process of fiber differentiation and structural changes inherent to activity, muscle cells need more Hsp60 than at rest, as we have determined in human primary fibroblasts in culture. Also there is a correlation between the expression of IL-6 and Hsp60, since both are expressed in greater quantities during training and their mRNAs are both targets of the same miR, i.e., miR-206 (Shan et al. 2010). This interaction has been demonstrated only in the heart muscle so far, but given the histological and molecular similarity between the two striated muscles, we cannot exclude that also in skeletal muscle the expression of these two molecules is somewhat co-regulated and correlated.

17.4 Conclusions

Hsp60 is constitutively expressed in many tissues of the body in basal conditions, upon stress and in pathological conditions. Its role as a mitochondrial chaperone is relatively well known, but new roles in the pathophysiology of the skeletal muscle are now emerging. We have for the first time demonstrated the presence of Hsp60 in exosomes of trained mice, and suggested a role for Hsp60 as a myokine. The pertinent intimate molecular mechanisms involved are currently under investigation.

Acknowledgements This study was funded by "Ministero dell'Istruzione, dell'Università e della Ricerca" (PRIN2012-prot. 2012N8YJC3- Prof. Daniela Caporossi/Prof. Felicia Farina). A.J.L.M was partially supported by IMET. This work was done under the agreement between IEMEST (Italy) and IMET (USA) (this is IMET contribution number IMET 19-005).

References

Barone R, Macaluso F, Sangiorgi C et al (2016) Skeletal muscle heat shock protein 60 increases after endurance training and induces peroxisome proliferator-activated receptor gamma coactivator 1 α1 expression. Sci Rep 27:19781

Barone R, Sangiorgi C, Marino Gammazza A et al (2017) Effects of conjugated linoleic acid associated with endurance exercise on muscle fibres and peroxisome proliferator-activated receptor γ coactivator 1 α isoforms. J Cell Physiol 232:1086–1094

Bodine SC (2013) Disuse-induced muscle wasting. Int J Biochem Cell Biol 45:2200–2208

Bornman LB, Polla S, Lotz BP, Gericke GS (1995) Expression of heat-shock/stress proteins in Duchenne muscular dystrophy. Muscle Nerve 18:23–31

Campanella C, Bucchieri F, Merendino AM et al (2012) The odyssey of Hsp60 from tumor cells to other destinations includes plasma membrane-associated stages and Golgi and exosomal protein-trafficking modalities. PLoS One 7:e42008

Cappello F, Conway de Macario E, Marasa L, Zummo G, Macario AJL (2008) Hsp60 expression, new locations, functions and perspectives for cancer diagnosis and therapy. Cancer Biol Ther 7:801–809
Czarnecka AM, Campanella C, Zummo G, Cappello F (2006) Mitochondrial chaperones in cancer: from molecular biology to clinical diagnostics. Cancer Biol Ther 5:714–720
Desplanches D (1997) Structural and functional adaptations of skeletal muscle to weightlessness. Int J Sports Med 18:259–264
Flück M, Hoppeler H (2003) Molecular basis of skeletal muscle plasticity – from gene to form and function. Rev Physiol Biochem Pharmacol 146:159–216
Folkesson M, Mackey AL, Langberg H et al (2013) The expression of heat shock protein in human skeletal muscle: effects of muscle fibre phenotype and training background. Acta Physiol 209:26–33
Georgopoulos C, Welch WJ (1993) Role of the major heat shock proteins as molecular chaperones. Ann Rev Cell Biol 9:601–634
Henderson B, Calderwood SK, Coates AR et al (2010) Caught with their PAMPs down? The extracellular signalling actions of molecular chaperones are not due to microbial contaminants. Cell Stress Chaperones 15:123–141
Hood DA, Takahashi M, Connor MK, Freyssenet D (2000) Assembly of the cellular powerhouse: current issues in mitochondrial biogenesis. Exerc Sport Sci Rev 28:68–73
Hoppeler H (1986) Exercise-induced ultrastructural changes in skeletal muscle. Int J Sports Med 7:187–204
Hoppeler H, Klossner S, Vogt M (2008) Training in hypoxia and its effects on skeletal muscle tissue. Scand J Med Sci Sports 18:38–49
Kleinridders A, Lauritzen HP, Ussar S et al (2013) Leptin regulation of Hsp60 impacts hypothalamic insulin signaling. J Clin Invest 123:4667–4680
Macario AJL (1995) Heat-shock proteins and molecular chaperones: implications for pathogenesis, diagnostics, and therapeutics. Int J Clin Lab Res 25:59–70
Macario AJL, Conway de Macario E (2019) Chaperone proteins and chaperonopathies. In: Fink G (ed) Stress physiology, biochemistry, and pathology. Handbook of stress. Elsevier, Amsterdam
Mattson JP, Ross CR, Kilgore JL, Musch TI (2000) Induction of mitochondrial stress proteins following treadmill running. Med Sci Sport Exerc 32:365–369
Moresi V, Garcia-Alvarez G, Pristerà A et al (2009) Modulation of caspase activity regulates skeletal muscle regeneration and function in response to vasopressin and tumor necrosis factor. PLoS One 4:e5570
Morici G, Frinchi M, Pitruzzella A et al (2017) Mild aerobic exercise training hardly affects the diaphragm of mdx mice. J Cell Physiol 232:2044–2052
Morton JP, Maclaren DP, Cable NT et al (2008) Trained men display increased basal heat shock protein content of skeletal muscle. Med Sci Sports Exerc 40:1255–1262
Morton JP, Kayani AC, McArdle A, Drust B (2009) The exercise-induced stress response of skeletal muscle, with specific emphasis on humans. Sports Med 39:643–662
Moura CS, Lollo PC, Morato PN, Nisishima LH, Carneiro EM, Amaya-Farfan J (2014) Whey protein hydrolysate enhances HSP90 but does not alter HSP60 and HSP25 in skeletal muscle of rats. PLoS One 20:e83437
Nonaka K, Une S, Komatsu M, Yamaji R, Akiyama J (2018) Heat stress prevents the decrease in succinate dehydrogenase activity in the extensor digitorum longus of streptozotocin-induced diabetic rats. Physiol Res 16:117–126
Ogata T, Oishi Y, Higashida K, Higuchi M, Muraoka I (2009) Prolonged exercise training induces long-term enhancement of HSP70 expression in rat plantaris muscle. Am J Physiol Regul Integr Comp Physiol 296:1557–1563
Ornatsky OI, Connor MK, Hood DA (1995) Expression of stress proteins and mitochondrial chaperonins in chronically stimulated skeletal muscle. Biochem J 311:119–123
Samelman TR (2000) Heat shock protein expression is increased in cardiac and skeletal muscles of Fischer 344 rats after endurance training. Exp Physiol 85:92–102

Shan ZX, Lin QX, Deng CY et al (2010) miR-1/miR-206 regulate Hsp60 expression contributing to glucose-mediated apoptosis in cardiomyocytes. FEBS Lett 584:3592–3600
Spangenburg EE, Booth FW (2003) Molecular regulation of individual skeletal muscle fibre types. Acta Physiol Scand 178:413–424
Tang H, Chen Y, Liu X et al (2016) Downregulation of HSP60 disrupts mitochondrial proteostasis to promote tumorigenesis and progression in clear cell renal cell carcinoma. Oncotarget 21:38822–38834
Vogt M, Puntschart A, Howald H et al (2003) Effects of dietary fat on muscle substrates, metabolism, and performance in athletes. Med Sci Sports Exerc 35:952–960
Voos W (2013) Chaperone-protease networks in mitochondrial protein homeostasis. Biochim Biophys Acta 1833:388–399
Zierath JR, Hawley JA (2004) Skeletal muscle fiber type: influence on contractile and metabolic properties. PLoS Biol 2:e348

Chapter 18
Heat Shock Protein 60 (HSP60): Role in Skeletal Muscle Diseases and Novel Prospects for Therapy

Richa Rathor, Geetha Suryakumar, Som Nath Singh, and Bhuvnesh Kumar

Abstract Eukaryotic Hsp60 is also known as mitochondrial chaperones (Chaperonin, Cpn60) as earlier it was considered to be present in mitochondria only. Last few years it has become clear that it is also present in cytosol, cell surface, extracellular space and in the peripheral blood. Hsp60 plays a vital role in quality control of proteins. It interacts with Hsp10 (resides in mitochondria, also named as Cpn10) to prepare native conformational protein from nascent polypeptides in the presence of ATP. Some other newly identified functions of Hsp60 include cell survival and proliferation. Hsp60 has significant role in various skeletal muscle wasting diseases like sarcopenia, cancer cachexia, sepsis, denervation, burns, and chronic obstructive pulmonary disease. The present chapter describes a brief representative research efforts aimed to establish the role of Hsp60 in various skeletal muscle wasting conditions with the purpose to illustrate possible protective and therapeutic implications for developing novel approach to rectify them.

Keywords Cytoprotection · High altitude · Hsp60 · Proteostasis · Quality control · Skeletal muscle

Abbreviations

HSBP1	Heat shock protein binding factor 1
HSEs	Heat shock elements
HSF1	Heat shock factor transcription factor
HSP	Heat shock family

R. Rathor (✉) · G. Suryakumar · S. N. Singh · B. Kumar
Cellular Biochemistry Group, Defense Institute of Physiology and Allied Sciences, Delhi, India
e-mail: richa110in@dipas.drdo.in

A. A. A. Asea, P. Kaur (eds.), *Heat Shock Protein 60 in Human Diseases and Disorders*, Heat Shock Proteins 18,
https://doi.org/10.1007/978-3-030-23154-5_18

Hsp	Heat shock proteins
Hsp60	Heat shock protein60
IGF1	Insulin growth factor-1
LMF	Lipid mobilising factor
MAPK	Mitogen-activated protein kinase
mTOR	Mammalian target of rapamycin
TLRs	Toll like receptors
UPS	Unfolded protein response

18.1 Introduction

Heat Shock proteins (Hsp) comprises a heterogeneous group of molecules involved with various functions such as response to stress and protein folding. In past, Hsp were considered stress proteins which usually involved in assisting quality control for protein folding of nascent polypeptides and refolding cellular stress induced partially denatured proteins (Kregel 2002; Brodsky and Chiosis 2006; Macario and de Macario 2007) and these Hsp are also known as molecular chaperones. Recently, a number of cellular functions attributed to Hsp like protein translocation, protein degradation, dissolution of pathologic protein aggregates, regulation of gene expression, cell differentiation, DNA replication, signal transduction, programmed cell death, cellular senescence and carcinogenesis. The first heat shock protein was discovered by Feruccio Ritossa (1962, 1996) in salivary gland cells of *Drosophila buskii* after a serendipitous heat shock. Followed by, it was clearly characterized later in 1974 (Tissieres et al. 1974). Previously it was assumed that they induced by heat shock stress but afterwards it was detailed that a large variety of stimuli or environmental stresses lead to produce heat shock proteins (Fig. 18.1).

18.1.1 Heat Shock Proteins, Molecular Chaperones

Occasionally, the term Hsp and molecular chaperone are used interchangeably. But, in actual few Hsp does not contain chaperoning function and some chaperones do not belong to Hsp group. As per the definition, chaperones are defined as "proteins that bind to and stabilize an otherwise unstable conformer of another protein – and, by controlled binding and release, facilitate its correct fate in vivo: be it folding, oligomeric assembly, transport to a particular subcellular compartment, or disposal by degradation" (Hartl 1996). Basically chaperones could be divided into two categories: small heat shock proteins and 90kD heat shock protein (Hsp90 family) (Table 18.1).

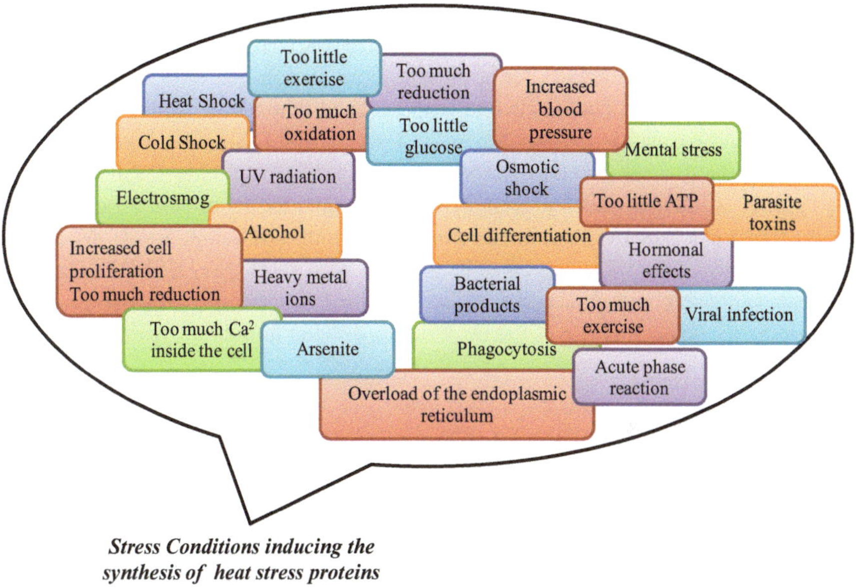

Fig. 18.1 Numerous adverse stimuli or environmental stress conditions lead to produce heat shock proteins

Table 18.1 Major families of molecular chaperones and their functions

Molecular chaperone families	Major functions of chaperone families
Small heat shock (e.g. Hsp27) proteins	Prevent the aggregation of other proteins
Hsp60 family	Involved in protein folding and re-folding
Hsp70 family	Involved in protein folding and re-folding
Hsp90 family	Involved in stabilizing substrate proteins interaction; maintain their active or inactive state; prevent the aggregation of other proteins
Hsp100 family	Involved in desegregation of proteins
Protein disulfide isomerases	Promoted correct disulfide-bridge formation and involved with reorganization of disulfide bridges
Peptidyl prolyl cis/trans isomerases	Catalysis of the cis/trans isomerization of peptide-bonds

18.1.2 Synthesis of Heat Shock Proteins

Synthesis of heat shock protein is dependent on the interaction between heat shock factor transcription factor (HSF1) and heat shock elements (HSEs), present on the heat shock protein gene promoter regions (Voellmy 1994; Morimoto et al. 1994). During unstressed condition, HSF1 is present in latent monomeric molecule which doesn't have capability to bind with DNA. While, during stressed conditions, HSF1 is hyperphosphorylated in a ras-dependent manner via mitogen-activated protein kinase (MAPK) subfamilies (e.g. ERK1, JNK/SAPK, p38 protein kinase) (Knauf et al. 1996; Kim et al. 1997). After phosphorylation, it converts into phosphorylated trimers, translocates to nucleus from cytoplasm and binds with promoter region as trimer phosphorylated HSF1 has capability to bind with DNA. Consequently, transcription of heat shock proteins starts (Shi et al. 1998). The other mechanism that also regulates the heat shock protein synthesis, is the binding of heat shock protein binding factor 1 (HSBP1) and active trimeric form of HSF1 and Hsp70. Once this binding occurs, HSF1 lost the capacity to bind with DNA (14) HSBP1 is mainly present in nucleus and unaffected by heat shock (Satyal et al. 1998) (Fig. 18.2). According to the occurrence of the Hsp60, it is divided into three types:

1. Intracellular Hsp60
2. Surface Hsp60
3. Extracellular Hsp60

18.1.2.1 Intracellular Hsp60

Hsp60, a molecular chaperone is known for its protein folding function in prokaryotics and eukaryotics cell organelles. In eukaryotes, it is mainly present in mitochondria (chaperonin 60 or Cpn60) and chloroplasts (in plants) and interacts with Hsp10 (or Cpn10) when involved with the function of chaperoning of nascent polypeptide to functional conformation or native protein (Ranford and Henderson 2002;

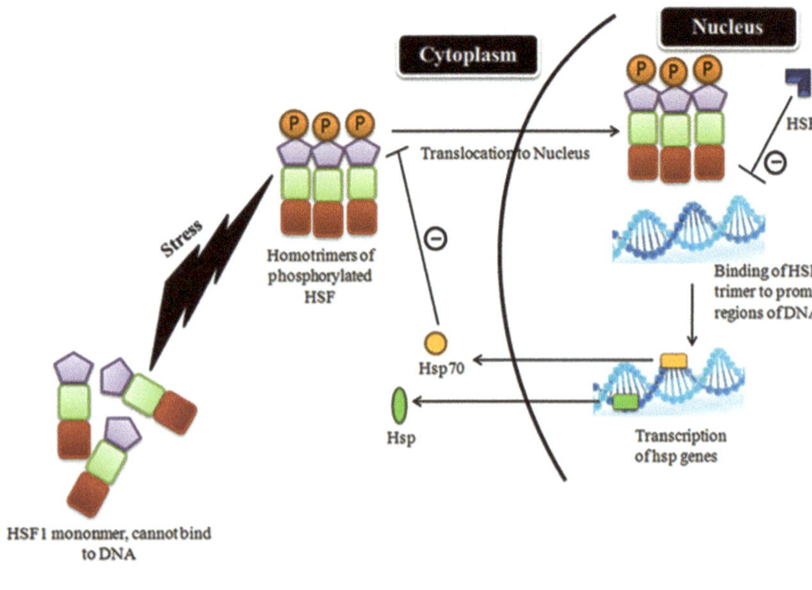

Fig. 18.2 Regulation of synthesis of heat shock proteins

Naylor and Hartl 2001). Along with its recognised role in protein folding, Hsp60 has also important role in intracellular protein trafficking (Deocaris et al. 2006; Czarnecka et al. 2006) and in peptide-hormone signalling (Sigal et al. 2001). Previously, Hsp60 was known as intracellular chaperone but few years back evidences suggested that Hsp60 is found in cytosol too and both, mitochondrial and cytosolic forms of Hsp60 may function in pro-survival or pro-apoptotic pathways as per the cellular situation (Chandra et al. 2007).

18.1.2.2 Surface Hsp60

Recent studies suggested that Hsp60 is also present on the surface of normal (Soltys and Gupta 1997) and tumor cells (Piselli et al. 2000; Feng et al. 2002; Shin et al. 2003) but its presence at cell surface was considered as danger signal which lead to activation of immune system, consisting activation and maturation of dendritic cells and generation of antitumor T-cell response (Osterloh et al. 2004).

18.1.2.3 Extracellular Hsp60

Hsp60 is also found in extracellular space and in circulation along with its intracellular and pericellular locations. However, the mechanism comprises its secretion and exosomal functions are still unanswered (Chen et al. 2006; Gupta and Knowlton 2007). Reports depicted, extracellular Hsp60 act as pro-inflammatory agent as it interacts with various cell-surface receptors like CD14, CD40 and toll like receptors (TLRs) and generate pro- and anti-inflammatory effects (Pockley et al. 2007). Some other studies reported its role in release of TNF-α, production of nitric oxide (NO) and induction of Th1- promoting cytokines IL-12 and IL-15 in macrophages (Chen et al. 1999). Whilst, as anti-inflammatory agent, it is involved with antitumoral immunosuppression activity (Atre et al. 2006).

Extracellular Hsp60 is also found in plasma and blood stream in the range of 1–1000 ng/ml (Shamaei Tousi et al. 2007). Figure 18.3 depicts the working process of molecular chaperone, Hsp60. First, Hsp60 binds its target protein to internal cavity of oligomeric protein. Hsp10-heptamer acts as a cap which closes the Hsp60 protein and this protected environment of Hsp60 cavity restricts the aggregation of protein completely (Hartl 1996). Then, binding and hydrolysis of ATP leads to conformational changes in Hsp60 and further help to loosen the hydrophobic core of the target protein to partial unfolding stage. Due to this multidirectional pulling of target protein, water can enter the hydrophobic sites and help to reorganizing the target protein (Csermely 1999). Ergo, hydrolysis of ATP and binding of new target protein to the other side of Hsp60 resultant into release of Hsp10 cap and target protein. Single round Hsp60-assisted folding is not enough to correct the folding of target protein while multiple folding and refolding cycles are required to commence the correct folding of target protein.

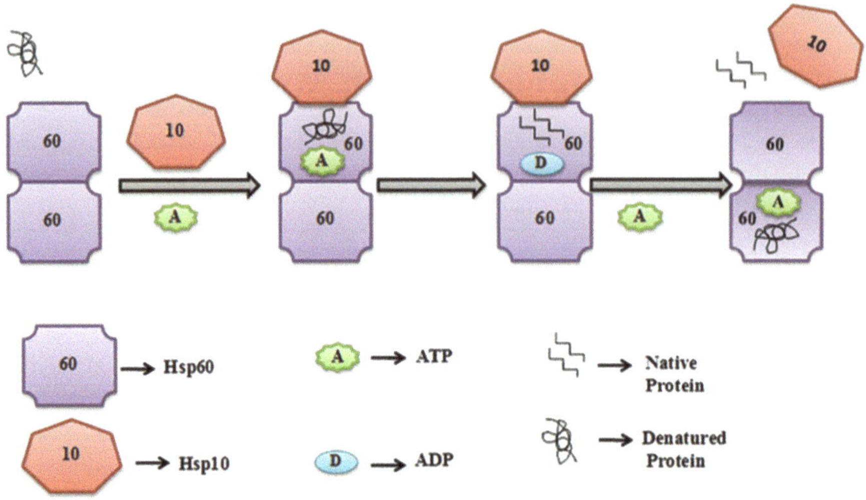

Fig. 18.3 Working Process of molecular chaperone, Hsp60

18.1.3 Quality Control and Heat Shock Proteins

In a normal phenomenon, eucaryotic cells operate the protein quality control (PQC) system for removing cytotoxic agents in time dependent manner as any chemical or physical stress lead to misfolding of proteins. Molecular chaperones are considered as the essential component of PQC as they are involved with refolding of misfolded protein and also inhibit protein aggregation (Kim et al. 2013). Kastle and Grune (2012) reported the molecular chaperones plays an pivotal role in proteostasis under normal and stress conditions and about 10% part of its constituted by proteome. Basically, molecular chaperones performed via three known mode of actions.

1. Some molecular chaperones like Hsp70 hold the unfolded protein until they achieve spontaneous folded proteins (Hartl et al. 2011; Kastle and Grune 2012).
2. Some molecular chaperones such as Hsp60 and Hsp70 assist the refolding of misfolded protein into natively protein species in the presence of ATP (Itoh et al. 2002; Tutar and Tutar 2010).
3. Some chaperones such as Hsp40 and Hsp110 performs as "disaggegases" which forcefully unfold and solubilize protein aggregates into natively refolded proteins (DeSantis et al. 2012).

Recently Wiechmann (2017) currently identified mitochondrial Hsp60 as a direct target of myrtucommulone (MC) and it is ATP-dependent chaperonin which assembles with co-chaperonin Hsp10 and this assembly is responsible for mitochondrial proteostasis by promoting folding of newly imported mitochondrial protein and preventing protein aggregation.

18.1.3.1 Role of Hsp60 in Skeletal Muscle Diseases

Numerous conditions are characterized by muscle wasting, including sarcopenia, cancer cachexia, sepsis, denervation, burns, and chronic obstructive pulmonary disease. Muscle trauma and loss of mass and physical capacity can significantly compromise quality of life. Heat shock proteins play a vital role during these adverse conditions.

18.1.3.2 Sarcopenia and Hsp60

Sarcopenia term refer to age-related loss of muscle mass, resulting into loss of strength (Morley et al. 2001). In simple words, the human body age increment is related to decrease in muscle mass which is associated with gain in fat mass and abdominal circumference. This age related muscle mass loss is termed as sarcopenia. As per definition of sarcopenia (from Greek sarx: flesh, penia: poverty), it is defined as "progressive loss of muscle mass and strength with a risk of adverse outcomes such as disability, poor quality of life and death" (Rosenberg 1997; Santilli et al. 2014). In view of public health, age related sarcopenia is a major clinical problem of older people with numerous adverse outcomes like disability, poor quality of life, falls, hospitalisation, nursing home admission and increased risk of death (Lauretani et al. 2003; Rizzoli et al. 2013; Rantanen 2003). The prevalence of sarcopenia is rising day by day all over the world with increase in world's older population. The meta analysis based results described the overall prevalence of sarcopenia is 10%. The prevalence of sarcopenia is higher in non-Asian countries as compared to Asian countries. Gender based studies provided the highest prevalence of sarcopenia in older men (50%) as compared to older females (43.8%) of the age of 80 years (Iannuzzi-Sucich et al. 2002; Kirchengast and Huber 2009). On average, 5–13% of 60–70 years old and 11–50% of ≥80 years old people suffers from sarcopenia with high prevalence of 68% who are also residing in nursing home (Landi et al. 2012).

Insulin sensitivity is impaired during sarcopenia, which leads to reduction in insulin growth factor-1 (IGF-1) release and decrease expression of Akt and mammalian target of rapamycin (mTOR) that lead to decrease protein synthesis. On the other hand, increment in protein degradation was reported via increase in calpain activity and unfolded protein response (UPS) (Dargelos et al. 2007). Whereas, autophagy declines and apoptosis increase with age (McMulen et al. 2009).

Only few reports detailed the expression of Hsp60 in skeletal muscle during rest and stress conditions and their results are also contradictory. Colotti et al. (2005) detected lower amount of Hsp60 present in mitochondria in aged ventricles and in aged rat hearts. Joseph et al. (2012) also evidenced an age-related decrement in Hsp60 protein levels. On the other hand, few studies reported a significant increase in heat shock proteins such as Hsp27, Hsp60 and inducible Hsp70 in skeletal muscle which associated with enhance apoptosis regulatory proteins like p53, Bcl-2,

Bax, Apaf-1, cleaved caspase-9, pro-caspase-12, pro-caspase-7 that ultimately lead to increase apoptosis process and muscle atrophy (Chung and Ng 2006).

18.1.3.3 Cancer Cachexia and Hsp60

Cancer cachexia is a multiorgan, multifactorial and irreversible wasting syndrome showing symptoms such as marked body weight loss, asthenia, anorexia, anaemia and biochemical alternations (reactive protein, CRP, albumin, haemoglobin (Bosaeus et al. 2001; Argilés et al. 2005; Fearon et al. 2006). It is a serious condition associated with cancer and other serious, chronic illnesses including AIDS, chronic heart failure, chronic kidney disease and chronic obstructive pulmonary disease (Graul et al. 2016). Cachexia results in poor performance status, poor quality of life with high mortality rate in cancer patients. A longitudinal study depicted a 2.5 kg weight change over 6–8 weeks which led to changes in performance status (O'Gorman et al. 1999) and death usually happened in the case of 30% weight loss (Tisdale 2002).

As per the previous definition of cachexia, it is described "a wasting syndrome involving loss of muscle and fat directly caused by tumour factors, or indirectly caused by an aberrant host response to tumour presence" (MacDonald et al. 2003). While recent definition depicts "a complex metabolic syndrome associated with underlying illness and characterised by loss of muscle with or without loss of fat mass" (Evans et al. 2008). In cachexia, increase muscle protein catabolism resilient into muscle mass loss while protein synthesis may be increased or unchanged (McMillan et al. 1994). The pivotal catabolic pathways involved with loss of muscle mass loss are proteolytic pathways such as lysosomal pathway (cathepsins B, H, D, and L) (Bosutti et al. 2002), calpain pathway (Busquets et al. 2000) and ATP ubiquitin-dependent proteolytic pathway (Khal et al. 2005; DeJong et al. 2005). Another important factor of cachexia is loss of adipose tissue mass or lipolysis (Tisdale 2002) which driven by lipid mobilising factor (LMF) and tumor factor, zinc-alpha-2 glycoprotein (Tisdale 2010). Other factor involved with lipolysis is dysregulation of energy metabolism which lead to increase resting energy expenditure (Tisdale 2002).

Heat shock proteins play an important role in cancer cachexia. Zhang et al. (2017) found surprisingly that cancer cachexia related muscle catabolic activity was associated with high level of Hsp70 and Hsp90 and another interesting finding of their research was high levels extracellular vesicles (EVs) which served as a carrier of tumor-released Hsp70 and Hsp90. Conclusively, the research group demonstrated that tumor cell-released EV-associated HSp70 and Hsp90 are required and sufficient to induce cancer cachexia induce muscle wasting. On the other hand, intracellular Hsp70/90 plays cytoprotective role as molecular chaperones against stressful conditions (Wang 2011). Hence, aiming these Hsp could become promising therapeutic targets for defeating cancer cachexia (Zhang et al. 2017). Along with this, role of Hsp60 was also established with exercise and cachexia and the results concluded that Hsp60 over expression could improve muscle performance and reduce cachexia (Barone et al. 2016).

18.1.3.4 Duchenne Muscular Dystrophy (DMD) and Hsp60

Dystrophin is a rod shaped cytoplasmic protein which primarily located in muscles and used for the movement of skeletal and cardiac muscles. DMD is a known human gene, provides instructions for translating dystrophin protein. Mutations in dystrophin gene are the cause of duchenne muscular dystrophy (DMD) that is a lethal X-linked recessive disorder (Guiraud et al. 2015). The disorder is so severe, characterized by progressive muscle wasting which lead to loss of ambulation by 8–12 years of age and affects 1 in 5000 boys and also results in death due to cardiorespiratory failure (Mendell et al. 2012; Emery 1993; Bach et al. 1987). A crucial link occurs between dystrophin associated protein complex (DAPC) present at the sarcolemma and the cytoskeleton. Dystrophin is required for maintaining strength, flexibility and stability in skeletal muscle (Davies and Nowak 2006). Absence of dystrophin results in increase myofibres susceptibility, leading to contraction-induced injury, muscle wasting and premature death (Emery 1990). Morici et al. (2016) reported the epithelial Hsp60 expression was progressively decreased and inversely correlated with epithelial apoptosis ($r = -0.66$, $P = 0.01$) in mdx mice. Further, the same group also observed low expression of Hsp60 in mdx mice which indicates progressive exhaustion of preventing epithelium integrity. Recently, Guiraud et al. (2017) identified Hsp60 as a new potential putative serum dystrophic marker.

18.1.3.5 High Altitude Associated Muscle Atrophy and Hsp60

High altitude is defined as an altitude exceeding 2438 meters or 8000 feet above sea level. Traditionally, according to the height, it is divided into the following categories (Fig. 18.4):

1. High altitude: 2438–3658 m (8000–12,000 ft)
2. Very high altitude: 3658–5487 m (12,000–18,000 ft)
3. Extreme high altitude: 5500+ m (18,000+ feet)

High altitude presents numerous adverse and challenging environmental conditions for human survival and its exposure sometimes unavoidable due to military, pilgrimages, sports and tourism activities. High velocity winds, low humidity, extreme low ambient temperature, high intensity solar radiation, reduced atmospheric pressure, reduce oxygen availability are variety of stresses faced at high altitude and these stress conditions leads to compromise physiological function and physical performance (Verratti et al. 2015). One of the main hassles at high altitude is hypobaric hypoxia which leads to disturb redox homeostasis and modified protein structure in skeletal muscle. Our recent studies established the detail overview of

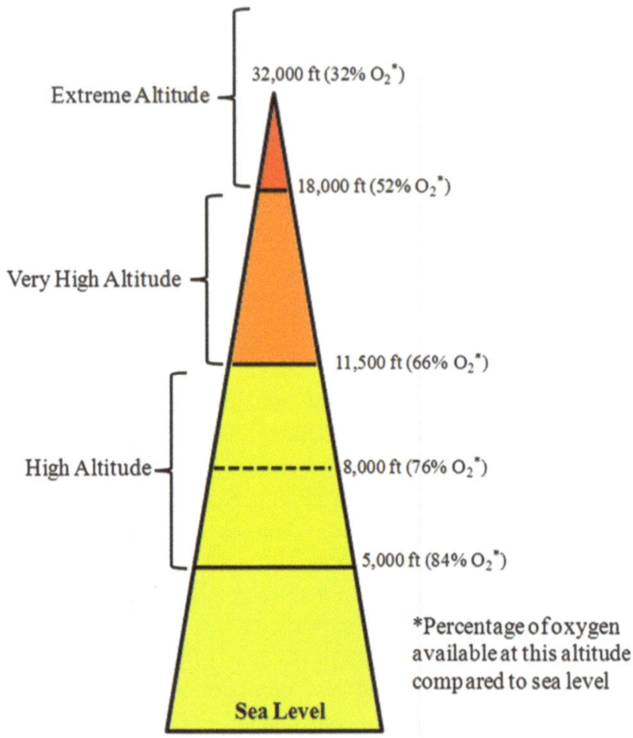

Fig. 18.4 Altitude Categories

oxidative protein modification in response to hypobaric hypoxia that altered protein homeostasis and skeletal muscle protein loss (Agrawal et al. 2017, 2018) that ultimately lead to compromised physical performance.

Ironically, the proposal has been made that hypobaric hypoxia provokes alternations in mitochondrial activity which lead to accumulation of reactive oxygen species (ROS) and oxidative stress (Barbieri and Sestili 2012; Barbieri et al. 2014). It has been further demonstrated the role of oxidative stress in the damage of lipids, proteins and nucleic acids (Beccafico et al. 2007; Pietrangelo et al. 2009). Paradoxically, accumulation of ROS leads to oxidation of proteins and oxidative protein modification which altered proteostasis (Chaudhury et al. 2012; Agrawal et al. 2017). The oxidative protein modification leads to the activation of ER stress in which GRP78, GRP94, Hsp70, Hsp60, Hsp90 gets activated to rectify the misfolded proteins (Jain et al. 2014; Rathor et al. 2015). When the cells protein folding machinery was not able to cope up with the stress situation; there were a marked upregulation of transcription factor C/EBP homologous protein (CHOP), associated with ER stress initiated that lead to apoptosis and cell death.

The well known fact regarding heat shock proteins is its crucial cytoprotective role in relation to environmental and metabolic stress. Jain et al. (2013) observed an enhancement of heat shock proteins during hypobaric hypoxia exposure which suggested its better chaperone activity. This chaperone activity prevents misfolding of protein structure and impaired protein function. Mestril and Dillmann (1995) also proposed a positive correlation between production of heat shock proteins and protection against myocardial, hepatic and muscle damage (Jain et al. 2014; Rathor et al. 2015).

18.1.4 Protective or Therapeutic Efficacy of Hsp60

Hsp60 depicts vital activities in angiogenesis, transformation and metastasis due to its unique anti-apoptotic property (Wu et al. 2017). The cell survival activity prompted via the pro-proliferative and pro-survival activity of HSP60. Recently, Hsp60 also emerging as a novel regulator of cytoprotective chaperone network via its ability to inhibit CypD-dependent tumor cell death (Haslbeck and Vierling 2015). Two pathways also proposed via Hsp60 shows its anti-apoptotic activity. First, via modulating mitochondrial survivin (SVV) stability and second, via control of p53 expression. Survivin (SVV) and p53 is well known for the activation of apoptosis pathway. During normal conditions, Hsp60 binds with SVV and p53 proteins and make complex with these proteins. But during stress conditions, Hsp60 silence from these proteins that result in loss of chaperonage function. Once SVV release from Hsp60, lead to mitochondrial apoptotic pathway. On the other hand, silencing of Hsp60 from p53 leads to p53-dependent transcription of apoptotic factors like BAX that ultimately promote cell death (Ghosh et al. 2008) (Fig. 18.5).

Hsp also has cellular degeneration activity, along with its major role in protein folding, refolding, transportation and translocation. It also comprises molecular chaperones activity of thermotolerance against stresses such as reactive oxygen species (Auluck et al. 2002). Andersen 2004 also reported its free radical scavenging activity and it also able increase catalase and superoxide dismutase activity during heat shock stress. Recently, Veereshwarayya et al. (2016) also depicted that overexpression of Hsp60 alone and in combination with Hsp70 and Hsp90 protect brain cell cultures via reducing cytotoxicity. Presently, a number of available information indicates that Hsp60 has key role in maintaining protein homeostasis and cell survival. It also consist an important role in progression of aging-related diseases. Aging is the condition in which chaperone levels and its functionality generally decreases. Barone et al. (2016) depicted that endurance training leads to increase Hsp60 level which further increase the levels of mitochondrial DNA compared to nuclear DNA. This process indicates mitochondrial biogenesis and release of Hsp60 protein in blood.

Cellular heat shock protein increment also showed cytoprotection not only against ischaemia and reperfusion but also against calcium paradox or the oxygen paradox (Marber 1994; Plumier and Currie 1996). Additionally, hyperthermic condition and excessive exercise result into increment in Hsp content in skeletal muscle which also related to cytoprotection of skeletal muscle damage (Febbraio and

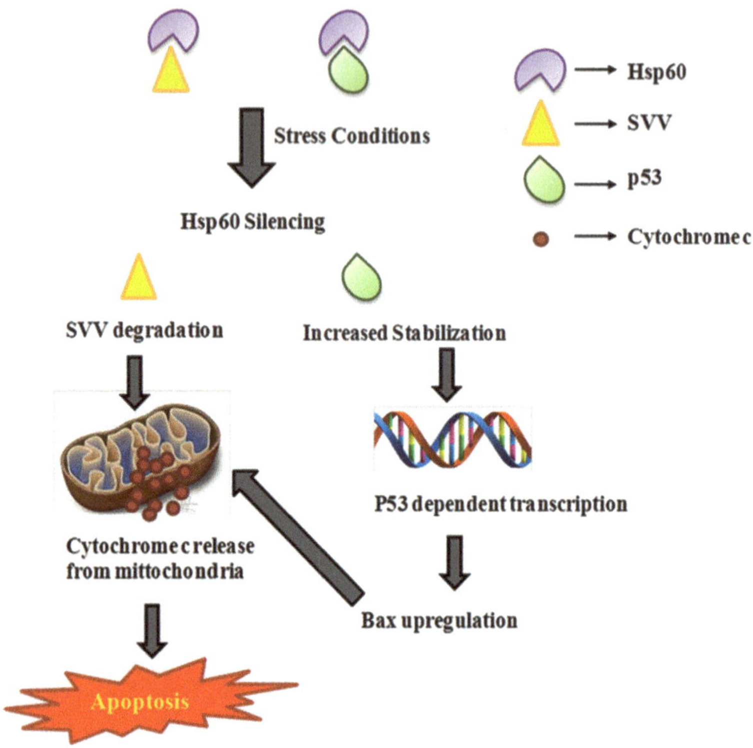

Fig. 18.5 Mechanisms of HSP60 for cytoprotection

Koukoulas 2000; Khassaf et al. 2001; McArdle et al. 2001; Maglara et al. 2003). Hsp60 also engaged as a key protein in a crosstalk between metabolic stress and inflammation. Interestingly during this crosstalk, it participates in both the ways like pro-inflammatory and anti-inflammatory way which could further induce both types of cytokines (Quintana and Cohen 2011). Moreover, Hsp60 also showed cytoprotective efficacy against tolerance to high altitude induced hypobaric stress (Jain et al. 2013). Increase Hsp60 level was also observed in skeletal muscle, liver and heart in multiple stress condition, CHR (cold-hypoxia-restraint) which predict its adaptogenic activity via restoring protein homeostasis (Rathor et al. 2015).

18.2 Conclusions

A number of studies demonstrated the inimitable anti-apoptotic function of Hsp60 by which it explain its crucial role in angiogenesis, transformation, metastasis and cell survival activity. The present chapter established an insight into different pathophysiological conditions of muscular atrophy and the probable role of Hsp60 into it. Understanding the role of Hsp60 in these diseases, might further offer its pharmacological interventions function so that various muscular stress conditions like sarcopenia, muscular fatigue and muscular dystrophy could be benefited.

Acknowledgements The authors are thankful to Dr. Bhuvnesh Kumar, Director, DIPAS, for his constant support and encouragement. The study was supported by the Defence Research and Development Organisation, Ministry of Defence, Government of India.

References

Agrawal A, Rathor R, Suryakumar G (2017) Oxidative protein modification alters proteostasis under acute hypobaric hypoxia in skeletal muscles: a comprehensive in vivo study. Cell Stress Chaperones 22(3):429–443

Agrawal A, Rathor R, Kumar R, Suryakumar G, Ganju L (2018) Role of altered proteostasis network in chronic hypobaric hypoxia induced skeletal muscle atrophy. PLoS One 13(9):e0204283

Andersen KJ (2004) Oxidative stress in neurodegeneration: cause or consequence? Nat Rev Neurosci 5:S18–S25

Argilés JM, Busquets S, García-Martínez C, López-Soriano FJ (2005) Mediators involved in the cancer anorexia-cachexia syndrome: past, present, and future. Nutrition 21:977–985

Atre N, Thomas L, Mistry R, Pathak K, Chiplunkar S (2006) Role of nitric oxide in heat shock protein induced apoptosis of gamma delta T cells. Int J Cancer 119:1368–1376

Auluck PK, Chan HY, Trojanowski JQ, Lee VM, Bonini NM (2002) Chaperone suppression of alpha-synuclein toxicity in a Drosophila model for Parkinson's disease. Science 295:865–868

Bach JR, O'Brien J, Krotenberg R, Alba AS (1987) Management of end stage respiratory failure in Duchenne muscular dystrophy. Muscle Nerve 10:177–182

Barbieri E, Sestili P (2012) Reactive oxygen species in skeletal muscle signaling. J Sig Transduct 2012:982794

Barbieri E, Sestili P, Vallorani L, Guescini M, Calcabrini C, Gioacchini AM, Annibalini G, Lucertini F, Piccoli G, Stocchi V (2014) Mitohormesis in muscle cells: a morphological, molecular, and proteomic approach. Muscle Ligaments Tendons J 3(4):254–266

Barone R, Macalusa F, Sangiogi C, D'Amico D, Gammazza AM, Campanella C, Cappello F, Zummo G, Farina F (2016) Hsp60 and interleukins expression in the skeletal muscle and its implications in exercise and cachexia. Ital J Anat Embryol 121(1):26

Beccafico S, Puglielli C, Pietrangelo T, Bellomo R, Fanò G, Fulle S (2007) Age-dependent effects on functional aspects in human satellite cells. Ann N Y Acad Sci 1100:345–352

Bosaeus I, Daneryd P, Svanberg E, Lundholm K (2001) Dietary intake and resting energy expenditure in relation to weight loss in unselected cancer patients. Int J Cancer 93:380–383

Bosutti A, Toigo G, Ciocchi B, Situlin R, Guarnieri G, Biolo G (2002) Regulation of muscle cathepsin B proteolytic activity in protein-depleted patients with chronic diseases. Clin Nutr 21(5):373–378

Brodsky JL, Chiosis G (2006) Hsp70 molecular chaperones: emerging roles in human disease and identification of small molecule modulators. Curr Top Med Chem 6:1215–1225

Busquets S, García-Martínez C, Alvarez B, Carbó N, López-Soriano FJ, Argilés JM (2000) Calpain-3 gene expression is decreased during experimental cancer cachexia. Biochim Biophys Acta 1475(1):5–9

Chandra D, Choy G, Tang DG (2007) Cytosolic accumulation of HSP60 during apoptosis with or without apparent mitochondrial release: evidence that its pro-apoptotic or pro-survival functions involve differential interactions with caspase-3. J Biol Chem 282:31289–31301

Chaudhury P, Suryakumar G, Prasad R, Singh SM, Ali S, Ilavazhagan G (2012) Chronic hypobaric hypoxia mediated skeletal muscle atrophy: role of ubiquitin–proteasome pathway and calpains. Mol Cell Biochem 364:101–113

Chen W, Syldath U, Bellmann K, Burkart V, Kolb H (1999) Human 60-kDa heat-shock protein: a danger signal to the innate immune system. J Immunol 162:3212–3219
Chen W, Wang J, Shao C, Liu S, Yu Y, Wang Q, Cao X (2006) Efficient induction of antitumor T cell immunity by exosomes derived from heat-shocked lymphoma cells. Eur J Immunol 36:1598–1607
Chung L, Ng YC (2006) Age-related alternations in expression of apoptosis regulatory proteins and heat shock proteins in rat skeletal muscle. BBA-Mol Basis Dis 1762(1):103–109
Colotti C, Cavallini G, Vitale RL, Donati A, Maltinti M, Del Ry S, Bergamini E, Giannessi D (2005) Effects of aging and anti-aging caloric restrictions on carbonyl and heat shock protein levels and expression. Biogerontology 6:397–406
Csermely P (1999) The "chaperone-percolator" model: a possible molecular mechanism of Anfinsen-cage type chaperone action. BioEssays 21:959–965
Czarnecka AM, Campanella C, Zummo G, Cappello F (2006) Heat shock protein 10 and signal transduction: a "capsula eburnea" of carcinogenesis? Cell Stress Chaperones 11:287–294
Dargelos E, Brulé C, Combaret L, Hadj-Sassi A, Dulong S, Poussard S, Cottin P (2007) Involvement of the calcium-dependent proteolytic system in skeletal muscle aging. Exp Gerontol 42(11):1088–1098
Davies KE, Nowak KJ (2006) Molecular mechanisms of muscular dystrophies: old and new players. Nat Rev Mol Cell Biol 7:762–773
DeJong CH, Busquets S, Moses AG, Schrauwen P, Ross JA, Argiles JM, Fearon KC (2005) Systemic inflammation correlates with increased expression of skeletal muscle ubiquitin but not uncoupling proteins in cancer cachexia. Oncol Rep 14(1):257–263
Deocaris CC, Kaul SC, Wadhwa R (2006) On the brotherhood of the mitochondrial chaperones mortalin and heat shock protein 60. Cell Stress Chaperones 11:116–128
DeSantis ME, Leung EH, Sweeny EA, Jackrel ME, Cushman-Nick M, Neuhaus-Follini A, Vashist S, Sochor MA, Knight MN, Shorter J (2012) Operational plasticity enables hsp104 to disaggregate diverse amyloid and nonamyloid clients. Cell 151:778–793
Emery AE (1990) Dystrophin function. Lancet 335:1289
Emery AE (1993) Duchenne muscular dystrophy–Meryon's disease. Neuromuscul Disord 3:263–266
Evans WJ, Morley JE, Argilés J, Bales C, Baracos V, Guttridge D, Jatoi A, Kalantar-Zadeh K, Lochs H, Mantovani G, Marks D, Mitch WE, Muscaritoli M, Najand A, Ponikowski P, Rossi Fanelli F, Schambelan M, Schols A, Schuster M, Thomas D, Wolfe R, Anker SD (2008) Cachexia: a new definition. Clin Nutr 27(6):793–799
Fearon KC, Voss AC, Hustead DS, Cancer Cachexia Study Group (2006) Definition of cancer cachexia: effect of weight loss, reduced food intake, and systemic inflammation on functional status and prognosis. Am J Clin Nutr 83:1345–1350
Febbraio MA, Koukoulas I (2000) HSP72 gene expression progressively increases in human skeletal muscle during prolonged, exhaustive exercise. J Appl Physiol 89:1055–1060
Feng H, Zeng Y, Graner MW, Katsanis E (2002) Stressed apoptotic tumor cells stimulate dendritic cells and induce specific cytotoxic T cells. Blood 100:4108–4115
Ghosh JC, Dohi T, Kang BH, Altieri DC (2008) Hsp60 regulation of tumor cell apoptosis. J Biol Chem 283:5188–5194. https://doi.org/10.1074/jbc.M705904200
Graul AI, Stringer M, Sorbera LC (2016) Drugs Today 52(9):519
Guiraud S, Aartsma-Rus A, Vieira NM, Davies KE, van Ommen GJ, Kunkel LM (2015) The pathogenesis and therapy of muscular dystrophies. Ann Rev Genom Hum Genet 16:281–308
Guiraud S, Edwards B, Squire SE, Babbs A, Shah N, Berg A, Chen H, Davies KE (2017) Identification of serum protein biomarkers for utrophin based DMD therapy. Sci Rep 7:43697
Gupta S, Knowlton AA (2007) Hsp60 trafficking in adult cardiac myocytes: role of exosomal pathway. Am J Physiol Heart Circ Physiol 292:3052–3056
Hartl FU (1996) Molecular chaperones in cellular protein folding. Nature 381:571–580
Hartl FU, Bracher A, Hayer-Hartl M (2011) Molecular chaperones in protein folding and proteostasis. Nature 475:324–332

Haslbeck M, Vierling E (2015) A first line of stress defense: small heat shock proteins and their function in protein homeostasis. J Mol Biol 427:1537–1548
Iannuzzi-Sucich M, Prestwood KM, Kenny AM (2002) Prevalence of sarcopenia and predictors of skeletal muscle mass in healthy, older men and women. J Gerontol A Biol Sci Med Sci 57(12):M772–M777
Itoh H, Komatsuda A, Ohtani H, Wakui H, Imai H, Sawada K, Otaka M, Ogura M, Suzuki A, Hamada F (2002) Mammalian HSP60 is quickly sorted into the mitochondria under conditions of dehydration. Eur J Biochem 269:5931–5938
Jain K, Suryakumar G, Prasad R, Ganju L (2013) Differential activation of myocardial ER stress response: a possible role in hypoxic tolerance. Int J Cardiol 168(5):667–677
Jain K, Suryakumar G, Ganju L, Singh SB (2014) Differential hypoxic tolerance is mediated by activation of heat shock response and nitric oxide pathway. Cell Stress Chaperones 19(6):801–812
Joseph AM, Adhihwtty PJ, Buford TW, Wohlgemuth SE, Lees HA, Nguyen LMD, Aranda JM, Sandesara BD, Pahor M, Manini TM, Marzetti E, Leeuwenburgh C (2012) The impact of aging on mitochondrial function and biogenesis pathways in skeletal muscle of sedentary high- and low-functioning elderly individuals. Aging Cell 11(5):801–809
Kastle M, Grune T (2012) Interactions of the proteasomal system with chaperones: protein triage and protein quality control. Prog Mol Biol Transl Sci 109:113–160
Khal J, Hine AV, Fearon KCH, Dejong CHC, Tisdale MJ (2005) Increased expression of proteasome subunits in skeletal muscle of cancer patients with weight loss. Int J Biochem Cell Biol 37(10):2196–2206
Khassaf M, Child RB, McArdle A, Brodie DA, Esanu C, Griffiths RD, Jackson MJ (2001) Time course of responses of human skeletal muscle to oxidative stress induced by nondamaging exercise. J Appl Physiol 90:1031–1035
Kim J, Nueda A, Meng YH, Dynan WS, Mivechi NF (1997) Analysis of the phosphorylation of human heat shock transcription factor-1 by MAP kinase family members. J Cell Biochem 67:43–54
Kim YE, Hipp MS, Bracher A, Hayer-Hartl M, Hartl FU (2013) Molecular chaperone functions in protein folding and proteostasis. Annu Rev Biochem 82:323–355
Kirchengast S, Huber J (2009) Gender and age differences in lean soft tissue mass and sarcopenia among healthy elderly. Anthropol Anz 67(2):139–151
Knauf U, Newton EM, Kyriakis J, Kingston RE (1996) Repression of human heat shock factor 1 activity at control temperature by phosphorylation. Genes Dev 10:2782–2793
Kregel KC (2002) Heat shock proteins: modifying factors in physiological stress responses and acquired thermotolerance. J Appl Physiol 92:2177–2186
Landi F, Liperoti R, Fusco D, Mastropaolo S, Quattrociocchi D, Proia A, Tosato M, Bernabei R, Onder G (2012) Sarcopenia and mortality among older nursing home residents. J Am Med Dir Assoc 13:121–126
Lauretani F, Russo CR, Bandinelli S, Bartali B, Cavazzini C, Di Iorio A, Corsi AM, Rantanen T, Guralnik JM, Ferrucci L (2003) Age-associated changes in skeletal muscles and their effect on mobility: an operational diagnosis of sarcopenia. J Appl Physiol 95(5):1851–1860
Macario AJL, de Macario CE (2007) Molecular chaperones: multiple functions, pathologies and potential applications. Front Biosci 12:2588–2600
MacDonald N, Easson AM, Mazurak VC, Dunn GP, Baracos VE (2003) Understanding and managing cancer cachexia. J Am Coll Surg 197(1):143–161
Maglara AA, Vasilaki A, Jackson MJ, McArdle A (2003) Damage to developing mouse skeletal muscle myotubes in culture: protective effect of heat shock proteins. J Physiol 548:837–846
Marber MS (1994) Stress proteins and myocardial protection. Clin Sci (Lond) 86:375–381
McArdle A, Pattwell D, Vasilaki A, Griffiths RD, Jackson MJ (2001) Contractile activity-induced oxidative stress: cellular origin and adaptive responses. Am J Physiol Cell Physiol 280:C621–C627

McMillan DC, Preston T, Fearon KCH, Burns HJG, Slater C, Shenkin A (1994) Protein synthesis in cancer patients with inflammatory response: investigations with [N]glycine. Nutrition 10(3):232–240

McMulen CA, Ferry AL, Gamboa JL, Andrade FH, Dupont-Versteegden EE (2009) Age-related changes of cell death pathways in rat extraocular muscle. Exp Gerontol 44:20–425

Mendell JR, Shilling C, Leslie ND, Flanigan KM, Al-Dahhak R, Gastier-Foster J, Kneile K, Dunn DM, Duval B, Aoyagi A, Hamil C, Mahmoud M, Roush K, Bird L, Rankin C, Lilly H, Street N, Chandrasekar R, Weiss RB (2012) Evidence-based path to newborn screening for Duchenne muscular dystrophy. Ann Neurol 71:304–313. 2012

Mestril R, Dillmann WH (1995) Heat shock proteins and protection against myocardial ischemia. J Mol Cell Cardiol 27(1):45–52

Morici G, Rappa F, Cappello F, Pace E, Pace A, Mudò G, Crescimanno G, Belluardo N, Bonsignore MR (2016) Lack of dystrophin affects bronchial epithelium in mdx mice. J Cell Physiol 231(10):2218–2223

Morimoto RI, Jurivich DA, Kroger PE, Mathur SK, Murphy SP, Nakai A, Sarge AK, Abravaya K, Sistonen LT (1994) Regulation of heat shock gene transcription by a family of heat shock factors. In: Morimoto R, Tissières A, Georgopoulos C (eds) The biology of heat shock proteins and molecular chaperones. Cold Spring Harbor Laboratory Press, Cold Spring Harbor, pp 417–455

Morley JE, Baumgartner RN, Roubenoff R, Mayer J, Nair KS (2001) Sarcopenia. J Lab Clin Med 137:231–24310

Naylor DJ, Hartl FU (2001) Contribution of molecular chaperones to protein folding in the cytoplasm of prokaryotic and eukaryotic cells. Biochem Soc Symp 68:45–68

O'Gorman P, McMillan DC, McArdle CS (1999) Longitudinal study of weight, appetite, performance status, and inflammation in advanced gastrointestinal cancer. Nutr Can 35(2):127–129

Osterloh A, Meier Stiegen F, Veit A, Fleischer B, von Bonin A, Breloer M (2004) Lipopolysaccharide- free heat shock protein 60 activates T cells. J Biol Chem 279:47906–47911

Pietrangelo T, Puglielli C, Mancinelli R, Beccafico S, Fanò G, Fulle S (2009) Molecular basis of the myogenic profile of aged human skeletal muscle satellite cells during differentiation. Exp Gerontol 44:523–531

Piselli P, Vendetti S, Vismara D, Cicconi R, Poccia F, Colizzi V, Delpino A (2000) Different expression of CD44, ICAM-1 and HSP60 on primary tumor and metastases of a human pancreatic carcinoma growing in scid mice. Anticancer Res 20:825–831

Plumier JC, Currie RW (1996) Heat shock-induced myocardial protection against ischemic injury: a role for HSP70? Cell Stress Chaperones 1:13–17

Pockley AG, Fairburn B, Mirza S, Slack LK, Hopkinson K, Muthana M (2007) A non-receptor-mediated mechanism for internalization of molecular chaperones. Methods 43:238–244

Quintana FJ, Cohen IR (2011) The HSP60 immune system network. Trends Immunol 32:89–95

Ranford JC, Henderson B (2002) Chaperonins in disease: mechanisms, models and treatments. Mol Pathol 55:209–213

Rantanen T (2003) Muscle strength, disability and mortality. (2003). Scand J Med Sci Sports 13(1):3–8

Rathor R, Sharma P, Suryakumar G, Ganju L (2015) A pharmacological investigation of *Hippophae salicifolia* (HS) and *Hippophae rhamnoides turkestanica* (HRT) against multiple stress (C-H-R): an experimental study using rat model. Cell Stress Chaperones 20(5):821–831

Ritossa F (1962) A new puffing pattern induced by temperature shock and DNP in Drosophila. Experientia 18:571–573

Ritossa F (1996) Discovery of the heat shock response. Cell Stress Chaperones 1(2):97–98

Rizzoli R, Reginster JY, Arnal JF, Bautmans I, Beaudart C, Bischoff-Ferrari H, Biver E, Boonen S, Brandi ML, Chines A, Cooper C, Epstein S, Fielding RA, Goodpaster B, Kanis JA, Kaufman JM, Laslop A, Malafarina V, Mañas LR, Mitlak BH, Oreffo RO, Petermans J, Reid K, Rolland

Y, Sayer AA, Tsouderos Y, Visser M, Bruyère O (2013) Quality of life in sarcopenia and frailty. Calcif Tissue Int 93(2):101–120
Rosenberg IH (1997) Sarcopenia: origins and clinical relevance. J Nutr 127:990S–991S
Santilli V, Bernetti A, Mangone M, Paoloni M (2014) Clinical definition of sarcopenia. Clin Cases Miner Bone Metab 11(3):177–180
Satyal SH, Chen D, Fox SG, Kramer JM, Morimoto RI (1998) Negative regulation of the heat shock transcriptional response by HSBP1. Genes Dev 12:1962–1974
Shamaei Tousi A, Steptoe A, O'Donnell K, Palmen J, Stephens JW, Hurel SJ, Marmot M, Homer K, D'Aiuto F, Coates AR, Humphries SE, Henderson B (2007) Plasma heat shock protein 60 and cardiovascular disease risk: the role of psychosocial, genetic and biological factors. Cell Stress Chaperones 12:384–392
Shi Y, Mosser DD, Morimoto RI (1998) Molecular chaperones as HSF1-specific transcriptional repressors. Genes Dev 12:654–666
Shin BK, Wang H, Yim AM, Le Naour F, Brichory F, Jang JH, Zhao R, Puravs E, Tra J, Michael CW, Misek DE, Hanash SM (2003) Global profiling of the cell surface proteome of cancer cells uncovers an abundance of proteins with chaperone function. J Biol Chem 278:7607–7616
Sigal LH, Williams S, Soltys B, Gupta R (2001) H9724, a monoclonal antibody to Borreliaburgdorferi's flagellin, binds to heat shock protein 60 (HSP60) within live neuroblastoma cells: a potential role for HSP60 in peptide hormone signaling and in an autoimmune pathogenesis of the neuropathy of Lyme disease. Cell Mol Neurobiol 21:477–495
Soltys BJ, Gupta RS (1997) Cell surface localization of the 60 kDa heat shock chaperonin protein (hsp60) in mammalian cells. Cell Biol Int 21:315–320
Tisdale MJ (2002) Cachexia in cancer patients. Nat Rev Cancer 2(11):862–871
Tisdale MJ (2010) Cancer cachexia. Curr Opin Gastroenterol 26(2):146–151
Tissieres A, Mitchell HK, Tracy UM (1974) Protein synthesis in salivary glands of Drosophila melanogaster: relation to chromosome puffs. J Mol Biol 84(3):389–398
Tutar L, Tutar Y (2010) Heat shock proteins; an overview. (2010). Curr Pharm Biotechnol 11(2):216–222
Veereshwarayya V, Kumar P, Rosen KM, Mestril R, Querfurt HW (2016) Differential effects of mitochondrial heat shock protein 60 and related molecular chaperones to prevent intracellular β-amyloid-induced inhibition of complex IV and limit apoptosis. J Biol Chem 281:29468–29478
Verratti V, Falone S, Doria C, Pietrangelo T, Di Giulio C (2015) Kilimanjaro Abruzzo expedition: effects of high-altitude trekking on anthropometric, cardiovascular and blood biochemical parameters. Sport Sci Health 11(3):271–278
Voellmy R (1994) Transduction of the stress signal and mechanisms of transcriptional regulation of heat shock/stress protein gene expression in higher eukaryotes. Crit Rev Eukaryot Gene Expr 4:357–401
Wang RE (2011) Targeting heat shock proteins 70/90 and proteasome for cancer therapy. Curr Med Chem 18:4250–4264
Wiechmann K, Müller H, Konig S, Wielsch N, Svatos A, Jauch J, Werz O (2017) Cell Chem Biol 24(5):614–623
Wu J, Liu T, Rios Z, Mei Q, Lin X, Cao S (2017) Heat shock proteins and cancer. Trends Pharmacol Sci 38:226–256
Zhang G, Liu Z, Ding H, Zhou Y, Doan HA, Sin KWT, Zhu ZJ, Flores R, Wen Y, Gong X, Liu O, Li YP (2017) Tumor induces muscle wasting in mice through releasing extracellular Hsp70 and Hsp90. Nat Commun 8:589

Chapter 19
Heat Shock Protein 60 Regulation of Skeletal Tissue Integrity

Feng-Sheng Wang, Wei-Shiung Lian, Chung-Wen Kuo, Yu-Shan Chen, and Pei-Rong Chang

Abstract Osteoporosis and osteoarthritis are the most prevalent degenerative skeletal diseases in the elderly. Deregulated osteoblast and chondrocyte behavior are prominent cellular features of these disorders. Organelle dysfunction disturbs cell survival and differentiation capacity, accelerating bone mass and articular cartilage loss. Heat shock protein 60 (HSP60) is a mitochondrial chaperonin essential to mitochondrial integrity and proteostasis. Its function to skeletal tissue homeostasis and degeneration warrants systemic characterization. Here, we highlight the merging evidence in regard to the involvement of this chaperonin in mitochondrial biogenesis, autophagy, and post-translational modification of bioactive proteins, contributing to tissue homeostasis, deterioration, and tumorigenesis in various physiological and pathological contexts. This article sheds a new light on the beneficial actions of HSP60 to osteoblast autophagy that protects bone tissue against osteoporosis development. We also offer a productive insight into how this chaperone protein sustains chondrocyte function to facilitate cartilage development and slow down osteoarthritis progression.

Keywords Autophagy · Chondrocytes · Heat shock protein 60 · Osteoarthritis · Osteoblasts · Osteoporosis

---

Authors Feng-Sheng Wang and Wei-Shiung Lian are equally contributed to this chapter.

F.-S. Wang (✉)
Core Laboratory for Phenomics & Diagnostics, Department of Medical Research, Kaohsiung Chang Gung Memorial Hospital, Kaohsiung, Taiwan

Graduate Institute of Clinical Medical Sciences, Chang Gung University College of Medicine, Kaohsiung Chang Gung Memorial Hospital, Kaohsiung, Taiwan
e-mail: wangfs@ms33.hinet.net

W.-S. Lian · C.-W. Kuo · Y.-S. Chen
Core Laboratory for Phenomics & Diagnostics, Department of Medical Research, Kaohsiung Chang Gung Memorial Hospital, Kaohsiung, Taiwan

P.-R. Chang
Department of Nursing, Fooyin University, Kaohsiung, Taiwan

A. A. A. Asea, P. Kaur (eds.), *Heat Shock Protein 60 in Human Diseases and Disorders*, Heat Shock Proteins 18,
https://doi.org/10.1007/978-3-030-23154-5_19

Abbreviations

μCT	microcomputed tomography
Acetyl CoA	acetyl coenzyme A
ADP	adenosine diphosphate
AMPK	AMP-activated protein kinase
Atg	autophagy related proteins
ATP	adenosine triphosphate
Bax	BCL2 associated X apoptosis regulator
BNIP3	BCL2 interacting protein 3
DKK-1	dickkopf-1
ERK	extracellular regulated MAP kinase
FOXO	forkhead box subgroup O
HDAC1	histone deacetylase 1
HSP60	heat shock protein 60
HSPd1	heat shock protein family D member 1
IGF-1	insulin-like growth factor-1
IL-1β	interleukin-1β
LC3	microtubule associates protein 1 light chain 3 alpha
MMP	matrix metalloproteinases
MSC	mesenchymal stem cells
mtDNA	mitochondrial DNA
mTOR	mechanistic target of rapamycin kinase
NAD+	nicotinamide adenine dinucleotide
NRF	nuclear respiratory factor
OA	osteoarthritis
O-GlcNAc	*O*-linked *N*-acetylaglucosamine
PGC-1α	peroxisome proliferator-activated receptor γ coactivator-1α
PGK	phosphoglycerates kinase
PINK1	PTEN induced kinase 1
PKA	protein kinase A
PP2A	serine-threonine protein phosphatase 2A
PTM	post-translational modification reactions
RANKL	receptor activator nuclear factor-κ ligand
REDD1	DNA damage inducible transcript 4
RPTOR	regulator-associated protein of mTOR complex 1
SOX9	SRY-box 9
TCA	tricarboxylic acid
SUMO	small ubiquitin like modifier proteins
TFAM	mitochondrial transcription factor A
TGF-β1	transforming growth factor-β1
TNF-α	tumor necrosis factor-α
ULK1	unc-51 like autophagy activating kinase

19.1 Introduction

Osteoporosis accounts for the most prevalent bone disorder in aged people. Extremely poor bone mass and fragile microstructure of affected skeletons undermine that of biomechanical strength incompetent against the impact of accidental falling, increasing the risk of osteoporotic fracture (Sattui and Saag 2014). Aging, estrogen deficiency, glucocorticoid excess, renal disorders, and diabetes, etc. are shown to accelerate the bone disease progression (Rachner et al. 2011). Osteoporosis has become an urgent healthcare issue to tackle down disability and even premature mortality of the aged people. To this end, having a productive insight into the molecular events underlying skeletal tissue deterioration will facilitate us to manage osteoporotic disorders.

Bone mass homeostasis is a dynamic status harmonized by bone mineral accretion and resorption reactions. Osteoblasts are responsible for piling up mineralized matrices, whereas osteoclastic cells resorb bone matrices to develop a well-woven three-dimensional microarchitecture (Favus 2010). In excessive bone turnover conditions, deregulated osteoblast behavior is a well-recognized hallmark of osteoporotic bone development. It slows down bone mineralization or overproduces osteoclastogenesis-promoting factors, like receptor activator nuclear factor-κ ligand (RANKL) and inflammatory cytokines, to activate osteoclasts and ultimately causes skeletal tissue to lose bone mineral density, microarchitectural integrity, and biomechanical properties (Canalis 2013).

Emerging evidence has shown that deregulated differentiation potential of bone-marrow mesenchymal stem cells (MSC) also contributes to osteoporosis development (Seibel et al. 2013). Loss of WNT and β-catenin signaling components suppresses osteogenic differentiation capacity of bone-marrow MSC, but upregulates adipocyte formation of marrow compartment in osteoporotic bone tissue (Wang et al. 2005). Upregulated WNT inhibitors, like glycogen synthase kinase-3β (GSK-3β) and dickkopf-1 (DKK-1), increase osteogenic cell apoptosis and fatty marrow accumulation in osteoporotic skeletal microenvironment (Wang et al. 2009, 2008). Neuropeptide and cannabinoid receptor pathways also shift MSC into adipogenic lineages rather than osteoblastic cells during osteoporosis development (Ko et al. 2012; Wang et al. 2016). Little has been characterized how intracellular disintegration causes osteoblastic cells to lose bone formation activity in the course of osteoporosis.

Heat shock proteins (HSP) belong to the chaperone family, ubiquitously existing in mammalian cells. These highly-conserved molecules along with E3 ubiquitin and adenosine triphosphate (ATP) have been shown to escort unfolded native proteins in various organelles to gain biological stability, facilitate transcription factor entry into nuclei, and sustain histone assembly, etc. (Arlet et al. 2014). HSP also plays an important role in modulating intracellular homeostasis as cells encounter adverse conditions, like heat, chemical, mechanical stress, hypoxia, nutrient depletion, and inflammation, etc. (Nisemblat et al. 2015).

This article highlights the biological roles of heat shock protein 60 (HSP60) in osteoporosis development and how HSP60 wards off detrimental stress-induced osteoblast dysfunction. We center on its actions to organelle function and post-translational modification of bioactive proteins in bone cells. We also shed light on HSP60 function to chondrocyte metabolism and cartilage integrity in osteoarthritis and arthritic diseases.

19.1.1 Organelle Dysfunction in Stressed Osteoblasts

Increasing studies reveal that organelle malfunction impairs survival and differentiation capacity of osteoblasts upon deleterious stresses. For example, a supraphysiologic level of dexamethasone impairs endoplasmic reticulum integrity, interrupting protein transportation and mitochondrial ATP synthesis in osteoblasts (Sato et al. 2015). Ageing hinders mitochondria function, increasing hydrogen peroxide and superoxide burst in osteoblasts and osteocytes, reducing mineralized matrix synthesis, and ultimately accelerating trabecular bone loss (Kobayashi et al. 2015). Hyperglycemic derivatives advanced glycation end products induce endoplasmic reticulum stress, which ramps up osteoblast apoptosis (Tanaka et al. 2013). Defective autophagy escalates endoplasmic reticulum dysfunction in osteoblasts, suppressing bone mineral density and trabecular integrity (Omari et al. 2018). The molecular mechanism underlying the disintegrated mitochondria, endoplasmic reticulum, and autophagy in intracellular microenvironment that deteriorate survival and bone matrix production of osteoblastic cells during osteoporosis progression has become an attractive topic.

19.1.1.1 Mitochondrial Function in Bone Tissue

Mitochondria, the largest organelle in cytosolic compartment, is responsible for ATP synthesis and fatty acid metabolism through a sophisticated respiratory reaction in cells (Bertero and Maack 2018; Rambold and Pearce 2018). Mitochondrial energy metabolism enables cells to deal with a plethora of biochemical reactions, like lipid metabolism, amino acid synthesis (Herzig and Shaw 2018), molecular pumps (Madeira 2018; Wu et al. 2016), and apoptotic program (Chen et al. 2018). Aerobic oxidation process of tricarboxylic acid (TCA) in mitochondrial micro-compartment plays an important role in consuming fatty acids, pyruvate, amino acid, and glutamine, etc. to produce adenosine diphosphate (ADP) and ATP. Several molecules within mitochondrial matrix, like acetyl coenzyme A (CoA), succinyl CoA, nicotinamide adenine dinucleotide (NAD^+), and NADH participate in electron transfer during the respiratory complex reaction (Spinelli and Haigis 2018).

In addition, this organelle is an enclosed bilayer membrane structure containing genetic materials with a circular form of mitochondrial DNA (mtDNA) distinct from genomic DNA. mtDNA is maternally inherited and composes of two initiated subunits with specific genetic codes, which are replicated by a unique genome-encoded RNA polymerase (Falkenberg 2018). This DNA component dynamically integrates metabolic processes and enzymatic reactions within cellular microenvironment under the physiological, extracellular stress, and pathological conditions. Accumulating evidence reveals that perturbed mtDNA is relevant to the disorganized extracellular matrix and Unstable genomic statuses in various human diseases.

With regard to the relevance of mitochondrial function to osteoporosis and osteoarthritis development, increased matrix metalloproteinases (MMP), interleukin-1β (IL-1β), and tumor necrosis factor-α (TNF-α) are shown to provoke oxidative stress in osteoporotic bone and osteoarthritic cartilage. Impaired respiratory chain reaction, ATP synthesis, calcium influx, and redox status also occur in osteoblasts and chondrocytes (Blanco et al. 2018; Ucer et al. 2017; Zarrouk et al. 2014). While shreds of emerging evidence reveal the correlation of poor mitochondrial integrity with osteoporosis and osteoarthritis development, the contribution of mtDNA to these skeletal diseases remains elusive.

19.1.1.2 Autophagy in Osteoblasts and Chondrocytes

Autophagy is an important intracellular reaction to dispose of unwanted macromolecules, like damaged organelles, pathogens, and unfolded proteins, through autophagosome inclusion and lysosomal breakdown processes (Memme et al. 2016). Beclin, autophagy related proteins (Atg), and microtubule associates protein 1 light chain 3 alpha (LC3) are essential builders during autophagic body formation, removing unnecessary molecules within cells upon encountering adverse conditions to sustain microenvironment homeostasis (Doria et al. 2013; Goutas et al. 2018). In bone and cartilage tissues, osteoblast-specific Atg5 knockout mice exhibit poor bone mass and trabecular morphology and those of primary bone-marrow MSC cultures exhibit low osteogenic differentiation capacity (Nollet et al. 2014). Mice deficient in Atg5 in chondrocytes show cartilage underdevelopment and defective skeletal morphology (Vuppalapati et al. 2015). In the starvation or nutrient depletion condition, autophagy upstream regulator serine/threonine kinase mechanistic target of rapamycin kinase (mTOR) pathway is shown to induce unc-51 like autophagy activating kinase (ULK1) and Atg13 phosphorylation and thus controls autophagic programs (Corona Velazquez and Jackson 2018). On the contrary, mTOR pathway is suppressed by forkhead box subgroup O (FOXO), Akt/AMP-activated protein kinase (AMPK), and DNA damage inducible transcript 4 (REDD1) signaling to maintain survival and extracellular matrix synthesis of osteoblasts and chondrocytes for retaining bone and cartilage matrix homeostasis (Alvarez-Garcia et al. 2017; Cai et al. 2018).

19.1.1.3 Autophagic Degradation of Mitochondria

Autophagy is also involved in maintaining mitochondrial matrix homeostasis through mitophagy. For example, dysfunctional mitochondrial components or damaged mtDNA are subjected to undergoing a mitophagic degradation process to keep mitochondrial ultrastructure and microarchitecture of cell cultures to stay intact upon oxidative stress, DNA damage, and endoplasmic reticulum stress (Vyas et al. 2016). Two pathways PTEN induced kinase 1 (PINK1) /Parkin and BCL2 interacting protein 3 (BNIP3) /Nix are key regulators, acting as "receptors" of the mitophagic reactions. Once autophagosome recognizes these molecules, an ubiquitination process is initiated to selectively get rid of unwanted mitochondrial components (Ashrafi and Schwarz 2013; Yamano et al. 2016).

19.1.1.4 Endoplasmic Reticulum Function and Chaperone Proteins

When mitochondrial biogenesis is impaired, an endoplasmic reticulum-mitochondrial network is important to maintain the autophagosome formation process (Sasaki and Yoshida 2015). The lumen compartment of endoplasmic reticulum acts as a factory for protein assembling, folding, and shaping *via* various post-translational modification processes. Several chaperone pathways within cytoplasmic compartment are found to secure a wide spectrum of protein unfolding/folding mechanism to facilitate proteostasis and stabilize biological function of newly-synthetized proteins. These molecules are shown to interact with unfolded proteins, in a ATP-dependent way, coordinating chaperonins to prevent the unstable proteins from aggregation. The chaperone family are highly conserved molecules, also participating in protein assembly, secretion, trafficking, and degradation. Of chaperonins, heat shock proteins (HSP), depending on types and sizes, exert various physiological and pathological activities (Table 19.1).

HSP60 exists in the mitochondrial compartment and forms an intricate hollow ultrastructure for securing and embedding misfolded or unfolded proteins into a properly folded state. It plays an important role in quality control of proteins. HSP60 is constitutively expressed by the majority of mammalian cells and also acts as a stress-inducible protein responsible for cellular and mitochondrial homeostasis. This molecule is shown to work with co-chaperonin HSP10, in an ATP-dependent fashion, to regulate protein assembly and facilitate the transportation/synthesis of proteins from cytoplasmic compartment toward mitochondria.

19.1.2 HSP60 Function to Tissue Integrity

HSP60 exerts various effects on tissue morphogenesis and deterioration. For example, mice lacking heat shock protein family D member 1 (HSPd1) gene that encodes HSP60 exhibit spastic paraplegia-like neurodegenerative disorders, embryo

Table 19.1 Biological actions and relevance of chaperone proteins to various diseases

HSP types	Cellular location	Biological functions	Related diseases
Chaperonins			
HSP60	Mitochondria	Importing newly formed proteins into mitochondrial from cytosol; protein folding	Osteoarthritis, rheumatoid arthritis, autoimmune disorders, aging, embryonic lethal, metabolic syndrome, and stem cells differentiation, etc.
HSP10	Cytosol	Co-chaperone with HSP60 for folding protein	
Molecular chaperones			
HSP40	Cytosol	Co-chaperone with HSP70, regulating ATPase activity	Degenerative neural disorders, dementia, Huntington's disease, Parkinson's disease, and mechanical damage of tissue integrity, etc.
HSP70	Cytosol	Protein folding/unfolding; regulating protein post-translational modification. Protecting unfolding proteins from aggregation.	
Bip/Grp78	Endoplasmic reticulum	Co-chaperone with GRP94; modulating ER stress, regulating protein unfolding/aggregation	Malignant melanoma and leukemic diseases
HSP90	Cytosol	Co-chaperone with HSP70, folding and conformational regulation of signaling proteins	Osteoporotic diseases

lethality (Christensen et al. 2010), and hypothalamus insulin resistance (Kleinridders et al. 2013). Mice deficient in epithelial-specific HSP60 show severe mitochondrial dysfunction and stemness loss of intestinal epithelial stem cells (Berger et al. 2016). However, HSP60 transgenic mice driven by CAGGS promoter show neonatal mortality with severe cardiac muscle dystrophy (Chen et al. 2015). Intriguing evidence is that mice ubiquitously overexpressing HSP60 driven by phosphoglycerates kinase (PGK) promoter are fertile and gain body weight (Ko et al. 2016), which is indicative of different in vivo HSP60 gene manipulation strategies may generate various phenotypes. While the contribution of HSP60 to tissue development is still contentious, increasing studies have revealed the involvement of mitochondrial HSP60 in various pathological conditions, like chronic metabolic syndrome, carcinogenesis, immune responses, and inflammation (Berger et al. 2016; Jeong et al. 2017; Swaroop et al. 2016).

19.1.2.1 HSP60 Control of Cell Survival

Chaperones are shown to facilitate protein folding within cell microenvironment upon intra or extracellular stress. Various molecular chaperones exist in nuclei, cytoplasm, endoplasmic reticulum, and mitochondrial compartments. HSP60 is a well-known chaperone that primarily appears in the mitochondrial matrix. Increasing evidence also uncovers the existence and function of this molecule outside the mitochondria (Cechetto et al. 2000; Choi et al. 2015; Deniset et al. 2018; Padma Priya et al. 2015). In mitochondrial compartment, HSP60 forms a complex with HSP10 to modulate mitochondria-dependent cell death pathway and mtDNA replication (Cheng et al. 2016; Jeffery 2018).

The biological role of HSP60 in cell apoptosis appears to depend on several factors, like cell, stress and disease types, etc., but still remains inconclusive. Cytoplasmic HSP60 is found to interact with pro-apoptotic regulators to form a complex, preventing cells from apoptosis (Choi et al. 2015); however, some groups suggest that HSP60 accumulates in the cytoplasm without changing mitochondrial integrity in apoptotic cells (Chandra et al. 2007). In yeast, cytoplasmic HSP60 is shown to change 6-phosphofructokinase activity, controlling the glycolysis pathway (Kalderon et al. 2015). This chaperonin also exists in endoplasmic reticulum in apoptotic breast cancer cells (Arya et al. 2015). Forced HSP60 expression in various human tumor cells enables HSP60 to accumulate in secretory vesicles and plasma membrane and thus shield cancer cells from deleterious stress (Marino Gammazza et al. 2017; Zhang et al. 2016), whereas other groups report that loss of HSP60 function increases the viability and behavior of tumor cells (Gorska et al. 2013; Wadhwa et al. 2016). In addition to the cytoplasmic compartment, various cell types are shown to secrete HSP60 and that of interacts with receptors of immune cells to change immune responses. HSP60 circulating in bloodstream also acts as a chaperokine to regulate cell fate and behavior (Gazali 2012; Osterloh et al. 2007; Suragani et al. 2013; Xie et al. 2010).

19.1.2.2 Post-Translational Modification Capacity of HSP60

Post-translational modification reactions (PTM), like phosphorylation, ubiquitination, sumoylation, acetylation, and glycosylation, etc. convert native proteins into biologically active and stable ones. Several studies have revealed the role of PTM in HSP60 function. Protein kinase A (PKA) and serine-threonine protein phosphatase 2 A (PP2A) are shown to phosphorylate HSP60 to dissociate that of protein from histone 2B in plasma membrane (Kotlo et al. 2014; Vilasi et al. 2017). With respect to the ubiquitination process, proteolytic enzymes breakdown the protein targets of interest and thus change the structure and function of the protein. Ubiquitin and proteasome are common regulators responsible for protein degradation reactions. HSP60 is ubiquitinated and subsequently sequestrated in mitochondrial compartment in hydrogen peroxide-stressed human monocytes (Tang et al. 2013). On the other hand, small ubiquitin like modifier proteins (SUMO)-mediated sumoylation is also a well-known PTM reaction similar to ubiquitination process to influence protein stability, transportation, and activity. HSP60 sumoylation is found to occur mainly in the mutated sumoylation regions of HSP60 in fungi and enables the microorganisms sensitive to a thermal stress (Leach et al. 2011).

Lysine residues of proteins are inclined to be targets of acetylation that functionally modulates cell biology and mitochondrial function (Lu et al. 2015; Merendino et al. 2010). Mass spectrometric analysis reveals that histone deacetylase 1 (HDAC1) deacetylates HSP60 (Choudhary et al. 2009). Mdm-2 antagonist nutlin-3 induces HSP60 hypoacetylation in myeloid leukemia cells (Haaland et al. 2014). Deacetylated HSP60 causes p53 to dissociate from the chaperonin and thus activates p53 to worsen cell apoptosis (Marino Gammazza et al. 2017). HSP60 hyperacetylation is also found to decrease HSP60 activity in osteosarcoma cells upon geldanamycin treatment (Gorska et al. 2013).

The carboxyl groups of lysine residue of proteins are also putatively reactive to malonylation or succinylation process, which alters cellular stress response and metabotic activities (Park et al. 2013; Peng et al. 2011). High performance liquid chromatography/tandem mass spectrometric analysis shows that Lys133 and Lys355 of HSP60 are the putative malonylation and succinulation sites; however, the contribution of these two PTM processes to HSP60 function is still unclear (Park et al. 2013; Peng et al. 2011). Methylation of lysine residue by specific lysine methyltransferases also occurs in HSP60. While HSP60 exists mainly in mitochondria, this molecule is also secreted to interact with immune cells via lipid raft-exosome system (Gupta and Knowlton 2007). On the other hand, *N*-glycosylated HSP60 is secreted through a endoplasmic reticulum-Golgi system in malignant fibrosarcoma cells (Hayoun et al. 2012). In the hyperglycemic condition, *O*-linked *N*-acetylaglucosamine (*O*-GlcNAc) modification of HSP60 occurs in β-cells in rat pancreas and dissociate HSP60/BCL2 associated X apoptosis regulator (Bax) complex to activate caspase-3 that accelerates cell apoptotic program (Kim et al. 2006).

19.1.3 HSP60 Function to Bone Tissue

Bone cells in skeletal microenvironment adapt extracellular biophysical or biochemical stresses through altering intracellular signaling pathways (Plotkin and Bellido 2016). Accumulating studies reveal that HSP in bone cells actively respond to various deleterious stresses, affecting cell fate or metabolic activity. HSP22 impairs migration capacity of osteoblastic cells upon the stimulation by osteogenic factor transforming growth factor-β1 (TGF-β1) (Yamamoto et al. 2016). Forced HSP27 expression mitigates the reactive oxidative stress-induced osteoblast apoptosis. Grafting HSP27-transfected osteoblasts prolongs bone cell survival in calvarial bone defects (Kitami et al. 2016). On the contrary, forced HSP27 expression causes osteoblasts to lose osteogenic matrix osteocalcin synthesis (Kato et al. 2011). HSP90 expression along with mineralized matrix production is increased in osteoblast cultures upon low-intensity pulsed ultrasound stimulation (Miyasaka et al. 2015). HSP90 inhibitor reduces the prostaglandin-mediated inflammatory cytokine production and the glucocorticoid excess-induced bone mass loss (Chen et al. 2017). While the contribution of HSP to osteoblast function may depend on extracellular stimulation or stress types, whether these HSP influence organelle function in osteoblasts is not defined.

19.1.3.1 HSP60 Is Essential to Tissue Integrity

Chaperonin HSP60 is indispensable in mitochondrial integrity and embryo survival. Thus, loss of HSP60 function results in fused sarcoma-linked neural disorders (Magnoni et al. 2013). This molecule is shown to increase energy metabolism in mitochondrial compartment in skeletal muscle and promote immune responses of macrophages in ischemia/perfusion lung (Liu et al. 2017). HSP60 also exerts inhibitory actions to some cancer cell types. For example, it acts as a tumor suppressor, slowing down the invasion activity of hepatocellular carcinoma (Zhang et al. 2016). HSP60 loss blocks mitochondrial function, exaggerates the epithelial to mesenchymal transition capacity, and exacerbates malignancy of clear renal cell carcinoma (Tang et al. 2016). These studies unravel the beneficial function of HSP60 to tissue homeostasis.

19.1.3.2 HSP60 Alleviates Bone Mass Loss

With regard to skeletal tissue homeostasis, comparative proteomic analysis shows that a low HSP60 abundance along with decreased bone mineral density and meager trabecular bone morphometry occurs in osteoporotic skeletons in rats upon long-term glucocorticoid treatment (Wang et al. 2011). In vivo knocking down HSP60 causes bone tissue to show porous bone structure and weak mechanical property lower than the baseline, which suggests that HSP60 is required for bone mass homeostasis. Striking findings are that gain of HSP60 action by adenovirus HSP60 gene therapy offsets the severity of glucocorticoid-mediated loss of bone mass and microstructure, as well as sustains biomechanical strength of the skeletal tissue (Wang et al. 2011).

Increasing HSP60 sustains mineral acquisition reaction and represses osteoblast apoptosis in bone microenvironment, whereas knocking down this chaperonin downregulates bone formation rate and bone cell survival below the baseline. The molecular events underlying the bone-protective effects are that HSP60 activates multiple cell survival regulators, like extracellular regulated MAP kinase (ERK), p38, and Atk pathways, and subsequently reverses the glucocorticoid-induced mitochondrial cytochrome c release and Bax-Caspase 3 activation (Wang et al. 2011). In contrast to the HSP60 protection against osteoblast dysfunction, a study of ovariectomized mice shows that increasing HSP60 expression reportedly accelerates osteoclast formation of macrophage precursor cells, bone resorption, and estrogen deficiency-induced bone loss (Koh et al. 2009). Exogenous HSP60 recombinant proteins also activate toll-like receptor pathways, provoking osteoblastic cell apoptosis (Kim et al. 2009). These findings hint that the effect of HSP60 on bone mass status may depend on osteoporosis types, endogenous/exogenous HSP60, and bone cell lineage.

19.1.3.3 HSP60 Retains Osteoblast Autophagy

Of interest, this chaperone protein is important to stabilize autophagy in osteoblasts upon high concentration of glucocorticoid stress as forced HSP60 expression attenuates the glucocorticoid-induced loss of autophagic puncta formation probed by fluorescent monodansylcadaverin (Fig. 19.1a). It maintains autophagic builders Atg4, Atg12, and LC3-II conversion, protecting from autophagosome loss to reduce apoptotic reactions in glucocorticoid-stressed osteoblasts (Lian et al. 2018). The high throughput tandem mass spectrometric analysis uncovers an interaction between HSP60 and autophagy upstream modulator regulator-associated protein of mTOR complex 1 (RPTOR), which is indispensable in sustaining autophagic program in osteoblastic cells and bone mass metabolism. Mechanistically, HSP60 stabilizes RPTOR function through repressing the glucocorticoid-deregulated multiple post-translational modification reactions, like de-phosphorylation, ubiquitinization, and aggregation of this autophagy regulator (Lian et al. 2018).

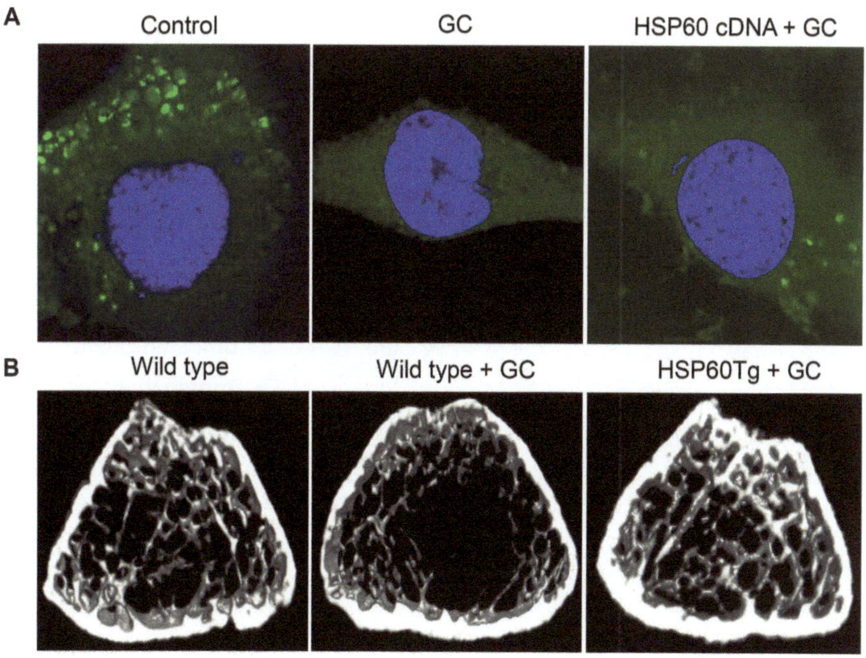

Fig. 19.1 Protective effects of HSP60 on osteoblast autophagy and bone microstructure. (**a**) Forced HSP60 expression represses the glucocorticoid-induced loss of osteoblast autophagy. Osteoblasts with or with HSP60 cDNA transfection were treated with 1 μM dexamethasone for 24 h. Autophagic puncta is probed by fluorescent monodansylcadaverin (green) and DAPI counterstaining (blue). (**b**) Mice overexpressing HSP60 (HSP60Tg) show a minor response to glucocorticoid-induced loss of trabecular bone. Mice were given 10 mg/kg/day methylprednisolone for 2 weeks. Trabecular bone microarchitecture was scanned by micro-CT. GC, glucocorticoid

19.1.3.4 HSP60 Overexpression Compromises Osteoporosis

The HSP60 promotion of autophagy enables osteoblasts to accumulate abundant mineralized matrices. Consistent with the analysis of in vitro model, mice overexpressing HSP60 driven by PGK promoter show high bone formation rate and bone mass, as well as increased osteogenic differentiation potential of primary bone-marrow MSC. These beneficial effects on bone tissue improve the glucocorticoid-mediated RPTOR loss, osteoblast dysfunction, like impaired autophagy and survival, and osteoporotic skeleton development as evident from micro-CT imaging (Fig. 19.1b) (Lian et al. 2018). The robust in vitro and in vivo evidence underpins the beneficial actions of this mitochondrial chaperonin to autophagosomal integrity important to maintain osteoblastic activity for bone formation (Fig. 19.2).

19.1.3.5 HSP60 Action to Osteosarcoma

In addition to osteoporosis, HSP60 is also relevant to osteosarcoma development, its action to the cancer cell behavior remains inconclusive though. For example, the differentially expressed gene array analysis reveals that HSPd1 is a candidate gene during the tumor development of osteosarcoma stem cells (Selvarajah et al. 2013).

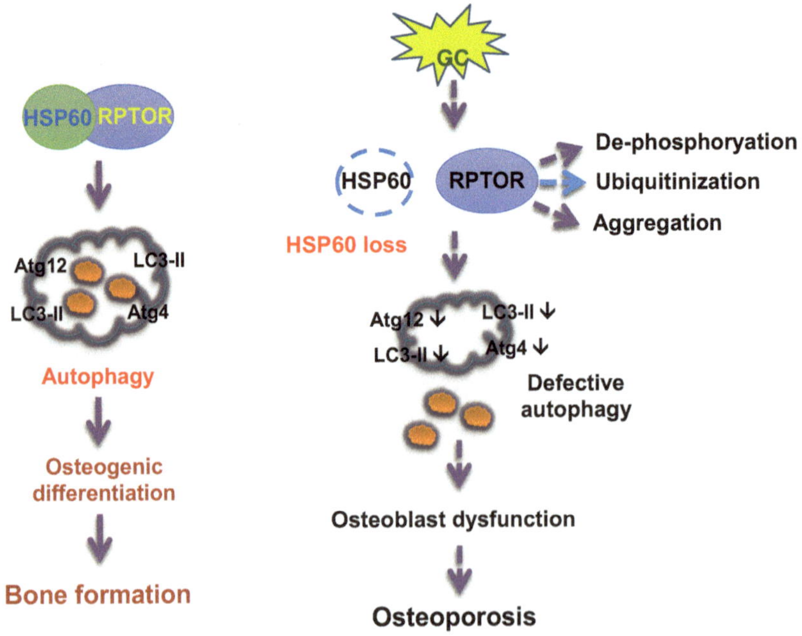

Fig. 19.2 Schematic drawings for HSP60 stabilization of autophagy in osteoblasts. HSP60 is indispensable in protecting from glucocorticoid-induced loss of RPTOR action to osteoblast autophagy and osteoporosis development

Low-grade osteosarcoma cells highly express this chaperone protein. Strong HSP60 immunostaining in human and canine osteosarcoma is correlated with poor prognosis. Interfering with HSP60 decreases survival of highly metastatic osteosarcoma cells. Chemotherapy with geldanamycin induces HSP60 hyperacetylation and ramps up apoptosis processes in osteosarcoma cell cultures (Gorska et al. 2013; Moon et al. 2010).

19.1.3.6 HSP60 Action in Other Osteopenic Conditions

Increasing evidence also links HSP60 action to the extracellular stress-deregulated MSC behavior. MSC incubated in a simulated microgravity status strongly express HSP60 along with high osteogenic maker expression (Cazzaniga et al. 2016). HSP60 expression in dental pulp MSC is decreased with age (Feng et al. 2014). A low HSP60 level is correlated with bone metastasis of laryngeal squamous cell carcinoma (Bodnar et al. 2016). While HSP60 expression is changed in progenitor cells upon these extracellular stresses, its biological function to cell fate is not yet identified.

19.1.4 HSP60 Action to Cartilage Integrity

Osteoarthritis (OA) is an irreversible joint degeneration provoked by progressive articular cartilage degradation. This joint disorder impacts patients' daily life quality badly, especially the aged people. A chronic inflammation status in articular microenvironment is shown to devastate chondrocyte function, like viability loss, extracellular matrix underproduction, and excessive matrix breakdown, that accelerates disease progression (Robinson et al. 2016). Accumulating studies have revealed the relevance of HSP to OA development. Systemic microarray profiles uncover that HSPd1 gene encoding HSP60 along with cartilage matrix production is increased in chondrocytes. HSPd1 knockdown decreases matrix expression, which is indicative of HSP60 action essential to cartilage metabolism (Suwanwela et al. 2011). The biological function of HSP60 signaling to cartilage and osteoarthritis development warrants mention.

19.1.4.1 HSP60 Represses Mitochondrial Dysfunction in Osteoarthritic Chondrocytes

In human end-stage osteoarthritis, weak HSP60 expression occurs in osteoarthritic cartilage (Fig. 19.3a). Forced HSP60 expression downregulates the IL-1β-mediated loss of mitochondrial biogenesis activity as mitochondrial transcription factor A (TFAM), nuclear respiratory factor (NRF), and peroxisome proliferator-activated receptor γ coactivator-1α (PGC-1α) are improved in chondrocyte cultures. Increasing

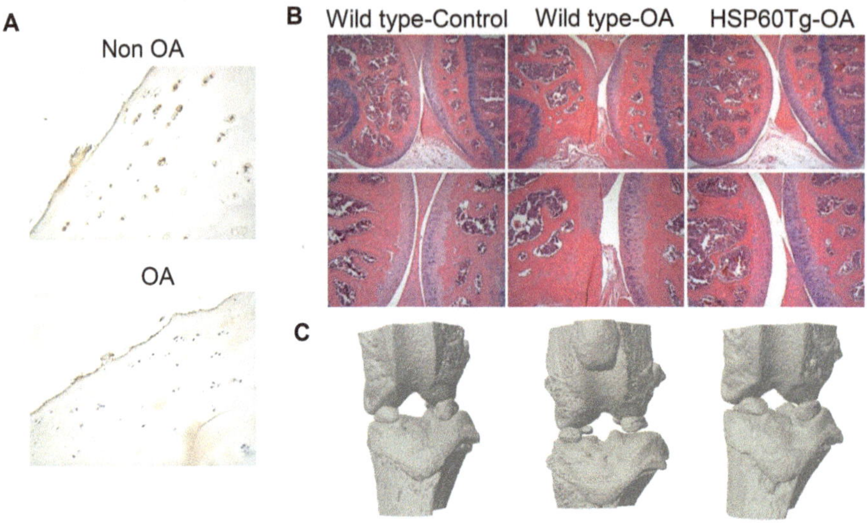

Fig. 19.3 HSP60 protection against osteoarthritis development. (**a**) Chondrocytes in human osteoarthritic cartilage weakly show HSP60 immunostaining as compared to cartilage from femoral neck fracture. Collagenase-induced OA signs, including (**b**) articular cartilage erosion and (**c**) osteophyte formation are compromised in HSP60 transgenic mice. OA was induced by an intra-articular injection of collagenase. Articular cartilage morphology and osteophyte formation were detected by histology and μCT imaging

this chaperonin thus restores chondrocytic activity (Ko et al. 2016). These studies confirm the indispensability of HSP60 signaling in maintaining mitochondrial integrity in cells upon experiencing deleterious conditions.

19.1.4.2 HSP60 Overexpression Compromises OA Progression

Consistent with the in vitro analysis, mild cartilage overgrowth, like increased chondrocyte proliferation, chondrocytic marker expression, and cartilage thickness, appears in transgenic mice overexpressing HSP60 driven by PGK promoter. Of interest, these transgenic mice show a minor response to collagenase-induced OA signs, like cartilage degradation, synovial hypertrophy, and osteophyte formation (Fig. 19.3b, c). These effects also protect joints from irregular walking patterns. These beneficial actions stand true for a further study that an intra-articular HSP60 gene therapy also lessens the severity of OA signs (Ko et al. 2016). Molecular events underlying these cartilage-promoting effects are that HSP60 chaperones chondrogenic transcription factor SRY-box 9 (SOX9) from ubiquitination and thus increases chondrocyte-promoting factor insulin growth factor-1 (IGF-1) expression (Chen et al. 2015) (Fig. 19.4).

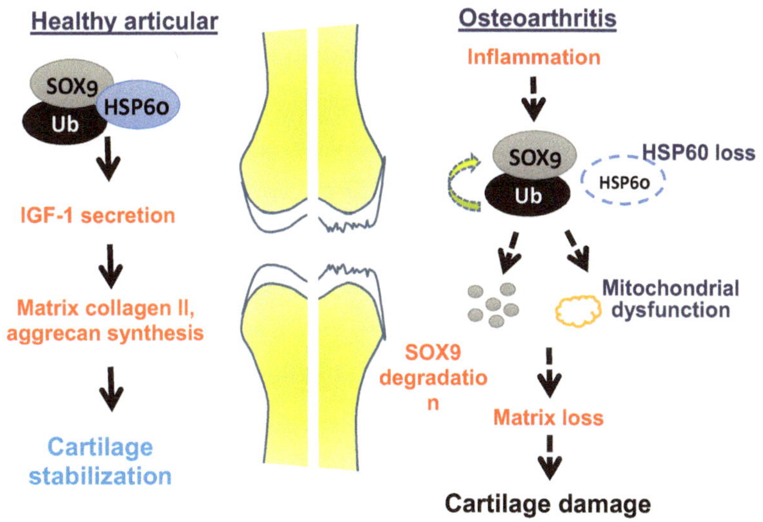

Fig. 19.4 Schematic drawings for HSP60 protection from chondrocyte dysfunction. HSP60 sustained chondrocyte metabolism by escorting chondrogenic transcription factor SOX9 from ubiquitination

19.1.4.3 HSP60 in Other Joint Disorders

HSP60 is also relevant to human arthritic diseases. For example, HSP60 gene polymorphism is strongly correlated with gouty arthritis (Ren et al. 2018). Significantly increased the abundances of HSP60 in peripheral blood exist in patients with rheumatoid arthritis (Sedlackova et al. 2011). The role this chaperone protein in organelle function or joint remodeling during the disease progress is still unknow.

19.1.5 Epigenetic Regulation of HSP60

HSPd1 gene is shown to encode HSP60 in the majority of cells types and that of mutation is relevant to mitochondria-associated diseases in human (Bross et al. 2012). Emerging evidence has uncovered epigenetic pathways, like small non-coding microRNAs and histone modification that regulate HSP60 gene expression without changing DNA sequence. For example, microRNA let-7 is found to target HSP60 mRNA expression in Müller glia cells (Ramachandran et al. 2010). MicroRNA-1 directly silences HSP60 mRNA expression through binding 3′-untranslated region of HSP60 in cardiomyocytes (Neumann et al. 2016). Given that HSP60 expression is affected during osteoporosis and osteoarthritis, the epigenetic actions to HSP60 expression appear to shine a new landscape of molecular mechanisms underlying HSP60 stabilization of bone and cartilage homeostasis.

19.2 Conclusions

Mitochondrial HSP60 exerts multifunction to organelle integrity, regulating cell behavior and fate in various physiological, pathological, and oncological conditions. This chaperone protein plays an important role in escorting molecules essential to sustain autophagy, mitochondrial function, endoplasmic reticulum integrity, and differentiation capacity, shielding bone and cartilage tissues from osteoporosis and osteoarthritis. This article offers profound insights into the HSP60 action to bone and joint metabolism and highlights the protective effects of this mitochondrial chaperonin on osteoporotic and osteoarthritic disorders in the future.

Acknowledgements This work was in part supported by grants [NHRI-EX107-10736SI] from the National Health Research Institute, [MOST107-2314-B-182A-038MY3; MOST107-2314-B-182A-012-MY2] from Ministry of Science and Technology, and [CMRPG8H0211-3; CLRPG8G0701-3] from Chang Gung Memorial Hospital, Taiwan.

References

Alvarez-Garcia O, Matsuzaki T, Olmer M, Plate L, Kelly JW, Lotz MK (2017) Regulated in development and DNA damage response 1 deficiency impairs autophagy and mitochondrial biogenesis in articular cartilage and increases the severity of experimental osteoarthritis. Arthritis Rheumatol 69:1418–1428

Arlet JB, Ribeil JA, Guillem F et al (2014) HSP70 sequestration by free alpha-globin promotes ineffective erythropoiesis in beta-thalassaemia. Nature 514:242–246

Arya RK, Singh A, Yadav NK et al (2015) Anti-breast tumor activity of Eclipta extract in-vitro and in-vivo: novel evidence of endoplasmic reticulum specific localization of Hsp60 during apoptosis. Sci Rep 5:18457

Ashrafi G, Schwarz TL (2013) The pathways of mitophagy for quality control and clearance of mitochondria. Cell Death Differ 20:31–42

Berger E, Rath E, Yuan D et al (2016) Mitochondrial function controls intestinal epithelial stemness and proliferation. Nat Commun 7:13171

Bertero E, Maack C (2018) Metabolic remodelling in heart failure. Nat Rev Cardiol 15:457–470

Blanco FJ, Valdes AM, Rego-Perez I (2018) Mitochondrial DNA variation and the pathogenesis of osteoarthritis phenotypes. Nat Rev Rheumatol 14:327–340

Bodnar M, Luczak M, Bednarek K et al (2016) Proteomic profiling identifies the inorganic pyrophosphatase (PPA1) protein as a potential biomarker of metastasis in laryngeal squamous cell carcinoma. Amino Acids 48:1469–1476

Bross P, Magnoni R, Bie AS (2012) Molecular chaperone disorders: defective Hsp60 in neurodegeneration. Curr Top Med Chem 12:2491–2503

Cai ZY, Yang B, Shi YX et al (2018) High glucose downregulates the effects of autophagy on osteoclastogenesis via the AMPK/mTOR/ULK1 pathway. Biochem Biophys Res Commun 503:428–435

Canalis E (2013) Wnt signalling in osteoporosis: mechanisms and novel therapeutic approaches. Nat Rev Endocrinol 9:575–583

Cazzaniga A, Maier JAM, Castiglioni S (2016) Impact of simulated microgravity on human bone stem cells: new hints for space medicine. Biochem Biophys Res Commun 473:181–186

Cechetto JD, Soltys BJ, Gupta RS (2000) Localization of mitochondrial 60-kD heat shock chaperonin protein (Hsp60) in pituitary growth hormone secretory granules and pancreatic zymogen granules. J Histochem Cytochem 48:45–56
Chandra D, Choy G, Tang DG (2007) Cytosolic accumulation of HSP60 during apoptosis with or without apparent mitochondrial release: evidence that its pro-apoptotic or pro-survival functions involve differential interactions with caspase-3. J Biol Chem 282:31289–31301
Chen TH, Liu SW, Chen MR et al (2015) Neonatal death and heart failure in mouse with transgenic HSP60 expression. Biomed Res Int 2015:539805
Chen H, Xing J, Hu X et al (2017) Inhibition of heat shock protein 90 rescues glucocorticoid-induced bone loss through enhancing bone formation. J Steroid Biochem Mol Biol 171:236–246
Chen X, Li X, Zhao W, Li T, Ouyang Q (2018) Parameter sensitivity analysis for a stochastic model of mitochondrial apoptosis pathway. PLoS One 13:e0198579
Cheng Y, Sun J, Chen H et al (2016) Expression and location of HSP60 and HSP10 in the heart tissue of heat-stressed rats. Exp Ther Med 12:2759–2765
Choi B, Choi M, Park C et al (2015) Cytosolic Hsp60 orchestrates the survival and inflammatory responses of vascular smooth muscle cells in injured aortic vessels. Cardiovasc Res 106:498–508
Choudhary C, Kumar C, Gnad F et al (2009) Lysine acetylation targets protein complexes and co-regulates major cellular functions. Science 325:834–840
Christensen JH, Nielsen MN, Hansen J et al (2010) Inactivation of the hereditary spastic paraplegia-associated Hspd1 gene encoding the Hsp60 chaperone results in early embryonic lethality in mice. Cell Stress Chaperones 15:851–863
Corona Velazquez AF, Jackson WT (2018) So many roads: the multifaceted regulation of autophagy induction. Mol Cell Biol 38
Deniset JF, Hedley TE, Hlavackova M et al (2018) Heat shock protein 60 involvement in vascular smooth muscle cell proliferation. Cell Signal 47:44–51
Doria A, Gatto M, Punzi L (2013) Autophagy in human health and disease. N Engl J Med 368:1845
Falkenberg M (2018) Mitochondrial DNA replication in mammalian cells: overview of the pathway. Essays Biochem 62:287–296
Favus MJ (2010) Bisphosphonates for osteoporosis. N Engl J Med 363:2027–2035
Feng X, Xing J, Feng G et al (2014) p16(INK4A) mediates age-related changes in mesenchymal stem cells derived from human dental pulp through the DNA damage and stress response. Mech Ageing Dev 141–142:46–55
Gazali A (2012) Conference scene: taking the heat out of chaperokine function. Immunotherapy 4:773–775
Gorska M, Marino Gammazza A, Zmijewski MA et al (2013) Geldanamycin-induced osteosarcoma cell death is associated with hyperacetylation and loss of mitochondrial pool of heat shock protein 60 (hsp60). PLoS One 8:e71135
Goutas A, Syrrou C, Papathanasiou I, Tsezou A, Trachana V (2018) The autophagic response to oxidative stress in osteoarthritic chondrocytes is deregulated. Free Radic Biol Med 126:122–132
Gupta S, Knowlton AA (2007) HSP60 trafficking in adult cardiac myocytes: role of the exosomal pathway. Am J Physiol Heart Circ Physiol 292:H3052–H3056
Haaland I, Opsahl JA, Berven FS et al (2014) Molecular mechanisms of nutlin-3 involve acetylation of p53, histones and heat shock proteins in acute myeloid leukemia. Mol Cancer 13:116
Hayoun D, Kapp T, Edri-Brami M et al (2012) HSP60 is transported through the secretory pathway of 3-MCA-induced fibrosarcoma tumour cells and undergoes N-glycosylation. FEBS J 279:2083–2095
Herzig S, Shaw RJ (2018) AMPK: guardian of metabolism and mitochondrial homeostasis. Nat Rev Mol Cell Biol 19:121–135
Jeffery CJ (2018) Protein moonlighting: what is it, and why is it important? Philos Trans R Soc Lond Ser B Biol Sci 373
Jeong DE, Lee D, Hwang SY et al (2017) Mitochondrial chaperone HSP-60 regulates anti-bacterial immunity via p38 MAP kinase signaling. EMBO J 36:1046–1065

Kalderon B, Kogan G, Bubis E, Pines O (2015) Cytosolic Hsp60 can modulate proteasome activity in yeast. J Biol Chem 290:3542–3551
Kato K, Adachi S, Matsushima-Nishiwaki R et al (2011) Regulation by heat shock protein 27 of osteocalcin synthesis in osteoblasts. Endocrinology 152:1872–1882
Kim HS, Kim EM, Lee J et al (2006) Heat shock protein 60 modified with O-linked N-acetylglucosamine is involved in pancreatic beta-cell death under hyperglycemic conditions. FEBS Lett 580:2311–2316
Kim YS, Koh JM, Lee YS et al (2009) Increased circulating heat shock protein 60 induced by menopause, stimulates apoptosis of osteoblast-lineage cells via up-regulation of toll-like receptors. Bone 45:68–76
Kitami M, Kaku M, Rocabado JM, Ida T, Akiba N, Uoshima K (2016) Prolonged survival of transplanted osteoblastic cells does not directly accelerate the healing of calvarial bone defects. J Cell Physiol 231:1974–1982
Kleinridders A, Lauritzen HP, Ussar S et al (2013) Leptin regulation of Hsp60 impacts hypothalamic insulin signaling. J Clin Invest 123:4667–4680
Ko JY, Wu RW, Kuo SJ et al (2012) Cannabinoid receptor 1 mediates glucocorticoid-induced bone loss in rats by perturbing bone mineral acquisition and marrow adipogenesis. Arthritis Rheum 64:1204–1214
Ko JY, Sun YC, Li WC, Wang FS (2016) Chaperonin 60 regulation of SOX9 ubiquitination mitigates the development of knee osteoarthritis. J Mol Med (Berl) 94:755–769
Kobayashi K, Nojiri H, Saita Y et al (2015) Mitochondrial superoxide in osteocytes perturbs canalicular networks in the setting of age-related osteoporosis. Sci Rep 5:9148
Koh JM, Lee YS, Kim YS et al (2009) Heat shock protein 60 causes osteoclastic bone resorption via toll-like receptor-2 in estrogen deficiency. Bone 45:650–660
Kotlo K, Xing Y, Lather S et al (2014) PR65A phosphorylation regulates PP2A complex signaling. PLoS One 9:e85000
Leach MD, Stead DA, Argo E, Brown AJ (2011) Identification of sumoylation targets, combined with inactivation of SMT3, reveals the impact of sumoylation upon growth, morphology, and stress resistance in the pathogen Candida albicans. Mol Biol Cell 22:687–702
Lian WS, Ko JY, Chen YS et al (2018) Chaperonin 60 sustains osteoblast autophagy and counteracts glucocorticoid aggravation of osteoporosis by chaperoning RPTOR. Cell Death Dis 9:938
Liu X, Cao H, Li J et al (2017) Autophagy induced by DAMPs facilitates the inflammation response in lungs undergoing ischemia-reperfusion injury through promoting TRAF6 ubiquitination. Cell Death Differ 24:683–693
Lu Z, Chen Y, Aponte AM, Battaglia V, Gucek M, Sack MN (2015) Prolonged fasting identifies heat shock protein 10 as a Sirtuin 3 substrate: elucidating a new mechanism linking mitochondrial protein acetylation to fatty acid oxidation enzyme folding and function. J Biol Chem 290:2466–2476
Madeira VMC (2018) Overview of mitochondrial bioenergetics. Methods Mol Biol 1782:1–6
Magnoni R, Palmfeldt J, Christensen JH et al (2013) Late onset motoneuron disorder caused by mitochondrial Hsp60 chaperone deficiency in mice. Neurobiol Dis 54:12–23
Marino Gammazza A, Campanella C, Barone R et al (2017) Doxorubicin anti-tumor mechanisms include Hsp60 post-translational modifications leading to the Hsp60/p53 complex dissociation and instauration of replicative senescence. Cancer Lett 385:75–86
Memme JM, Oliveira AN, Hood DA (2016) Chronology of UPR activation in skeletal muscle adaptations to chronic contractile activity. Am J Physiol Cell Physiol 310:C1024–C1036
Merendino AM, Bucchieri F, Campanella C et al (2010) Hsp60 is actively secreted by human tumor cells. PLoS One 5:e9247
Miyasaka M, Nakata H, Hao J, Kim YK, Kasugai S, Kuroda S (2015) Low-intensity pulsed ultrasound stimulation enhances heat-shock protein 90 and mineralized nodule formation in mouse calvaria-derived osteoblasts. Tissue Eng Part A 21:2829–2839
Moon A, Bacchini P, Bertoni F et al (2010) Expression of heat shock proteins in osteosarcomas. Pathology 42:421–425

Neumann E, Brandenburger T, Santana-Varela S et al (2016) MicroRNA-1-associated effects of neuron-specific brain-derived neurotrophic factor gene deletion in dorsal root ganglia. Mol Cell Neurosci 75:36–43

Nisemblat S, Yaniv O, Parnas A, Frolow F, Azem A (2015) Crystal structure of the human mitochondrial chaperonin symmetrical football complex. Proc Natl Acad Sci U S A 112:6044–6049

Nollet M, Santucci-Darmanin S, Breuil V et al (2014) Autophagy in osteoblasts is involved in mineralization and bone homeostasis. Autophagy 10:1965–1977

Omari S, Makareeva E, Roberts-Pilgrim A et al (2018) Noncanonical autophagy at ER exit sites regulates procollagen turnover. Proc Natl Acad Sci U S A 115:E10099–E1E108

Osterloh A, Kalinke U, Weiss S, Fleischer B, Breloer M (2007) Synergistic and differential modulation of immune responses by Hsp60 and lipopolysaccharide. J Biol Chem 282:4669–4680

Padma Priya P, Grover M, Tatu US, Natarajan V (2015) Characterization of precursor PfHsp60 in Plasmodium falciparum cytosol during its asexual development in human erythrocytes. PLoS One 10:e0136401

Park J, Chen Y, Tishkoff DX et al (2013) SIRT5-mediated lysine desuccinylation impacts diverse metabolic pathways. Mol Cell 50:919–930

Peng C, Lu Z, Xie Z et al (2011) The first identification of lysine malonylation substrates and its regulatory enzyme. Mol Cell Proteomics 10:M111 012658

Plotkin LI, Bellido T (2016) Osteocytic signalling pathways as therapeutic targets for bone fragility. Nat Rev Endocrinol 12:593–605

Rachner TD, Khosla S, Hofbauer LC (2011) Osteoporosis: now and the future. Lancet 377:1276–1287

Ramachandran R, Fausett BV, Goldman D (2010) Ascl1a regulates Muller glia dedifferentiation and retinal regeneration through a Lin-28-dependent, let-7 microRNA signalling pathway. Nat Cell Biol 12:1101–1107

Rambold AS, Pearce EL (2018) Mitochondrial dynamics at the interface of immune cell metabolism and function. Trends Immunol 39:6–18

Ren D, Fan M, Sun C, Zhou C, Li Y (2018) Capillary electrophoresis with laser induced fluorescence detection for study of the association of HSP60 gene polymorphism with gouty arthritis. J AOAC Int 102(3):810–814

Robinson WH, Lepus CM, Wang Q et al (2016) Low-grade inflammation as a key mediator of the pathogenesis of osteoarthritis. Nat Rev Rheumatol 12:580–592

Sasaki K, Yoshida H (2015) Organelle autoregulation-stress responses in the ER, Golgi, mitochondria and lysosome. J Biochem 157:185–195

Sato AY, Tu X, McAndrews KA, Plotkin LI, Bellido T (2015) Prevention of glucocorticoid induced-apoptosis of osteoblasts and osteocytes by protecting against endoplasmic reticulum (ER) stress in vitro and in vivo in female mice. Bone 73:60–68

Sattui SE, Saag KG (2014) Fracture mortality: associations with epidemiology and osteoporosis treatment. Nat Rev Endocrinol 10:592–602

Sedlackova L, Sosna A, Vavrincova P et al (2011) Heat shock protein gene expression profile may differentiate between rheumatoid arthritis, osteoarthritis, and healthy controls. Scand J Rheumatol 40:354–357

Seibel MJ, Cooper MS, Zhou H (2013) Glucocorticoid-induced osteoporosis: mechanisms, management, and future perspectives. Lancet Diabetes Endocrinol 1:59–70

Selvarajah GT, Bonestroo FA, Kirpensteijn J et al (2013) Heat shock protein expression analysis in canine osteosarcoma reveals HSP60 as a potentially relevant therapeutic target. Cell Stress Chaperones 18:607–622

Spinelli JB, Haigis MC (2018) The multifaceted contributions of mitochondria to cellular metabolism. Nat Cell Biol 20:745–754

Suragani M, Aadinarayana VD, Pinjari AB et al (2013) Human resistin, a proinflammatory cytokine, shows chaperone-like activity. Proc Natl Acad Sci U S A 110:20467–20472

Suwanwela J, Farber CR, Haung BL et al (2011) Systems genetics analysis of mouse chondrocyte differentiation. J Bone Miner Res 26:747–760

Swaroop S, Sengupta N, Suryawanshi AR, Adlakha YK, Basu A (2016) HSP60 plays a regulatory role in IL-1beta-induced microglial inflammation via TLR4-p38 MAPK axis. J Neuroinflammation 13:27
Tanaka K, Yamaguchi T, Kaji H, Kanazawa I, Sugimoto T (2013) Advanced glycation end products suppress osteoblastic differentiation of stromal cells by activating endoplasmic reticulum stress. Biochem Biophys Res Commun 438:463–467
Tang H, Tian E, Liu C, Wang Q, Deng H (2013) Oxidative stress induces monocyte necrosis with enrichment of cell-bound albumin and overexpression of endoplasmic reticulum and mitochondrial chaperones. PLoS One 8:e59610
Tang H, Chen Y, Liu X et al (2016) Downregulation of HSP60 disrupts mitochondrial proteostasis to promote tumorigenesis and progression in clear cell renal cell carcinoma. Oncotarget 7:38822–38834
Ucer S, Iyer S, Kim HN et al (2017) The effects of aging and sex steroid deficiency on the murine skeleton are independent and mechanistically distinct. J Bone Miner Res 32:560–574
Vilasi S, Bulone D, Caruso Bavisotto C et al (2017) Chaperonin of group I: oligomeric spectrum and biochemical and biological implications. Front Mol Biosci 4:99
Vuppalapati KK, Bouderlique T, Newton PT et al (2015) Targeted deletion of autophagy genes Atg5 or Atg7 in the chondrocytes promotes caspase-dependent cell death and leads to mild growth retardation. J Bone Miner Res 30:2249–2261
Vyas S, Zaganjor E, Haigis MC (2016) Mitochondria and cancer. Cell 166:555–566
Wadhwa R, Priyandoko D, Gao R et al (2016) Stress chaperone mortalin regulates human melanogenesis. Cell Stress Chaperones 21:631–644
Wang FS, Lin CL, Chen YJ et al (2005) Secreted frizzled-related protein 1 modulates glucocorticoid attenuation of osteogenic activities and bone mass. Endocrinology 146:2415–2423
Wang FS, Ko JY, Yeh DW, Ke HC, Wu HL (2008) Modulation of Dickkopf-1 attenuates glucocorticoid induction of osteoblast apoptosis, adipocytic differentiation, and bone mass loss. Endocrinology 149:1793–1801
Wang FS, Ko JY, Weng LH, Yeh DW, Ke HJ, Wu SL (2009) Inhibition of glycogen synthase kinase-3beta attenuates glucocorticoid-induced bone loss. Life Sci 85:685–692
Wang FS, Wu RW, Ko JY et al (2011) Heat shock protein 60 protects skeletal tissue against glucocorticoid-induced bone mass loss by regulating osteoblast survival. Bone 49:1080–1089
Wang FS, Lian WS, Weng WT et al (2016) Neuropeptide Y mediates glucocorticoid-induced osteoporosis and marrow adiposity in mice. Osteoporos Int 27:2777–2789
Wu M, Gu J, Guo R, Huang Y, Yang M (2016) Structure of mammalian respiratory supercomplex I1III2IV1. Cell 167:1598–609 e10
Xie J, Zhu H, Guo L et al (2010) Lectin-like oxidized low-density lipoprotein receptor-1 delivers heat shock protein 60-fused antigen into the MHC class I presentation pathway. J Immunol 185:2306–2313
Yamamoto N, Tokuda H, Kuroyanagi G et al (2016) Heat shock protein 22 (HSPB8) limits TGF-beta-stimulated migration of osteoblasts. Mol Cell Endocrinol 436:1–9
Yamano K, Matsuda N, Tanaka K (2016) The ubiquitin signal and autophagy: an orchestrated dance leading to mitochondrial degradation. EMBO Rep 17:300–316
Zarrouk A, Vejux A, Mackrill J et al (2014) Involvement of oxysterols in age-related diseases and ageing processes. Ageing Res Rev 18:148–162
Zhang J, Zhou X, Chang H et al (2016) Hsp60 exerts a tumor suppressor function by inducing cell differentiation and inhibiting invasion in hepatocellular carcinoma. Oncotarget 7:68976–68989

Part VII
HSP60 in Human Health

Chapter 20
Role of HSP60 in Steroidogenesis and Reproduction

Sofia Olvera-Sanchez, Jessica Monreal-Flores, Oscar Flores-Herrera, and Federico Martinez

Abstract Heat shock proteins (HSP) play an important role in cellular homeostasis; however, they have recently been involved also in diverse relevant biological processes, like cellular proliferation and differentiation. It has been suggested that they could play an important role in some pathologies, such as cancer and atherosclerosis, making them an attractive therapeutic target. The high degree of HSP' conservation along evolution suggests that these proteins provide intrinsic defense mechanisms to cells against stress conditions. This chapter provides general information on heat shock proteins, as well as on their role in some pathophysiological processes related to reproduction. It also approaches the association of HSP with hormones and their participation in steroidogenesis, because hormones are critical mediators in regulating development, reproduction, and homeostasis in general. This chapter emphasizes the Hsp60, which has been suggested to have a relevant role in steroid hormones synthesis; besides, this protein is located in mitochondrial contact sites favoring progesterone synthesis in the human placenta.

Keywords Cholesterol transport · Human reproduction · Mitochondrial contact sites · Pregnancy · Steroidogenesis · Syncytiotrophoblast

Abbreviations

3β-HSD	3 beta hydroxy steroid dehydrogenase
ACTH	adrenocorticotropic hormone
AKT	protein kinase B (PKB, Akt)
APAF-1	apoptotic protease activating factor 1
APG-1 protein	ATP and peptide-binding protein in germ cells

S. Olvera-Sanchez · J. Monreal-Flores · O. Flores-Herrera · F. Martinez (✉)
Departamento de Bioquímica, Facultad de Medicina, Universidad Nacional Autónoma de México, Mexico City, Mexico
e-mail: fedem@bq.unam.mx

A. A. A. Asea, P. Kaur (eds.), *Heat Shock Protein 60 in Human Diseases and Disorders*, Heat Shock Proteins 18,
https://doi.org/10.1007/978-3-030-23154-5_20

BAD	Bcl2-associated agonist of cell death (Bcl2-L-8)
Bax	apoptosis regulator (Bcl2-L-4)
ERK	extracellular signal-regulated kinases
FSH	follicle-stimulating hormone
HDL	high-density lipoproteins
HspB1	heat shock protein beta-1
LH	luteinizing hormone
LPS	lipopolysaccharide
MLN64	metastatic lymph node 64 protein (STARD3)
P450scc	cholesterol side-chain cleavage enzyme
PI3K	phosphoinositide 3-kinase
PKA	protein kinase A
RAF kinase	rapidly accelerated fibrosarcoma kinase
SMAC	second mitochondria-derived activator of caspases
StAR	steroidogenic acute protein (STARD1)
TIM	translocase of the inner membrane
TNF	tumor necrosis factor
TOM	translocase of the outer membrane
TSPO	translocator protein
VDAC	voltage-dependent anion channel

20.1 Introduction

Heat shock proteins are a family of proteins that were originally associated with a specialized mechanism of cellular protection. Their "chaperone" activity contributes to avoid the structural and functional changes of certain proteins, conferring them stress resistance to maintain cell viability in the presence of chemical or physical agents, such as changes in temperature. These enzymes are called chaperones because their activity is to prevent the irreversible aggregation of non-native conformations of housekeeping proteins, thus maintaining their vital structure and function.

HSP are practically distributed in all cellular compartments and have been associated with several biological processes different to chaperonin function. The classification of HSP in different families involves among other aspects, specific structural characteristics, their cellular functions such as the transport of proteins to specific cellular sites, and the proliferation or cell signaling.

Through time, it has been observed that the function of a type of HSP was not linked to a single activity. Even more, it was reported that certain HSP could change their subcellular location in response to alterations produced by diseases or specific metabolic conditions. Different functions and diverse subcellular locations of the same HSP, suggests that the cell has acquired an adaptive strategy to maintain its integrity and survival with a limited number of proteins that have a broad spectrum of functions. Therefore, the term "moonlight proteins" is appropriate to describe the multiple functionality of the HSP. Additionally, the cellular requirements allow HSP to perform functions both inside and outside of cells, which makes them potential candidates as markers of physiological processes or diseases such as pregnancy or atherosclerosis, respectively. In this chapter, a review of the functions described for several HSP in the processes of fertilization and pregnancy is done, emphasizing on the role of Hsp60.

20.1.1 General Aspects of Heat Shock Proteins (HSP)

Heat shock proteins (HSP) are highly conserved from bacteria to eukaryotes (Seigneuric et al. 2011). In the 1960-decade, Ritossa (1996) described, for the first time, a group of proteins whose expression is regulated under stress conditions, like the increase of temperature. Posterior studies described other stress factors that also regulate the expression of HSP, such as UV radiation (Keyse and Tyrrell 1989), anoxia (Li and Shrieve 1982), ischemia (Rani et al. 2013), bacterial infections (Zhang et al. 2011), exposure to heavy metals (Burdon et al. 1982), and cancer (Whitesell and Lindquist 2005), among others. The first function described for HSP was proteins folding; however, currently, it has been described that they participate in diverse processes, such as proliferation, cellular differentiation and apoptosis (Omori et al. 2013).

The HSP are classified in six families according to their molecular weight: (1) Hsp110 (HspH), (2) Hsp90 (HspC), (3) Hsp70 (HspA), (4) Hsp60 (HspD), (5) Hsp40 (DNAJ), and (6) small Hsp (HspB); in turn, each family is constituted by various members (Seigneuric et al. 2011). The HSP are located in different cell compartments such as cytoplasm (Chun et al. 2010), mitochondrion (Miyata et al. 2013), endoplasmic reticulum (Miyata et al. 2013), and nucleus (Whitesell and Lindquist 2005). The human genome encodes four genes for the Hsp110 family (1, 2b, 3b, and 4b) (Kampinga et al. 2009), which is expressed practically in all vertebrates, although its characterization has been made mostly in mammals (mouse, hamster, and humans) (Levinson et al. 1980; Hightower 1980). It is localized mainly in the cytoplasm, whereas under stress conditions like heat shock, as well as in proliferative stage cells, the protein is localized in the nucleus (Easton et al. 2000). Two isoforms of Hsp110 have been identified: The APG-1 protein that is the product of *Hsp110-3* gene and protein APG-2, derived from *Hsp110-2* gene (Kampinga et al. 2009). Protein APG-1 is expressed in germ cells and has been suggested to play an important role in the maturation of these cells, as it increases their expression while maturation is underway (Kaneko et al. 1997a). Isoform APG-2 is mainly expressed in gonads, testicles, germ cells (Kaneko et al. 1997b), and, in contrast to APG-1, is not heat inducible (Yagita et al. 1999). Overexpression of APG-2 increases the cellular proliferation, and this action makes it the primary therapeutic target against carcinogenesis (Li et al. 2010).

Likewise, it has been described that Hsp110 is overexpressed in colorectal cancer, so that it has been suggested as a possible marker for this disease (Slaby et al.

2009; Kim et al. 2013). Other studies determined that the loss of expression of APG-2 causes fibrosis and cardiac hypertrophy, because denaturalized proteins are accumulated inducing loss of proteasome function, which indicates that these proteins are key elements in cell homeostasis (Mohamed et al. 2012).

The Hsp90 is the most abundant protein in the cytoplasm of eukaryote cells and corresponds to 1% of all soluble proteins under normal conditions (Omori et al. 2013). There are five isoforms encoded by five genes, being TRAP1 isoform a protein of mitochondrial localization, which has been proposed as an alternative for cancer treatment (Harris et al. 2004; Lettini et al. 2017). In contrast to the other chaperone proteins, the Hsp90 does not participate in the folding of newly synthesized; they rather interact with folded proteins in almost native state. Currently it is known that the Hsp90 regulates the maturation of more than 300 proteins (Rohl et al. 2013). The first proteins identified to interact with Hsp90 were the kinase Src 2 and 3 (Xu and Lindquist 1993). Later on, it was found that Hsp90 is essential in the activation of steroid hormone receptors (Joab et al. 1984), and that it participates in telomere maintenance (Woo et al. 2009), transduction of signals (He et al. 2012), vesicular transport (Taylor et al. 2010), immune response, viral infection, signaling of steroids (Pratt et al. 2006), and cancer development (Rohl et al. 2013). In this regard, it has been reported that Hsp90 is overexpressed in various tumors, suggesting its involvement in the regulation of processes related to cancer through its interaction and maturation of kinase proteins like AKT, the TNF receptor, and the transcription factor NF-κB (Becker et al. 2004; Cheng et al. 2012). In addition, overexpression of Hsp90 participates in the development of neurodegenerative diseases such as Alzheimer, promoting accumulation of protein aggregates, such as the Tau protein, which are toxic for nervous system cells (Luo et al. 2010).

The Hsp70 is inducible by thermal stress (Seigneuric et al. 2011), and participates in the folding of newly synthesized proteins, preventing their aggregation, has anti-apoptotic properties, and translocates target organelles to specific proteins (Shi and Thomas 1992; Gambill et al. 1993). The Hsp70 is of organelle-specific location according to its isoforms; Hsp70-5 is located in the endoplasmic reticulum, Hsp70-0 in mitochondria, and other six isoforms are located in the cytoplasm and nucleus (Vos et al. 2008). It has been suggested that Hsp70 plays a relevant role in cancer because its overexpression promote tumorigenesis and cell growth (Jaattela 1995; Nanbu et al. 1998). This effect is accomplished by the inhibition of the intrinsic and extrinsic apoptotic pathways when sequestering BAX in the cytoplasm and avoiding its translocation to the mitochondrion (Gotoh et al. 2004), thereby impeding its interaction with the apoptosome when recruiting APAF-1 (Saleh et al. 2000). It also binds to the apoptosis inducing factor, inhibiting it and producing chromatin condensation (Ravagnan et al. 2001; Murphy 2013).

The human genome encodes 49 genes for the Hsp40 family (Hageman and Kampinga 2009), which participates in the recruiting and binding of specific substrates for Hsp70 (Kelley 1998). Isoforms of Hsp40 have an important role in cancer development, through their role as co-chaperone with Hsp70 and Hsp90, which are overexpressed in breast and ovary cancers (Sullivan and Pipas 2002; Mitra et al. 2009; Sterrenberg et al. 2011). Elevated expression of Hsp40 has been associated with the increase in aggregation of the huntingtin protein (poliQ), inducing the formation of inclusion bodies and giving rise to Huntington's disease, which is characterized by the presence of muscular coordination disorders and cognitive deterioration (Wyttenbach et al. 2000).

The Hsp60 is encoded by gene HspD1 in chromosome 2q33.1, and it correlates with Hsp10, since the genes of both proteins are localized in the same chromosome and are placed contiguously (Hansen et al. 2003). Although expression of Hsp60 is constitutive, its overexpression is regulated by factors like high temperatures (42 °C) (Nadin et al. 2003), hypoxia (Chandra et al. 2007), sodium arsenite (Somji et al. 2000), or cadmium chloride (Somji et al. 2000). The Hsp60 is translated in the rough endoplasmic reticulum and is transported to the mitochondria, as it contains a signal peptide in the amino terminal region (Trapasso et al. 2008). However, it is also located in the cytoplasm, nucleus, plasmatic membrane and, in some cases, in the extracellular space (Soltys and Gupta 1996; Alard et al. 2007). As will be discussed later, the 60 kDa proteins take up different structural associations, they can be found as tetradecamer, tetramer, or dimer, apparently under some particular conditions of the cell, as for example during cell growth (Henderson et al. 2013). The Hsp60 is considered a "moonlighting" protein, as it has more than one biological activity, independently from the activity for which it was characterized. These functions can be divided as follows: (A) when the protein acts in the interior of the cell, (B) when Hsp60 acts as a ligand/receptor for other molecules in the plasmatic membrane, and (C) when it acts exogenously as an intercellular signal (Henderson et al. 2013). Some of its reported actions are binding to the surface of human cells (Hickey et al. 2010), it stimulates the synthesis of pro-inflammatory cytokines (Friedland et al. 1993), it binds to HDL in the cells surface (Bocharov et al. 2000), it stimulates the proliferation of vascular cells of the smooth muscle (de Graaf et al. 2006), it regulates apoptosis, as it can have and anti-apoptotic as well as a pro-apoptotic effect. It exerts its anti-apoptotic effect by forming a macromolecular complex at the cytosolic level with Bak and Bax, preventing apoptosis (Czarnecka et al. 2006; Deocaris et al. 2006); the pro-apoptotic effect is exerted by promoting maturation of the procaspase 3 (Arya et al. 2007), among many other described functions (Henderson et al. 2013). The Hsp60 plays an important role in regulating some types of cancer; while the increase in its expression inhibits apoptosis by interacting with Bax, which suggests its participation in cancer progression in neuroepithelial cells (Rappa et al. 2013). Other studies indicate that the interaction of Hsp60 with p53 also inhibits apoptosis, favoring survival of tumor cells (Ghosh et al. 2008). Besides, the interaction of Hsp60 with β-catenin promotes metastases (Tsai et al. 2009). It has been reported that Hsp60 is involved in cardiovascular diseases because it has been found at high concentrations in its soluble form in blood plasma (Xiao et al. 2005). Likewise, Hsp60 induces the production of pro-inflammatory cytokines promoting the development of atherosclerotic plaques. An association between hypercholesterolemia and high anti-Hsp60 antibodies levels has been reported, which has allowed suggesting that this protein possesses properties relating it with the distribution of cholesterol in the organism (Mandal et al. 2005).

Expression of small heat-shock proteins is induced by stress (Lund 2001) and they participate in diverse cellular processes such as signals transduction and translation. The HspB1 are considered anti-apoptotic proteins which interferes with the activation of procaspase 9 and 3 inhibiting the release of pro-apoptotic factors like cytochrome c and the Smac/Diablo complex. Because of this anti-apoptotic role, HspB1 can exert tumorigenic effects; besides, it has been found in some types of cancer (Sun et al. 2007a, b; Kase et al. 2009; Boncoraglio et al. 2012).

20.1.2 HSP in General Cell Functions

In a general way, several proteins exert functions or activities that are different from those originally assigned; therefore, they have been termed as "moonlighting proteins". It has been proposed that the human proteome is more complex than the genome. Thus, providing proteins with several activities becomes relevant, because the number of genes that produce proteins needed for the adequate control of cellular biological function is not enough (Henderson and Martin 2014). This information is based on data obtained from the human proteome, which contains 3×10^9 base pairs that produce 19,000 proteins, whereas the water flea *Daphnia pulex* has 0.2×10^9 base pairs coding for at least 31,000 genes. This has been interpreted as that the number of proteins expressed by the human genome is low to cover totally all the cell functions. Hence, the fact that a protein has several activities, which depend on its association with another protein(s) or the milieu in that it is found, allows the cells to respond adequately and to maintain their vital functions.

In this context, Hsp60 is one of the proteins that have been studied about its different activities and it has been termed as a moonlight protein (Jeffery 1999). The Hsp60s have been classified in two groups (Skjærven et al. 2015). Group I consisting of bacterial and mitochondrial variants whereas group II includes archaea and eukaryote variants of the cytoplasm. It has been suggested that the activities described as "moonlighting" are not due to a fusion of genes or multiple fragments of proteins, but to native form of the proteins; even, they have been named "multitasking proteins". On the other side, the alternate activities are not necessarily performed where their function was originally described. In several occasions, their functions are performed inside the cell in compartments where it is not frequent to find them, like the nucleus (Monaghan and Whitmarsh 2015) or outside cells, or associated to the plasmatic membrane (Ellis 1999; Jeffery 2017), as occurs in bacteria. In the latter case, some proteins of the bacterial surface that participate in infection, virulence, or invasion processes shown: (1) identity with intracellular proteins, like chaperons, and (2) some lack the signal peptides for their translocation to outside the cells (Wang and Jeffery 2016). As a whole, these data suggest that cells adopt mechanisms that allow them to conform to the different metabolic and regulatory conditions required to maintain their vital functions.

Similarly to what occurs in bacteria, in several eukaryotic cells Hsp60 and other chaperones can be localized in different compartment inside and outside cells. In a study with electron microscopy using different gold-marked antibodies against Hsp60, it was found that between 80% and 85% of the gold-marked antibodies were localized associated to mitochondria. However, the rest of the Hsp60 protein was detected, depending on the cell line, in the endoplasmic reticulum, vesicles, peroxisomes, and cell surface (Soltys and Gupta 1996, 1997). Recently, it was reported that Hsp60 could be transported to the plasmatic membrane via exosomes and the Golgi apparatus, because inhibition with brefeldin A, a specific inhibitor of the Golgi apparatus, diminishes significantly the presence of extracellular Hsp60 (Campanella et al. 2012). Results revealed that part of the Hsp60 extracellular release is associated with the formation of multivesicles that converge with lipid rafts where cholesterol concentration is high. Hence, it is proposed that part of the adverse effects of circulating Hsp60s or bound to plasma membranes could be due to this type of associations, as will be discussed further ahead (Fig. 20.1). Hsp60 becomes phosphorylated in spermatic cells to be translocated from the internal to the external membrane of the acrosome during capacitation in the mouse. These data suggest that Hsp60 in spermatozoids could play a relevant role in oocyte fertilization (Naaby-Hansen and Herr 2010).

In a review, Henderson et al. (2013) describe more than 50 different actions reported for Hsp60, which shows the versatility of this protein in modulating different activities in different cell locations. Likewise, it describes the complexity of these actions which differ from those assigned to them at the beginning of their discovery that opens a door for more detailed studies. Among the functions described in bacteria are aggregation, association with neurotoxins, formation of biofilms, association with toxins, and membrane adhesion. In mammalian cells, the functions presented by Hsp60 occurs both inside and outside of cells, which are classified in four groups. Among them are worthwhile mentioning the intracellular functions that differ from those originally described, i.e., association to ligands in the plasmatic membrane and signaling actions among cells. These functions as a moonlight protein are not only exclusive for Hsp60, but are for other HSP, in which pleiotropic functions have been described, like those described for Hsp90 (Hartl et al. 2011). There is even a specific database for moonlight proteins that describes the reported activities (Mani et al. 2015; Chen et al. 2018).

Although the Hsp60 functions have been associated with pathological processes, its presence has been demonstrated in the peripheral circulation of healthy individuals (Pockley et al. 1999). Results revealed that its concentration is 3.5 times higher in women than in men, suggesting that Hsp60 is necessary to fulfill different essential functions, as proposed for tolerance in the immune system. Up to the present, the role played by Hsp60 in human sera is unknown; however, some studies have related the presence of Hsp60 with some diseases. For example, this protein activates the innate and adaptive responses of the immune system, as well as its binding to membrane receptors participating in both pathological and immunity protection

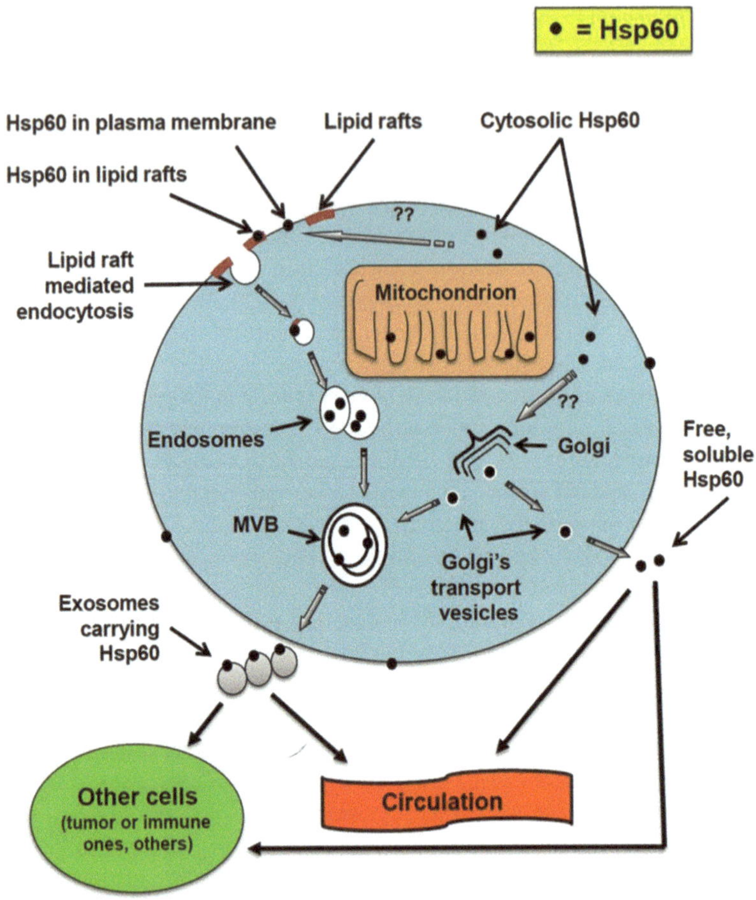

Fig. 20.1 Proposed Hsp60 secretion pathways in a tumor cell. Hsp60 (black dots) in normal cells localizes mainly in mitochondria, while in tumor cells it accumulates in the cytoplasm and, for unknown reasons (post-translational modifications?), it reaches the cell membrane and the Golgi. At the membrane, Hsp60 binds to lipid rafts internalized (endocytose) toward multivesicular bodies (MVB), later it is secreted via exosomes. In these, it is located in the membrane and probably also inside. Hsp60-loaded exosomes thus would reach other cells near and far through the circulation. The Golgi may also participate in Hsp60 secretion via transport vesicles moving to both MVB and the extracellular space. Hsp60 released in the extracellular space by Golgi vesicles (free Hsp60) can thus reach other cells in the vicinity and distant via blood circulation. (Figure is from Campanella et al. 2012)

mechanisms (Landstein et al. 2015). In this regard, the actions exerted by Hsp60 in the immune system with an intra- and extra-cellular location have been described; at the same time, it could be a marker of the disease process that would also allow assessing the effects of treatment on some diseases (Quintana and Cohen 2011). Two isomers of Hsp60 of *Mycobacterium tuberculosis* (Cpn60.1 and Cpn60.2) bind to different membrane receptors contributing thereby to the infectious process of mycobacteria (Cehovin et al. 2010). On the other side, Hsp60 can bind to certain proteins to contribute with a biological effect, as the protein ULBP2 that associates with Hsp60, which, in this case, has to be phosphorylated in the amino acid tyrosine (Leung et al. 2015). Similarly, Hsp60 associates with ATP synthase in mitochondria (Prasad et al. 1990), an association that has also been observed in endothelial cells between Hsp60 and the β subunits of the ATP synthase in membranes containing lipid rafts (Alard et al. 2011). Other diseases related to Hsp60 are cardiovascular disorders (Shamaei-Tousi et al. 2007), particularly, atherosclerosis where Hsp60 is associated with lipoproteins (Milioti et al. 2008).

Apoprotein-II (Apo-II) resulted to be an Hsp60 which binds to HDL. In this case, the plasmatic membrane-located Hsp60 would help in the recognition of HDL and apoproteins. The mechanism proposed for this process could occur through two ways; one is that the enrichment of glycines and methionines in the C-terminal end would facilitate their insertions in the lipid bilayer, favoring the transport of peptides inside and outside of cells, as occurs with apoproteins, that has been described to be under retro-endocytosis, which would generate an important mechanism in the reverse cholesterol transport route (Bocharov et al. 2000). The other mechanism proposed is related to changes between the cycles of folded protein and a molten globule at the membrane level, which modifies the capacity of Hsp60 to dissociate from the membrane and allow the lipid to leak out. Notwithstanding, the possibility of binding and acting with other proteins with greater impact, such as the participation in signaling cascades, is another possibility on how HSP accomplish their functions. For example, heat shock proteins 90 and 70 can bind to RAF kinases that are associated with mitochondria.

Given the role played by Hsp60 in several cell functions, both under normal and disease conditions, it has been suggested that Hsp60 could become a therapeutic target for some inflammatory type and infectious diseases or cancer (Meng et al. 2018). In the human brain, HSP are expressed constitutively and, in particular, are located in lipid rafts that are rich in cholesterol in association with the protein flotillin-1, which could contribute to the stabilization of membrane proteins or to signal transduction complexes (Chen et al. 2005). Besides, Hsp60s from different biological sources do not necessarily express the same actions. For example, the effect of the Hsp60 from Gram-negative bacteria stimulates bone resorption, whereas that from mycobacteria is unable to do so (Meghji et al. 2003). It was also observed, that mutants of the N-terminal had no effect. Apparently, the action mechanism is related with signal transduction processes in the presence of immunological system stimulation (Ranford and Henderson 2002). To this regard, leptin modifies the expression of Hsp60 in the hypothalamus, and this causes insulin resistance; the insulin itself is capable of inducing the presence of Hsp60, suggesting that the latter could be acting as an integrating molecule in the crosstalk between leptin and insulin (Kleinridders et al. 2013). The results of these authors revealed that the deficiency of Hsp60 by itself induces relevant changes in mitochondrial morphology.

To this regard, determinations have been made on which are the Hsp60 epitopes for its biological effects. For example, in the effect of Hsp60 in the activation of macrophages by LPS, a specific binding of Hsp60 for LPS was observed and that this binding lies in the region of clone 4B9/89 (aa 335–366). The important observation is that these peptides exert their effect without the need to modify the structure of heptamer or decatetramer, which is the functional conformation of Hsp60. Data on the binding to LPS showed a relation of approximately two molecules of LPS per mole of Hsp60. Results also revealed that when making an analogy between Hsp60 and GroEL of *E. coli*, the region that binds LPS would lie in the apical zone of Hsp60. In this zone, the amino acids of clone 4B9/89 do not interact with the intramolecular contacts nor with the adjacent monomers, suggesting that the binding site is accessible for LPS in both the monomeric and oligomeric states of Hsp60 (Habich et al. 2005). Besides, the recognition of Hsp60 for other proteins does not lie in the same region, so that the interactions become specific. Thus, the region that recognizes LPS is specific, since these authors characterized that the epitope that reacts with macrophages lies between amino acids 481 and 500 (Habich et al. 2004). In the case of diabetes type 1, a destruction of pancreatic cells that synthetize insulin has occurred due to an autoimmune action. Aiming at cancelling this activity, a peptide named DiaPep277 was synthetized from Hsp60, which covers the largest recognition region of T cells. This peptide is being investigated as a possible alternative in the initial treatment of diabetes type 1, which reveals the relevant role of Hsp60 in the physiology of the human organism (Tuccinardi et al. 2011).

20.1.3 Hsp60 in Mitochondrial Functions

The mitochondrion is a subcellular structure that performs multiple functions. Although, frequently it has been restricted to the production of energy in the form of ATP as its main function, its activities go far beyond this concept. For example, its participation in the events of apoptosis, calcium regulation, transformation of cholesterol into pregnenolone for the synthesis of steroid hormones, and its relation to mechanisms of cell signaling are some of the activities in which mitochondria participate actively in cell metabolism. The outer mitochondrial membrane participates in related signaling processes such as cell death, apoptosis, and regulated necrosis. It has even been proposed that mitochondria produce risk signals to alert against perturbations in cell homeostasis. In addition, it has been suggested that mitochondria are the center of a complex system of sensors that detect alterations in intracellular homeostasis, including oxidative stress, abstinence of the growth factor, and viral infection. Mitochondria not only decode input risk signals, activating adaptative responses or promoting cell death, but they also emit risk signals that alert the cell against mitochondrial stress conditions. The additional mitochondrial factors that can act as risk signals include, among others, ATP, cardiolipin, SMAC, Hsp60, β-subunit of the F_1-ATPase. Both the Hsp60 and the β-subunit of the F_1-ATPase are directed by antibodies that contribute to (or at least accompany) some diseases like atherosclerosis and vasculitis. Interestingly, an ectopic variant of the β-subunit of the F_1-ATPase can act as a receptor for the coupling of Hsp60 in the surface of endothelial cells, which contributes to the action mechanism of Hsp60

(Galluzzi et al. 2012). In this sense, it has been suggested that this association could favor the phosphorylation of antigens to special T cells (Mookerjee-Basu et al. 2010); it would have to be evaluated whether this could be the mechanism used by other cell lines.

As mentioned upper, the location of HSP are not restricted to mitochondria, which opens a wide range of possibilities in which HSP participate, such as cell signaling cascades, among other actions described in the former section (Gupta et al. 2008). In this regard, Hsp60 and the disulfide isomerase are required by the hepatic growth factor as inducer or ERK activator and cell transformation to hepatocellular cancer (Lin et al. 2016). Some kinases associated with mitochondria, like RAF, could interact with HSP (Galmiche and Fueller 2007). In this way, the RAF kinase could phosphorylate the BAD protein, promote indirectly the phosphorylation of two other kinases, as PKA and PKB, and could recruit a small but functionally relevant group of MEK, which reveals the value of the relations of this kinase with HSP. Particularly, it was demonstrated that the overexpression of the domain 283–288 of Hsp60 increases the levels of the subunit 110 of PI3K. A similar mechanism was observed for the induction of c-Myc expression. Besides, an increase in the levels of Akt phosphorylation was also observed (Tsai et al. 2008; Yan et al. 2015). These results show that Hsp60 participates in cell signaling processes.

20.1.4 The Role of Hsp60 in Steroidogenesis, Generalities

As described in the previous sections, HSP are implicated in several events like folding and translocation of mitochondrial proteins, assembling of oligomeric proteins, rupture and formation of supramolecular complexes associated with transport, intracellular catabolism of proteins, signal transduction mechanisms, regulation of genetic expression, cell protection in stress conditions, among others. The HSP constitute a complex system, as they interact physically among themselves forming multifunctional complexes with other proteins, whose joint action allows modulating or regulating the transport through membranes. In this sense, incorporation of specific proteins at the level of mitochondria becomes relevant because most proteins that constitute them are encoded in the nucleus and are synthetized in the cytoplasm with a pre-sequence or a signal peptide that allows their transportation into mitochondria. This process requires complex systems, like the TIM and the TOM, where HSP are required (Marom et al. 2011). For example, for the incorporation of the translocator of adenine nucleotides into the inner mitochondrial membrane, the presence of Hsp60 is indispensable (Mahlke et al. 1990). In this case, it has been proposed that contact points are formed among mitochondrial membranes that recognize the signal peptide and, once localized in the matrix, it is processed undergoing a retrotranslocation by interacting with Hsp60 toward the internal membrane to exert its functions in the transport of ADP/ATP. Another example is the incorporation of adrenodoxin, which has been shown to require a 60 kDa-chaperonin for its incorporation into mitochondria (Grunau et al. 1995). Although HSP perform multiple functions, the studies carried out in relation to steroidogenesis have been limited. Probably, several of the alterations observed respect to HSP could be due to direct alterations in the steroidogenic processes, in the modulation or expression of enzymes related to cholesterol transport and transformation, or rather due to alterations of receptors associated with steroidogenesis, aside from the already demonstrated effects. In this way, steroids synthesis would be affected indirectly; however, future research is needed to determine the role of HSP in different physiological processes and/or diseases, and to relate them to the production of steroid hormones.

20.1.5 Steroidogenesis

Steroidogenic tissues transform cholesterol in steroid hormones, like the suprarenal glands, ovaries, testicles, and placenta, being mitochondria the site where this transformation occurs (Papadopoulos and Miller 2012). The sterol hormones are critical mediators in processes such as regulation of development, reproduction, and homeostasis, in general. Under physiological conditions, steroidogenic cells have low amounts of steroid hormones; however, when there is an external stimulus (ACTH, LH, and FSH) large amounts of glucocorticoids (cortisol, corticosterone), mineralocorticoids (aldosterone), progestins (progesterone), androgens (testosterone, dihydrotestosterone) or estrogens (estradiol) are produced (Miller 2013; Issop et al. 2013). Synthesis of steroid hormones starts in mitochondria, where cholesterol is transferred from the cytoplasm to the outer membrane, and then to the internal mitochondrial membrane, where it is metabolized to pregnenolone by cytochrome P450scc, and finally transformed into progesterone by the 3β-HSD. There are two isoforms of enzyme 3β-HSD: type II localized in the smooth endoplasmic reticulum and expressed in the suprarenal glands and gonads; and type I, localized in the mitochondrion and expressed in the placenta (Sanderson 2006; Miller and Bose 2011; Elustondo et al. 2017).

20.1.6 Cholesterol Transport in Mitochondria Is Mediated by HSP-Associated Proteins

Steroidogenesis can be modified by altering some of the steps that participate in its production. For example, it has been proposed that VDAC contributes to cholesterol transport by forming multiprotein complexes with the TSPO and the StAR at mitochondrial binding sites, which allows cholesterol to reach cytochrome P450scc (Campbell and Chan 2008; Graham 2015). At least 700 proteins are capable of binding cholesterol; among them are receptors, enzymes, and transport proteins like VDACs and several HSP, like 10, 70, and 90 (Hulce et al. 2013). Several of the proteins that bind cholesterol are not directly related to its metabolism and/or transformation, revealing the relevance of cholesterol in cell functions. On the other side, associations between proteins and HSP could be a factor to promote some cell activities, as more than 150 proteins bind to Hsp60 in *Methanosarcina mazei* (Hirtreiter et al. 2009). Based on the aforementioned paragraphs, it could be suggested that the interaction of proteins with HSP is necessary for a biological activity to occur, as is the case of membrane recognition of some T lymphocytes or the association with proteins for their transport to mitochondria. To this regard, results from our laboratory suggest that there are several proteins involved in cholesterol transport in the human placenta (Navarrete et al. 1999; Espinosa-Garcia et al. 2000; Flores-Herrera et al. 2002). It was observed that in mitochondria isolated from the placenta and treated with trypsin, synthesis of pregnenolone is not stimulated in the presence of exogenous cholesterol, whereas synthesis of pregnenolone in trypsin-treated mitochondria was stimulated in the presence of 22-OH cholesterol, a metabolite that crosses freely the membranes. These data show that there are proteins of the external mitochondrial membrane that participate in cholesterol transfer to cytochrome P450scc located in the internal mitochondrial membrane (Espinosa-Garcia et al. 2000; Martinez et al. 2015).

20.1.7 Participation of HSP in Receptors for Steroidogenesis

Other examples are the reports that suggest that Hsp40 and Hsp70 participate in folding of the receptor for glucocorticoids, which is associated with the synthesis of steroid hormones. On the other side, Hsp90 regulates the activation and translocation of the GR to the nucleus, where it modulates the expression of target genes involved in the synthesis of steroid hormones (Mangelsdorf et al. 1995; Grad and Picard 2007). In this same sense, mitochondria make contact with the endoplasmic reticulum forming associations named mitochondria-associated ER membrane (MAM), where some proteins are processed for their incorporation to mitochondria. Protein StAR is associated with the glucose regulatory protein 78 (Grp78), which is a chaperone of the endoplasmic reticulum that allows protein StAR to perform its function of cholesterol transport into the mitochondrion (Prasad et al. 2017). Knockdown of Grp78 resulted in an important diminution of StAR activity, suggesting that this protein is a regulator of steroidogenesis at the MAM's level.

A direct effect of some HSP has been shown on steroidogenesis. For example, overexpression of Hsp70 inhibits the synthesis of steroid hormones in luteal cells. It has been suggested that the action mechanism by which Hsp70 inhibits steroids' synthesis is by interfering with cholesterol translocation to the mitochondrial cytochrome P450scc (Khanna et al. 1994). Similarly, testosterone synthesis in a model of cryptorchidism in mice diminished significantly along time. Results revealed that the presence of Hsp70 in Leydig cells was constant, whereas Hsp60 was detected in both germinal and Leydig cells associated with an increase in the heat shock factor 1 and a diminution in the StAR protein, with the consequent diminution in testosterone production (Oka et al. 2017). Authors proposed that the heat shock factor 1 regulates testicular steroidogenesis by stabilizing cholesterol transport through StAR.

In steroidogenic tissues, such as the corpus luteum of rats and the suprarenal cortex, incubation of these cells during 10 min at 45 °C induce the synthesis of a 70-kDa protein, which was identified as a heat shock protein. It was also found that this protein modifies the levels of progesterone sensitive to the LH. However, the mechanism by which these HSP modify the synthesis of steroids is complex. Besides inhibiting the synthesis of progesterone when temperature increases, steroidogenesis is blocked in response to 8-bromo-cAMP and forskolin. On the other side, the heat shock increases eight to ten times the synthesis of progesterone with respect to control cells. Authors suggest that one of the effects could be the modification of cholesterol transport at the mitochondrial level (Khanna et al. 1994, 1995a, 1995b).

20.1.8 Role of Hsp60 in the Steroidogenic Process

Although, it has been shown that Hsp60 has diverse effects on cell functions, its repercussion on blood circulation has been associated to several diseases. An association between high levels of Hsp60 and hypercholesterolemia has been reported in the serum of atherosclerotic patients (Ohashi et al. 2004), suggesting that Hsp60 can play a relevant role in the regulation of cholesterol distribution in the organism. It has also been reported that heat of the summer produces stress in bovines increasing infertility. In this sense it has been reported that all HSP increase, particularly Hsp60 increases 2.6 times, and Hsp70 type 3 was the highest with a 4.9-times increase. Although levels of StAR and P450scc diminished, this did not affect the synthesis of progesterone, but 17β-estradiol did, which, in turn, induced apoptosis. There is no explanation for this lack of effects on P4 synthesis (Li et al. 2016). Similarly, overexpression of Hsp60 in Leydig cells induced by human chorionic gonadotropin and the follicle stimulating hormone is associated with the increase in testosterone secretion and regulation of steroidogenesis (Miller 2013; Issop et al. 2013).

As mentioned, association of Hsp60 with other proteins could be a mechanism for cholesterol transport. In MA-10 mouse tumor Leydig cells, as well as in HeLa cells, the possible participation of Hsp60 in the incorporation or insertion of the translocator protein (TPSO) was determined in the outer mitochondrial membrane. TPSO has been associated to the cholesterol transport in these cells. However, contradictory results were reported where the Hsp60 was identified as a soluble form in the mitochondrial matrix and not associated to TPSO, whereas TPSO bound specifically to Hsp70 (Rone et al. 2009). The effect on steroidogenesis is not clear, since by using MA-10 cells subjected to heat stress, where Hsp70 levels increased between 6 and 20 times, depending on the treatment time, the inhibition of steroidogenesis was induced, without affecting the P450scc or 3β-HSD activities; however, it was suggests that the observed effect was due to a diminution in the expression of the StAR protein (Liu and Stocco 1997; Murphy et al. 2001). One of the effects of Hsp60 would be on the signaling mechanisms. To this regard, it was demonstrated that RhoA, a GTPase of the family of Rho GTPases, participates in the transmission of signals for diverse cell processes, including the female reproductive system. Deletion of RhoA in the corpus luteum of mice reduced significantly the mitochondrial Hsp60, among others (El Zowalaty et al. 2017). It was proposed that RoA deletion produced important changes in the cell's cytoskeleton, causing accumulation of lipid droplets, and avoiding mitochondrial fusion/fission processes, thereby impeding steroidogenesis by decreasing mitochondrial Hsp60.

20.1.9 Hsp60 and the StAR Effect

In follicular cells of the ovary, the induction of protein StAR synthesis was accompanied by an increase of Hsp60 and several proteases when stimulated with hCG (Bahat et al. 2014). However, transfection of HEK293 cells with StAR and their overproduction did not modify the synthesis of Hsp60. Transfection with the whole electron transport chain of P450scc, which is needed for cholesterol transformation into pregnenolone, increased significantly the expression of Hsp60 with the concomitant hormonal increase (Monreal-Flores et al. 2017). As mentioned, it has been proposed that part of the mitochondrial proteins involved in steroidogenesis is associated to contact sites. In our laboratory, we isolated this mitochondrial contact sites from the human placenta (Uribe et al. 2003). They contained the proteins of the cytochrome P450scc chain, several HSP, among them Hsp90, Hsp40, and Hsp27, and Hsp72 at a lower concentration, as well as Hsp60 (Olvera-Sanchez et al. 2011; Monreal-Flores et al. 2017). These contact sites were defined as steroidogenic contact sites, since their capacity to transform cholesterol into progesterone in the presence of isocitrate and $NADP^+$ without the addition of exogenous cholesterol (Uribe et al. 2003). The results allowed demonstrate that mitochondrial integrity is not necessary to perform steroidogenesis, that cholesterol concentration of mitochondrial membranes as substrate was enough and that the proteins present in these contact sites let to transport cholesterol for steroidogenesis.

Complementarily, in human placenta mitochondria, we identified the Hsp60 using an antibody that recognizes the cholesterol binding domain (called the START domain) to the protein MLN64, which is located in the C-terminal region of the protein and has been involved in cholesterol transport to mitochondria in steroidogenic cells (Fig. 20.2). Accordingly, amino acid sequence alignment of Hsp60 with the START domain of the StAR and MLN64 showed an identity of 18.6% and of 26.1%, respectively (Fig. 20.3). This result allowed suggesting that Hsp60 presents a common region with the START domain in the C-terminal. To this regard, it has been proposed the Hsp60 could be associated with the movement of cholesterol among mitochondrial membranes of the human placenta (Olvera-Sanchez et al. 2011). Based on these antecedents, the role of Hsp60 in mitochondrial steroidogen-

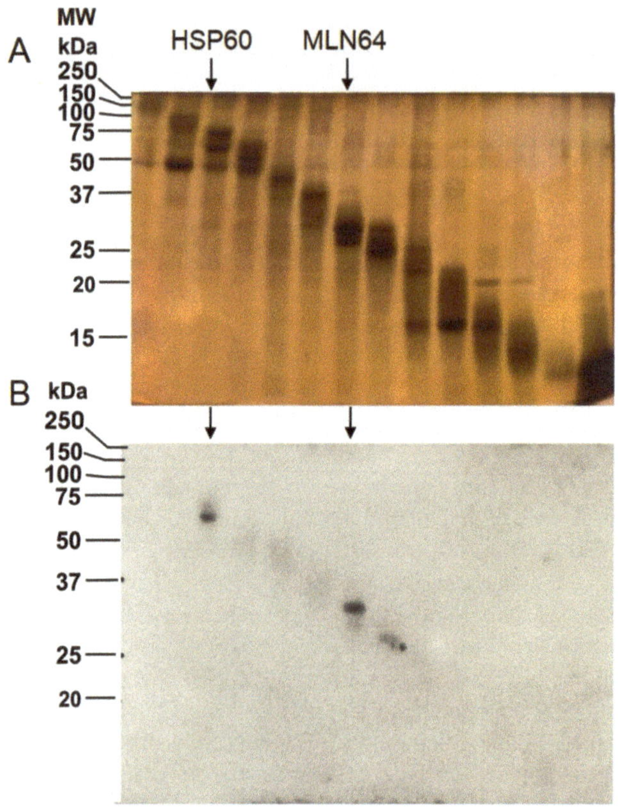

Fig. 20.2 Electroelution of proteins SDS-PAGE and Western blot analysis. The syncytiotrophoblast mitochondria were incubated for progesterone synthesis and recovered by centrifugation. Protein from mitochondria was separated by SDS-PAGE, electroeluted and separated again by SDS-PAGE, and stained with silver (**a**). The proteins were electro transferred to PVDF membranes and blotted against MLN64 (**b**) bu. using a rabbit anti-MLN64 antibody raised against recombinant MLN64 protein recognizing the START domain (ABR Inc., USA). Arrows indicate the 60 and 30 kDa positions on the MLN64-like protein

esis in the human placenta was characterized. Previous studies demonstrated that the human Hsp60 has a Cys-442 in its active site, which is crucial for its chaperone activity. Both *in vitro* and *in situ* studies revealed that N-ethylmaleimide and ETB (epolactaene tertiary butyl ester), a metabolite derived from epolactaene that binds to Cys-442, inhibit the activity of Hsp60 (Nagumo et al. 2005; Lin et al. 2016). The afore mentioned allowed using them to determine the Hsp60 activity in mitochondria of the placenta.

Steroidogenesis from human syncytiotrophoblast mitochondria was inhibited by fluorescein-5-maleimide and *N*-ethyl maleimide, similar to observed in the JEG-3 cells model (Olvera-Sanchez et al. 2011). Given that these maleimides bind to cysteines 442 and 447, inhibiting the Hsp60 activity (Nagumo et al. 2005; Lin et al.

```
MLN64           ----------MSKLPRELTRDLERSLPAVASLGSSLSHSQSLSSHLLP------------ 38
HSP60           MLRLPTVFRQMRPVSRVLAPHLTRAYAKDVKFGADARALMLQGVDLLADAVAVTMGPKGR 60
                          *  :.* *: .* *: .  ..:*:.         . .**.
MLN64           ---------PPEKRRAISDVRRTFCLFVTFDLLFISLLWIIELNTNTGIRKN------LE 83
HSP60           TVIIEQSWGSPKVTKDGVTVAKSIDLKDKYKNIGAKLVQDVANNTNEEAGDGTTTATVLA 120
                         .*:  :    * ::: *  .:. :  .*:  :  ***     ..      *
MLN64           QEIIQYNFKTS---------------FFDIFVLAFFRFS--------------------- 107
HSP60           RSIAKEGFEKISKGANPVEIRRGVMLAVDAVIAELKKQSKPVTTPEEIAQVATISANGDK 180
                :.* : .*:.                .* .:  : : *
MLN64           --GLLLGYAVLRLRHWWVIALLSKGAFGYLLPIVS---FVLAWLETWFLDFKVLPQEAEE 162
HSP60           EIGNIISDAMKKVGRKGVITVKDGKTLNDELEIIEGMKFDRGYISPYFINTSKGQKCEFQ 240
                  * ::. *: :: :  **:: .  ::.  * *:.   *  .::..:*:: .   :   :
MLN64           ERWYLAAQVAVARGPLLFSGALSEGQFYSPPESFAGSDNESDEE-------------VAG 209
HSP60           DAYVLLSEKKISSIQSIVP-ALEIANAHRKPLVIIAEDVDGEALSTLVLNRLKVGLQVVA 299
                : : * ::  ::    :.. **. .: :  *  : ..* :.:                *..
MLN64           KKSFSAQEREYIRQGKEATAVVDQILAQE------ENWKFEKNNEYGDTVYTIEVPFHG- 262
HSP60           VKAPGFGDNRKNQLKDMAIATGGAVFGEEGLTLNLEDVQPHDLGKVGEVIVTKDDAMLLK 359
                 *: .  :..  :  . * *. . ::.:*      *: : .. .: *:.: * : .:
MLN64           ----KTFILKTFLPCPAELVYQEVILQPERMVLWNKTVTACQILQRVEDNTLISYDVSAG 318
HSP60           GKGDKAQIEKRIQEIIEQLDVTTSEYEKEKLNERLAKLSDGVAVLKVGGTSDVEVNEKKD 419
                    *: * * :     :*         : *::     .::     : :* ..: :. : . .
MLN64           ------------------AAGGVVSPRDFVNVRRIERRRDRYLSSGIATSHSAKPPTHKY 360
HSP60           RVTDALNATRAAVEEGIVLGGGCALLRCIPALDSLTPANEDQKIGIEIIKRTLKIPAMTI 479
                                  .** .  * :  :  :   .:    .    .:: * *: .
MLN64           VRGENGPGGFIVLKSASNPR-------VCTFVWILNT---DLKGRLPRYLIHQSLAATMF 410
HSP60           AKNAGVEGSLIVEKIMQSSSEVGYDAMAGDFVNMVEKGIIDPTKVVRTALLDAAGVASLL 539
                .:. .  *.:** *  ...        .  ** :::.   * .  :   *:. : .*:::
MLN64           EFAFHLRQRIS------ELGARA----------- 427
HSP60           TTAEVVVTEIPKEEKDPGMGAMGGMGGGMGGGMF 573
                  *  :  .*.      :** .
```

Fig. 20.3 Amino acids alignment of MLN64 and Hsp60. Sequences were obtained from https://www.uniprot.org/uniprot/. Amino acids sequences of Metastatic lymph node gene 64 protein (MLN64 or StARD3; UniprotKB-Q14849) and Heat shock protein 60 (Hsp60; UniprotKB-P10809) were aligned with BioEdit CLUSTALW multiple sequence alignment program. (∗) Identical residues, (:) Conserved residues, (•) Semi-conserved residues. The START domain of MLN64 is in yellow

2016), the interpretation was that the mitochondrial Hsp60 of the human placenta is directly related with the transport of cholesterol. Furthermore, the presence of maleimides did not modify the oxidative phosphorylation in these mitochondria, confirming that inhibition of progesterone synthesis is related with blocking the active cysteines of Hsp60 (Olvera-Sanchez et al. 2011). On the other side, immunoprecipitation assays with antibodies against MLN64 allowed identifying Hsp60, supporting the proposal that there is an association between both proteins (Monreal-Flores et al. 2017). Similar results were obtained with immunohistological assays in cotyledons of human placenta, demonstrating that Hsp60 and MLN64 co-localize in syncytiotrophoblast cells (Olvera-Sanchez et al. 2011). This suggests that Hsp60 can present a new activity within the steroidogenic metabolism of the human placenta (Papadopoulos and Miller 2012).

Confirmation of Hsp60 participation in steroidogenesis of the placenta was obtained with the overexpression of Hsp60 in choriocarcinoma cells (Monreal-Flores et al. 2017). Furthermore, a reproducible model of non-steroidogenic cells (HEK293) transformed into steroidogenic with the transfection of P450scc system, and the protein 3β-HSD confirmed that overexpression of Hsp60 increased the

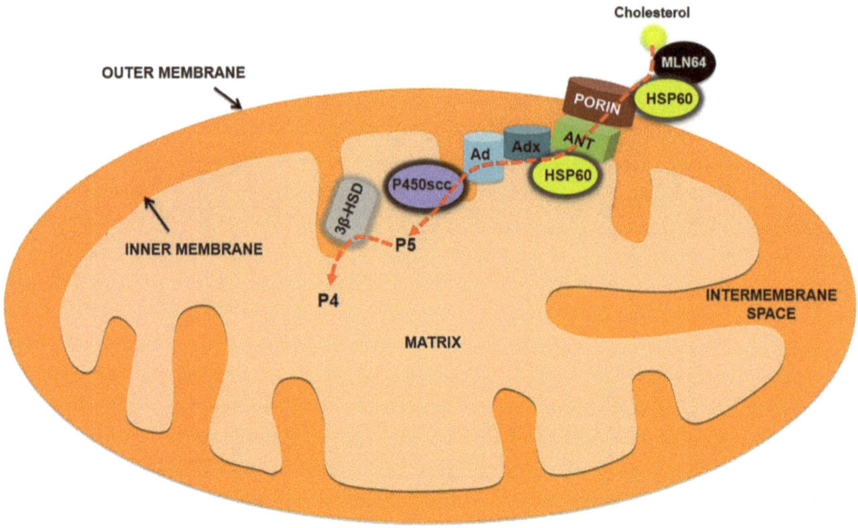

Fig. 20.4 Proposed participation of HSP in the placental steroidogenesis. The Hsp60 is located in both sides of the mitochondrial membranes. An association of Hsp60 with the MLN64 protein is required to transport the cholesterol to outer mitochondrial membrane, although experimental research will be necessary to determine the molecular mechanism. Also, the presence of Hsp60 inside the mitochondria contribute to present the cholesterol to the electron transport chain coupled to cytochrome P450scc. ANT, adenine nucleotide translocase; Adx, adrenodoxin reductase; Ad, adrenodoxin; P5, pregnenolone; P4, progesterone; 3β-HSD, 3 beta hydroxy steroid dehydrogenase

transformation of cholesterol into progesterone (Monreal-Flores et al. 2017). Consistent with these data, in vitro assays demonstrated that the recombinant His10-Hsp60 protein favored progesterone synthesis in mitochondria isolated from JEG-3 cells, confirming its participation in steroidogenesis. Immunoprecipitation also revealed that Hsp60 is associated to cytochrome P450scc and to MNL64, suggesting that a multiprotein is formed to accomplish the transformation of cholesterol into progesterone (Monreal-Flores et al. 2017) (Fig. 20.4).

Since there are antecedents suggesting that Hsp60 can bind to cholesterol-rich regions, like lipid rafts (Ohashi et al. 2004) or studies in which Hsp60 has been involved in cholesterol metabolism at a systemic level and its association to atherosclerosis (Bocharov et al. 2000; Mandal et al. 2005; Ohashi et al. 2004), the cholesterol binding capacity of Hsp60 was investigated as a possible action mechanism of Hsp60 in steroidogenesis. Data from our laboratory (Monreal-Flores et al. 2017) showed that Hsp60 could bind cholesterol similarly to what has been reported for the StAR protein in MA10 cells (Roostaee et al. 2009). As a whole, data suggest that, in the placenta, Hsp60 plays a relevant role in steroidogenesis, aside from the folding functions associated with mitochondria (Monreal-Flores et al. 2017). Although, Hsp60 is an important protein in mitochondrial cholesterol transport in the human placenta, the mechanism by which this effect is accomplished must still be determined.

20.1.10 Participation of Hsp60 in Obstetric and Developmental Alterations

Based on the shown information, the detailed study about HSP and their participation in reproductive biology is evident, as well as to find, as has been suggested, markers and standards that will allow to monitor pregnancy in a safe and objective manner. Studies have even made to establish possible associations with some obstetric and gynecological alterations. In general, it has been reported that HSP are the first proteins to be produced in in vitro culture of embryos, and that they are immunodominant antigens of some pathogens like *Chlamydia trachomatis*, which have been associated to infertility problems (Neuer et al. 1999). The presence of practically all HSP has been described in the endometrium along the menstrual cycle. Of these, Hsp60 shows variation during that cycle, increasing during the proliferative phase and decreasing during the secretory phase. It has been proposed that the presence of HSP, in particular of Hsp70, could avoid the cytotoxic effects produced by TNFα, which induces activation of phospholipase A2 with the subsequent activation of mediators that induce inflammation and mitochondrial formation of reactive oxygen species (Tabibzadeh et al. 1996; Tabibzadeh and Broome 1999). Finally, it is proposed that these modifications can be modulated, like the menstrual cycle, by the effects of steroid hormones. Apparently, in the oviduct, epithelial cells release Hsp60 and Grp78, which have the capacity of binding to spermatic cells. This relation favors the viability, mobility, and integrity of the acrosome. The presence of added Hsp60 modulated spermatic motility, tyrosine phosphorylation, and intracellular calcium concentration during capacitation of spermatozoids; although the mechanisms to produce these effects are still unknown (Lachance et al. 2007).

Therefore, it is convenient to mention that pregnancy is a fundamental biological process to ensure the subsistence of humans. Efforts have been made to count upon better indicators to provide information on the advancement and state of the fetus and the mother during pregnancy. In this sense, with the help of modern technologies, both the placenta and its cells have been studied to know their proteome and gather information that will allow distinguishing between normal pregnancy and pre-eclampsia. A difference of 34 proteins has been reported between the cells isolated from the trophoblast of normal placentas and those with pre-eclampsia, of which, seven spots were identified. Expression of the following proteins diminished, disulfide isomerase ER-60, peroxiredoxin 2, and $\Delta^{3,5}$-$\Delta^{2,4}$-dienoyl-CoA isomerase; whereas four proteins (protein disulfide isomerase precursor, endoplasmic reticulum resident protein, dihydrolipoyl dehydrogenase, and TIM21-like protein) increased significantly (Sun et al. 2007a, b). Data suggest that mitochondria could play a relevant role in pre-eclampsia. In complete placentas, more than 1000 spots were observed, of which 650 coincided, allowing to differentiate 21 spots that showed a 140% increase or a 50% diminution. The identified proteins were divided in the following groups; structural, antioxidant, and detoxicant, related to stress, apoptosis, $NADP^+$ regeneration through the enzyme isocitrate dehydrogenase, glycolytic and immunomodulator proteins. Of the proteins that increased their presence in the placentas of patients with pre-eclampsia are: Hsp60, catepsin D, GST, VDAC, and ERp29 (Kim et al. 2007). In a similar study, 2636 proteins were detected, of these, 171 differed between normal placentas and those with pre-eclampsia, among them stand out peroxiredoxin, cytochrome P450scc, and 3-βHSD (Wang et al. 2013). Finally, in another study, using normal and placentas from patients with pre-eclampsia, 110 proteins were characterized, among them were Hsp27, 60, and 70, as well as peroxiredoxin 1 and 2 (Mine et al. 2007).

Using placental trophoblasts obtained through laser capture micro-dissection, 962 deficient proteins were observed in patients with pre-eclampsia. Proteins identities were classified in the following groups: membrane (17%), nucleus (15%), cytoplasm (14%), and mitochondrion (11%). Of these, the proteins of interest that were classified are: the 14-3-3, which participates in signal transduction processes, the heat shock protein 27, and anexins II and V, which are involved in the response to stress (Jin et al. 2008). In another study using complete placentas, 1275 proteins were identified, in which 520 different spots were detected, 17 spots were expressed differentially, and 11 proteins were characterized. Four proteins increased: chloride intracellular channel 3, apolipoprotein A-I, transthyretin, and protein disulphide isomerase. Seven diminished their expression in the placenta from patients with pre-eclampsia: peroxiredoxin 2, peroxiredoxin 3, Hsp70. Cu/Zn-superoxide dismutase, actin gamma 1 propetide, chain A of enoyl-coenzyme A hydratase, and Hsp-gp90 precursor (Gharesi-Fard et al. 2010). Using the central core portion of the placenta, changes in Hsp27, Hsp70, vimentin, and peroxiredoxin II, among others, were considered relevant (Shin et al. 2011; Yang et al. 2015). Similar results were obtained in isolated mitochondria from human placenta, where Hsp10 and Hsp70 increase their expression whereas the cytochrome P450scc diminishes. Results suggest that mitochondria are related to pre-eclampsia (Shi et al. 2013). Based on this information, Hsp70 has been proposed as a possible marker of pre-eclampsia (Cuffe et al. 2017).

In a study, in which the presence of Hsp60 was particularly determined in control patients and in patients with pre-eclampsia, the values of serum antibodies against Hsp60 and Hsp70 were not different between both studied groups, suggesting that they could be the normal antibodies occurring in humans (Molvarec et al. 2009). Although results from proteomic studies show differences among them, it is clear that the stress conditions in which the placenta is during pre-eclampsia would allow suggesting that Hsp60 could become a marker or indicator of the damage occurring in that disease. In spite of all the previous data and the importance of Hsp60 in the different metabolic processes described, efforts are underway to find a possible relation between the presence of Hsp60 and certain obstetric pathologies. It has been reported that sera obtained from both control pregnant women and the umbilical cord of neonates presented anti-HSP antibodies, concluding that these are part of the natural antibodies, being part of a particular maternal response with the production of IgA antibodies, and fetal response with IgM antibodies. Authors suggest that the natural autoimmunity during pregnancy starts in the uterus (Merbl et al. 2007). For example, in mice, a relation between the increase in anti-Hsp60 and the arrest of in vitro embryos was observed with respect to controls (Neuer et al. 1997).

Similar results were obtained in women with *Chlamydia trachomatis* infections, in which high levels of anti-Hsp60 were related with a diminution in in vitro fertilization (Witkin 1999), suggesting that this type of infection could be part of the infertility problems, or even of abortions (Eggert-Kruse et al. 2014). In endometriosis, Hsp60 levels increase in response to activation of macrophages, T lymphocytes, and the release of cytokine kinase, which can lead to infertility (Kligman et al. 1996).

Likewise, the stress during pregnancy gives rise to increases in HSP levels, being Hsp60 which showing the highest increment during fetal hypoxia (Hromadnikova et al. 2015), suggesting that determination of the circulating levels of the mRNA of HSP as biomarkers could be a non-invasive control measure to avoid complications during pregnancy, like the premature rupture of membranes. In the premature rupture of membranes and in the preterm labor with intact membranes, an increase in both Hsp60 and Hsp27 was observed (Dvorakova et al. 2017). Likewise, the formation of antibodies against Hsp60 and Hsp70 seem to modify the functions during pregnancy, leading to preterm births (Ziegert et al. 1999). Alterations occurring during pregnancy also affect the fetus. For example, the presence of IgG and IgM antibodies against Hsp60 was demonstrated in the sera of small fetuses according to their gestational age, being IgM antibodies the only one generated by the fetus. IgM antibodies against Hsp60 increased significantly in small fetuses as compared to controls, thus, it was suggested that this response could risk the life of fetuses, as occurred in three of the clinical cases studied (Belhia et al. 2010). Similarly, in patients with recurrent pregnancy losses, the levels of anti-Hsp60 and 70 antibodies increase, apparently associated with maternal vascular disorders (Matsuda et al. 2017). There are even evidences that the high levels of anti-Hsp70 could also be associated with other diseases during pregnancy (Molvarec et al. 2010). Interestingly, immunoidentification and distribution of HSP in the placenta shows changes in their concentration and a particular distribution along the pregnancy (Shah et al. 1998).

The presented information shows that HSP have multiple functions during pregnancy and there is interest in determining the effects produced by their presence in different sites inside and outside of cells, as well as the immunological reactions that become associated with infections during gestation and their repercussion on the development of the fetus. It is clear, that research has been aimed at searching for the possible role that these proteins would have in the circulation and their antibodies, and even of other proteins that are modified by the presence of Hsp60, to establish biometric markers that would allow making a follow up of the maternal-fetal health status. Data of the literature have also suggested that acting against the heat shock proteins could be a therapeutic alternative for some diseases, and it is not discarded that these strategies could also have implications in the realm of reproductive biology in the future. There are still many subjects to be studied regarding HSP, in particular Hsp60, given the multiple interactions they display and, above all, their participation as one more element in the signaling cascades, which opens new horizons in the field of hormonal control within the physiology and biochemistry of reproductive biology, among other biomedical areas.

20.2 Conclusions

Procreation is a harmonic process that requires strategies that allow organisms, in this case the human, to transmit their genetic background to the next generation and preserve the specie. Collectively, the mechanisms of maturation of the gametes, the relationships generated during implantation and the hormonal production to keep the fetus in the womb and not be expelled, require a series of steps and strategies to protect both the mother and the fetus. As mentioned, the role played by HSP at different levels acquires relevance given the complexity of the reproductive process. In this sense, intense studies have been carried out in order to find possible markers that allow the monitoring of pregnancy and to prevent probable complications or alterations. It is interesting that in the human placenta, Hsp60 is associated with the steroidogenic process by participating in the transport of cholesterol at the mitochondrial level where is transformed into progesterone by the P450scc chain. Although up to now, it has not been possible to have a HSP as a biomarker of pregnancy, it is possible that in the near future, the identification and characterization of HSP in the maternal serum could be an alternative for monitoring the course of pregnancy, as well as the fetal well-being.

Acknowledgements This work was partially supported by Grants IN211912, IN211715 and IN215518 (FM), IN222617 (OFH) from Dirección General de Apoyo al Personal Académico (UNAM), and 168025 (FM) from Consejo Nacional de Ciencia y Tecnología, México.

References

Alard JE, Dueymes M, Youinou P, Jamin C (2007) Modulation of endothelial cell damages by anti-Hsp60 autoantibodies in systemic autoimmune diseases. Autoimmun Rev 6:438–443

Alard JE, Hillion S, Guillevin L, Saraux A, Pers J-O, Youinou P, Jamin C (2011) Autoantibodies to endothelial cell surface ATP synthase, the endogenous receptor for Hsp60, might play a pathogenic role in vasculatides. PLoS One 6:e14654

Arya R, Mallik M, Lakhotia SC (2007) Heat shock genes – integrating cell survival and death. J Biosci 32:595–610

Bahat A, Perlberg S, Melamed-Book N, Lauria I, Langer T, Orly J (2014) StAR enhances transcription of genes encoding the mitochondrial proteases involved in its own degradation. Mol Endocrinol 28:208–224

Becker B, Multhoff G, Farkas B, Wild PJ, Landthaler M, Stolz W, Vogt T (2004) Induction of Hsp90 protein expression in malignant melanomas and melanoma metastases. Exp Dermatol 13:27–32

Belhia F, Gremlich S, Muller-Brochut AC, Damnon F, Hohlfeld P, Witkin SS, Gerber S (2010) Anti-60-kDa heat shock protein antibodies in fetal serum: a biomarker for unexplained small for gestational age fetuses. Gynecol Obstet Investig 70:299–305

Bocharov AV, Vishnyakova TG, Baranova IN, Remaley AT, Patterson AP, Eggerman TL (2000) Heat shock protein 60 is a high-affinity high-density lipoprotein binding protein. Biochem Biophys Res Commun 277:228–235

Boncoraglio A, Minoia M, Carra S (2012) The family of mammalian small heat shock proteins (HSPBs): implications in protein deposit diseases and motor neuropathies. Int J Biochem Cell Biol 44:1657–1669

Burdon RH, Slater A, McMahon M, Cato AC (1982) Hyperthermia and the heat-shock proteins of HeLa cells. Br J Cancer 45:953–963

Campanella C, Bucchieri F, Merendino AM, Fucarino A, Burgio G, Corona DF, Barbieri G, David S, Farina F, Zummo G, de Macario EC, Macario AJ, Cappello F (2012) The odyssey of Hsp60 from tumor cells to other destinations includes plasma membrane-associated stages and Golgi and exosomal protein-trafficking modalities. PLoS One 7:e42008

Campbell AM, Chan SH (2008) Mitochondrial membrane cholesterol, the voltaje dependent anion channel (VDAC), and the Warburg effect. J Bioenerg Biomembr 40:193–197

Cehovin A, Coates AR, Hu Y, Riffo-Vasquez Y, Tormay P, Botanch C, Altare F, Henderson B (2010) Comparison of the moonlighting actions of the two highly homologous chaperonin 60 proteins of Mycobacterium tuberculosis. Infect Immunol 78:3196–3206

Chandra D, Choy G, Tang DG (2007) Cytosolic accumulation of HSP60 during apoptosis with or without apparent mitochondrial release: evidence that its pro-apoptotic or pro-survival functions involve differential interactions with caspase-3. J Biol Chem 282:31289–31301

Chen S, Bawa D, Besshoh S, Gurd JW, Brown IR (2005) Association of heat shock proteins and neuronal membrane components with lipid rafts from the rat brain. J Neurosci Res 81:522–529

Chen C, Zabad S, Liu H, Wang W, Jeffery C (2018) MoonProt 2.0: an expansion and update of the moonlighting proteins database. Nucleic Acids Res 46:D640–D644

Cheng Q, Chang JT, Geradts J, Neckers LM, Haystead T, Spector NL, Lyerly HK (2012) Amplification and high-level expression of heat shock protein 90 marks aggressive phenotypes of human epidermal growth factor receptor 2 negative breast cancer. Breast Cancer Res 14:R62

Chun JN, Choi B, Lee KW, Lee DJ, Kang DH, Lee JY, Song IS, Kim HI, Lee SH, Kim HS, Lee NK, Lee SY, Lee SY, Lee KJ, Kim J, Kang SW (2010) Cytosolic Hsp60 is involved in the NF-kappaB-dependent survival of cancer cells via IKK regulation. PLoS One 5:e9422

Cuffe JSM, Holland O, Salomon C, Rice GE, Perkins AV (2017) Review: placental derived biomarkers of pregnancy disorders. Placenta 54:104–110

Czarnecka AM, Campanella C, Zummo G, Capello F (2006) Mitochondrial chaperones in cancer: from molecular biology to clinical diagnostics. Cancer Biol Ther 5:714–720

de Graaf R, Kloppenburg G, Kitslaar PJ, Bruggeman CA, Sassen F (2006) Human heat shock protein 60 stimulates vascular smooth muscle cell proliferation through toll-like receptors 2 and 4. Microbes Infect 8:1859–1865

Deocaris CC, Kaul SC, Wadhwa R (2006) On the brotherhood of the mitochondrial chaperones mortalin and heat shock protein 60. Cell Stress Chaperones 11:116–128

Dvorakova L, Ivankova K, Krofta L, Hromadnikova I (2017) Expression profile of heat shock proteins in placental tissues of patients with preterm prelabor rupture of membranes and spontaneous preterm labor with intact membranes. Am J Reprod Immunol 78:e12698

Easton DP, Kaneko Y, Subjeck JR (2000) The hsp110 and Grp1 70 stress proteins: newly recognized relatives of the Hsp70s. Cell Stress Chaperones 5:276–290

Eggert-Kruse W, Scholz S, Kirschfink M, Strowitzki T (2014) Recurrent miscarriages, innate immunity, and autoimmune reaction to chlamydial 60-kDa heat shock protein—is there an association? Fertil Steril 101:1675–1680

El Zowalaty AE, Li R, Zheng Y, Lydon JP, DeMayo FJ, Ye X (2017) Deletion of RhoA in progesterone receptor-expressing cells leads to luteal insufficiency and infertility in female mice. Endocrinology 158:2168–2178

Ellis RJ (1999) Chaperonins. Curr Biol 9:R352

Elustondo P, Martin LA, Karten B (2017) Mitochondrial cholesterol import. Biochim Biophys Acta 1862:90–101

Espinosa-Garcia MT, Strauss JFIII, Martinez F (2000) A trypsin-sensitive protein is required for utilization of exogenous cholesterol for pregnenolone synthesis by placental mitochondria. Placenta 21:654–660

Flores-Herrera O, Uribe A, Garcia-Perez C, Milan R, Martinez F (2002) 5′-p-Fluorosulfonylbenzoyl adenosine inhibits progesterone synthesis in human placental mitochondria. Biochim Biophys Acta 1585:11–18
Friedland JS, Shattock R, Remick DG, Griffin GE (1993) Mycobacterial 65-kD heat shock protein induces release of proinflammatory cytokines from human monocytic cells. Clin Exp Immunol 91:58–62
Galluzzi L, Kepp O, Kroemer G (2012) Mitochondria: master regulators of danger signalling. Nat Rev Mol Cell Biol 13:780–788
Galmiche A, Fueller J (2007) RAF kinases and mitochondria. Biochim Biophys Acta 1773:1256–1262
Gambill BD, Voos W, Kang PJ, Miao B, Langer T, Craig EA, Pfanner N (1993) A dual role for mitochondrial heat shock protein 70 in membrane translocation of preproteins. J Cell Biol 123:109–117
Gharesi-Fard B, Zolghadri J, Kamali-Sarvestani E (2010) Proteome differences of placenta between pre-eclampsia and normal pregnancy. Placenta 31:121–125
Ghosh JC, Dohi T, Kang BH, Altieri DC (2008) Hsp60 regulation of tumor cell apoptosis. J Biol Chem 283:5188–5194
Gotoh T, Terada K, Oyadomari S, Mori M (2004) hsp70-DnaJ chaperone pair prevents nitric oxide- and CHOP-induced apoptosis by inhibiting translocation of Bax to mitochondria. Cell Death Differ 11:390–402
Grad I, Picard D (2007) The glucocorticoid responses are shaped by molecular chaperones. Mol Cell Endocrinol 275:2–12
Graham A (2015) Mitochondrial regulation of macrophage cholesterol homeostasis. Free Radic Biol Med 89:982–992
Grunau C, Dettmer R, Behlke J, Bernhardt R (1995) Bovine adrenodoxin – a mitochondrial iron-sulphur protein--binds to chaperonin GroEL. Biochem Biophys Res Commun 210:1001–1008
Gupta RS, Ramachandra NB, Bowes T, Singh B (2008) Unusual cellular disposition of the mitochondrial molecular chaperones Hsp60, Hsp70 and Hsp10. Novartis Found Symp 291:59–68
Habich C, Kempe K, Burkart V, Van Der Zee R, Lillicrap M, Gaston H, Kolb H (2004) Identification of the heat shock protein 60 epitope involved in receptor binding on macrophages. FEBS Lett 568:65–69
Habich C, Kempe K, van der Zee R, Rümenapf R, Akiyama H, Kolb H, Burkart V (2005) Heat shock protein 60: specific binding of lipopolysaccharide. J Immunol 174:1298–1305
Hageman J, Kampinga HH (2009) Computational analysis of the human HSPH/HSPA/DNAJ family and cloning of a human HSPH/HSPA/DNAJ expression library. Cell Stress Chaperones 14:1–21
Hansen JJ, Bross P, Westergaard M, Nielsen MN, Eiberg H, Borglum AD, Mogensen J, Kristiansen K, Bolund L, Gregersen N (2003) Genomic structure of the human mitochondrial chaperonin genes: HSP60 and HSP10 are localised head to head on chromosome 2 separated by a bidirectional promoter. Hum Genet 112:71–77
Harris SF, Shiau AK, Agard DA (2004) The crystal structure of the carboxy-terminal dimerization domain of htpG, the *Escherichia coli* Hsp90, reveals a potential substrate binding site. Structure 12:1087–1097
Hartl FU, Bracher A, Hayer-Hartl M (2011) Molecular chaperones in protein folding and proteostasis. Nature 475:324–332
He F, Kumar A, Song ZH (2012) Heat shock protein 90 is an essential molecular chaperone for CB2 cannabinoid receptor-mediated signaling in trabecular meshwork cells. Mol Vis 18:2839–2846
Henderson B, Martin AC (2014) Protein moonlighting: a new factor in biology and medicine. Biochem Soc Trans 42:1671–1678
Henderson B, Fares MA, Lund PA (2013) Chaperonin 60: a paradoxical, evolutionarily conserved protein family with multiple moonlighting functions. Biol Rev Camb Philos Soc 88:955–987

Hickey TB, Ziltener HJ, Speert DP, Stokes RW (2010) *Mycobacterium tuberculosis* employs Cpn60.2 as an adhesin that binds CD43 on the macrophage surface. Cell Microbiol 12:1634–1647

Hightower LE (1980) Cultured animal cells exposed to amino acid analogues or puromycin rapidly synthesize several polypeptides. J Cell Physiol 102:407–427

Hirtreiter AM, Calloni G, Forner F, Scheibe B, Puype M, Vandekerckhove J, Mann M, Hartl FU, Hayer-Hartl M (2009) Differential substrate specificity of group I and group II chaperonins in the aechaeon *Methanosarcina mazei*. Mol Microbiol 74:1152–1168

Hromadnikova I, Dvorakova L, Kotlabova K, Kestlerova A, Hympanova L, Novotna V, Doucha J, Krofta L (2015) Assessment of placental and maternal stress responses in patients with pregnancy related complications via monitoring of heat shock protein mRNA levels. Mol Biol Rep 42:625–637

Hulce JJ, Cognetta AB, Niphakis MJ, Tully SE, Cravatt BF (2013) Proteome-wide mapping of cholesterol-interacting proteins in mammalian cells. Nat Methods 10:259–264

Issop L, Rone MB, Papadopoulos V (2013) Organelle plasticity and interactions in cholesterol transport and steroid biosynthesis. Mol Cell Endocrinol 371:34–46

Jaattela M (1995) Over-expression of hsp70 confers tumorigenicity to mouse fibrosarcoma cells. Int J Cancer 60:689–693

Jeffery CJ (1999) Moonlighting proteins. Trends Biochem Sci 24:8–11

Jeffery CJ (2017) Protein moonlighting: what is it, and why is it important? Phil Trans R Soc B 373:20160523

Jin H, Ma KD, Hu R, Chen Y, Yang F, Yao J, Li XT, Yang PY (2008) Analysis of expression and comparative profile of normal placental tissue proteins and those in preeclampsia patients using proteomic approaches. Anal Chim Acta 629:158–164

Joab I, Radanyi C, Renoir M, Buchou T, Catelli MG, Binart N, Mester J, Baulieu EE (1984) Common non-hormone binding component in non-transformed chick oviduct receptors of four steroid hormones. Nature 308:850–853

Kampinga HH, Hageman J, Vos MJ, Kubota H, Tanguay RM, Bruford EA, Cheetham ME, Chen B, Hightower LE (2009) Guidelines for the nomenclature of the human heat shock proteins. Cell Stress Chaperones 14:105–111

Kaneko Y, Nishiyama H, Nonoguchi K, Higashitsuji H, Kishishita M, Fujita J (1997a) A novel hsp110-related gene, APG-1, that is abundantly expressed in the testis responds to a low temperature heat shock rather than the traditional elevated temperatures. J Biol Chem 272:2640–2645

Kaneko Y, Kimura T, Kishishita M, Noda Y, Fujita J (1997b) Cloning of apg-2 encoding a novel member of heat shock protein 110 family. Gene 189:19–24

Kase S, Parikh JG, Rao NA (2009) Expression of heat shock protein 27 and alpha-crystallins in human retinoblastoma after chemoreduction. Br J Ophthalmol 93:541–544

Kelley WL (1998) The J-domain family and the recruitment of chaperone power. Trends Biochem Sci 23:222–227

Keyse SM, Tyrrell RM (1989) Heme oxygenase is the major 32-kDa stress protein induced in human skin fibroblasts by UVA radiation, hydrogen peroxide, and sodium arsenite. Proc Natl Acad Sci U S A 86:99–103

Khanna A, Aten RF, Behrman HR (1994) Heat shock protein induction blocks hormone-sensitive steroidogenesis in rat luteal cells. Steroids 59:4–9

Khanna A, Aten RF, Behrman HR (1995a) Heat shock protein-70 induction mediates luteal regression in the rat. Mol Endocrinol 9:1431–1440

Khanna A, Aten RF, Behrman HR (1995b) Physiological and pharmacological inhibitors of luteinizing hormone-dependent steroidogenesis induce heat shock protein-70 in rat luteal cells. Endocrinology 136:1775–1781

Kim YN, Kim HK, Warda M, Kim N, Park WS, Prince AB, Jeong DH, Lee DS, Kim KT, Han J (2007) Toward a better understanding of preeclampsia: comparative proteomic analysis of preeclamptic placentas. Proteomics Clin Appl 1:1625–1636

Kim JH, Kim KJ, Rhee YY, Oh S, Cho NY, Lee HS, Kang GH (2013) Expression status of wild-type HSP110 correlates with HSP110 T deletion size and patient prognosis in microsatellite-unstable colorectal cancer. Mod Pathol 27:443–453
Kleinridders A, Lauritzen HPMM, Ussar S, Christensen JH, Mori MA, Bross P, Kahn CR (2013) Leptin regulation of Hsp60 impacts hypothalamic insulin signaling. J Clin Invest 123:4667–4680
Kligman L, Grifo JA, Witkin SS (1996) Expression of the 60 kDa heat shock protein in peritoneal fluids from women with endometriosis: implications for endometriosis-associated infertility. Hum Reprod 11:2736–2738
Lachance C, Bailey JL, Leclerc P (2007) Expression of Hsp60 and Grp78 in the human endometrium and oviduct, and their effect on sperm functions. Hum Reprod 22:2606–2614
Landstein D, Ulmansky R, Naparstek Y (2015) HSP60: a double edge swoed in autimmunity. Oncotarget 6:32299–32300
Lettini G, Maddalena F, Sisinni L, Condelli V, Matassa DS, Costi MP, Simoni D, Esposito F, Landriscina M (2017) TRAP1: a viable therapeutic target for future cancer treatments? Expert Opin Ther Targets 21:805–815
Leung WH, Vong QP, Lin W, Bouck D, Wendt S, Sullivan E, Li Y, Bari R, Chen T, Leung W (2015) PRL-3 mediates the protein maturation of ULBP2 by regulating the tyrosine phosphorylation of HSP60. J Immunol 194:2930–2941
Levinson W, Oppermann H, Jackson J (1980) Transition series metals and sulfhydryl reagents induce the synthesis of four proteins in eukaryotic cells. Biochim Biophys Acta 606:170–180
Li GC, Shrieve DC (1982) Thermal tolerance and specific protein synthesis in Chinese hamster fibroblasts exposed to prolonged hypoxia. Exp Cell Res 142:464–468
Li C, Liu D, Yuan Y, Huang S, Shi M, Tao K, Feng W (2010) Overexpresion of APG-2 increases cell proliferation and protects from oxidative damage in BaF3-BCR/ABL cells. Int J Oncol 36:899–904
Li L, Wu J, Luo M, Sun Y, Wang G (2016) The effect of heat stress on gene expression, synthesis of steroids, and apoptosis in bovine granulosa cells. Cell Stress Chaperones 21:467–475
Lin CY, Hu CT, Cheng CC, Lee MC, Pan SM, Lin TY, Wu WS (2016) Oxidation of heat shock protein 60 and protein disulfide isomerase activates ERK and migration of human hepatocellular carcinoma HepG2. Oncotarget 7:11067–11082
Liu Z, Stocco DM (1997) Heat shock-induced inhibition of acute steroidogenesis in MA-10 cells is associated with inhibition of the synthesis of the steroidogenic acute regulatory protein. Endocrinology 138:2722–2728
Lund PA (2001) Microbial molecular chaperones. Adv Microb Physiol 44:93–140
Luo W, Sun W, Taldone T, Rodina A, Chiosis G (2010) Heat shock protein 90 in neurodegenerative diseases. Mol Neurodegener 5:24
Mahlke K, Pfanner N, Martin J, Horwich AL, Hartl FU, Neupert W (1990) Sorting pathways of mitochondrial inner membrane proteins. Eur J Biochem 192:551–555
Mandal K, Foteinos G, Jahangiri M, Xu Q (2005) Role of antiheat shock protein 60 autoantibodies in atherosclerosis. Lupus 14:742–746
Mangelsdorf DJ, Thummel C, Beato M, Herrlich P, Schütz G, Umesono K, Blumberg B, Kastner P, Mark M, Chambon P, Evans RM (1995) The nuclear receptor superfamily: the second decade. Cell 83:835–839
Mani M, Chen C, Amblee V, Liu H, Mathur T, Zwicke G, Zabad S, Patel B, Thakkar J, Jeffery CJ (2015) MoonProt: a database for proteins that are known to moonlight. Nucleic Acids Res 43:D277–D282
Marom M, Dayan D, Demishtein-Zohary K, Mokranjac D, Neupert W, Azem A (2011) Direct interaction of mitochondrial targeting presequences with purified components of the TIM23 protein complex. J Biol Chem 286:43809–43815
Martinez F, Olvera-Sanchez S, Esparza-Perusquia M, Gomez-Chang E, Flores-Herrera O (2015) Multiple functions of syncytiotrophoblast mitocondria. Steroids 103:11–22
Matsuda M, Sasaki A, Shimizu K, Kamada Y, Noguchi S, Hiramatsu Y, Nakatsuka M (2017) Increased anti-HSP60 and anti-HSP70 antibodies in women with unexplained recurrent pregnancy loss. Acta Med Okayama 71:201–208

Meghji S, Lillicrap M, Maguire M, Tabona P, Gaston JS, Poole S, Henderson B (2003) Human chaperonin 60 (Hsp60) stimulates bone resorption: structure/function relationships. Bone 33:419–425

Meng Q, Li BX, Xiao X (2018) Toward developing chemical modulators of Hsp60 as potential therapeutics. Front Mol Biosci 5:35

Merbl Y, Zucker-Toledano M, Quintana FJ, Cohen IR (2007) Newborn humans manifest autoantibodies to defined self molecules detected by antigen microarray informatics. J Clin Invest 117:712–718

Milioti N, Bermudez-Fajardo A, Penichet ML, Oviedo-Orta E (2008) Antigen-induced immunomodulation in the pathogenesis of atherosclerosis. Clin Dev Immunol 2008:723539

Miller WL (2013) Steroid hormone synthesis in mitochondria. Mol Cell Endocrinol 379:62–73

Miller WL, Bose HS (2011) Early steps in steroidogenesis: intracellular cholesterol trafficking. J Lipid Res 52:2111–2135

Mine K, Katayama A, Matsumura T, Nishino T, Kuwabara Y, Ishikawa G, Murata T, Sawa R, Otsubo Y, Shin S, Takeshita T (2007) Proteome analysis of human placentae: pre-eclampsia versus normal pregnancy. Placenta 28:676–687

Mitra A, Shevde LA, Samant RS (2009) Multi-faceted role of HSP40 in cancer. Clin Exp Metastasis 26:559–567

Miyata S, Mizuno T, Koyama Y, Katayama T, Tohyama M (2013) The endoplasmic reticulum-resident chaperone heat shock protein 47 protects the Golgi apparatus from the effects of O-glycosylation inhibition. PLoS One 8:e69732

Mohamed BA, Barakat AZ, Zimmermann WH, Bittner RE, Muhlfeld C, Hunlich M, Engel W, Maier LS, Adham IM (2012) Targeted disruption of Hspa4 gene leads to cardiac hypertrophy and fibrosis. J Mol Cell Cardiol 53:459–468

Molvarec A, Derzsy Z, Kocsis J, Bőze T, Nagy B, Balogh K, Makó V, Cervenak L, Mézes M, Karádi I, Prohászka Z, Rigó JJ (2009) Circulating anti-heat-shock-protein antibodies in normal pregnancy and preeclampsia. Cell Stress Chaperones 14:491–498

Molvarec A, Tamasi L, Losonczy G, Madach K, Prohaszka Z, Rigo JJ (2010) Circulating heat shock protein 70 (HSPA1A) in normal and pathological pregnancies. Cell Stress Chaperones 15:237–247

Monaghan RM, Whitmarsh AJ (2015) Mitochondrial proteins moonlighting in the nucleus. Trends Biochem Sci 40:728–735

Monreal-Flores J, Espinosa-Garcia MT, Garcia-Regalado A, Arechavaleta-Velasco F, Martinez F (2017) The heat shock protein 60 promotes progesterone synthesis in mitochondria of JEG-3 cells. Reprod Biol 17:154–161

Mookerjee-Basu J, Vantourout P, Martinez LO, Perret B, Collet X, Périgaud C, Peyrottes S, Champagne E (2010) F_1-adenosine triphosphatase displays properties characteristic of an antigen presentation molecule for Vgamma9Vdelta2 T cells. J Immunol 184:6920–6928

Murphy ME (2013) The HSP70 family and cancer. Carcinogenesis 34:1181–1188

Murphy BD, Lalli E, Walsh LP, Liu Z, Soh J, Stocco DM, Sassone-Corsi P (2001) Heat shock interferes with steroidogenesis by reducing transcription of the steroidogenic acute regulatory protein gene. Mol Endocrinol 15:1255–1263

Naaby-Hansen S, Herr JC (2010) Heat shock proteins on the human sperm surface. J Reprod Immunol 84:32–40

Nadin SB, Vargas-Roig LM, Cuello-Carrion FD, Ciocca DR (2003) Deoxyribonucleic acid damage induced by doxorubicin in peripheral blood mononuclear cells: possible roles for the stress response and the deoxyribonucleic acid repair process. Cell Stress Chaperones 8:361–372

Nagumo Y, Kakeya H, Shoji M, Hayashi Y, Dohmae N, Osada H (2005) Epolactaene binds human Hsp60 Cys442 resulting in the inhibition of chaperone activity. Biochem J 387:835–840

Nanbu K, Konishi I, Mandai M, Kuroda H, Hamid AA, Komatsu T, Mori T (1998) Prognostic significance of heat shock proteins HSP70 and HSP90 in endometrial carcinomas. Cancer Detect Prev 22:549–555

Navarrete J, Flores-Herrera O, Uribe A, Martínez F (1999) Differences in cholesterol incorporation into mitochondria from hepatoma AS-30D and human term placenta. Placenta 20:285–291
Neuer A, Spandorfer S, Giraldo P, Mele C, Liu HC, Marzusch K, Kneissl D, Witkin SS (1997) Immune sensitization to the 60 kD heat shock protein and pregnancy outcome. Infect Dis Obstet Gynecol 5:154–157
Neuer A, Spandorfer SD, Giraldo P, Dieterle S, Rosenwaks Z, Witkin SS (1999) The role of heat shock proteins in reproduction. Hum Reprod Update 6:149–159
Ohashi R, Mu H, Yao Q, Chen C (2004) Atherosclerosis: immunopathogenesis and immunotherapy. Med Sci Monit 10:RA255–RA260
Oka S, Shiraishi K, Fujimoto M, Katiyar A, Takii R, Nakai A, Matsuyama H (2017) Role of heat shock factor 1 in conserving cholesterol transportation in leydig cell steroidogenesis via steroidogenic acute regulatory protein. Endocrinology 158:2648–2658
Olvera-Sanchez S, Espinosa-Garcia MT, Monreal J, Flores-Herrera O, Martinez F (2011) Mitochondrial heat shock protein participates in placental steroidogenesis. Placenta 32:222–229
Omori H, Otsu M, Suzuki A, Nakayama T, Akama K, Watanabe M, Inoue N (2013) Effects of heat shock on survival, proliferation and differentiation of mouse neural stem cells. Neurosci Res 79:13–21
Papadopoulos V, Miller WL (2012) Role of mitochondria in steroidogenesis. Best Pract Res Clin Endocrinol Metab 26:771–790
Pockley AG, Bulmer J, Hanks BM, Wright BH (1999) Identification of human heat shock protein 60 (Hsp60) and anti-Hsp60 antibodies in the peripheral circulation of normal individuals. Cell Stress Chaperones 4:29–35
Prasad TK, Hack E, Hallberg RL (1990) Function of the maize mitochondrial chaperonin hsp60: specific association between hsp60 and newly synthesized F1 ATPase alpha subunits. Mol Cell Biol 10:3979–3986
Prasad M, Pawlak KJ, Burak WE, Perry EE, Marshall B, Whittal RM, Bose HS (2017) Mitochondrial metabolic regulation by GRP78. Sci Adv 3:e1602038
Pratt WB, Morishima Y, Murphy M, Harrell M (2006) Chaperoning of glucocorticoid receptors. Handb Exp Pharmacol 172:111–138
Quintana FJ, Cohen IR (2011) The HSP60 immune system network. Trends Immunol 32:89–95
Ranford J, Henderson B (2002) Chaperonins in disease: mechanisms, models and therapies. J Clin Pathol Mol Pathol 55:209–213
Rani N, Bharti S, Manchanda M, Nag TC, Ray R, Chauhan SS, Kumari S, Arya DS (2013) Regulation of heat shock proteins 27 and 70, p-Akt/p-eNOS and MAPKs by Naringin dampens myocardial injury and dysfunction in vivo after ischemia/reperfusion. PLoS One 8:e82577
Rappa F, Unti E, Baiamonte P, Cappello F, Scibetta N (2013) Different immunohistochemical levels of Hsp60 and Hsp70 in a subset of brain tumors and putative role of Hsp60 in neuroepithelial tumorigenesis. Eur J Histochem 57:e20
Ravagnan L, Gurbuxani S, Susin SA, Maisse C, Daugas E, Zamzami N, Mak T, Jaattela M, Penninger JM, Garrido C, Kroemer G (2001) Heat-shock protein 70 antagonizes apoptosis-inducing factor. Nat Cell Biol 3:839–843
Ritossa F (1996) Discovery of the heat shock response. Cell Stress Chaperones 1:97–98
Rohl A, Rohrberg J, Buchner J (2013) The chaperone Hsp90: changing partners for demanding clients. Trends Biochem Sci 38:253–262
Rone MB, Liu J, Blonder J, Ye X, Veenstra TD, Young JC, Papadopoulos V (2009) Targeting and insertion of the cholesterol-binding translocator protein into the outer mitochondrial membrane. Biochemistry 48:6909–6920
Roostaee A, Barbar E, Lavigne P, LeHoux JG (2009) The mechanism of specific binding of free cholesterol by the steroidogenic acute regulatory protein: evidence for a role of the C-terminal α-helix in the gating of the binding site. Biosci Rep 29:89–101
Saleh A, Srinivasula SM, Balkir L, Robbins PD, Alnemri ES (2000) Negative regulation of the Apaf-1 apoptosome by Hsp70. Nat Cell Biol 2:476–483
Sanderson JT (2006) The steroid hormone biosynthesis pathway as a target for endocrine-disrupting chemicals. Toxicol Sci 94:3–21

Seigneuric R, Mjahed H, Gobbo J, Joly AL, Berthenet K, Shirley S, Garrido C (2011) Heat shock proteins as danger signals for cancer detection. Front Oncol 1:37
Shah M, Stanek J, Handwerger S (1998) Differential localization of heat shock proteins 90, 70, 60 and 27 in human decidua and placenta during pregnancy. Histochem J 30:509–518
Shamaei-Tousi A, Steptoe A, O'Donnell K, Palmen J, Stephens JW, Hurel SJ, Marmot M, Homer K, D'Aiuto F, Coates AR, Humphries SE, Henderson B (2007) Plasma heat shock protein 60 and cardiovascular disease risk: the role of psychosocial, genetic, and biological factors. Cell Stress Chaperones 12:384–392
Shi Y, Thomas JO (1992) The transport of proteins into the nucleus requires the 70-kilodalton heat shock protein or its cytosolic cognate. Mol Cell Biol 12:2186–2192
Shi Z, Long W, Zhao C, Guo X, Shen R, Ding H (2013) Comparative proteomics analysis suggests that placental mitochondria are involved in the development of pre-eclampsia. PLoS One 8:e64351
Shin JK, Baek JC, Kang MY, Park JK, Lee SA, Lee JH, Choi WS, Paik WY (2011) Proteomic analysis reveals an elevated expression of heat shock protein 27 in preeclamptic placentas. Gynecol Obstet Investig 71:151–157
Skjærven L, Cuellar J, Martinez A, Valpuesta JM (2015) Dynamics, flexibility, and allostery in molecular chaperonins. FEBS Lett 589:2522–2532
Slaby O, Sobkova K, Svoboda M, Garajova I, Fabian P, Hrstka R, Nenutil R, Sachlova M, Kocakova I, Michalek J, Smerdova T, Knoflickova D, Vyzula R (2009) Significant overexpression of Hsp110 gene during colorectal cancer progression. Oncol Rep 21:1235–1241
Soltys BJ, Gupta RS (1996) Immunoelectron microscopic localization of the 60-kDa heat shock chaperonin protein (Hsp60) in mammalian cells. Exp Cell Res 222:16–27
Soltys BJ, Gupta RS (1997) Cell surface localization of the 60 kDa heat shock chaperonin protein (hsp60) in mammalian cells. Cell Biol Int 21:315–320
Somji S, Todd JH, Sens MA, Garrett SH, Sens DA (2000) Expression of heat shock protein 60 in human proximal tubule cells exposed to heat, sodium arsenite and CdCl(2). Toxicol Lett 115:127–136
Sterrenberg JN, Blatch GL, Edkins AL (2011) Human DNAJ in cancer and stem cells. Cancer Lett 312:129–142
Sullivan CS, Pipas JM (2002) T antigens of simian virus 40: molecular chaperones for viral replication and tumorigenesis. Microbiol Mol Biol Rev 66:179–202
Sun X, Fontaine JM, Bartl I, Behnam B, Welsh MJ, Benndorf R (2007a) Induction of Hsp22 (HspB8) by estrogen and the metalloestrogen cadmium in estrogen receptor-positive breast cancer cells. Cell Stress Chaperones 12:307–319
Sun LZ, Yang NN, De W, Xiao YS (2007b) Proteomic analysis of proteins differentially expressed in preeclamptic trophoblasts. Gynecol Obstet Investig 64:17–23
Tabibzadeh S, Broome J (1999) Heat shock proteins in human endometrium throughout the menstrual cycle. Infect Dis Obstet Gynecol 7:5–9
Tabibzadeh S, Kong QF, Satyaswaroop PG, Babaknia A (1996) Heat shock proteins in human endometrium throughout the menstrual cycle. Hum Reprod 11:633–640
Taylor M, Navarro-Garcia F, Huerta J, Burress H, Maseey S, Ireton K, Teter K (2010) Hsp90 is required for transfer of the cholera toxin A1 subunit from the endoplasmic reticulum to the cytosol. J Biol Chem 285:31261–31267
Trapasso F, Pichiorri F, Gaspari M, Palumbo T, Aqeilan RI, Gaudi E, Okumura H, Luliano R, Di Leva G, Fabbri M, Birk DE, Raso C, Green-Church K, Spagnoli LG, Venuta S, Huebner K, Croce CM (2008) Fhit interaction with ferredoxin reductase triggers generation of reactive oxygen species and apoptosis of cancer cells. J Biol Chem 283:13736–13744
Tsai YP, Teng SC, Wu KJ (2008) Direct regulation of HSP60 expression by c-MYC induces transformation. FEBS Lett 582:4083–4088
Tsai YP, Yang MH, Huang CH, Chang SY, Chen PM, Liu CJ, Teng SC, Wu KJ (2009) Interaction between HSP60 and beta-catenin promotes metastasis. Carcinogenesis 30:1049–1057

Tuccinardi D, Fioriti E, Manfrini S, D'Amico E, Pozzilli P (2011) DiaPep277 peptide therapy in the context of other immune intervention trials in type 1 diabetes. Expert Opin Biol Ther 11:1233–1240

Uribe A, Strauss JF III, Martínez F (2003) Contact sites from human placental mitochondria: characterization and role in progesterone synthesis. Placenta 413:172–181

Vos MJ, Hageman J, Carra S, Kampinga HH (2008) Structural and functional diversities between members of the human HSPB, HSPH, HSPA, and DNAJ chaperone families. Biochemistry 47:7001–7011

Wang W, Jeffery CJ (2016) An analysis of surface proteomics results reveals novel candidates for intracellular/surface moonlighting proteins in bacteria. Mol BioSyst 12:1420–1431

Wang F, Shi Z, Wang P, You W, Liang G (2013) Comparative proteome profile of human placenta from normal and preeclamptic pregnancies. PLoS One 8:e78025

Whitesell L, Lindquist SL (2005) HSP90 and the chaperoning of cancer. Nat Rev Cancer 5:761–772

Witkin SS (1999) Immunity to heat shock proteins and pregnancy outcome. Infect Dis Obstet Gynecol 7:35–38

Woo SH, An S, Lee HC, Jin HO, Seo SK, Yoo DH, Lee KH, Rhee CH, Choi EJ, Hong SI, Park IC (2009) A truncated form of p23 down-regulates telomerase activity via disruption of Hsp90 function. J Biol Chem 284:30871–30880

Wyttenbach A, Carmichael J, Swartz J, Furlong RA, Narain Y, Rankin J, Rubinsztein DC (2000) Effects of heat shock, heat shock protein 40 (HDJ-2), and proteasome inhibition on protein aggregation in cellular models of Huntington's disease. Proc Natl Acad Sci U S A 97:2898–2903

Xiao Q, Mandal K, Schett G, Mayr M, Wick G, Oberhollenzer F, Willeit J, Kiechl S, Xu Q (2005) Association of serum-soluble heat shock protein 60 with carotid atherosclerosis: clinical significance determined in a follow-up study. Stroke 36:2571–2576

Xu Y, Lindquist S (1993) Heat-shock protein hsp90 governs the activity of pp60v-src kinase. Proc Natl Acad Sci U S A 90:7074–7078

Yagita Y, Kitagawa K, Taguchi A, Ohtsuki T, Kuwabara K, Mabuchi T, Matsumoto M, Yanagihara T, Hori M (1999) Molecular cloning of a novel member of the HSP110 family of genes, ischemia-responsive protein 94 kDa (irp94), expressed in rat brain after transient forebrain ischemia. J Neurochem 72:1544–1551

Yan FQ, Wang JQ, Tsai YP, Wu KJ (2015) HSP60 overexpression increases the protein levels of the p110a subunit of phosphoinositide 3-kinase and c-Myc. Clin Exp Pharmacol Physiol 42:1092–1097

Yang JI, Kong TW, Kim HS, Kim HY (2015) The proteomic analysis of human placenta with preeclampsia and normal pregnancy. J Korean Med Sci 30:770–778

Zhang X, Pang H, Wu Z, Jian J (2011) Molecular characterization of heat shock protein 70 gene transcripts during Vibrio harveyi infection of humphead snapper, *Lutjanus sanguineus*. Fish Physiol Biochem 37:897–910

Ziegert M, Witkin SS, Sziller I, Alexander H, Brylla E, Hirtig W (1999) Heat shock proteins and heat shock protein-antibody complexes in placental tissues. Infect Dis Obstet Gynecol 7:180–185

Chapter 21
Heat Shock Protein 60 in Skin Diseases

Papapit Tuchinda and Visith Thongboonkerd

Abstract In addition to serving as a stress regulatory/response protein, heat shock protein 60 (Hsp60) also plays important roles in disease mechanism and progression. This chapter summarizes all aspects of the current knowledge on Hsp60 related to various skin diseases, including acne, atopic dermatitis (AD), dermatitis herpetiformis (DH), vasculitis, Behçet's disease (BD), microscopic polyangiitis (MPA), systemic sclerosis (SSc), dermatomyositis, lichen planus (LP), and psoriasis. The data have shown that not only human Hsp60 but also its homologs in bacteria or microbes (e.g., GroEL) are involved in immune response and inflammatory cascade of these skin diseases. Furthermore, Hsp60 expression is also associated with severity of some diseases. Therefore, Hsp60 can be considered as a potential target for future development of a useful biomarker for diagnostics and prognostics in skin diseases. Moreover, it may also serve as a new therapeutic target for better treatment outcome.

Keywords Chaperone · Dermatology · Hsp60 · Immune response · Inflammation · Skin disorders

Abbreviations

AD	atopic dermatitis
AECA	anti-endothelial cell antibodies
ANCA	anti-neutrophil cytoplasmic antibodies
BD	Behçet's disease

P. Tuchinda
Department of Dermatology, Faculty of Medicine Siriraj Hospital, Mahidol University, Bangkok, Thailand

V. Thongboonkerd (✉)
Medical Proteomics Unit, Office for Research and Development,
Faculty of Medicine Siriraj Hospital, Mahidol University, Bangkok, Thailand

A. A. A. Asea, P. Kaur (eds.), *Heat Shock Protein 60 in Human Diseases and Disorders*, Heat Shock Proteins 18,
https://doi.org/10.1007/978-3-030-23154-5_21

DH	dermatitis herpetiformis
Hsp60	heat shock protein 60
ICAM-1	intercellular adhesion molecule-1
IFN-γ	interferon-γ
IL	interleukin
LP	lichen planus
MPA	microscopic polyangiitis
MPO	myeloperoxidase
PBMC	peripheral blood mononuclear cells
SLE	systemic lupus erythematosus
SSc	systemic sclerosis
TGF-β	transforming growth factor-β
TLR	toll-like receptors
TNF-α	tumor necrosis factor-α

21.1 Introduction

Heat shock proteins are generally considered as the stress regulatory/response molecules induced by various types of stimuli and are involved in several inflammatory and autoimmune disorders. In human skin, heat shock protein 60 (Hsp60) has been demonstrated to play roles in various stress conditions. For example, ultraviolet A and B radiation, which frequently induces cellular apoptosis, causes increased level of Hsp60 in keratinocytes in a dose-dependent manner (Wang et al. 2013). In addition, hyperthermia can increase nuclear expression of Hsp60 in the normal human skin (Subjeck et al. 1982). The increased Hsp60 expression has been also observed in cellular compartment of skin fibroblasts during an early stage of senescence (Di Felice et al. 2005). Such increases in expression of Hsp60 following stress conditions have been thought to be involved in the refolding process of cellular proteins to protect human keratinocytes and other skin cells from stress-induced damage (Wilson et al. 2000).

In addition to human Hsp60, its homologs in bacteria or microbes (e.g., GroEL) also play crucial roles in cutaneous response against pathogens. A previous study has demonstrated that human keratinocytes increased secretion of tumor necrosis factor-α (TNF-α), interleukin-1α (IL-1α), IL-6 and soluble intercellular adhesion molecule-1 (ICAM-1) after 48-h exposure to GroEL from *Escherichia coli* (Marcatili et al. 1997). Hsp60 purified from *Actinobacillus actinomycetemcomitans*, the oral pathogen, could induce proliferation and migration of HaCaT human keratinocytes through activation of ERK1/2 MAP kinase pathway, whereas exogenous recombinant human Hsp60 showed no such effect (Zhang et al. 2001). This process might be involved in wound repair of the skin and oral mucosa after exposure to bacterial pathogens. However, prolonged exposure of HaCaT cells to bacterial Hsp60 caused decrease in cell viability (Zhang et al. 2004).

Hsp60 not only serves as the stress regulatory/response protein but also plays important roles in pathogenic mechanisms of several skin disorders. This chapter summarizes all aspects of the current knowledge on Hsp60 related to various skin diseases, including acne, atopic dermatitis (AD), dermatitis herpetiformis (DH), vasculitis, Behçet's disease (BD), microscopic polyangiitis (MPA), systemic sclerosis (SSc), dermatomyositis, lichen planus (LP), and psoriasis.

21.2 Hsp60 in Acne

Acne (also known as acne vulgaris) is a common skin problem associated with sebaceous follicles and *Propionibacterium acnes* infection (Das and Reynolds 2014). Acne has a varied spectrum of manifestations, including non-inflammatory comedones, inflammatory papules, pustules, nodules and cysts. Common sites of acne include face, chest, upper back and upper arms (Eichenfield et al. 2013). The pathophysiology of acne involves follicular epithelial hyperproliferation, inflammatory processes, sebum overproduction and proliferation of *P. acnes*, which is a Gram-positive rod bacterium that resides in pilosebaceous follicles (Das and Reynolds 2014).

P. acnes can induce inflammatory processes leading to diffuse infiltration of inflammatory cells around the hair follicle and rupture of follicular wall (Beylot et al. 2014; Das and Reynolds 2014; Harper 2004). In addition, *P. acnes* may induce a cell-mediated inflammatory response to physiological stress and increase production of heat shock proteins. Wilcox et al. investigated the proliferative response of peripheral blood mononuclear cells (PBMC) to *P. acnes*, Hsp60/Hsp70 derived from *P. acnes* and Hsp65 derived from *Mycobacterium bovis* BCG isolated from acne patients, resolved acne subjects and healthy controls (Wilcox et al. 2007). The data have indicated that PBMC stimulated by mycobacterial Hsp65 from acne patients showed significantly higher proportion of positive responders (as determined by purified $CD4^+$ T-cells) than those derived from the resolved acne subjects and healthy controls (Wilcox et al. 2007). However, the proportion of such positive response was comparable between stimulation with *P. acnes* vs. Hsp60/Hsp70 derived from *P. acnes*. Although these results could not conclude that *P. acnes* Hsp60 plays a crucial role in the pathogenesis of acne, the limiting dilution analysis showed a significantly lower proportion of the resolved acne subjects responded to *P. acnes* fitting the single-hit kinetic model than acne patients and controls, suggesting an association of spontaneous resolution of acne and negative regulation of the $CD4^+$ T-cell response to *P. acnes* (Wilcox et al. 2007).

21.3 Hsp60 in Atopic Dermatitis (AD)

AD is a common chronic inflammatory skin disorder affecting approximately 20% of children and 10% of adults (Flohr and Mann 2014). AD patients usually have personal or family history of other atopic symptoms, such as asthma and allergic rhinitis, and frequently have xerosis, pruritus and eczema at flexural areas (Weidinger and Novak 2016). Distribution of AD skin lesions varies according to the patient age. In infants, the commonly affected areas are cheeks, forehead and scalp, whereas flexures, neck and hands are more common in older children (Akdis et al. 2006; Wollenberg et al. 2016). Lichenification may occur from repeated scratching and rubbing (Akdis et al. 2006; Wollenberg et al. 2016). The pathogenesis of AD is considerably complex and is associated with genetic factors, skin barrier disruption and immune dysregulation. Mutation of gene encoding filaggrin (a filament aggregating protein) has been reported in AD patients, leading to disruption of the skin barrier (Thyssen and Kezic 2014). The impaired epidermal barrier then increases transepidermal permeability, water loss and penetration of external antigens that subsequently activate innate immune response (Weidinger and Novak 2016).

Roles of Hsp65 in the pathogenesis of AD have been investigated. Ghoreishi et al. studied the expression of Hsp65 in AD skin lesion compared with contact dermatitis lesion and normal skin (Ghoreishi et al. 2000). Hsp65 expression was more intense in keratinocytes of the whole epidermis in AD lesion than in contact dermatitis lesion and normal skin (Ghoreishi et al. 2000). Infiltrating lymphocytes in the dermis of approximately half of AD patients showed Hsp65 expression, which was not observed in contact dermatitis, suggesting that Hsp65-expressed lymphocytes may play a role in the pathogenic processes of AD (Ghoreishi et al. 2000).

Subsequent studies also support roles of Hsp60 in AD (Jassies-van der Lee et al. 2008; Kapitein et al. 2013). Kapitein et al. have demonstrated increased Hsp60 expression in AD lesion as compared to the non-lesional skin and suggested that Hsp60-specific T-cell response might affect local inflammation found in AD (Kapitein et al. 2013). Another study in dogs has shown that intradermal injection of recombinant human Hsp60 at AD lesion could induce regulatory cytokines (i.e., IL-10, transforming growth factor-β (TGF-β) and IL-12p40) and Th1 phenotype in the skin (Kapitein et al. 2013).

21.4 Hsp60 in Dermatitis Herpetiformis (DH)

DH is an autoimmune blistering disease associated with gluten sensitivity and celiac disease. These patients usually present with pruritic papules or vesicles mainly on extensor surfaces of the elbows, buttocks, knees, back and scalp, but frequently spare the mucosal areas (Nicolas et al. 2003). Pathogenic mechanism of DH is not completely understood but multiple factors (e.g., genetic and environmental factors) have been demonstrated to get involved in the disease process. Following the ingestion of gluten-containing foods, tissue transglutaminase can modify gliadin (a fraction of gluten) into an autoantigen, leading to activation of the gluten-responsive T-cells (Nicolas et al. 2003). IgA1 is the main type of autoantibody detected in DH and granular deposition of IgA1 in dermal papillae and along the basement membranes serves as one of the hallmarks for the diagnosis of DH (Bolotin and Petronic-Rosic 2011a; Nicolas et al. 2003). Neutrophilic microabscess can be found in the area of IgA deposition, suggesting that IgA may trigger inflammatory cascade (Bolotin and Petronic-Rosic 2011a; Nicolas et al. 2003).

In celiac disease, cross reaction of autoantibodies to the celiac peptide and Hsp60 may induce intestinal mucosal damage and increase intestinal permeability (Tukaj et al. 2017; Zanoni et al. 2006). Serum anti-tissue transglutaminase and anti-Hsp60 IgA antibodies were detected in active celiac patients (Tukaj et al. 2017; Zanoni et al. 2006). These antibodies were no longer detectable after elimination of gluten from the diet (Zanoni et al. 2006). Because DH is closely related with gluten-sensitive disease, Hsp60 may also play such important role in the pathogenesis of DH (Bolotin and Petronic-Rosic 2011a, b).

Kasperkiewicz et al. studied the role of heat shock proteins in autoimmune vesiculobullous diseases, including DH, bullous pemphigoid and pemphigus vulgaris (Kasperkiewicz et al. 2014). They have demonstrated that serum anti-Hsp60 IgG antibody was significantly increased only in patients with active DH, not in those with other active autoimmune vesiculobullous diseases (Kasperkiewicz et al. 2014). In addition to dapsone therapy, gluten-free diet is the mainstay treatment for DH (Bolotin and Petronic-Rosic 2011b; Nicolas et al. 2003). The strict gluten-free diet leads to resolution of the skin lesion and improvement of gastrointestinal symptoms in association with reduction of anti-Hsp60 IgG autoantibody (Kasperkiewicz et al. 2014).

21.5 Hsp60 in Vasculitis

Anti-endothelial cell antibodies (AECA) are the circulating autoantibodies targeting to endothelial cells (Guilpain and Mouthon 2008). There is evidence suggesting the role of AECA in the pathogenesis of various vasculitides (Alard et al. 2008; Guilpain and Mouthon 2008). Nevertheless, the AECA-targeting antigens are not well characterized. Hsp60 is commonly localized in the cytoplasm of human endothelial cells. When endothelial cells are exposed to heat stress, Hsp60 can be translocated to the cell membranes and thus be accessible to antibodies (Jamin et al. 2005). An interaction between AECA and Hsp60 has been reported to play pathogenic role in vasculitis-associated systemic autoimmune diseases (Alard et al. 2008, 2011; Jamin et al. 2005). Furthermore, anti-Hsp60 could trigger an inflammatory response of vasculitis by inducing apoptosis of endothelial cells (Jamin et al. 2005). However, apoptosis could be inhibited only partially by pre-incubating recombinant Hsp60 with purified IgG, indicating that there should be other antibodies associated with AECA-induced apoptosis (Jamin et al. 2005). Details of roles for Hsp60 in specific types of vasculitis are discussed below.

21.5.1 Behçet's Disease (BD)

BD is a rare chronic systemic vasculitis of unknown etiology that may affect multiple parts of the body. Common manifestations include recurrent aphthous and genital ulcers, uveitis and cutaneous lesions (Pineton et al. 2012). Etiopathogenesis of BD is highly complex and remains unclear. Genetic predisposition, immune dysregulation, infections, and environmental factors have been implicated in the pathogenic mechanisms of BD (Greco et al. 2018; Mendoza-Pinto et al. 2010; Pineton et al. 2012). Increasing evidence has supported the role for Hsp60 in the etiopathogenesis of BD (Kaneko et al. 2011; Shaker et al. 2007; Shimizu et al. 2012). In addition, Hsp60 level has been found to increase in peripheral blood lymphocytes, intestinal tissues, mucocutaneous lesions and plasma of patients with BD (Ergun et al. 2001; Imamura et al. 2005; Shaker et al. 2007). Moreover, it has been reported that excessive Th1 immune response and Hsp60-reactive T-cells also play roles in active BD (Greco et al. 2018; Imamura et al. 2005).

A precipitating factor of BD is infection (Lule et al. 2017; Mendoza-Pinto et al. 2010; Pineton et al. 2012). Cho et al. have demonstrated that GroEL from *Streptococcus sanguinis* is a target for anti-*S. sanguinis* IgA antibody reactivity, which is higher in BD patients than in healthy controls (Cho et al. 2013). Interestingly, T-cell immune response to Hsp60 peptide was markedly increased in patients with BD comparing to those with rheumatoid arthritis and healthy controls (Kaneko et al. 1997). Furthermore, Hsp60 peptide up-regulated mRNA expression of proinflammatory cytokines, i.e., IL-8, TNF-α and TNF-β (Kaneko et al. 1997). Hsp60 can also activate immune response through toll-like receptors (TLR). A previous study has shown that TLR were involved in the pathogenesis of BD and level of TLR-6 expressing granulocytes of BD patients was significantly increased after Hsp60 stimulation (Yavuz et al. 2008).

21.5.2 Microscopic Polyangiitis (MPA)

MPA is an autoimmune disease affecting small vessels and characterized by necrotizing vasculitis, but without paucity of immune deposits within blood vessel walls. MPA can affect several organs, including skin, lungs and kidneys. Although the etiology of MPA is still unclear, the autoimmune process has been thought to play a critical role in the pathogenesis of MPA. Anti-neutrophil cytoplasmic antibodies (ANCA) are positive in most of the MPA patients and targets mainly to myeloperoxidase (MPO) (Kallenberg 2014). A study in 58 patients with MPO-ANCA positive MPA has shown significantly higher frequency and titer of anti-human Hsp60 antibody in these patients than in those with rheumatoid arthritis, systemic lupus erythematosus (SLE) and healthy controls (Komiya et al. 2011).

21.6 Hsp60 in Systemic Sclerosis (SSc)

SSc or diffuse scleroderma is an uncommon autoimmune disease characterized by extensive fibrosis of skin and internal organs, as well as vasculopathy (Denton and Khanna 2017). Other common clinical manifestations of SSc include sclerodactyly of the fingers, digital tip ulcers, telangiectasia, Raynaud's phenomenon, interstitial lung disease, and renal involvement (Denton and Khanna 2017). Histopathology commonly shows excessive collagen accumulation, vascular injury, and autoimmune activation (Yazawa et al. 2007). Danieli et al. studied serum levels of antibodies against *M. tuberculosis* Hsp65 in 53 SSc patients, 36 patients with primary Raynaud's phenomenon, and 36 SLE patients (Danieli et al. 1992). The data showed that 47% and 38% of patients with SSc and primary Raynaud's phenomenon, respectively, had serum antibodies against Hsp65, whereas such antibodies were detected only in 5% of SLE patients (Danieli et al. 1992). While the role of immunity to bacterial Hsp65 has been implicated in rheumatoid arthritis, its role in SSc is controversial (Gaston et al. 1989, 1990).

21.7 Hsp60 in Dermatomyositis

Dermatomyositis is a systemic autoimmune disease affecting mainly children who present with rash and proximal muscle weakness. Both genetic and environmental conditions have been reported as the pathogenic factors affecting dermatomyositis. However, its mechanism and autoantigen(s) remain poorly defined (Quartier and Gherardi 2013; Thompson et al. 2018). Expression of Hsp60 has been reported to increase in muscle tissues of inflammatory myositis patients (Hohlfeld and Engel 1992). In addition, Elst et al. studied the expression of Hsp60 in juvenile dermatomyositis tissues and found that all of these muscle tissues had increased Hsp60 expression in both degenerating and regenerating muscle fibers and in the mural layer of small blood vessels (Elst et al. 2008). Moreover, muscle tissue-derived mononuclear cells and PBMC from juvenile dermatomyositis patients could activate T-cell proliferation (Elst et al. 2008). In vitro activation of PBMC isolated from juvenile dermatomyositis patients with human and microbial Hsp60 significantly induced secretion of IL-1β, TNF-α, and IL-10 (Elst et al. 2008). This study has suggested that Hsp60 induced both effector and regulatory T-cell response to control inflammation in juvenile dermatomyositis (Elst et al. 2008).

21.8 Hsp60 in Lichen Planus (LP)

LP is an idiopathic chronic inflammatory disorder that is mediated through T-cell immune response. The most frequently affected site is oral mucosa (found in approximately 70% of cases), followed by genital mucosa and skin (Farhi and

Dupin 2010; Kurago 2016). Cutaneous LP presents as polygonal, pruritic, flat-topped violaceous papules on the trunk or extremities overlying with whitish lacy lesions known as Wickham striae. In the oral cavity, the commonly involved areas include buccal mucosa, tongue and gingiva, characterized by multiple papules typically with Wickham striae (Alrashdan et al. 2016; Olson et al. 2016).

Multiple factors, including genetic background, infections, dental materials, medications and autoimmunity, can affect the pathogenesis of oral LP (Olson et al. 2016). A meta-analysis has demonstrated the association between hepatitis C seropositivity and oral LP in certain populations, such as Mediterranean and Japan (Shengyuan et al. 2009). Some medications are also associated with this disease, including beta blockers, angiotensin-converting enzyme inhibitors, anti-inflammatory drugs, diuretics and dapsone (Alrashdan et al. 2016; Roopashree et al. 2010). Contact hypersensitivity to dental materials, such as amalgam, dental acrylics, cobalt, composite and nickel has been also reported as the priming cause of LP. Moreover, replacement of such dental materials lead to resolution of the LP oral lesions (Ismail et al. 2007). Histopathology includes liquefaction degeneration of the basal layer, saw-tooth rete pegs, band-like infiltration of lymphocytes in subepithelial layer, necrotic keratinocytes, hyperkeratosis and acanthosis (Fernandez-Gonzalez et al. 2011). The immunoreaction of oral LP is mediated through T-cells, whereas antigen that can induce such immune process can be exogenous antigen or autoantigen (Alrashdan et al. 2016; Roopashree et al. 2010). Activation of $CD4^+$ and $CD8^+$ T-lymphocytes can then induce secretion of inflammatory cytokines, such as IL-2, interferon-γ (IFN-γ) and TNF-α, leading to apoptosis of basal layer due to a cytotoxic reaction (Olson et al. 2016; Roopashree et al. 2010).

Hsp60 has been implicated in the pathogenesis of oral LP but with unclear mechanism. It may serve as an autoantigen that can induce T-cell mediated immune response or associated with an autoimmune response to basal cell antigens (Bayramgurler et al. 2004; Bramanti et al. 1995). Hsp60 expression was found in the basal layer (Bramanti et al. 1995; Chaiyarit et al. 1999) and epithelial-connective tissue interface of oral LP tissues (Bramanti et al. 1995). A study by Chaiyarit et al. has suggested that Hsp60 expression in basal keratinocytes might be up-regulated by cytokines produced from T-lymphocytes in the subepithelial layer (i.e., IL-6, granulocyte-macrophage colony stimulating factor (GM-CSF) and TNF-α) (Chaiyarit et al. 1999).

21.9 Hsp60 in Psoriasis

Psoriasis is a common chronic inflammatory skin disease, which can be classified into five subtypes based on historic descriptions of its underlying histology and morphology (Menter et al. 2008). The most common form is psoriasis vulgaris, in which well demarcated erythematous plaques are covered by thick silvery scales. The second subtype, guttate psoriasis, classically presents as small circumscribed erythematous scaly lesions, which usually occur in young adults. The third subtype, inverse psoriasis, is the uncommon form localized at flexural and intertriginous areas in which scales are rarely present due to the local moist environment. The forth subtype is (localized and generalized) pustular psoriasis, which is characterized by pustules without or with fever and systemic symptoms. The last form is erythrodermic psoriasis with diffuse erythema and scaling of the skin covering nearly the entire body. Nails and scalp are commonly involved in psoriasis, whereas systemic symptoms (i.e., fever and malaise) may also present (Boehncke and Schon 2015; Menter et al. 2008).

Although its etiology is unclear, the pathogenic mechanism of psoriasis has been thought to involve genetic and environmental factors, as well as infections, stress, drugs and trauma (Menter et al. 2008). Both innate and adaptive immune processes, particularly cell-mediated adaptive immune response triggered by environmental factors, are also considered to be involved in the disease mechanism (Gaspari 2006). Infiltrating $CD4^+$ and $CD8^+$ T-lymphocytes are commonly found in the affected skins (Griffiths and Barker 2007). Additionally, many cytokines have been shown to be up-regulated in psoriatic lesions, including IL-12, IL-17, IL-22, IL-23, TNF-α and IFN-γ (Gaspari 2006; Kim and Krueger 2015).

The role of Hsp60 has been implicated in the pathogenic mechanism of psoriasis. A previous study has demonstrated that Hsp60 expression was significantly greater in epidermal keratinocytes of plaque psoriasis and guttate psoriasis than those of the normal skin (Seung et al. 2007). A study in severe combined immunodeficient (SCID) mice transplanted with skin-grafts from pustulosis palmaris et plantaris patients has demonstrated the strong expression of Hsp60 in epidermal keratinocytes of the animals (Hayashi et al. 2009). The role of association between Hsp60 and TLR has been also suggested in the innate immune response of psoriasis. Zanin-Zhorov et al. have shown that soluble Hsp60 regulated response of T-cells by interacting with TLR2 (Zanin-Zhorov et al. 2003). Ohashi et al. have demonstrated that TLR4-defective macrophages isolated from C3H/HeJ mice did not response to Hsp60 (Ohashi et al. 2000).

Human immunodeficiency virus (HIV)-infected patients are frequently associated with more severe form of psoriasis and refractory skin lesions (Mallon and Bunker 2000). Puig et al. has demonstrated intense expression of Hsp65 in psoriatic skins and lesions of AIDS-associated psoriasiform dermatitis (Puig et al. 1995). This study has suggested that immunodysregulation background in HIV-infected individuals may be associated with severity of psoriasis through modified or increased expression of Hsp65 in the skin (Puig et al. 1995).

21.10 Conclusions

Increasing evidence has indicated that Hsp60 is involved in pathogenic mechanisms of a broad spectrum of skin diseases. The data have shown that not only human Hsp60 but also its bacterial or microbial homologs (e.g., GroEL) play important roles in immune response and inflammatory cascade of these skin diseases.

In addition, Hsp60 expression is associated with severity of some diseases. Therefore, Hsp60 can be considered as a potential target for future development of a useful biomarker for diagnostics and prognostics in skin diseases. Moreover, it may also serve as a new therapeutic target for better treatment outcome.

Acknowledgements This work was supported by Mahidol University research grant and the Thailand Research Fund (IRN60W0004 and IRG5980006).

References

Akdis CA, Akdis M, Bieber T, Bindslev-Jensen C, Boguniewicz M, Eigenmann P, Hamid Q, Kapp A, Leung DY, Lipozencic J et al (2006) Diagnosis and treatment of atopic dermatitis in children and adults: European Academy of Allergology and Clinical Immunology/American Academy of Allergy, Asthma and Immunology/PRACTALL consensus report. Allergy 61:969–987

Alard JE, Dueymes M, Youinou P, Jamin C (2008) HSP60 and anti-HSP60 antibodies in vasculitis: they are two of a kind. Clin Rev Allergy Immunol 35:66–71

Alard JE, Hillion S, Guillevin L, Saraux A, Pers JO, Youinou P, Jamin C (2011) Autoantibodies to endothelial cell surface ATP synthase, the endogenous receptor for hsp60, might play a pathogenic role in vasculatides. PLoS One 6:e14654

Alrashdan MS, Cirillo N, McCullough M (2016) Oral lichen planus: a literature review and update. Arch Dermatol Res 308:539–551

Bayramgurler D, Ozkara SK, Apaydin R, Ercin C, Bilen N (2004) Heat shock proteins 60 and 70 expression of cutaneous lichen planus: comparison with normal skin and psoriasis vulgaris. J Cutan Pathol 31:586–594

Beylot C, Auffret N, Poli F, Claudel JP, Leccia MT, Del Giudice P, Dreno B (2014) *Propionibacterium acnes*: an update on its role in the pathogenesis of acne. J Eur Acad Dermatol Venereol 28:271–278

Boehncke WH, Schon MP (2015) Psoriasis. Lancet 386:983–994

Bolotin D, Petronic-Rosic V (2011a) Dermatitis herpetiformis. Part I. Epidemiology, pathogenesis, and clinical presentation. J Am Acad Dermatol 64:1017–1024

Bolotin D, Petronic-Rosic V (2011b) Dermatitis herpetiformis. Part II. Diagnosis, management, and prognosis. J Am Acad Dermatol 64:1027–1033

Bramanti TE, Dekker NP, Lozada-Nur F, Sauk JJ, Regezi JA (1995) Heat shock (stress) proteins and gamma delta T lymphocytes in oral lichen planus. Oral Surg Oral Med Oral Pathol Oral Radiol Endod 80:698–704

Chaiyarit P, Kafrawy AH, Miles DA, Zunt SL, Van Dis ML, Gregory RL (1999) Oral lichen planus: an immunohistochemical study of heat shock proteins (HSPs) and cytokeratins (CKs) and a unifying hypothesis of pathogenesis. J Oral Pathol Med 28:210–215

Cho SB, Zheng Z, Ahn KJ, Choi MJ, Cho S, Kim DY, Lee HS, Bang D (2013) Serum IgA reactivity against GroEL of Streptococcus sanguinis and human heterogeneous nuclear ribonucleoprotein A2/B1 in patients with Behcet disease. Br J Dermatol 168:977–983

Danieli MG, Candela M, Ricciatti AM, Reginelli R, Danieli G, Cohen IR, Gabrielli A (1992) Antibodies to mycobacterial 65 kDa heat shock protein in systemic sclerosis (scleroderma). J Autoimmun 5:443–452

Das S, Reynolds RV (2014) Recent advances in acne pathogenesis: implications for therapy. Am J Clin Dermatol 15:479–488

Denton CP, Khanna D (2017) Systemic sclerosis. Lancet 390:1685–1699

Di Felice V, Ardizzone N, Marciano V, Bartolotta T, Cappello F, Farina F, Zummo G (2005) Senescence-associated HSP60 expression in normal human skin fibroblasts. Anat Rec A Discov Mol Cell Evol Biol 284:446–453

Eichenfield LF, Krakowski AC, Piggott C, Del Rosso J, Baldwin H, Friedlander SF, Levy M, Lucky A, Mancini AJ, Orlow SJ et al (2013) Evidence-based recommendations for the diagnosis and treatment of pediatric acne. Pediatrics 131(Suppl 3):S163–S186

Elst EF, Klein M, de Jager W, Kamphuis S, Wedderburn LR, van der ZR, Albani S, Kuis W, Prakken BJ (2008) Hsp60 in inflamed muscle tissue is the target of regulatory autoreactive T cells in patients with juvenile dermatomyositis. Arthritis Rheum 58:547–555

Ergun T, Ince U, Eksioglu-Demiralp E, Direskeneli H, Gurbuz O, Gurses L, Aker F, Akoglu T (2001) HSP 60 expression in mucocutaneous lesions of Behcet's disease. J Am Acad Dermatol 45:904–909

Farhi D, Dupin N (2010) Pathophysiology, etiologic factors, and clinical management of oral lichen planus, part I: facts and controversies. Clin Dermatol 28:100–108

Fernandez-Gonzalez F, Vazquez-Alvarez R, Reboiras-Lopez D, Gandara-Vila P, Garcia-Garcia A, Gandara-Rey JM (2011) Histopathological findings in oral lichen planus and their correlation with the clinical manifestations. Med Oral Patol Oral Cir Bucal 16:e641–e646

Flohr C, Mann J (2014) New insights into the epidemiology of childhood atopic dermatitis. Allergy 69:3–16

Gaspari AA (2006) Innate and adaptive immunity and the pathophysiology of psoriasis. J Am Acad Dermatol 54:S67–S80

Gaston JS, Life PF, Bailey LC, Bacon PA (1989) In vitro responses to a 65-kilodalton mycobacterial protein by synovial T cells from inflammatory arthritis patients. J Immunol 143:2494–2500

Gaston JS, Life PF, Jenner PJ, Colston MJ, Bacon PA (1990) Recognition of a mycobacteria-specific epitope in the 65-kD heat-shock protein by synovial fluid-derived T cell clones. J Exp Med 171:831–841

Ghoreishi M, Yokozeki H, Hua WM, Nishioka K (2000) Expression of 27 KD, 65 KD and 72/73 KD heat shock protein in atopic dermatitis: comparison with those in normal skin and contact dermatitis. J Dermatol 27:370–379

Greco A, De Virgilio A, Ralli M, Ciofalo A, Mancini P, Attanasio G, De Vincentiis M, Lambiase A (2018) Behcet's disease: new insights into pathophysiology, clinical features and treatment options. Autoimmun Rev 17:567–575

Griffiths CE, Barker JN (2007) Pathogenesis and clinical features of psoriasis. Lancet 370:263–271

Guilpain P, Mouthon L (2008) Antiendothelial cells autoantibodies in vasculitis-associated systemic diseases. Clin Rev Allergy Immunol 35:59–65

Harper JC (2004) An update on the pathogenesis and management of acne vulgaris. J Am Acad Dermatol 51:S36–S38

Hayashi M, Fujihara K, Beder LB, Yamamoto Y, Hotomi M, Yamanaka N (2009) Pathogenic role of tonsillar lymphocytes in associated with HSP60/65 in Pustulosis palmaris et plantaris. Auris Nasus Larynx 36:578–585

Hohlfeld R, Engel AG (1992) Expression of 65-kd heat shock proteins in the inflammatory myopathies. Ann Neurol 32:821–823

Imamura Y, Kurokawa MS, Yoshikawa H, Nara K, Takada E, Masuda C, Tsukikawa S, Ozaki S, Matsuda T, Suzuki N (2005) Involvement of Th1 cells and heat shock protein 60 in the pathogenesis of intestinal Behcet's disease. Clin Exp Immunol 139:371–378

Ismail SB, Kumar SK, Zain RB (2007) Oral lichen planus and lichenoid reactions: etiopathogenesis, diagnosis, management and malignant transformation. J Oral Sci 49:89–106

Jamin C, Dugue C, Alard JE, Jousse S, Saraux A, Guillevin L, Piette JC, Youinou P (2005) Induction of endothelial cell apoptosis by the binding of anti-endothelial cell antibodies to Hsp60 in vasculitis-associated systemic autoimmune diseases. Arthritis Rheum 52:4028–4038

Jassies-van der Lee A, Rutten V, van Kooten P, van der Zee R, Willemse T (2008) Intradermal injection of Hsp60 induces cytokine responses in canine atopic and healthy skin. Cell Stress Chaperones 13:387–391

Kallenberg CG (2014) The diagnosis and classification of microscopic polyangiitis. J Autoimmun 48-49:90–93

Kaneko S, Suzuki N, Yamashita N, Nagafuchi H, Nakajima T, Wakisaka S, Yamamoto S, Sakane T (1997) Characterization of T cells specific for an epitope of human 60-kD heat shock protein (hsp) in patients with Behcet's disease (BD) in Japan. Clin Exp Immunol 108:204–212

Kaneko F, Togashi A, Saito S, Sakuma H, Oyama N, Nakamura K, Yokota K, Oguma K (2011) Behcet's disease (Adamantiades-Behcet's disease). Clin Dev Immunol 2011:681956

Kapitein B, Aalberse JA, Klein MR, de Jager W, Hoekstra MO, Knol EF, Prakken BJ (2013) Recognition of self-heat shock protein 60 by T cells from patients with atopic dermatitis. Cell Stress Chaperones 18:87–95

Kasperkiewicz M, Tukaj S, Gembicki AJ, Sillo P, Gorog A, Zillikens D, Karpati S (2014) Evidence for a role of autoantibodies to heat shock protein 60, 70, and 90 in patients with dermatitis herpetiformis. Cell Stress Chaperones 19:837–843

Kim J, Krueger JG (2015) The immunopathogenesis of psoriasis. Dermatol Clin 33:13–23

Komiya I, Arimura Y, Nakabayashi K, Yamada A, Osaki T, Yamaguchi H, Kamiya S (2011) Increased concentrations of antibody against heat shock protein in patients with myeloperoxidase anti-neutrophil cytoplasmic autoantibody positive microscopic polyangiitis. Microbiol Immunol 55:531–538

Kurago ZB (2016) Etiology and pathogenesis of oral lichen planus: an overview. Oral Surg Oral Med Oral Pathol Oral Radiol 122:72–80

Lule S, Colpak AI, Balci-Peynircioglu B, Gursoy-Ozdemir Y, Peker S, Kalyoncu U, Can A, Tekin N, Demiralp D, Dalkara T (2017) Behcet disease serum is immunoreactive to neurofilament medium which share common epitopes to bacterial HSP-65, a putative trigger. J Autoimmun 84:87–96

Mallon E, Bunker CB (2000) HIV-associated psoriasis. AIDS Patient Care STDs 14:239–246

Marcatili A, Cipollaro dl, Galdiero M, Folgore A, Petrillo G (1997) TNF-alpha, IL-1 alpha, IL-6 and ICAM-1 expression in human keratinocytes stimulated in vitro with *Escherichia coli* heat-shock proteins. Microbiology 143(Pt 1):45–53

Mendoza-Pinto C, Garcia-Carrasco M, Jimenez-Hernandez M, Jimenez HC, Riebeling-Navarro C, Nava ZA, Vera RM, Espinosa G, Jara QJ, Cervera R (2010) Etiopathogenesis of Behcet's disease. Autoimmun Rev 9:241–245

Menter A, Gottlieb A, Feldman SR, Van Voorhees AS, Leonardi CL, Gordon KB, Lebwohl M, Koo JY, Elmets CA, Korman NJ et al (2008) Guidelines of care for the management of psoriasis and psoriatic arthritis: section 1. Overview of psoriasis and guidelines of care for the treatment of psoriasis with biologics. J Am Acad Dermatol 58:826–850

Nicolas ME, Krause PK, Gibson LE, Murray JA (2003) Dermatitis herpetiformis. Int J Dermatol 42:588–600

Ohashi K, Burkart V, Flohe S, Kolb H (2000) Cutting edge: heat shock protein 60 is a putative endogenous ligand of the toll-like receptor-4 complex. J Immunol 164:558–561

Olson MA, Rogers RS III, Bruce AJ (2016) Oral lichen planus. Clin Dermatol 34:495–504

Pineton dC, Wechsler B, Geri G, Cacoub P, Saadoun D (2012) New insights into the pathogenesis of Behcet's disease. Autoimmun Rev 11:687–698

Puig L, Fernandez-Figueras MT, Ferrandiz C, Ribera M, de Moragas JM (1995) Epidermal expression of 65 and 72 kd heat shock proteins in psoriasis and AIDS-associated psoriasiform dermatitis. J Am Acad Dermatol 33:985–989

Quartier P, Gherardi RK (2013) Juvenile dermatomyositis. Handb Clin Neurol 113:1457–1463

Roopashree MR, Gondhalekar RV, Shashikanth MC, George J, Thippeswamy SH, Shukla A (2010) Pathogenesis of oral lichen planus--a review. J Oral Pathol Med 39:729–734

Seung NR, Park EJ, Kim CW, Kim KH, Kim KJ, Cho HJ, Park HR (2007) Comparison of expression of heat-shock protein 60, toll-like receptors 2 and 4, and T-cell receptor gammadelta in plaque and guttate psoriasis. J Cutan Pathol 34:903–911

Shaker O, Ay El-Deen MA, El Hadidi H, Grace BD, El Sherif H, Abdel HA (2007) The role of heat shock protein 60, vascular endothelial growth factor and antiphospholipid antibodies in Behcet disease. Br J Dermatol 156:32–37

Shengyuan L, Songpo Y, Wen W, Wenjing T, Haitao Z, Binyou W (2009) Hepatitis C virus and lichen planus: a reciprocal association determined by a meta-analysis. Arch Dermatol 145:1040–1047

Shimizu J, Izumi T, Suzuki N (2012) Aberrant activation of heat shock protein 60/65 reactive T cells in patients with Behcet's disease. Autoimmune Dis 2012:105205

Subjeck JR, Sciandra JJ, Johnson RJ (1982) Heat shock proteins and thermotolerance; a comparison of induction kinetics. Br J Radiol 55:579–584

Thompson C, Piguet V, Choy E (2018) The pathogenesis of dermatomyositis. Br J Dermatol 179:1256–1262

Thyssen JP, Kezic S (2014) Causes of epidermal filaggrin reduction and their role in the pathogenesis of atopic dermatitis. J Allergy Clin Immunol 134:792–799

Tukaj S, Gorog A, Kleszczynski K, Zillikens D, Karpati S, Kasperkiewicz M (2017) Autoimmunity to heat shock proteins and vitamin D status in patients with celiac disease without associated dermatitis herpetiformis. J Steroid Biochem Mol Biol 173:23–27

Wang XY, Tao CJ, Wu QY, Yuan CD (2013) Protein extract of ultraviolet-irradiated human skin keratinocytes promote the expression of mitogen-activated protein kinases, nuclear factor-kappaB and interferon regulatory factor-3 in Langerhans cells via toll-like receptor 2 and 4. Photodermatol Photoimmunol Photomed 29:41–48

Weidinger S, Novak N (2016) Atopic dermatitis. Lancet 387:1109–1122

Wilcox HE, Farrar MD, Cunliffe WJ, Holland KT, Ingham E (2007) Resolution of inflammatory acne vulgaris may involve regulation of CD4+ T-cell responses to *Propionibacterium acnes*. Br J Dermatol 156:460–465

Wilson N, McArdle A, Guerin D, Tasker H, Wareing P, Foster CS, Jackson MJ, Rhodes LE (2000) Hyperthermia to normal human skin in vivo upregulates heat shock proteins 27, 60, 72i and 90. J Cutan Pathol 27:176–182

Wollenberg A, Oranje A, Deleuran M, Simon D, Szalai Z, Kunz B, Svensson A, Barbarot S, von Kobyletzki L, Taieb A et al (2016) ETFAD/EADV eczema task force 2015 position paper on diagnosis and treatment of atopic dermatitis in adult and paediatric patients. J Eur Acad Dermatol Venereol 30:729–747

Yavuz S, Elbir Y, Tulunay A, Eksioglu-Demiralp E, Direskeneli H (2008) Differential expression of toll-like receptor 6 on granulocytes and monocytes implicates the role of microorganisms in Behcet's disease etiopathogenesis. Rheumatol Int 28:401–406

Yazawa N, Fujimoto M, Tamaki K (2007) Recent advances on pathogenesis and therapies in systemic sclerosis. Clin Rev Allergy Immunol 33:107–112

Zanin-Zhorov A, Nussbaum G, Franitza S, Cohen IR, Lider O (2003) T cells respond to heat shock protein 60 via TLR2: activation of adhesion and inhibition of chemokine receptors. FASEB J 17:1567–1569

Zanoni G, Navone R, Lunardi C, Tridente G, Bason C, Sivori S, Beri R, Dolcino M, Valletta E, Corrocher R et al (2006) In celiac disease, a subset of autoantibodies against transglutaminase binds toll-like receptor 4 and induces activation of monocytes. PLoS Med 3:e358

Zhang L, Pelech SL, Mayrand D, Grenier D, Heino J, Uitto VJ (2001) Bacterial heat shock protein-60 increases epithelial cell proliferation through the ERK1/2 MAP kinases. Exp Cell Res 266:11–20

Zhang L, Pelech S, Uitto VJ (2004) Long-term effect of heat shock protein 60 from Actinobacillus actinomycetemcomitans on epithelial cell viability and mitogen-activated protein kinases. Infect Immun 72:38–45

Index

A. A. A. Asea, P. Kaur (eds.), *Heat Shock Protein 60 in Human Diseases and Disorders*, Heat Shock Proteins 18,
https://doi.org/10.1007/978-3-030-23154-5

FSC
www.fsc.org
MIX
Papier | Fördert
gute Waldnutzung
FSC® C083411

Kolibri 360 GmbH
Ferdinand-Jühlke-Straße 7
99095 Erfurt, Deutschland
produktsicherheit@kolibri360.de